Biogeography: An Ecological Approach

Biogeography: An Ecological Approach

Editor: Neil Griffin

www.callistoreference.com

Callisto Reference,
118-35 Queens Blvd., Suite 400,
Forest Hills, NY 11375, USA

Visit us on the World Wide Web at:
www.callistoreference.com

ISBN: 978-1-63239-938-0 (Hardback)

Cataloging-in-Publication Data

Biogeography : an ecological approach / edited by Neil Griffin.
        p. cm.
Includes bibliographical references and index.
ISBN 978-1-63239-938-0
1. Biogeography. 2. Ecology. I. Griffin, Neil.
QH84 .B56 2018
577--dc23

Table of Contents

Permissions

List of Contributors

Index

Preface

The purpose of the book is to provide a glimpse into the dynamics and to present opinions and studies of some of the scientists engaged in the development of new ideas in the field from very different standpoints. This book will prove useful to students and researchers owing to its high content quality.

Biogeography is the study of the distribution of plant and animal species on Earth over a period of time. It has two branches, which are phytogeography and zoogeography. This field contributes essential knowledge about the composition of various ecosystems as well as the evolutionary processes and migratory patterns occurring within species over time. Habitat adaptation is an important phenomenon that all living organisms exhibit. Conservation and protection of vulnerable regions is vital for the survival of the various species found on our planet. This book is meant for students who are looking for an elaborate reference text on biogeography. For all readers who are interested in this subject, the case studies included in this book will serve as an excellent guide to develop a comprehensive understanding.

At the end, I would like to appreciate all the efforts made by the authors in completing their chapters professionally. I express my deepest gratitude to all of them for contributing to this book by sharing their valuable works. A special thanks to my family and friends for their constant support in this journey.

Editor

High Levels of Diversity Uncovered in a Widespread Nominal Taxon: Continental Phylogeography of the Neotropical Tree Frog *Dendropsophus minutus*

Marcelo Gehara[1,2]*, Andrew J. Crawford[3,4], Victor G. D. Orrico[5], Ariel Rodríguez[1], Stefan Lötters[6], Antoine Fouquet[7], Lucas S. Barrientos[3], Francisco Brusquetti[8], Ignacio De la Riva[9], Raffael Ernst[10], Giuseppe Gagliardi Urrutia[11], Frank Glaw[12], Juan M. Guayasamin[13], Monique Hölting[10], Martin Jansen[14], Philippe J. R. Kok[15], Axel Kwet[16], Rodrigo Lingnau[17], Mariana Lyra[18], Jiří Moravec[19], José P. Pombal Jr.[20], Fernando J. M. Rojas-Runjaic[21], Arne Schulze[22], J. Celsa Señaris[23], Mirco Solé[24], Miguel Trefaut Rodrigues[25], Evan Twomey[26], Celio F. B. Haddad[18], Miguel Vences[1], Jörn Köhler[22]

1 Division of Evolutionary Biology, Zoological Institute, Technical University of Braunschweig, Braunschweig, Germany, 2 Pós-graduação em Sistemática e Evolução, Centro de Biociências, Universidade Federal do Rio Grande do Norte, Campus Universitário Lagoa Nova, Natal, RN, Brasil, 3 Departamento de Ciencias Biológicas, Universidad de los Andes, Bogotá, Colombia, 4 Smithsonian Tropical Research Institute, Panamá, Republic of Panama, 5 Universidade de São Paulo, Instituto de Biociências, Departamento de Zoologia, São Paulo, Brasil, 6 Trier University, Biogeography Department, Trier, Germany, 7 CNRS-Guyane - USR3456, Immeuble Le Relais - 2, Cayenne, French Guiana, 8 Departamento de Zoologia, Instituto de Biociências, UNESP, Rio Claro, São Paulo, Brasil; Instituto de Investigación Biológica del Paraguay, Asunción, Paraguay, 9 Museo Nacional de Ciencias Naturales, Madrid, Spain, 10 Museum of Zoology, Senckenberg Natural History Collections Dresden, Dresden, Germany, 11 Peruvian Center for Biodiversity and Conservation (PCRC), Nanay, Iquitos, Peru, 12 Zoologische Staatssammlung München, München, Germany, 13 Universidad Tecnológica Indoamérica, Centro de Investigación de la Biodiversidad y el Cambio Climático (BioCamp), Cotocollao, Quito, Ecuador, 14 Senckenberg Gesellschaft für Naturforschung, Frankfurt am Main, Germany, 15 Amphibian Evolution Lab, Department of Biology, Vrije Universiteit Brussel, Brussels, Belgium, 16 German Herpetological Society (DGHT), Mannheim, Germany, 17 Universidade Tecnológica Federal do Paraná, Francisco Beltrão, PR, Brasil, 18 Departamento de Zoologia, Instituto de Biociências, UNESP, Rio Claro, São Paulo, Brasil, 19 Department of Zoology, National Museum, Prague, Czech Republic, 20 Departamento de Vertebrados, Museu Nacional, Universidade Federal do Rio de Janeiro, Rio de Janeiro, Brazil, 21 Fundación La Salle de Ciencias Naturales, Museo de Historia Natural La Salle, Caracas, Venezuela, 22 Hessisches Landesmuseum Darmstadt, Department of Zoology, Darmstadt, Germany, 23 Laboratorio de Ecología y Genética de Poblaciones, Centro de Ecología, Instituto Venezolano de Investigaciones Científicas, Caracas, Venezuela, 24 Universidade Estadual de Santa Cruz, Departamento de Ciências Biológicas, Rodovia Ilhéus-Itabuna, Bahia, Brasil, 25 Universidade de São Paulo, Instituto de Biociências, Departamento de Zoologia, São Paulo, Brasil, 26 Department of Biology, East Carolina University, Greenville, North Carolina, United States of America

Abstract

Species distributed across vast continental areas and across major biomes provide unique model systems for studies of biotic diversification, yet also constitute daunting financial, logistic and political challenges for data collection across such regions. The tree frog *Dendropsophus minutus* (Anura: Hylidae) is a nominal species, continentally distributed in South America, that may represent a complex of multiple species, each with a more limited distribution. To understand the spatial pattern of molecular diversity throughout the range of this species complex, we obtained DNA sequence data from two mitochondrial genes, cytochrome oxidase I (COI) and the 16S rhibosomal gene (16S) for 407 samples of *D. minutus* and closely related species distributed across eleven countries, effectively comprising the entire range of the group. We performed phylogenetic and spatially explicit phylogeographic analyses to assess the genetic structure of lineages and infer ancestral areas. We found 43 statistically supported, deep mitochondrial lineages, several of which may represent currently unrecognized distinct species. One major clade, containing 25 divergent lineages, includes samples from the type locality of *D. minutus*. We defined that clade as the *D. minutus* complex. The remaining lineages together with the *D. minutus* complex constitute the *D. minutus* species group. Historical analyses support an Amazonian origin for the *D. minutus* species group with a subsequent dispersal to eastern Brazil where the *D. minutus* complex originated. According to our dataset, a total of eight mtDNA lineages have ranges >100,000 km^2. One of them occupies an area of almost one million km^2 encompassing multiple biomes. Our results, at a spatial scale and resolution unprecedented for a Neotropical vertebrate, confirm that widespread amphibian species occur in lowland South America, yet at the same time a large proportion of cryptic diversity still remains to be discovered.

Editor: Donald James Colgan, Australian Museum, Australia

Funding: AR was supported by a Post-Doctoral fellowship of the Alexander von Humboldt foundation. CFBH acknowledges support by FAPESP (proc. 2008/50928-1) and Conselho Nacional de Desenvolvimento Científico e Tecnológico - CNPq. FB thanks Programa Nacional de Incentivo a Investigadores (PRONII, Paraguay) and Coordenação de Aperfeiçoamento de Pessoal de Nível Superior (CAPES, Brazil) for financial support. JM was financially supported by the Ministry of Culture of the Czech Republic (DKRVO 2012 and DKRVO 2013/14; 00023272). MG was supported by a PhD fellowship of the KAAD. MTR acknowledges support by FAPESP (#s 03/10335-8, 10/51071-7, and 11/50146-6), CNPq, and NSF (DEB 1035184 and 1120487). MV acknowledges support by the Deutsche Forschungsgemeinschaft (grant VE247/7-1). PJRK was mainly supported by the Belgian Directorate-General of Development Cooperation, the King Léopold III Fund for Nature Exploration and Conservation, and the Percy Sladen Memorial Fund, with additional support by the non-profit organization "les Amis de l'Institut Royal des Sciences Naturelles". RE and MH were supported by a research grant from the German Research Foundation (DFG ER 589/2*1). VGDO acknowledges support by Fundação de Amparo à Pesquisa do Estado de São Paulo - FAPESP (#2012/12500-5). Laboratory work in Colombia was financed by grant #156-09 from Ecopetrol. Funding was provided by the Universidad Tecnológica Indoamérica through the project "Patrones de diversidad de los anfibios andinos del Ecuador". JPP is grateful to CNPq and FAPERJ for financial support. The funders had no role in study design, data collection and analysis, decision to publish, or preparation of the manuscript.

Competing Interests: The authors have declared that no competing interests exist.

* Email: marcelo.gehara@gmail.com

Introduction

The application of molecular methods has expedited tremendously the discovery and characterization of global biological diversity [1]. This is particularly true for amphibians, where the rate of species descriptions has accelerated enormously in the past 20 years [2–7]. Integrative approaches that combine multiple lines of evidence have allowed taxonomists to define and name many of these evolutionary independent lineages as proper species [8–11]. The improved delimitation of species diversity, transforming one widely distributed species into several species, each with a smaller range, in many cases has notable impact on conservation. For instance, the International Union for Conservation of Nature (IUCN) status of certain populations may change from 'Least Concern' to one of the various threat categories or simply 'Data Deficient' [12–14].

Cryptic genetic diversity is now so commonly reported in molecular studies of amphibian species that the existence of nominally widespread tropical species has been called into question [15,16]. However, supposedly widespread species occurring across multiple biomes and countries are rarely comprehensively sampled across their complete geographic range in screenings of genetic diversity [5,6] or phylogeographic studies [17–21]. Sampling of species from across vast continental areas and across political borders is often handicapped by financial, logistic and political factors.

In the Neotropics, nominal taxa such as *Rhinella margaritifera* (Bufonidae), *Leptodactylus fuscus* (Leptodactylidae), and *Scinax ruber* (Hylidae) are prominent examples of anuran species once considered to occur across nearly the entire tropical lowlands of South America. Evidence has accumulated that many such putatively widespread species could in fact be complexes of cryptic taxa (e.g. [20,22]). However, given limited genetic sampling and the difficulty in reviewing material from all countries hosting populations, their relationships and systematics remain in many cases as unclear as they were decades ago [23,24].

A further example of a putatively widespread Neotropical amphibian species is *Dendropsophus minutus* (Peters, 1872), a small hylid frog of 21–28 mm snout-vent length, distributed in Cis-Andean South America, including the Andean slopes, the Amazon Basin, the Guiana Shield, down to the Atlantic Forests of southeastern Brazil, with an elevational record from near sea level up to 2,000 m [25]. Variation in coloration, osteology, advertisement calls and larval morphology [6,26–29], along with molecular data from limited parts of the species' distribution [21] suggest that the nominal *D. minutus* might represent a species complex. However, the sheer size of its supposed geographical range along with nomenclatural and taxonomic complexity (six junior synonyms, [25]) and unresolved relationships in the *D.*

minutus species group [30] have so far made these frogs inaccessible to revision.

In this case study, we use *D. minutus* to understand to what degree a tropical, small-sized anuran has the potential to be continentally widespread with limited genetic structure within its range, as expected for a single species. In addition to conservation concerns, this question has important implications for South American biogeography in general and amphibian systematics and evolution in particular. Evidence is accumulating that body size in amphibians has a positive correlation with range size [31,32], but contrary to this trend many Holarctic amphibians occur with little genetic substructure across the vast ranges they colonized after the last glaciation, despite sometimes moderate to small body sizes (examples in [16]). Whether such patterns also exist across vast ranges in tropical regions, with their distinct historical climatic dynamics [33], is an open question. Deciphering possible cryptic diversity within the nominal *D. minutus* would also help inform conservation assessments which typically use species'geographic distributions as criteria for conservation status [13].

The present study is a multinational collaborative effort to sample nominal *D. minutus* across its entire range and at a spatial resolution unprecedented for a Neotropical vertebrate. Based on mitochondrial DNA sequences as a proxy for overall genetic diversity, we identify genealogical lineages currently subsumed within *D. minutus* and putative allies and assess their historical relationships and geographic ranges. Although there are some data on morphology and bioacoustics, we only partially discuss these here and refrain from making taxonomic decisions, but instead provide a roadmap for future integrative studies. Our focus, therefore, is on the biogeographical implications of the phylogeographic origins and evolutionary history of the *D. minutus* species group. We reveal here that this species complex exists as a mixture of both geographically widespread lineages and probable micro-endemic lineages.

Methods

Data collection and laboratory methods

No experiments were conducted using living animals. All field researches and collecting of specimens were approved by competent authorities, these being: Instituto Chico Mendes – ICMBio, Brazil, through collection permits granted to MG (21710-2), VGDO (19920), RL (26957-1), MTR (10126-1), CFBH (22511-1) and JPP (12600-2); Museo Nacional de Historia Natural – Colección Boliviana de Fauna, La Paz, Bolivia (permits: CBF CITE No. 02/2006 and No. 81/2007); DGB and the INRENA (granted permits to IDLR for collecting Bolivian and Peruvian material respectively); The Guyana Environmental Protection Agency and the Guyanese Ministry of Amerindian Affairs and Environmental Protection Agency of Guyana through research

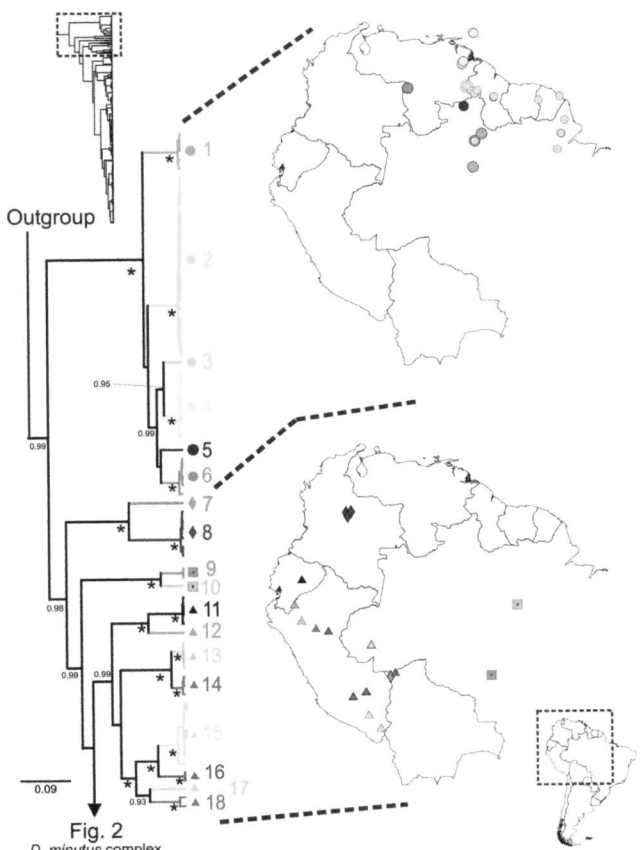

Figure 1. *Dendropsophus minutus* **tree with lineage distribution part 1.** 50% Maximum Clade Credibility tree and distribution maps of mtDNA lineages 1–18. Asterisks represent nodes with posterior probability equal to 1. Posterior probabilities lower than 0.9 are not shown.

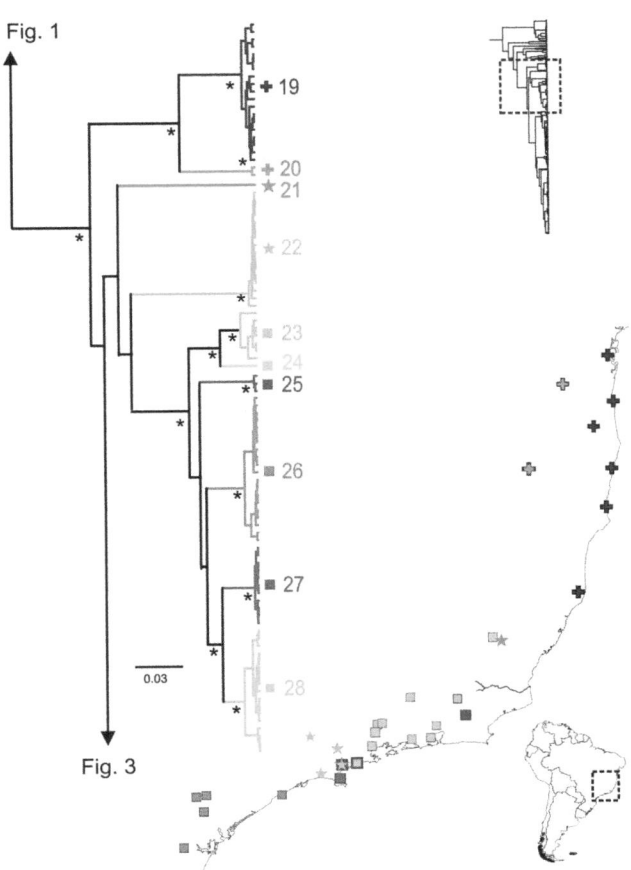

Figure 2. *Dendropsophus minutus* **tree with lineage distribution part 2.** 50% Maximum Clade Credibility tree and distribution maps of mtDNA lineages 19–28. Asterisks represent nodes with probability equals to 1. Probabilities lower than 0.9 are not shown.

permit granted to PJRK's (180609 BR 112); Ministerio de Ambiente, Vivienda y Desarrollo Territorial of Colombia through research and collecting permit granted to AJC (No. 15 of 26 July 2010 and access to genetic resources permit No. 44); Ministerio Venezolano del Poder Popular para el Ambiente through permits granted to FJMRR and MHNLS (collection permit #4156, period 2009-2010; access to genetic resources permit #0076 of 22 February 2011). Samples from Ecuador were obtained under permits MAE-DPP-2011-0691 and 05-2011-Investigación-B-DPMS/MAE. Voucher specimens were euthanized using methods that do not require approval by an ethics committee. None of the species collected for this study is listed in the Convention on International Trade in Endangered Species of Wild Fauna and Flora – CITES (www.cites.org).

We analyzed 407 samples of specimens identified as *Dendropsophus aperomeus*, *D. delarivai*, *D. minutus*, *D. stingi* or *D. xapuriensis* which we here consider together with *D. limai* (not included in our analysis), to constitute the *Dendropsophus minutus* species group (see Results, Figures 1–3 and Figures S1-2). The *D. minutus* species group was defined by Faivovich et al. (2005) [30] to comprise *D. delarivai*, *D. limai*, *D. minutus* and *D. xapuriensis*. These authors tentatively allocated *Dendropsophus aperomeus* to the *D. minimus* species group (in accordance with [34]). However, later molecular phylogenetic analyses suggested different positions for *D. aperomeus* [35–38]. *Dendropsophus stingi* has not been

associated with any species group so far [30], but shares morphological characters with *D. minutus* [28,39]. Because of their unsolved relationships, *D. aperomeus* and *D. stingi* were included in our study.

Genomic DNA was extracted in multiple laboratories using various routine methodologies. We used polymerase chain reaction and direct sequencing with PCR primers on automated Sanger sequencers to obtain DNA sequences of two mitochondrial gene fragments: the 16S ribosomal RNA gene and the 5′ portion of the cytochrome oxidase subunit I (COI) gene, the latter corresponding to the standard DNA barcode fragment [40]. See Supplementary Materials for detailed protocols and primers. Sequence alignment was performed using the MUSCLE algorithm [41] as implemented in the software MEGA version 5.0 [42]. 16S sequences were available for all samples (407), while COI sequences were only available for a subset of these (335). Gapped regions of the 16S alignment were treated as missing data. To fill the concatenated alignment missing COI sequences were also treated as missing data. All newly determined sequences were deposited in GenBank (accession numbers: KJ817824 - KJ817835, KJ833032 - KJ833585, KJ933533 - KJ933690, KJ940033 - KJ940049, see also Table S1 for detailed information of data collection).

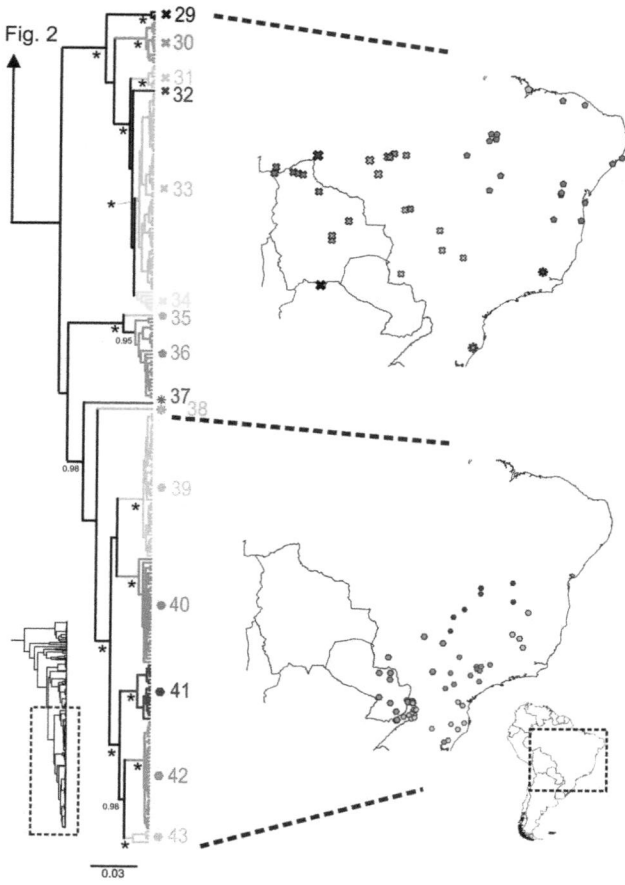

Figure 3. *Dendropsophus minutus* **tree with lineage distribution part 3.** 50% Maximum Clade Credibility tree and distribution maps of mtDNA lineages 29–43. Asterisks represent nodes with probability equals to 1. Probabilities lower than 0.9 are not shown.

Substitution rate estimation and monophyly of the *Dendropsophus minutus* species group

To evaluate the monophyly of the *D. minutus* group and to estimate a substitution rate that could be used for calibrating the phylogeographic analysis (see below), we constructed a time-calibrated 16S gene tree with node constraints based on fossil and geographic evidence [30,43,44], comprising sequences of 216 species of hylids (Table S2), including all *Dendropsophus* species available from GenBank, plus 28 sequences representing the main clades of the *D. minutus* group. Regions of the 16S DNA sequence data that did not overlap our 16S alignment were excluded to ensure that the fragments used in the substitution rate estimation and in the phylogeographic analysis were the same. Additionally we included one sequence representing each main clade of *D. minutus* found in this study.

The calibrated tree was estimated using Bayesian Inference with posterior probabilities via a Markov Chain Monte Carlo (MCMC) sampling as implemented in the software BEAST version 1.7.2. We used a temporal calibration scheme similar to what was used in previous studies of hylids and other amphibians [44,48–50], as follows. We constrained three nodes with uniform priors based on fossil evidence: 1) most recent common ancestor (MRCA) of *Acris* and *Pseudacris* to 15 million years (Ma) or older; 2) MRCA of *Hyla squirella* and *H. cinerea* to 15 Ma or older; and 3) MRCA of

all *Hyla* as 33 Ma or older [51]. Additionally, we assumed the MRCA of Phyllomedusinae and Pelodryadinae to be related to the separation between South America and Australia [52,53]. Therefore, we applied a fourth constraint to the respective MRCA of the two subfamilies using a normal prior with 40 Ma of mean and standard deviation of 6 Ma to allow a higher sampling probability around this value (quantiles: 5% = 30 Ma and 95% = 49.9 Ma) while avoiding hard boundaries. See Methods S1 for further information.

Substitution models were selected using the Akaike Information Criterion [45] in jModeltest version 0.1 [46] which suggested a GTR+I+Γ model for the 16S data set. A coalescent constant size tree prior was used with a lognormal uncorrelated relaxed clock model. The chain was run for 10^8 steps, sampling every 10^4 steps. We repeated the analysis five times to ensure convergence of posterior distributions. A maximum credibility clade tree was summarized using Tree Annotator version 1.7.2 provided in the BEAST 1.7.2 package. The first 20% of the trees were discarded as burn-in after empirical assessment of appropriate chain convergence and mixing with Tracer version 1.5 [47]. We used the median of the posterior density of the substitution rate (*ucld.mean*) obtained by this analysis as prior for phylogeographic analysis of the *D. minutus* complex. See Methods S1 for additional information.

Gene tree inference, GMYC analysis and genetic distance calculations

We estimated an ultrametric mitochondrial gene tree using BEAST software, and all available 16S and COI sequences of the *D. minutus* group (1,068 bp concatenated alignment). As outgroups we included samples of *Dendropsophus nanus*, *D. bipunctatus* and *D. microcephalus*. We subdivided the data set into three partitions: (1) 16S, (2) third codon positions of COI, and (3) combined first and second positions of COI, and implemented a GTR+I+Γ model for each of the partitions, as suggested by the software PartitionFinder [54]. Bayesian inference of an ultrametric phylogenetic tree, using BEAST software, followed the method described above for the taxonomically more inclusive 16S data set.

The resulting ultrametric consensus tree was used as an input to estimate the number of statistically (and presumed evolutionarily) distinct lineages using the generalized mixed Yule-coalescent (GMYC) algorithm. Mitochondrial DNA sequence variation can provide a preliminary yet objective estimate of the number of phenetic clusters or evolutionary lineages represented in a given dataset [55,56]. Of the many available algorithms [57,58], we applied the GMYC algorithm which identifies clusters of mtDNA haplotypes by testing for a transition between fast coalescent rates within clusters relative to slower times to common ancestry among clusters, i.e., as described by a stochastic birth-only Yule model in forward time [59]. GMYC is expected to perform best under dense spatial sampling and limited migration [60,61], two criteria which our study of the *D. minutus* species group would seem to meet (see below). Each statistically significant cluster identified by GMYC may correspond to a deep conspecific lineage or possibly an undescribed species, depending on support from other available taxonomic data [62]. The GMYC algorithm requires only an ultrametric tree as input, for which we used the concatenated 2-gene dataset (see above). We compared the likelihood of a single versus multiple transition threshold model via a χ^2 likelihood ratio test [63]. All GMYC calculations were performed using the 'splits' package, downloaded as 'gmyc.pkg.0.9.6.R' for the R statistical platform [64].

Using the clusters identified in the GMYC analysis, we calculated the uncorrected pairwise *p*-distances between these lineages and the mean *p*-distances within lineages using the 16S sequences only with pairwise deletion in the software MEGA5 [42]. Additionally, to determine the geographical extent of each lineage we used the occurrence data to generate minimum convex polygons representing the distribution range of each lineage observed in more than two localities. We then calculated the area (in km^2) of each polygon using a Winkel-tripel projection in ArcGIS 10 software (ESRI, Redlands, CA).

Phylogeographic analysis and connectivity surfaces

To reconstruct dispersal pathways within the *D. minutus* species group we applied a phylogeographic method that reconstructs geographical coordinates at the nodes of the genealogy using continuous trait reconstruction assuming a log-normal Relaxed Random Walk model implemented in BEAST software. The method estimates ancestral traits and topology simultaneously in a Bayesian framework, taking into account uncertainty of the topology [65]. For details of our implementation of this method, see Methods S1. Analyses assumed a coalescent prior with constant population size and an uncorrelated log-normal relaxed clock model [66], as well as an HKY+Γ model of substitution for each of the partitions used also in the gene tree estimation (see above). We ran three independent chains of 5×10^8 iterations with different random seeds sampling every 50,000 steps. In order to calibrate the tree we used the substitution rate previously estimated for the 16S fragment using a wider time tree analysis of *Dendropsophus* (see above). Mixing of parameter sampling, effective sample size (ESS) and convergence were checked in Tracer 1.5. The resulted tree was summarized with Tree Annotator 1.7.2. The software SPREAD [67] was used to generate a *kml* (Keyhole Markup Language) file which was plotted in a Google Earth map (http://earth.google.com).

To obtain independent support for these dispersal pathways derived from phylogeographic analysis, we constructed conductance maps for the *D. minutus* species group based on Species Distribution Models (SDM) using the software Circuitscape version 3.5.8 [68]. This approach does not incorporate phylogeographic information but uses circuit theory to predict connectivity between localities or areas in heterogeneous landscapes. The algorithm explicitly incorporates effects of limited and irregular habitat extent accounting for multiple pathways and wider habitat bands connecting populations [69]. SDMs can be taken as possible conductive surfaces where highly suitable areas would have high conductance (or low resistance to dispersal) and low suitability areas would have low conductance (or high resistance to dispersal). We thus used a correlative niche model approach and derived a SDM to be used as a base layer for constructing conductivity surfaces.

As our data might represent multiple cryptic species or lineages, a SDM generated on the basis of all samples could be affected by the magnitude of the differences in climatic niches among the different lineages. To assess the extent of climatic niche diversification in the *D. minutus* group, we plotted the first two PC scores of a Principal Component Analysis (PCA) calculated from the values of 19 bioclimatic variables associated with each locality. Given the strong overlap of bioclimatic niches, especially of the majority of lineages 19–43 (Figure S3), i.e., the *D. minutus* complex, we combined these for constructing a single SDM. A robust test of niche conservatism between the lineages (as presented by [70]) was not applicable in our study because only a low number of localities was available for most lineages and because these tests often overestimate niche differentiation and are highly sensitive to sampling bias [71]. We furthermore projected the SDM to climatic scenarios representing past warm and cold

extremes of the Late Pleistocene (Last Interglacial, 120 kyrBP, Last Glacial Maximum, 21 kyrBP). The SDM was constructed using six bioclimatic variables (see Methods S1) [72] using MaxEnt version 3.3.3a [73,74]. The maps of the SDMs obtained for the present and past climate scenarios (see Figure S4) were then used as the input for Circuitscape along with the localities of the whole *D. minutus* group to calculate a conductance map between all pairs of sampled localities. The four resulting conductance maps were averaged. In addition, to highlight areas that under different models maintained high conductance, we applied a 25% quartile threshold to generate binary conductance maps, keeping the grids with higher values of conductance. The maps were then superimposed and spatially summed to highlight the areas of high sustained conductance (stable corridors). All geospatial processing was performed in ArcGis 10 (ESRI; Redlands, CA). See Methods S1 for detailed methodology.

Results

The 16S tree containing all *Dendropsophus* for which sequences were available recovered the monophyly of the *D. minutus* species group (Figure S1). Within the group, the clade containing samples representing lineages 19–43 received a maximal posterior probability (1.0) and is defined here as the *D. minutus* complex (Figures 1–3, Figure S2B-D), given that lineage 25 contains samples from the type locality of *D. minutus* (Figures S2B). The substitution rate for the 16S fragment estimated from this analysis was 7.35×10^{-3}/site/Ma [median of *ucld.mean* parameter (95% HPD $= 6.1–8.7 \times 10^{-3}$)], or 1.47% total divergence per Ma. The exclusion of the third temporal constraint involving the MRCA of the genus *Hyla* (Methods S1) did not change substantially the substitution rate estimate.

The 16S sequences had an average of 477 base pairs (bp) across individuals (standard deviation: 9.6) while the COI sequences had an average of 586 bp (standard deviation: 19.8). The alignment of all 16S sequences of the *D. minutus* species group contained 407, while the COI fragment contained 335 samples. The GMYC analysis on the concatenated data identified 43 entities excluding the outgroup samples, consisting of 31 clusters and 12 'singletons', i.e., an entity consisting of a single concatenated mtDNA haplotype. The likelihood ratio test failed to reject the single-threshold GMYC model ($\chi^2_{9 \text{ d.f.}} = 6.09$, *P*-value $= 0.7308$), thus the following results are based on this simpler model, which tends to be more conservative in estimating number of statistically significant clusters.

Most of the mitochondrial lineages containing more than one sample received strong nodal support (Figures 1–3). The lineages splitting off from basal nodes of the tree (lineages 1–18) are distributed in the Guiana Shield, and in the Andean region of Peru, Ecuador and Colombia, with an eastern extralimital clade assembling disjunct localities in Mato Grosso and Pará (lineages 9–10; Figure 1).

The remaining lineages are in general more widely distributed in central and eastern South America (Figures 2–3). Lineages are largely allopatric but several cases of sympatry were observed (Figure S5). The uncorrected pairwise distances between lineages for the 16S gene ranged from 0.7 to 13%, while within-lineage *p*-distances ranged from 0.0 to 1.8% (Table S3).

Most of the lineages (45%) were found in only one or two localities. Fifty per cent of the lineages were only found in areas smaller than 10 km^2, and more than 70% have known ranges smaller than 10,000 km^2. Eight out of the 43 lineages have a distribution larger than 100,000 km^2 (Figure 4). Largest range sizes were found in northeastern Brazil (Caatinga domain;

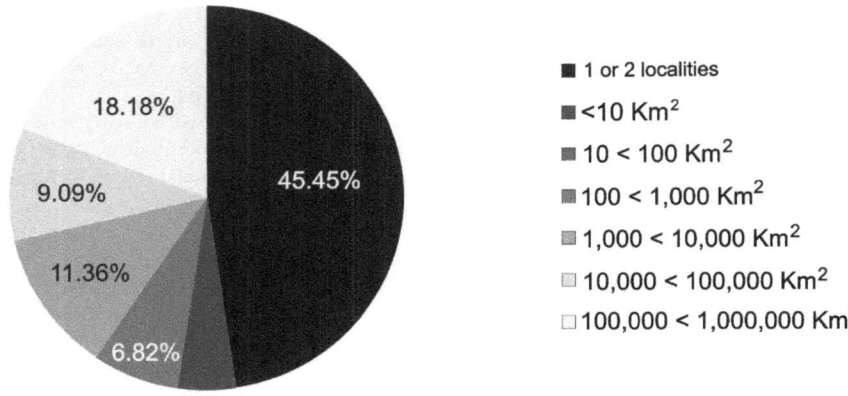

Figure 4. Comparative range size of lineages. Range size of lineages within the *Dendropsophus minutus* complex. Most lineages appear to be microendemic and are only recorded from one or two localities while eight lineages have ranges larger than 100,000 km².

997,262 km², lineage 36), eastern Bolivia and western Brazil (Cerrado, Chaco and Dry Forest domains; 293,321 km², lineage 33) and the Guiana Shield (269,741 km², lineage 2).

The phylogeographic analysis suggested an Amazonian origin for the *D. minutus* group (Figure 5A), with subsequent dispersal to the Atlantic Forest, Guiana shield and the Andean region

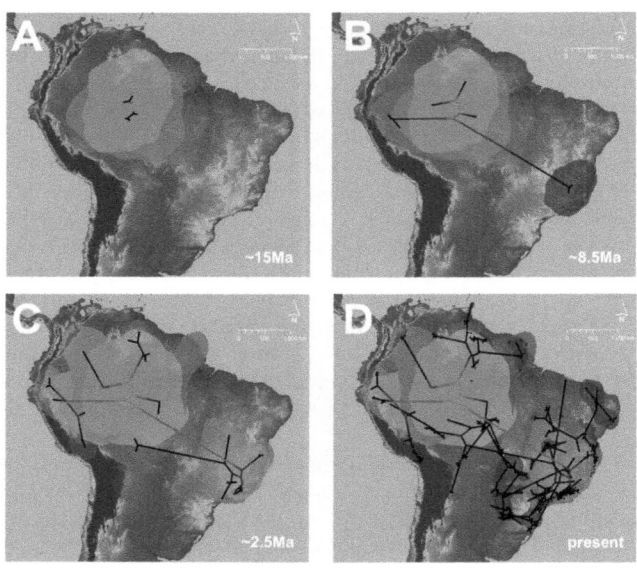

Figure 5. Phylogeographic reconstruction of *Dendropsophus minutus* group. Phylogeographical analysis of the *Dendropsophus minutus* group based on the 16S+COI mitochondrial dataset using a Relaxed Random Walk model for continuous trait reconstruction in Beast software. A) center of origin of the *D. minutus* group. B) Dispersal to west Amazonia, Guiana Shield, Andean region of Peru and eastern Brazil; polygon at the east represent the geographic origin of lineages representing the *D. minutus* complex. C) Dispersal from east Brazil to lowland of Bolivia; further dispersal to Guiana Shield, Peruvian and Colombian areas. D) Recent dispersals to northeast and south Brazil, east Paraguay and Guiana shield. Green polygons and red branches indicate relatively older events while dark polygons and black branches indicate later events. Maps were generated using google earth (earth.google.com).

(Figures 5B and 5C). From there, lineages then dispersed outward from the Atlantic Forest and Amazonia into the southern Atlantic Forest and eastern Paraguay, northeastern and central Brazil, and to the Guiana Shield (Figures 5C and 5D). The conductance analysis with Circuitscape software resulted in the most stable conductance areas along the Brazilian Atlantic coast, central Brazil, northeastern Argentina (Misiones region) and northern Bolivia. The analysis suggested three paths of connectivity between eastern Brazil and the Amazonian region: 1) across northeastern Brazil; 2) across central Brazil; and 3) across southern Brazil, the third being the most stable (Figure 6).

Discussion

Diversity in the *Dendropsophus minutus* species group

Given the numerous divergent lineages and high degree of differentiation revealed here, our molecular approach shows that more than one species could be hidden behind the name *Dendropsophus minutus*. At this stage, we refrain from any formal taxonomic action, as the clarification of the taxonomic status of the populations involved requires thorough integrative revision. In a few cases, there exists information on morphology and calls, but data are not sufficient to either support or reject species status for individual lineages, or allocate available names with certainty. In at least one example GMYC has been found to overestimate the number of species when compared to an integrative approach [75]. Thus, it is possible that several of the lineages identified by this method may not correspond to distinct species. Despite this qualification, some conclusions concerning species diversity can be derived from our results based on monophyly and GMYC results.

Concerning lineages in the *D. minutus* species group not belonging to the *D. minutus* complex (lineages 1–18; Figure 1), our analyses revealed multiple deeply differentiated lineages. Our results under the criteria of mitochondrial monophyly and statistically distinct genealogical lineage (GMYC) suggest recognition of the nominal taxa *D. aperomeus*, *D. delarivai*, *D. stingi* and *D. xapuriensis* as members of the *D. minutus* species group, whose monophyly has not previously been rigorously tested and the relationships among its species has not been studied. In two cases (*D. aperomeus*, *D. stingi*) the species were not previously allocated to the *D. minutus* species group [28,30,34,39,76]. Kaplan (1994) [28] acknowledged that *D. stingi* was morphologically similar to *D. minutus*, while Köhler and Lötters (2001) [39] pointed to similarities of *D. aperomeus* and *D. delarivai* (the latter species at

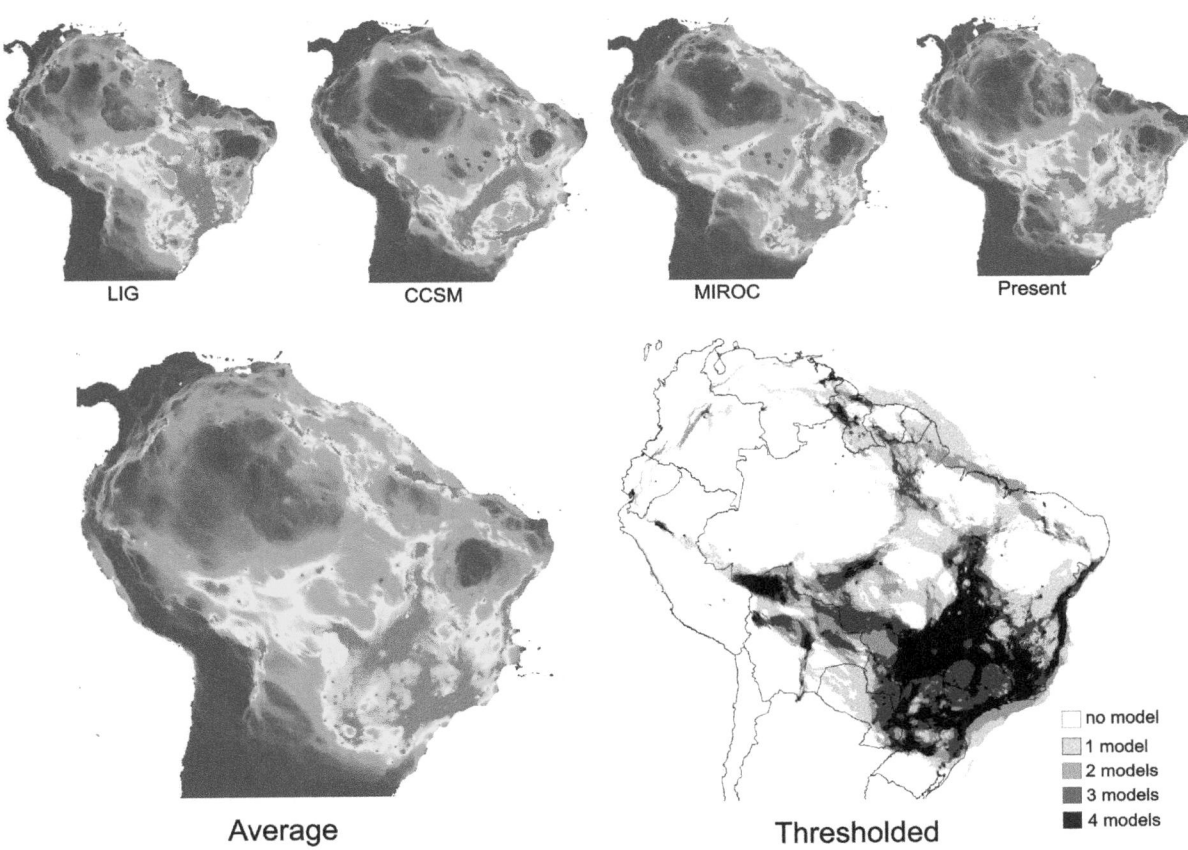

Figure 6. Estimated conductance maps for *Dendropsophus minuts* complex. Conductance maps constructed with the program Circuitscape. Conductances were estimated according to a spacial distribution modeling for the *D. minutus* complex projected to four different climatic scenarios. Conductance maps were averaged and thresholded to show stable dispersal corridors (see methods for details).

that time being tentatively allocated to the *D. minutus* group based on phenetic similarities), and Wiens et al. (2005) [35] found *D. aperomeus* to be related to *D. minutus*. The relationships of *D. aperomeus* and *D. minutus* also remained unresolved in the analysis of Pyron and Wiens (2011) [37].

Among the *D. minutus* species group members external to the *D. minutus* complex, lineages 1–6 are Guianan, while 7–18 are primarily distributed along the Andean foothills, and all show well-pronounced molecular differentiation and divergence. Among lineages 1–6, there is moderate genetic differentiation. Considering mitochondrial reciprocal monophyly and GMYC results as criteria, and being taxonomically conservative, one available name, *Hyla goughi* Boulenger, 1911 (type locality: Trinidad), should likely be removed from the synonymy of *D. minutus* and allocated to populations comprised by all or some of lineages 1–6. As a conservative estimate, we hypothesize that lineages 7–18 comprise seven distinct species, *i.e.*, five named taxa and two undescribed species (lineages 9+10 and 11+12).

Within the *D. minutus* complex (i.e., *D. minutus* sensu lato), GMYC analyses revealed 25 divergent clades (lineages 19–43 in Figures 2 and 3, Figure S2B-D). In several cases, these lineages occur in sympatry (e.g., lineages 19 and 20; 19 and 36; 22 and 27; 27 and 28; 34 and 42). In one case (lineages 30 and 33; Figure 3) genetic differentiation between allopatric lineages is concordant with consistant differences in larval morphology (A. Schulze unpubl. Data, [6]), suggesting species differentiation. However, larval morphology of *D. minutus* has not yet been explored

globally across its range. There are some descriptions of external morphology mainly based on single (or closely located) populations (e.g. [77,78–84]) and one description of internal larval buccal morphology [85]. While larval morphology may offer an alternative source of abundant taxonomic characters [86,87] it also can be highly plastic, suggesting that more data need to be collected and coded with care (e.g. [88,89]).

Advertisement calls among populations which, according to our analyses, would represent distinct lineages, were in some cases shown to be very similar or even identical [26,90]. The certain assignment of calls to any mitochondrial lineage may require sequencing of voucher specimens having recorded calls. On the other hand, call differences have been found among some lineages within *D. minutus* sensu lato identified here through molecular analyses [6], thus potentially providing further support for the presence of additional cryptic species.

Our analysis included topotypic material that may correspond to some of the available names currently regarded as synonyms of *D. minutus*. If additional integrative taxonomic studies are able to link unique diagnostic traits to our mtDNA clades, the following synonyms may require revalidation: *Hyla pallens* Lutz, 1925 (lineage 28), *H. velata* Cope, 1887 (lineage 33), and *H. bivittata* Boulenger, 1888 (lineage 39). On the other hand, *H. suturata* Miranda-Ribeiro, 1926 (lineage 28), and *H. emrichi* Mertens, 1927 (lineage 39) likely represent junior synonyms of *H. pallens* and *H. bivittata*, respectively. Samples from the type locality of *Hyla minuta* Peters, 1872 form a distinct clade (lineage 25) and do not cluster with samples from any other locality studied here (see Figures S2B-D). Results of the GMYC analyses should be interpreted with caution, however, as genealogical clusters identified by this method may correspond to species but may also simply be conspecific phylogeographic lineages [75].

At some localities our analyses revealed up to three independent lineages of the *D. minutus* species group occurring in sympatry or at least in very close proximity (e.g., lineages 21, 24, 37; lineages 22, 27, 28; lineages 7, 18, 30; lineages 2, 3, 5; Figures 2 and 3; Figure S5). Several of the sympatric yet highly divergent lineages are found in or close to areas recognized as Quaternary refugia (southern Bahia and southeastern São Paulo) [91,92]. Phenotypically or genetically distinct groups that are able to maintain their genetic integrity in sympatry may be interpreted as biological species [93]. However, divergent mtDNA lineages in sympatry could also be the product of recent secondary contact between previously isolated yet conspecific (and freely interbreeding) populations [94]. Therefore, we prefer to interpret these sympatric entities as Deep Conspecific Lineages pending additional evidence for possible species status [7,95].

Biogeographic origins of the *Dendropsophus minutus* species group

The region of likely origin of the *D. minutus* group as suggested by the mitochondrial history (Figure 5A) corresponds to the Amazon Basin. The estimated time frame of coalescence of all mtDNA lineages at 18.0 Ma (95% HPD, 14.6–21.4 Ma) falls in the Early to Middle Miocene, a period following the first peak of mountain building in the Andes (~23 Ma) and when, according to Hoorn et al. (2010) [33], western Amazonia was characterized by a large wetland formation of shallow lakes and swamps known as the Lake Pebas System (11–17 Ma). This was apparently an important period for the diversification of the Neotropical fauna and flora in the Amazonian region [33,96,97]. For instance, this period coincides with the diversification of other anurans such as the *Allobates trilineatus* complex 14.0–15.0 Ma [98], the *Rhinella marina* complex 10.7–17.2 Ma [99], the genus *Phyzelaphryne* 9.0–18 Ma [100] and the Amazonian *Adelophryne* clade 13.0–24.0 Ma [100]; as well as with the diversification of Amazonian gymnophthalmids of the genus *Leposoma* around 13.9 Ma [101].

The phylogeographic reconstruction of the *D. minutus* group suggests dispersal to the Andes and the Guiana Shield after its initial diversification between 8.0–11.0 Ma (Figures 5A and B), with additional recent dispersals (Figures 5C and D). This is in remarkable concordance with the pattern observed in dendrobatid frogs for which dispersals from the Amazon Basin into the same regions were reconstructed at 8.8–10.8 Ma, followed by another more recent wave 0.7–6.0 Ma [98]. Concomitant with the first dispersal wave there was a first long distance dispersal of *D. minutus* from the Amazonia to the Atlantic Forest.

An old relationship between Amazonia and an Atlantic Forest area at southeastern Bahia has long been proposed (the "Hiléia Bahiana" [102]). Independent molecular-based studies have recovered similar relationships (e.g. [103]) and studies with amphibians have provided additional evidence for this scenario [100,104–107]. Some authors proposed that climatic changes in the Eocene, leading to the formation of a diagonal band of open and dry biomes (currently Caatinga, Cerrado and Chaco biomes) separating the Atlantic Forest from Amazonia, likely caused ancient vicariance of forest dwelling amphibians [103,105,106]. Accordingly, only generalist species that tolerate both dry and humid conditions could disperse across this potential environmental barrier, and therefore would be distributed in both wet-forest biomes [100]. On the other hand, evidence of Miocene dispersal

of forest-dwelling frogs suggests the possibility of relatively more recent forested connections between Atlantic Forest and Amazonia [100,107]. Our analysis recovers a Late Miocene dispersal of *D. minutus* from Amazonia to eastern Brazil (8–13 Ma) (Figures 5B, C and D). However, frogs of the group are currently found in a variety of habitats, including savannas and deciduous forest. Thus, this first dispersal event could have occurred without the necessity of dense forest continuity and may be unrelated to ancient dispersal of forest dwelling species.

Diversification and dispersal within the *Dendropsophus minutus* complex

The polygon corresponding to the reconstructed area of origin of the *D. minutus* complex encompasses potentially climatically stable areas within the Atlantic Forest [108]. From here, the complex further diversified. Within the Atlantic Forest, sympatry of lineages was observed within or close to Pleistocene refugia (lineages 19 and 20; lineages 21 and 24; lineages 22, 23, 25, 26, 27 and 28; Figure 2) suggesting that these stable areas [91,92,108] might be related to the persistence of deeply divergent lineages. The inferred history of lineage diversification further recovers two discontinuities also found in other vertebrates, possibly representing suture zones [109]. The first of these is located at the southern border of the state of São Paulo, Brazil, represented by the southern limit of the distribution of lineages 26 and 40, and the northern limit of lineage 39, a discontinuity that has also been observed in pit vipers and toads [92,110,111]. The second possible suture zone is in the state of Espirito Santo, Brazil, represented by the northern limit of lineages 21, 23 and 24, and the southern limit of lineage 19 (Figs. 2–3; Fugure S6), coinciding with genetic breaks observed in toads, geckos, foxes and woodcreepers [92,112–114].

After the initial diversification of the *D. minutus* complex in the Atlantic Forest, our phylogeographic inference suggests subsequent dispersal to other areas of eastern South America, central Brazil and Amazonia, supporting a southern dispersal route between Amazonia and the Atlantic Forest (Figures 5C and D). Distribution patterns of different vertebrate groups, as well as climatic and floristic evidence, suggest recent dispersal corridors between these forested areas [103,107,114–116]. On the other hand, the phylogeography of tropical rattlesnakes adapted to dry biomes suggests past connectivity between open area formations that are currently isolated [117,118]. These results combined with paleoclimatic models of the Atlantic Forest and Cerrado [108,119] support the past existence of a dynamic interplay of dispersal corridors and temporary barriers across the South American lowlands. Our conductance analysis that took into account contrasting climatic scenarios suggests the existence of stable stretches of favorable habitat for *D. minutus* along northern and particularly southern Amazonia during the Pleistocene, which would have allowed recent dispersal between forested areas of South America, corroborating the phylogeographic results (Figure 6).

The stable habitat corridor inferred along southern Amazonia (Figure 6) was proposed previously as the main floristic connection between the Atlantic Forest and Amazonia [120] and phylogeographic analyses of some species of small mammals corroborate this hypothesis [103]. Moreover, the palynological record from different continents of the southern hemisphere suggest the existence of a band of moisture at this latitude (~23°S) during the Last Glacial Maximum (LGM) [121], which raises the possibility that this same corridor may have persisted even earlier. The presence of a mosaic of forest and open formations, and higher moisture, could have generated conditions necessary for the existence of a Pleistocenic southern dispersal route for *D. minutus*.

Widespread amphibian species in the Neotropics

Data presented herein provide conclusive evidence for a strong genetic subdivision of the nominal species *Dendropsophus minutus* as currently understood. Current taxonomy conservatively assumes a putatively widespread species encompassing a vast area of South America (from approximately latitude 11.0°N to 35.0°S), distributed across several biomes. Our results, however, reveal high genetic diversity within *D. minutus* that would suggest the existence of numerous distinct species, leading to an important increase in number of species. If this hypothesis is confirmed through further studies, the existence of an increased number of species with decreased range sizes would have important consequences for the definition of centers of endemism and for assessing conservation status.

Despite revealing a substantial amount of cryptic genetic diversity within *D. minutus* sensu lato, our results also confirm the existence of widespread Neotropical species of anurans. While we cannot yet confirm which of the mitochondrial lineages within the *D. minutus* complex will merit a status as separate species, we can inversely conclude that in most cases, all samples assigned to one mtDNA lineage should be conspecific. Although cases of distinct amphibian species with low mtDNA differentiation exist (e.g. [86]) and phenomena of mtDNA introgression can potentially blur species identities (e.g. [122,123]), such cases remain exceptional. Therefore, these factors are unlikely to substantially affect our calculation of range sizes according to which a total of eight mtDNA lineages have ranges >100,000 km^2 (lineages 2, 19, 33, 34, 36, 39 41, 42). In the most inflationist taxonomic scenario, with each of these lineages representing separate species, our dataset still provides evidence for a species of lowland Neotropical amphibian (lineage 36) occupying an area of almost one million km^2, encompassing multiple biomes across a distance of about 1,600 km between its two most distant populations.

The most widespread lineages within *D. minutus* sensu lato have distributions restricted to or centered in Brazil and occur within rather open habitats, while lineage 2 of the *D. minutus* group (with a range of almost 270,000 km^2) occurs in rainforest. Various of the lineages known from only few or single sites (e.g., lineages 8, 11, 12, 13, 14, 15, 16) occur in the Andean foothills or on mountain slopes. Nevertheless, in the Andean area, a higher sampling density is needed before it can be concluded with certainty that those lineages are restricted to small ranges. Hence, the distribution of mitochondrial lineages in the *D. minutus* group indicates that in open lowland areas of South America, small-sized species of anurans can be widespread.

A plea for collaboration in taxonomy and large scale phylogeography

Whether a researcher is interested in taxonomic or biogeographic questions concerning widespread species groups and complexes, this study points to several advantages of analysing spatialy dense and geographically complete datasets. Large-scale analyses are of course pivotal to understanding processes at a continental level. Patterns may be misinterpreted when looking at a limited geographical area and/or limited sample size. Widespread groups suchs as *D. minutus* group are suitable for continental analyses, but they are often associated with taxonomic difficulties because of the lack of comprehensive datasets. Also, certain biogeographic analyses, like those performed here, damand extensive data. Through collaboration we can increase the efficiency of data collection, thus overcoming the problem of incomplete datasets.

The taxonomic and conservation crisis we are currently facing, where some species will be extinct before the community becomes aware of their existence [124], underscores the urgency and importance of collaborative work. Species extinction represents the loss of the genealogical and biogeographic information embedded in its evolutionary history. Broad collaboration among scientists is necessary to rapidly tackle taxonomic and biogeographic questions. As in other areas of science, we feel that in taxonomy and biogeography the establishment of multi-investigator, multi-institutional and multi-national consortia dealing with widely distributed and taxonomically convoluted groups will improve quality and speed of taxonomic revision, consequently improving our understanding of biodiversity patterns, the evolutionary processes that generated them, and the conservation status of tropical organisms.

Supporting Information

Figure S1 16S genealogy of the genus *Dendropsophus*: partial view of the 50% Maximum Clade Credibility tree derived from Bayesian phylogenetic inference of 216 mitochondrial 16S sequences of Hylidae species plus 28 exemplars of the *D. minutus* group, that was performed for the substitution rate estimations using the program BEAST 1.7.2. The *Dendropsophus minutus* group is highlighted by the dashed line and the *Dendropsophus minutus* complex by the grey box. Node numbers indicate posterior probabilities which are only shown when higher than 0.8. Numbers between brakets indicate lineage number acording with the GMYC results or GenBank accession numbers.

Figure S2 A. *Dendropsophus minutus* tree with samples names and annotations Part 1. 50% Maximum Clade Credibility tree, lineages 1–18. Asterisks represent nodes with probability equals to 1. Probabilities lower than 0.9 are not shown. Annotations refer to samples of particular interest, mainly samples collected at or close to the type locality of certain nominal taxa. **B. *Dendropsophus minutus* tree with samples names and annotations Part 2.** 50% Maximum Clade Credibility tree, lineages 19–28. Asterisks represent nodes with probability equals to 1. Probabilities lower than 0.9 are not shown. Annotations refer to samples of particular interest, mainly samples collected at or close to the type locality of certain nominal taxa. **C. *Dendropsophus minutus* tree with samples names and annotations Part 3.** 50% Maximum Clade Credibility tree, lineages 29–36. Asterisks represent nodes with probability equals to 1. Probabilities lower than 0.9 are not shown. Annotations refer to samples of particular interest, mainly samples collected at or close to the type locality of certain nominal taxa. **D. *Dendropsophus minutus* tree with samples names and annotations Part 4.** 50% Maximum Clade Credibility tree, lineages 37–43. Asterisks represent nodes with probability equals to 1. Probabilities lower than 0.9 are not shown. Annotations refer to samples of particular interest, mainly samples collected at or close to the type locality of certain nominal taxa.

Figure S3 PCA of climatic variables. PCA plots showing the first two principal components separately for lineages 1–18 (left) and 19–43 (right). Symbols and colours match those in Figs. 1–3. Loadings of the first two principal components were as follows (separated by colon): annual mean temperature 0.324, 0; mean monthly temperature range −0.122, 0.274; isothermality 0.192, 0; temperature seasonality −0.245, −0.13; maximum temperature warmest month 0.269, 0; minimum temperature coldest month 0.336, 0; temperature annual range −0.214, 0.123; mean temperature wettest quarter 0.274, 0; mean temperature driest quarter 0.336, 0; mean temperature warmest quarter 0.282, 0; mean temperature coldest quarter 0.339, 0; annual precipitation 0.187, −0.292; precipitation wettest month 0.226, 0; precipitation driest month 0, −0.477; precipitation seasonality 0, 0.421; precipitation wettest quarter 0.224, 0; precipitation driest quarter 0, −0.482; precipitation warmest quarter −0.113, −0.184; precipitation coldest quarter 0.132, −0.331. Main result of this analysis: the first two principal components (PCs) accounted for 64% of the climatic variance, with highest loadings for temperature variables along PC 1 (minimum temperature coldest month, mean temperature driest quarter, mean temperature coldest quarter) and precipitation variables along PC 2 (precipitation driest month, precipitation seasonality, precipitation of driest quarter). The PCA suggests that within the *D. minutus* group, the climatic niches (i.e., 19 bioclimatic temperature and precipitation dimensions) are rather similar, even when the two main groups are compared (i.e. lineages 1–18 vs. 19–43). Some lineages from the periphery of the known geographic distribution of the group, including lineages 1–18, are weakly separated (lineages 9, 11–13 15, 16).

Figure S4 Spatial distribution models used as resistance layers in the Circuit Scape analysis.

Figure S5 Distribution of all mitochondrial lineages. Map showing the distribution of all mitochondrial lineages in the *Dendropsophus minutus* group as revealed by this study. Symbols refer to those used in Figs. 1–3.

Table S1 Complete data collection with genbank accession numbers, voucher numbers and locality information.

Table S2 Species used in the substitution rate estimation with respective Genbank accession numbers.

Table S3 Uncorrected pairwise *p*-distance among lineages and average within lineage *p*-distance.

Table S4 List of primers used in this study with respective anealing temperatures.

Table S5 Localities and coordinates used in the Spatial Distribution Modeling.

Methods S1 Supplementary methods.

Acknowledgments

We are grateful to numerous friends and colleagues who provided samples, information or helped in the field. Diego Baldo, Santiago Castroviejo-Fisher, and José M. Padial kindly provided sequences and/or tissues. Julian Faivovich (Museo Argentino de Ciencias Naturales "Bernardino Rivadavia"-CONICET) provided sequences of specimens from Argentina. PJRK warmly acknowledges D. Bruce Means (Coastal Plains Institute and Land Conservancy, USA) for the loan of additional critical tissue samples. Meike Kondermann and Gabi Keunecke helped with laboratory procedures in Braunschweig and Jakob Fahr provided usefull recommendations for GIS analysis. Diana Flores was in charge of laboratory procedures at BioCamb, Ecuador. Fieldwork in Guyana was made possible through the help of the Iwokrama International Centre, particularly R. Thomas-Caesar and C. I. Bovolo.

Author Contributions

Conceived and designed the experiments: MG MV JK. Performed the experiments: MG AJC VGDO SL AF LSB FB IR RE GGU FG JMG MH MJ PJRK AK RL ML JM JPP FJMR AS JCS MS MTR ET CFBH MV JK. Analyzed the data: MG AJC AR SL MV JK. Contributed reagents/materials/analysis tools: MG AJC VGDO SL AF LSB FB IR RE GGU FG JMG MH MJ PJRK AK RL ML JM JPP FJMR AS JCS MS MTR ET CFBH MV JK. Wrote the paper: MG AJC VGDO MV JK.

References

1. Bickford D, Lohman DJ, Sodhi NS, Ng PKL, Meier R, et al. (2007) Cryptic species as a window on diversity and conservation. Trends Ecol Evol 22: 148–155.
2. Köhler J, Vieites DR, Bonett RM, García FH, Glaw F, et al. (2005) New amphibians and global conservation: a boost in species discoveries in a highly endangered vertebrate group. BioScience 55: 693–696.
3. Vences M, Köhler J (2008) Global diversity of amphibians (Amphibia) in freshwater. Hydrobiologia 595: 569–580.
4. Fouquet A, Gilles A, Vences M, Marty C, Blanc M, et al. (2007) Underestimation of species richness in Neotropical frogs revealed by mtDNA analyses. PLoS One 2: e1109.
5. Funk WC, Caminer M, Ron SR (2012) High levels of cryptic species diversity uncovered in Amazonian frogs. Proc Biol Sci 279: 1806–1814.
6. Jansen M, Bloch R, Schulze A, Pfenninger M (2011) Integrative inventory of Bolivia's lowland anurans reveals hidden diversity. Zool Scripta 40: 567–583.
7. Vieites DR, Wollenberg KC, Andreone F, Köhler J, Glaw F, et al. (2009) Vast underestimation of Madagascar's biodiversity evidenced by an integrative amphibian inventory. Proc Natl Acad Sci U S A 106: 8267–8272.
8. Biju SD, van Bocxlaer I, Mahony S, Dinesh KP, Radhakrishnan C, et al. (2011) A taxonomic review of the Night Frog genus Nyctibatrachus Boulenger, 1882 in the Western Ghats, India (Anura: Nyctibatrachidae) with description of twelve new species. Zootaxa 3029.
9. Funk WC, Angulo A, Caldwell JP, Ryan MJ, Cannatella DC (2008) Comparison of morphology and calls of two cryptic species of Physalaemus (Anura: Leiuperidae). Herpetologica 64: 290–304.
10. Glaw F, Köhler J, De la Riva I, Vieites DR, Vences M (2010) Integrative taxonomy of Malagasy treefrogs: combination of molecular genetics, bioacoustics and comparative morphology reveals twelve additional species of Boophis. Zootaxa 2383: 1–82.
11. Padial AA, De la Riva I (2009) Integrative taxonomy reveals cryptic Amazonian species of Pristimantis (Anura: Strabomantidae). Zool J Linnean Soc 155: 97–122.
12. Maciel DB, Nunes I (2010) Comparison of morphology and calls of two cryptic species of Physalaemus (Anura: Leiuperidae) from the rock meadows of Espinhaço, Brazil. Zootaxa: 53–61.
13. Angulo A, Icochea J (2010) Cryptic species complexes, widespread species and conservation: lessons from Amazonian frogs of the Leptodactylus marmoratus group (Anura: Leptodactylidae). Syst Biodivers 8: 357–370.
14. Gehara M, Canedo C, Haddad CFB, Vences M (2013) From widespread to microendemic: molecular and acoustic analyses show that Ischnocnema guentheri (Amphibia: Brachycephalidae) is endemic to Rio de Janeiro, Brazil. Conservation Genetics 14: 973–982.
15. Wynn A, Heyer WR (2001) Do geographically widespread species of tropical amphibians exist? An estimate of genetic relatedness within the neotropical frog Leptodactylus fuscus (Schneider 1799) (Anura Leptodactylidae). Trop Zool 14: 255–285.
16. Zeisset I, Beebee TJC (2008) Amphibian phylogeography: a model for understanding historical aspects of species distribution. Heredity: 109–119.
17. Elmer KR, Davila JA, Lougheed SC (2007) Cryptic diversity and deep divergence in an upper Amazonian leaflitter frog, Eleutherodactylus ockendeni. BMC Evol Biol 7: 247.
18. Noonan BP, Gaucher P (2005) Phylogeography and demography of Guianan harlequin toads (Atelopus): diversification within a refuge. Mol Ecol 14: 3017–3031.
19. Symula R, Schulte R, Summers K (2003) Molecular systematics and phylogeography of Amazonian poison frogs of the genus Dendrobates. Mol Phylogenet Evol 26: 452–475.
20. Camargo A, De Sá RO, Heyer WR (2006) Phylogenetic analyses of mtDNA sequences reveal three cryptic lineages in the widespread neotropical frog Leptodactylus fuscus (Schneider, 1799) (Anura, Leptodactylidae). Biol J Linn Soc 87: 325–341.
21. Hawkins MA, Sites JW Jr, Noonan BP (2007) Dendropsophus minutus (Anura: Hylidae) of the Guiana Shield: using DNA barcodes to assess identity and diversity. Zootaxa 1540: 61–67.

22. Fouquet A, Vences M, Salducci MD, Meyer A, Marty C, et al. (2007) Revealing cryptic diversity using molecular phylogenetics and phylogeography in frogs of the *Scinax ruber* and *Rhinella margaritifera* species groups. Mol Phylogenet Evol 43: 567–582.

23. Duellman WE (1956) The frogs of the hylid genus *Phrynohyas* Fitzinger, 1843. Miscellaneous Publications Museum of Zoology, University of Michigan: 1–47.

24. Duellman WE (1971) A taxonomic review of South American hylid frogs, genus *Phrynohyas*. Occas pap Mus Nat Hist (Lawrence) 1–21.

25. Frost DR (2013) Amphibian Species of the World: an Online Reference. Version 5.6. Electronic Database accessible at http://research.amnh.org/vz/herpetology/amphibia/. American Museum of Natural History, New York, USA. Accessed 2013 Dec.

26. Cardoso AJ, Haddad CFB (1984) Variabilidade acústica em diferentes populações e interações agressivas de *Hyla minuta* (Amphibia, Anura). Ciência e Cultura: 1393–1399.

27. Donnelly MA, Myers CW (1991) Herpetological results of the 1990 Venezuelan expedition to the summit of Cerro Guaiquinima, with new Tepui reptiles. Am Mus Novit: 1–54.

28. Kaplan M (1994) A new species of frog of the genus *Hyla* from the Cordillera Oriental in northern Colombia with comments on the taxonomy of *Hyla minuta*. J Herpetol 28: 79–87.

29. Murphy JC (1997) Amphibians and reptiles of Trinidad and Tobago. Malabar: Krieger Publ.

30. Faivovich J, Haddad CFB, Garcia PCA, Frost DR, Campbell JA, et al. (2005) Systematic review of the frog family Hylidae, with special reference to Hylinae: phylogenetic analysis and taxonomic revision. Bull Am Mus Nat Hist 294: 240 pp.

31. Etienne RS, Olff H (2004) How dispersal limitation shapes species-body size distributions in local communities. Am Nat 163: 69–83.

32. Pabijan M, Wollenberg KC, Vences M (2012) Small body size increases the regional differentiation of populations of tropical mantellid frogs (Anura: Mantellidae). J Evol Biol 25: 2310–2324.

33. Hoorn C, Wesselingh FP, ter Steege H, Bermudez MA, Mora A, et al. (2010) Amazonia through time: Andean uplift, climate change, landscape evolution, and biodiversity. Science 330: 927–931.

34. Duellman WE (1982) A new species of small yellow *Hyla* from Perú (Anura: Hylidae). Amphibi Reptil 3: 153–160.

35. Wiens JJ, Fetzner JW, Parkinson CL, Reeder TW (2005) Hylid frog phylogeny and sampling strategies for speciose clades. Syst Biol 54: 778–807.

36. Wiens JJ, Kuczynski CA, Hua X, Moen DS (2010) An expanded phylogeny of treefrogs (Hylidae) based on nuclear and mitochondrial sequence data. Mol Phylogenet Evol 55: 871–882.

37. Pyron RA, Wiens JJ (2011) A large-scale phylogeny of Amphibia including over 2800 species, and a revised classification of extant frogs, salamanders, and caecilians. Mol Phylogenet Evol 61: 543–583.

38. Motta A, Castroviejo-Fisher S, Venegas P, Orrico V, Padial J (2012) A new species of the *Dendropsophus parviceps* group from the western Amazon Basin (Amphibia: Anura: Hylidae). Zootaxa: 18–30.

39. Köhler J, Lötters S (2001) Description of a small tree frog, genus *Hyla* (Anura, Hylidae), from humid Andean slopes of Bolivia. Salamandra 37: 175–184.

40. Hebert PD, Ratnasingham S, de Waard JR (2003) Barcoding animal life: cytochrome c oxidase subunit 1 divergences among closely related species. Proc Biol Sci 270: S96–S99.

41. Edgar RC (2004) MUSCLE: multiple sequence alignment with high accuracy and high throughput. Nucleic Acids Res 32: 1792–1797.

42. Tamura K, Peterson D, Peterson N, Stecher G, Nei M, et al. (2011) MEGA5: molecular evolutionary genetics analysis using Maximum Likelihood, evolutionary eistance, and Maximum Parsimony methods. Mol Biol Evol 28: 2731–2739.

43. Sanmartin I, Ronquist F (2004) Southern hemisphere biogeography inferred by event-based models: plant versus animal patterns. Syst Biol 53: 216–243.

44. Wiens JJ, Graham CH, Moen DS, Smith SA, Reeder TW (2006) Evolutionary and ecological causes of the latitudinal diversity gradient in hylid frogs: treefrog trees unearth the roots of high tropical diversity. Am Nat 168: 579–596.

45. Akaike H (1974) A new look at the statistical model identification. IEEE Transactions on Automatic Control 19: 716–723.

46. Posada D (2008) jModelTest: phylogenetic model averaging. Mol Biol Evol 25: 1253–1256.

47. Rambaut A, Drummond AJ (2009) Tracer v1.5 Available at http://beast.bio.ed.ac.uk/Tracer. Accessed 2012 July.

48. Lemmon EM, Lemmon AR, Cannatella DC (2007) Geological and climatic forces driving speciation in the continentally distributed trilling chorus frogs (Pseudacris). Evolution 61: 2086–2103.

49. San Mauro D, Vences M, Alcobendas M, Zardoya R, Meyer A (2005) Initial diversification of living amphibians predated the breakup of Pangaea. Am Nat 165: 590–599.

50. Smith SA, Stephens PR, Wiens JJ (2005) Replicate patterns of species richness, historical biogeography, and phylogeny in Holarctic treefrogs. Evolution 59: 2433–2450.

51. Holman JA (1998) Pleistocene amphibians and reptiles in Britain and Europe. New York: Oxford University Press. vi, 254 p. p.

52. McLoughlin S (2001) The breakup history of Gondwana and its impact on pre-Cenozoic floristic provincialism. Aust J Bot 49: 271–300.

53. Scher HD, Martin EE (2006) Timing and climatic consequences of the opening of Drake Passage. Science 312: 428–430.

54. Lanfear R, Calcott B, Ho SY, Guindon S (2012) Partitionfinder: combined selection of partitioning schemes and substitution models for phylogenetic analyses. Mol Biol Evol 29: 1695–1701.

55. Goldstein PZ, DeSalle R (2011) Integrating DNA barcode data and taxonomic practice: determination, discovery, and description. Bioessays 33: 135–147.

56. Nielsen R, Matz M (2006) Statistical approaches for DNA barcoding. Syst Biol 55: 162–169.

57. Boykin LM, Armstrong KF, Kubatko L, De Barro P (2012) Species delimitation and global biosecurity. Evol Bioinform Online 8: 1–37.

58. Paz A, Crawford AJ (2012) Molecular-based rapid inventories of sympatric diversity: a comparison of DNA barcode clustering methods applied to geography-based vs clade-based sampling of amphibians. J Biosci 37: 887–896.

59. Pons J, Barraclough TG, Gomez-Zurita J, Cardoso A, Duran DP, et al. (2006) Sequence-based species delimitation for the DNA taxonomy of undescribed insects. Syst Biol 55: 595–609.

60. Lohse K (2009) Can mtDNA barcodes be used to delimit species? A response to Pons et al. (2006). Systematic Biology 58: 439–442.

61. Papadopoulou A, Monaghan MT, Barraclough TG, Vogler AP (2009) Sampling error does not invalidate the Yule-coalescent model for species delimitation. A response to Lohse (2009). Syst Biol 58: 442–444.

62. Padial J, Miralles A, De la Riva I, Vences M (2010) The integrative future of taxonomy. Frontiers in Zoology 7: 16.

63. Monaghan MT, Wild R, Elliot M, Fujisawa T, Balke M, et al. (2009) Accelerated species inventory on Madagascar using coalescent-based models of species delineation. Systematic Biology 58: 298–311.

64. R Core Team (2014) R: A language and environment for statistical computing. R Foundation for Statistical Computing. Vienna, Austria: URL: http://www.R-project.org/. Accessed 2014 Jan.

65. Lemey P, Rambaut A, Welch JJ, Suchard MA (2010) Phylogeography takes a relaxed random walk in continuous space and time. Mol Biol Evol 27: 1877–1885.

66. Drummond AJ, Ho SYW, Phillips MJ, Rambaut A (2006) Relaxed phylogenetics and dating with confidence. PLoS Biol 4: e88.

67. Bielejec F, Rambaut A, Suchard MA, Lemey P (2011) SPREAD: spatial phylogenetic reconstruction of evolutionary dynamics. Bioinform 27: 2910–2912.

68. McRae BH, Shah VB (2008) Circuitscape 3.5.8.

69. McRae BH (2006) Isolation by resistance. Evolution 60: 1551–1561.

70. Warren DL, Glor RE, Turelli M (2008) Environmental niche equivalency versus conservatism: Quantitative approaches to niche evolution. Evolution 62: 2868–2883.

71. Peterson AT (2011) Ecological niche conservatism: a time-structured review of evidence. Journal of Biogeography 38: 817–827.

72. Busby JR (1991) BIOCLIM - a bioclimatic analysis and prediction system. Nature Conservation: cost effective biological surveys and data analysis. pp. 64–68.

73. Phillips SJ, Dudík M (2008) Modeling of species distributions with Maxent: new extensions and a comprehensive evaluation. Ecography 31: 161–175.

74. Phillips SJ, Anderson RP, Schapire RE (2006) Maximum entropy modeling of species geographic distributions. Ecol Model 190: 231–259.

75. Miralles A, Vences M (2013) New metrics for comparison of taxonomies reveal striking discrepancies among species delimitation methods in *Madascincus* lizards. PLoS One 8: e68242.

76. Martins M, Cardoso AJ (1987) Novas espécies de hilídeos do estado do Acre (Amphibia: Anura). Rev Bras Biol 47: 549–558.

77. Bokermann WCA (1963) Girinos de Anfíbios Brasileiros - I (Amphibia - Salientia). An Acad Bras Ciênc 35: 465–474.

78. Cei JM (1980) Amphibians of Argentina. Monitore Zoologico Italiano (N S), Monogr 2: 609 pp.

79. Duellman WE (1978) The biology of an equatorial herpetofauna in Amazonian Ecuador. Misc publ Univ Kans Mus Nat Hist 65: 1–352.

80. Hero J-M (1990) An illustrated key to tadpoles occurring in the Central Amazon rainforest, Manaus, Amazonas, Brasil. Amazoniana 11: 201–262.

81. Heyer WR, Rand AS, Cruz CAG, Peixoto OL, Nelson CE (1990) Frogs of Boracéia. Arq Zool 31: 231–410.

82. Kenny JS (1969) The Amphibia of Trinidad. Studies on the Fauna of Curaçao and other Caribbean Islands XXIX 29: 1–78.

83. Lescure J, Marty C (2000) Atlas des Amphibiens de Guyane. Patrimoines Naturels 45: 1–388.

84. Rossa-Feres DdC, Nomura F (2006) Characterization and taxonomic key for tadpoles (Amphibia: Anura) from the northwestern region of São Paulo State, Brazil. Biota Neotrop 6.

85. Echeverría DD (1997) Microanatomy of the buccal apparatus and oral cavity of *Hyla minuta* Peters, 1872 larvae (Anura, Hylidae), with data on feeding habits. Alytes 15: 26–36.

86. Vences M, Gehara M, Kohler J, Glaw F (2012) Description of a new Malagasy treefrog (*Boophis*) occurring syntopically with its sister species, and a plea for studies on non-allopatric speciation in tropical amphibians. Amphibi Reptil 33: 503–520.

87. Vences M, Köhler J, Crottini A, Glaw F (2010) High mitochondrial sequence divergence meets morphological and bioacoustic conservatism: *Boophis*

quasiboehmei sp. n., a new cryptic treefrog species from south-eastern Madagascar. Bonn zool Bul: 241–255.

88. Reading CJ (2010) The impact of environmental temperature on larval development and metamorph body condition in the common toad, *Bufo bufo*. Amphibi Reptil 31: 483–488.

89. Gomes MdR, Peixoto OL (1991) Considerações sobre os girinos de *Hyla senicula* (Cope, 1868) e *Hyla soaresi* (Caramaschi e Jim, 1983) (Amphibia, Anura, Hylidae). Acta Biologica Leopoldensia 13: 5–18.

90. Morais AR, Batista VG, Gambale PG, Signorelli L, Bastos RP (2012) Acoustic communication in a Neotropical frog (*Dendropsophus minutus*): vocal repertoire, variability and individual discrimination. Herpetol J 22: 249–257.

91. Carnaval AC, Hickerson MJ, Haddad CFB, Rodrigues MT, Moritz C (2009) Stability predicts genetic diversity in the Brazilian Atlantic forest hotspot. Science 323: 785–789.

92. Thomé MT, Zamudio KR, Giovanelli JG, Haddad CF, Baldissera FA Jr, et al. (2010) Phylogeography of endemic toads and post-Pliocene persistence of the Brazilian Atlantic Forest. Mol Phylogenet Evol 55: 1018–1031.

93. Mallet J (2008) Hybridization, ecological races and the nature of species: empirical evidence for the ease of speciation. Philos Trans R Soc Lond B Biol Sci 363: 2971–2986.

94. Hewitt GM (2011) Quaternary phylogeography: the roots of hybrid zones. Genetica 139: 617–638.

95. Vences M, Köhler J, Vieites DR, Glaw F (2011) Molecular and bioacoustic differentiation of deep conspecific lineages of the Malagasy treefrogs *Boophis tampoka* and *B. luteus*. Herpetol Notes 4: 239–246.

96. Antonelli A, Quijada-Mascareñas A, Crawford AJ, Bates JM, Velazco PM, et al. (2009) Molecular Studies and Phylogeography of Amazonian Tetrapods and their Relation to Geological and Climatic Models. Amazonia: Landscape and Species Evolution: Wiley-Blackwell Publishing Ltd. pp. 386–404.

97. Antonelli A, Nylander JA, Persson C, Sanmartin I (2009) Tracing the impact of the Andean uplift on Neotropical plant evolution. Proc Natl Acad Sci U S A 106: 9749–9754.

98. Santos JC, Coloma LA, Summers K, Caldwell JP, Ree R, et al. (2009) Amazonian amphibian diversity is primarily derived from late Miocene Andean lineages. PLoS Biol 7: e56.

99. Vallinoto M, Sequeira F, Sodré D, Bernardi JAR, Sampaio I, et al. (2010) Phylogeny and biogeography of the *Rhinella marina* species complex (Amphibia, Bufonidae) revisited: implications for Neotropical diversification hypotheses. Zoologica Scripta 39: 128–140.

100. Fouquet A, Loebmann D, Castroviejo-Fisher S, Padial JM, Orrico VGD, et al. (2012) From Amazonia to the Atlantic forest: Molecular phylogeny of Phyzelaphryninae frogs reveals unexpected diversity and a striking biogeographic pattern emphasizing conservation challenges. Mol Phylogenet Evol 65: 547–561.

101. Pellegrino KC, Rodrigues MT, Harris DJ, Yonenaga-Yassuda Y, Sites JW Jr (2011) Molecular phylogeny, biogeography and insights into the origin of parthenogenesis in the Neotropical genus *Leposoma* (Squamata: Gymnophthalmidae): Ancient links between the Atlantic Forest and Amazonia. Mol Phylogenet Evol 61: 446–459.

102. Andrade-Lima D (1966) Vegetação. IBGE, Atlas Nacional do Brasil: Rio de Janeiro (Conselho Nacional de Geografia).

103. Costa LP (2003) The historical bridge between the Amazon and the Atlantic Forest of Brazil: a study of molecular phylogeography with small mammals. J Biogeogr 30: 71–86.

104. Canedo C, Haddad CFB (2012) Phylogenetic relationships within anuran clade Terrarana, with emphasis on the placement of Brazilian Atlantic rainforest frogs genus *Ischnocnema* (Anura: Brachycephalidae). Mol Phylogenet Evol 65: 610–620.

105. Fouquet A, Recoder R, Teixeira M Jr, Cassimiro J, Amaro RC, et al. (2012) Molecular phylogeny and morphometric analyses reveal deep divergence between Amazonia and Atlantic Forest species of *Dendrophryniscus*. Mol Phylogenet Evol 62: 826–838.

106. Heinicke MP, Duellman WE, Hedges SB (2007) Major Caribbean and Central American frog faunas originated by ancient oceanic dispersal. Proc Natl Acad Sci U S A 104: 10092–10097.

107. de Sá RO, Streicher JW, Sekonyela R, Forlani MC, Loader SP, et al. (2012) Molecular phylogeny of microhylid frogs (Anura: Microhylidae) with emphasis on relationships among New World genera. BMC Evol Biol 12: 241.

108. Carnaval AC, Moritz C (2008) Historical climate modelling predicts patterns of current biodiversity in the Brazilian Atlantic forest. J Biogeogr 35: 1187–1201.

109. Remington CL (1968) Suture-zones of hybrid interaction between recently joined biotas. In: Dobzhansky T, Hecht MK, Steere WC, editors. Evolutionary Biology: Springer US. pp. 321–428.

110. Grazziotin FG, Monzel M, Echeverrigaray S, Bonatto SL (2006) Phylogeography of the *Bothrops jararaca* complex (Serpentes: Viperidae): past fragmentation and island colonization in the Brazilian Atlantic Forest. Mol Ecol 15: 3969–3982.

111. Amaro RC, Rodrigues MT, Yonenaga-Yassuda Y, Carnaval AC (2012) Demographic processes in the montane Atlantic rainforest: molecular and cytogenetic evidence from the endemic frog *Proceratophrys boiei*. Mol Phylogenet Evol 62: 880–888.

112. Pellegrino KCM, Rodrigues MT, Waite AN, Morando M, Yassuda YY, et al. (2005) Phylogeography and species limits in the *Gymnodactylus darwinii* complex (Gekkonidae, Squamata): genetic structure coincides with river systems in the Brazilian Atlantic Forest. Biol J Linn Soc 85: 13–26.

113. Tchaicka L, Eizirik E, De Oliveira TG, Candido JF Jr, Freitas TR (2007) Phylogeography and population history of the crab-eating fox (*Cerdocyon thous*). Mol Ecol 16: 819–838.

114. Cabanne GS, d'Horta FM, Sari EH, Santos FR, Miyaki CY (2008) Nuclear and mitochondrial phylogeography of the Atlantic forest endemic *Xiphorhynchus fuscus* (Aves: Dendrocolaptidae): biogeography and systematics implications. Mol Phylogenet Evol 49: 760–773.

115. Melo Santos AM, Cavalcanti DR, Silva JMCd, Tabarelli M (2007) Biogeographical relationships among tropical forests in north-eastern Brazil. J Biogeogr 34: 437–446.

116. Wang X, Auler AS, Edwards RL, Cheng H, Cristalli PS, et al. (2004) Wet periods in northeastern Brazil over the past 210kyr linked to distant climate anomalies. Nature 432: 740–743.

117. Adrian Quijada-Mascareñas J, Ferguson JE, Pook CE, Salomão MDG, Thorpe RS, et al. (2007) Phylogeographic patterns of trans-Amazonian vicariants and Amazonian biogeography: the Neotropical rattlesnake (*Crotalus durissus* complex) as an example. J Biogeogr 34: 1296–1312.

118. Wüster W, Ferguson JE, Quijada-Mascarenas JA, Pook CE, Salomao Mda G, et al. (2005) Tracing an invasion: landbridges, refugia, and the phylogeography of the Neotropical rattlesnake (Serpentes: Viperidae: *Crotalus durissus*). Mol Ecol 14: 1095–1108.

119. Werneck FP, Nogueira C, Colli GR, Sites JW, Costa GC (2012) Climatic stability in the Brazilian Cerrado: implications for biogeographical connections of South American savannas, species richness and conservation in a biodiversity hotspot. J Biogeogr 39: 1695–1706.

120. Por FD (1992) Sooretama: the Atlantic rain forest of Brazil: SPB Academic Publishing, The Hague.

121. Ledru MP, Rousseau DD, Cruz FW JR, Riccomini C, Karmann I, et al. (2005) Paleoclimate changes during the last 100,000 yr from a record in the Brazilian Atlantic rainforest region and interhemispheric comparison. Quaternary Res 64: 444–450.

122. Babik W, Szymura JM, Rafinski J (2003) Nuclear markers, mitochondrial DNA and male secondary sexual traits variation in a newt hybrid zone (*Triturus vulgaris* x *T. montandoni*). Mol Ecol 12: 1913–1930.

123. Wielstra B, Arntzen JW (2012) Postglacial species displacement in *Triturus* newts deduced from asymmetrically introgressed mitochondrial DNA and ecological niche models. BMC Evol Biol 12: 161.

124. Crawford AJ, Lips KR, Bermingham E (2010) Epidemic disease decimates amphibian abundance, species diversity, and evolutionary history in the highlands of central Panama. Proc Natl Acad Sci U S A 107: 13777–13782.

2

Genotypic Diversity Analysis of *Mycobacterium tuberculosis* Strains Collected from Beijing in 2009, Using Spoligotyping and VNTR Typing

Yi Liu[1,2], Miao Tian[2], Xueke Wang[2¤], Rongrong Wei[2], Qing Xing[3], Tizhuang Ma[2], Xiaoying Jiang[4], Wensheng Li[2], Zhiguo Zhang[5], Yu Xue[2], Xuxia Zhang[2], Wei Wang[2], Tao Wang[2], Feng Hong[3], Junjie Zhang[1]*, Sumin Wang[3]*, Chuanyou Li[2]*

1 The Key Laboratory for Cell Proliferation and Regulation Biology, Ministry of Education, College of Life Sciences, Beijing Normal University, Haidian District, Beijing, China, 2 Department of Bacteriology and Immunology, Beijing Key Laboratory on Drug-resistant Tuberculosis Research, Beijing Tuberculosis and Thoracic Tumor Research Institute/Beijing Chest Hospital, Capital Medical University, Tongzhou District, Beijing, PR China, 3 Central Laboratory, Beijing Research Institute for Tuberculosis Control, Xicheng District, Beijing, PR China, 4 Clinical Center on TB, China CDC, Beijing Tuberculosis and Thoracic Tumor Research Institute/Beijing Chest Hospital, Capital Medical University, Tongzhou District, Beijing, PR China, 5 Beijing Changping Center for Tuberculosis Control and Prevention, Changping District, Beijing, PR China

Abstract

Background: Tuberculosis (TB) is a serious problem in China. While there have been some studies on the nationwide genotyping of *Mycobacterium tuberculosis* (*M. tuberculosis*), there has been little detailed research in Beijing, the capital of China, which has a huge population. Here, *M. tuberculosis* clinical strains collected in Beijing during 2009 were genotyped by classical methods.

Methodology/Principal Findings: Our aim was to analyze the genetic diversity of *M. tuberculosis* strains within the Beijing metropolitan area. We characterized these strains using two standard methods, spoligotyping (n = 1585) and variable number of tandem repeat (VNTR) typing (n = 1053). We found that the most prominent genotype was Beijing family genotype. Other genotypes included the MANU, T and H families etc. Spoligotyping resulted in 137 type patterns, included 101 unclustered strains and 1484 strains clustered into 36 clusters. In VNTR typing analysis, we selected 12-locus (QUB-11b, MIRU10, Mtub21, MIRU 23, MIRU39, MIRU16, MIRU40, MIRU31, Mtub24, Mtub04, MIRU20, and QUB-4156c) and named it 12-locus (BJ) VNTR. VNTR resulted in 869 type patterns, included 796 unclustered strains and 257 strains clustered into 73 clusters. It has almost equal discriminatory power to the 24-locus VNTR.

Conclusions/Significance: Our study provides a detailed characterization of the genotypic diversity of *M. tuberculosis* in Beijing. Combining spoligotyping and VNTR typing to study the genotyping of *M. tuberculosis* gave superior results than when these techniques were used separately. Our results indicated that Beijing family strains were still the most prevalent *M. tuberculosis* in Beijing. Moreover, VNTR typing analyzing of *M. tuberculosis* strains in Beijing was successfully accomplished using 12-locus (BJ) VNTR. This method used for strains genotyping from the Beijing metropolitan area was comparable. This study will not only provide TB researchers with valuable information for related studies, but also provides guidance for the prevention and control of TB in Beijing.

Editor: Jonathan Hon-Kwan Chen, Queen Mary Hospital, the University of Hong Kong, Hong Kong

Funding: This work was supported financially by grants from the Capital Medical Development Foundation (2009-1055 and 2007-1020), and by a grant from the National Natural Science Foundation of China (31170064). The funders had no role in study design, data collection and analysis, decision to publish, or preparation of the manuscript.

Competing Interests: The authors have declared that no competing interests exist.

* Email: lichuanyou6688@hotmail.com (CL); suminwang085@126.com (SW); jjzhang@ bnu.edu.cn (JZ)

¤ Current address: International Trading Department, Beijing Clover Seed & Turf Co. Chaoyang District, Beijing, PR China

Introduction

Tuberculosis (TB), caused by *Mycobacterium tuberculosis* (*M. tuberculosis*), is one of the major causes of death in the world today [1]. It is well known that China was the one of the 22 TB high-burden countries in the world [1]. TB is a reemerging infectious disease and a substantial public health problem in Beijing [2,3], the capital of the People's Republic of China, a megacity containing many districts with large populations and high population mobility. It has been reported that the prevalence and incidence of TB is gradually becoming higher in Beijing [4], thus the prevention and control of TB is a great challenge for this city.

Figure 1. The ratio of Beijing genotype strains accounted for in each district of Beijing in 2009.

Although Beijing has been thought of as the headstream of the Beijing family strains of *M. tuberculosis*, only a few studies have been conducted on strains collected in the Beijing area. Jia et al. investigated the distribution and magnitude of TB cases in permanent residents and migrant populations of Beijing between 2000 and 2006 [5], but they did not carry out genotyping analyses. Wan et al. [6] also performed studies on the diversity of *M. tuberculosis* in Beijing using strains collected randomly at the Beijing TB control and cure Institutes, and TB hospitals during the period 2005 to 2007, but only selected a fraction of the strains acquired from Beijing city, and not all strains. The strains chosen were not sufficiently representative to reveal the true prevalence and genotypic diversity of TB in Beijing. As the molecular epidemiology of TB in Beijing has not previously been well studied, we designed a study to examine the genetic diversity of TB strains by genotyping strains collected in 2009 from different areas of Beijing.

Spacer oligonucleotide typing (Spoligotyping) is a rapid and convenient genotyping method that is well suited for the identification of Beijing family *M. tuberculosis* strains [5]. However, it has a fairly low discriminatory power [7]. Variable number of tandem repeat (VNTR) genotyping is technically easier and possesses high discriminatory power and can therefore complement spoligotyping. Therefore, we combined these two methods together in this study. Moreover, its digital results are easy to be compared between different laboratories [8,9]. Studying the molecular epidemiology of *M. tuberculosis* has been proven to

be a useful tool for TB study and control, as it can be used to tracking transmission chains and detect suspected outbreaks [10]. In 2008, Jiao established in 2008 that classical VNTR methods can differentiate Beijing strains from other strains [11].

In this study, we have combined spoligotyping and VNTR to map the genotypic and molecular epidemiology of TB strains collected in Beijing during 2009. A collection of 1585 representative *M. tuberculosis* strains isolates was first genotyped by spoligotyping, and 1053 *M. tuberculosis* strains isolates were then genotyped by VNTR. We found that the most prominent *M. tuberculosis* family in Beijing in 2009 was the Beijing family; other families included the MANU2, T1, T2 and H families et al. VNTR typing analysis selected 12 loci as the VNTR genotype locus applied to Beijing strains. Our findings should contribute to the design of more efficient strategies for TB prevention and control in Beijing.

Results

Spoligotyping-based diversity of *M. tuberculosis* strains collected from the Beijing metropolitan area

To demonstrate spoligotype diversity among *M. tuberculosis* strains in the Beijing metropolitan area, we collected 1585 strains from different districts of Beijing (Figure S1) during 2009. The strains number of four districts of Chaoyang, Haidian, Xicheng and Xuanwu occupied more than 1/2 of the total strains. Spoligotyping data profiles of the all strains isolates can be seen in

Table S4. Thirteen hundred (82%) strains were Beijing family strains, and 285(18.1%) strains belonging to non-Beijing family (Table 1). In all Beijing lineage strains, 1225 strains (77.3%) were typical Beijing genotype strains, and 75 (4.7%) were atypical Beijing genotype strains. The percentage of strains with a Beijing genotype exceeded 67% in all districts (Figure 1). Of the non-Beijing strains, 105 (6.6% of all strains) belonged to the poorly-defined T lineage, 69 (4.4%) being from the T1 (SIT53) lineage, 24 (1.6%) from the T2 (SIT52) lineage, 10 (0.6%) from the T3 (SIT37) lineage and 2 from the T4 lineage. In addition, 76 (4.8%) strains were from the MANU2 (SIT54) lineage, 12 (0.75%) from the U lineage and 84 (5.3%) were of an "Unknown" genotype (absent from the current global spoligotyping database: Spo1DB4). Two *Mycobacterium bovis* Bacillus Calmette-Guérin (BCG) strains, unique strains deserving careful follow-up study, were also found.

The numbers and frequencies of strains collected in Beijing in 2009, along with their spoligotypes were listed in Table 2. 1484 strains were been formed 36 clusters, only 101 strains were not been clustered. Thus the ratio of cluster of spoligotyping method was fairly high. As summarized in Table 3, spoligotyping clustered 1431(90%) strains into 8 maximum clusters. Among these 8 types, SIT1, SIT190, SIT265 and SIT269 were members of the Beijing family lineage, and included 1225 (77.3%), 23 (1.45%), 20(1.26%) and 11 (0.69%) of the strains, respectively. SIT37 (T3 lineage), SIT52 (T2 lineage), SIT53 (T1 lineage) and SIT54 (MANU2 lineage) belonged to the non-Beijing family. More data were summarized in Table S1, we list all spoligotyping shared types (n = 1585 strains) appeared in this research.

We also analyzed the relationship between the Beijing family and sex, age ($\leq$45 and >45), and the household address of patients (this city or other areas). We found a statistically significant difference (P = 0.01<0.05) in sex (Table S2), of which the proportion of male infections with Beijing family genotype strains higher than the proportion of female infections. There were no statistically significant associations with age and household registration.

The population clad structure of TB strains in Beijing, based on major *M. tuberculosis* complex (MTBC) clades, was as follows: East Asian (Beijing) (82%), Euro-American (9.5%), Indo-Oceanic (8.3%), *Mycobacterium bovis* BCG (0.1%) and unknown (0.1%) (Table S3). We also investigated the proportion of modern and ancient lineages [12,13] among these groups. In this study, modern lineages,including Beijing/T/H lineages, accounted for 88.8% of all strains. Ancestral lineages, including the MANU/U/ S and CAS lineages, accounted for only 11.2% of all strains. These results indicate modern lineage strains were the most epidemiological strains in the Beijing metropolitan area.

The selection of VNTR loci and the result of VNTR genotyping

To determine whether different VNTR loci possess discrepant discriminatory power when applied to these strains, we genotyped 1053 of the TB strains collected in Beijing during 2009 using the VNTR method (Table S4). As a preliminary experiment, we first selected 13 loci (QUB-11b, MIRU10, Mtub21, MIRU39, MIRU16, MIRU40, MIRU31, Mtub24, Mtub04, QUB-4156c, Mtub39, Rv-2372, and Qub-15) to genotype the 465 strains collected during 2008 (data not shown). Results indicated that several loci were not suitable for genotyping as their, PCR products were not obtained consistently despite repeated attempts. We therefore eliminated 4 genotype loci (Mtub24, Mtub04, Qub-15, and Rv-2372), and added 3 new loci (MIRU 23, ETR-A and Mtub30) to study the strains collected in 2009 by comparing allelic diversity of different VNTR loci from different areas, bringing the

Table 1. Spoligotyping patterns result of *M. tuberculosis* strains collected from Beijing in 2009.

Year	No.§	Beijing families*		non-Beijing families#											
		Typical Beijing	Atypical Beijing	T1	T2	T3	T4	MANU2	H3	H4	U	S	CAS1-Delhi	BCG	Unknown
2009	1585	1225 (77.28%)	75 (4.73%)	69 (4.35%)	24 (1.51%)	10 (0.63%)	2	76 (4.79%)	3	1	12 (0.75%)	1	1	2	84 (5.29%)

§The number of collected strains isolates.
*1300(82.01%) strains belonged to Beijing families.
#285 strains (include 84 unknown strains) belonged to non-Beijing families.

Table 2. Numbers and frequencies of strains clustered assorted by spoligotyping from 2009 years in Beijing.

Spoligotyping Parameter	Value
No. of strains studied	1585
No. of clusters	36
No. of unknown spoligotype	84
Mean No. of strains per cluster	41.22
No.(%) of clustered strains	1484 (93.62)
No.(%) of unclustered strains	101 (6.37)

total number of genotype loci included in this study to 12 (Table 4). QUB-4156c (VNTR 4156), MIRU23 (VNTR 2531), MIRU20 (VNTR 3007) and MIRU39 (VNTR 4348) presented low levels of allelic diversity but were useful for the confirmation of different lineages [14] and were thus included in the analysis.

The 12-locus (BJ) VNTR method differentiated the 1053 strains into 869 genotypes (Table 5). A total of 796 strains had unique patterns and the remaining 257 formed 73 clusters (2 to 20 strains per cluster). The allelic diversity for the 1053 strains of each VNTR locus was estimated using the Hunter-Gaston discriminatory index (HGDI) (Table 4). The discriminatory power of 5 loci (QUB-11b, Mtub21, MIRU39, MIRU16, and MIRU31) exceeded 0.5 and these were regarded as highly discriminatory. Three loci (ETRA, MIRU10 and MIRU40) showed moderately high discriminatory (0.3 to 0.5). Other loci (QUB-4156c, Mtub30, Mtub39 and MIRU23) were found to be less polymorphic, with HGDI within the range of 0.04 to 0.3. Of the 12 VNTR loci, QUB-11b had the greatest allelic diversity and the highest discriminatory power for all genotypes including Beijing genotype strains. All results can be seen in Table 4.

Comparison of spoligotyping and VNTR genotyping methods

Spoligotyping and VNTR genotyping results were summarized in Table 5 which showed the discriminatory power of the two genotyping methods. Spoligotyping (n = 1585) had a low discrim-inatory power (HGDI = 0.399), especially when applied to Beijing family strains (HGDI = 0.111) (Table 5); among 115 different types, spoligotyping identified 101 unique strains (8.4%) and grouped 1484 (91.6%) strains into 17 clusters, giving a clustering rate of 93.6%. The largest cluster was assigned to the typical

Beijing genotype and included 1225 strains. Compared to spoligotyping, the original set of 12-locus (BJ) VNTR alone (n = 1053) identified 796 unique strains and clustered 257 strains into 73 clusters, giving a significantly higher discriminatory power (HGDI = 0.9984). When spoligotyping and 12-VNTR methods were used together (n = 1053), the discriminatory power increased, giving an HGDI value of 0.9989, and the clustering rate was reduced to 19.6% (Table 5). This study showed that spoligotyping and VNTR are appropriate methods for constructing primary genotype libraries of these strains, and the combination of these methods provides high discriminatory power.

Comparison of the discriminatory power of kinds of VNTR loci

In this study, we did not use 15-locus and 24-locus VNTR primer sets [9], because these loci were not always suited to strains from the Beijing district. We found that the allelic diversity of the VNTR loci varied significantly used to Beijing family isolates in different countries and districts when the HGDIs of these loci were compared with those reported from other areas (Table 6).

To verify the discriminatory power of each locus and the cumulative discriminatory power of all 12-locus (BJ), we calculated the discriminatory power of the loci listed in Table 7 to get an indication of allelic diversity and cumulative HGDI discriminatory power. The respective discriminatory powers of 4 VNTRs methods, (15-locus (Supply) [9], 24-locus (Supply) [9,13], 12-locus (JATA) [15], and 16-locus (Gao) [16]) were compared (Table 8). The cumulative HGDI discriminatory power of these 12-locus (BJ) VNTR reached 0.9990 when applied to all strains, 0.9994 when applied to non-Beijing family and 0.9984 when applied to Beijing family (Table 7 & Table 8). These results show that our VNTR

Table 3. The description of these clusters containing the maximum 8 clusters of *M. tuberculosis* collected from Beijing in this study.

Sequence number	SIT *(Clade) Octal number spoligotype description	Number(%#) in study
1	1 (Typical Beijing) 000000000003771	1225(77.28)
2	37(T3) 777777777760700	6(0.26%)
3	52(T2) 777777777760731	24(1.51%)
4	53(T1) 777777777760771	69(4.35%)
5	54(MANU2) 777777777763771	76(4.79%)
6	190(Beijing like) 000000000003731	23(1.45%)
7	265(Beijing like) 000000000003371	20(1.26%)
8	269(Beijing like) 000000000000771	11(0.69%)

*spoligotype international type, SIT number from SpolDB4.0.
#Represents the percentage of strains with a common SIT among all strains in this study.

Table 4. The Hunter-Gaston discriminatory index of the 12 VNTR loci in *M. tuberculosis* strains from Beijing in 2009.

Order	VNTR[#] locus	VNTR alias	Number of alleles	Range of repeats	Allelic diversity (h*) for		All strains (n = 1053)
					Beijing family (n = 884)	non-Beijing family (n = 169)	
1	2163	QUB-11b	14	1-10	0.7262	0.7687	0.7637
2	1955	Mtub21	9	1-9	0.6881	0.7339	0.7196
3	4348	MIRU39	7	2-9	0.6848	0.5691	0.7172
4	1644	MIRU16	6	0-8	0.6923	0.7333	0.6962
5	3192	MIRU31	6	0-9	0.6544	0.7394	0.6518
6	0960	MIRU10	6	0-7	0.4990	0.7312	0.5478
7	0802	MIRU40	5	1-7	0.4007	0.5987	0.4725
8	2165	ETR A	5	0-6	0.3760	0.5749	0.4261
9	4156	QUB-4156c	8	0-7,11	0.2867	0.2938	0.2992
10	2401	Mtub30	5	1-10	0.1727	0.5409	0.2459
11	3690	Mtub39	6	1-7	0.2300	0.4136	0.2723
12	2531	MIRU23	7	0-7	0.0443	0.1386	0.0561

*Hunter-Gaston discriminatory index.
#Variable number tandem repeat.

Table 5. Different discriminatory power of different typing methods used in this study.

Method	Beijing Family % (No.)	No. of type patterns	No. of unique strains	No. of clustered strains	No. of clusters	Cluster size (No. of strains)	Clustering rate (%)	Maximum No. of strains in a cluster	Allelic diversity (h*) for		
									All strains	Beijing family	Other strains
Spoligotyping (1585)	82%(1300)	137	101	1484	36	2-1225	93.6	1225	0.3999	0.1110	0.9177
Spoligotyping (1053)	83.8%(884)	105	80	973	25	2-830	92.4	830	0.3759	0.1190	0.9234
12-locus VNTR(1053)	83.8%(884)	869	796	257	73	2-20	25.84	20	0.9990	0.9984	0.9994
VNTR and Spoligotyping(1053)	83.8%(884)	899	847	206	52	2-16	19.56	16	0.9994	0.9989	0.9996

*Hunter-Gaston discriminatory index.

Table 6. Allelic diversity (h*) of different VNTR loci in Beijing family isolates from different areas.

Order	VNTR locus	VNTR alias	Beijing, China(this study)	Shanghai, China [16]	Heilongjiang, China [40]	Hong kong, China [41]	Japan [15]	Russia [42]
1	2163	QUB-11b	0.7262	0.6548	0.644	-	0.815	0.205
2	1955	Mtub21	0.6881	0.5231	0.396	-	0.598	0.330
3	4348	MIRU39	0.6848	0.2856	0.290	0.320	0.156	0.000
4	1644	MIRU16	0.6923	0.2423	0.200	0.058	0.258	0.082
5	3192	MIRU31	0.6544	0.2461	0.599	None	0.270	0.160
6	0960	MIRU10	0.4990	0.1952	0.154	0.377	0.431	0.082
7	0802	MIRU40	0.4007	0.1471	0.292	0.196	0.229	0.122
8	2165	ETR A	0.3760	0.0308	0.238	0.201	0.223	0.158
9	4156	QUB-4156c	0.2867	0.4923	0.107	-	-	-
10	2401	Mtub30	0.1727	0.0909	0.133	-	0.379	0.042
11	3690	Mtub39	0.2300	0.0606	0.174	-	0.215	0.000
12	2531	MIRU23	0.0443	0.0606	-	-	0.158	0.000

*Hunter-Gaston discriminatory index.

system is superior to the reported 15-locus VNTR and has almost equal discriminatory power to the 24-locus VNTR, 12-locus (JATA) and 16-locus (Gao) (Table 8), especially when applied to Beijing family strains, although it only uses 12 loci. Therefore, our 12-locus (BJ) VNTR method and the methods used for strains from the Beijing metropolitan area are comparable.

Distribution of clustering shown by the minimum spanning tree using spoligotype and VNTR data

Another way to represent the strains graphically is to construct a minimum spanning tree, which shows the clustering of TB strains. To show the major clusters and map their genetic links map, we constructed minimum spanning trees using BioNumerics software.

Minimum spanning trees were first generated by using spoligotype data (Figure 2). Each nodal point represented a particular spoligotype type, and the size of each nodal point was related to the number of strains within that spoligotype. As shown in Figure 2, the biggest cluster represents the Beijing genotype, but clusters of atypical Beijing genotype were also seen. The Beijing and non-Beijing family clusters were divided into two large groups. Annotations in the figure represented the 8 most frequent spoligotypes found in the Beijing municipality. The largest cluster (ST1) corresponded to the Beijing strains, as confirmed by spoligotyping, and three clusters (ST190, ST265 and ST269) showed the signature of strains belonging to the Beijing-like strains. According to the minimum spanning tree, the new SIT type was more similar to non-Beijing spoligotypes.

Other clusters showed mutually recognizable with spoligotype signatures. However, for some strains, there was no concordance with the spoligotyping results. In particular, 10 newly-identified strains with a spoligotype profile corresponding to MANU2 were clustered with Beijing strains. These discordant findings may result from mixed strains.

In Figure 3, the minimum spanning tree showed the clustering of 1053 Beijing strains using the VNTR typing method; different clusters were shown in different colors. From the figure clearly shows that the polymorphism of these strains was very high, and that this VNTR typing method has high discriminatory power. Although it was possible to assign strains to different branches, it was difficult to discern the evolutionary subtype to which each strain belongs. Figure 3 shows that the Beijing family strains form the largest group (light red). Non-Beijing family groups were shown in orange, green, purple and blue etc. In this MST tree, Beijing family strains were separated from non-Beijing family strains. Cluster analysis show identified 73 clusters (2 to 20 strains per cluster). Almost all of the clustered strains belonged to Beijing family genotype; only two clusters contained 2 and 3 strains respectively that were from the MANU2 family.

Discussion

TB is a serious problem globally, and its prevalence is reaching epidemic proportions in countries and districts with large and mobile populations, including many megacities. DNA fingerprinting technology is now a commonly used tool to detect and investigate outbreaks of *M. tuberculosis*. Particular *M. tuberculosis* genotypes have been found to be related to the risk of transmission. Previous studies [6,16–18] have described the distribution of various genotypes of TB strains from many areas in China. However, a detailed analysis of the population structure of TB strains from the metropolitan area of Beijing, one of the largest cities in the world, with a high prevalence of TB, has not been conducted. Acquiring information on the genotype diversity of TB strains in Beijing and exploring their molecular character-

Table 7. Allelic diversity of the 12 VNTR loci in *M. tuberculosis* strains from Beijing 2009.

Order	VNTR locus	VNTR alias	Beijing family (n = 884)		non-Beijing family (n = 169)		All strains (n = 1053)	
			HGDI (Individual locus)	HGDI* (cumulative)	HGDI (Individual locus)	HGDI (cumulative)	HGDI (Individual locus)	HGDI (cumulative)
1	2163	QUB-11b	0.7262	0.7262	0.7687	0.7687	0.7637	0.7637
2	1955	Mtub21	0.6881	0.8382	0.7339	0.8871	0.7196	0.7974
3	4348	MIRU39	0.6848	0.9479	0.5691	0.9765	0.7172	0.9528
4	1644	MIRU16	0.6923	0.9766	0.7333	0.9904	0.6962	0.9786
5	3192	MIRU31	0.6544	0.9926	0.7394	0.9972	0.6518	0.9927
6	0960	MIRU10	0.4990	0.9958	0.7312	0.9976	0.5478	0.9940
7	0802	MIRU40	0.4007	0.9968	0.5987	0.9981	0.4725	0.9961
8	2165	ETR A	0.3760	0.9972	0.5749	0.9985	0.4261	0.9969
9	4156	QUB-4156c	0.2867	0.9975	0.2938	0.9988	0.2992	0.9976
10	2401	Mtub30	0.1727	0.9976	0.5409	0.9990	0.2459	0.9983
11	3690	Mtub39	0.2300	0.9983	0.4136	0.9993	0.2723	0.9989
12	2531	MIRU23	0.0443	0.9984	0.1386	0.9994	0.0561	0.9990

*Hunter-Gaston discriminatory index.

Table 8. The comparison of the discriminatory power combination of different VNTR loci in different country.

Typing methods	All strains				
	HGDI*	No. of types	No. of unique types	No. of clusters	Percentage of clustering
15-locus(Supply)	0.990	291	269	22	17.2
24-locus (Supply)	0.999	303	287	16	11.7
12-locus (JATA)	0.999	302	284	18	12.6
16-locus (Gao)	0.9983	183	–	27	14.9
12-locus (this paper)	0.999	869	796	73	25.84

*Hunter-Gaston discriminatory index.

istics is important for understanding and controlling the spread of TB.

To identify which strains of TB were prevalent strains occur in Beijing, we genotyped strains using spoligotyping and VNTR typing. Spoligotyping is a classical method used for TB genotype analysis, especially for the identification of Beijing family strains [19], although its discriminatory power is not high. VNTR has been used extensively around the world [20], and was recommended as a standard genotyping method. The characteristics of prevalent strains in different regions are easy to compare using the digital features available with VNTR genotyping.

A previous study in which we genotyped on clinical samples of Beijing TB strains collected from several areas of China provided us with baseline information on the distribution of dominant TB genotypes in China [17]. However, the strains selected in that study do not accurately represent the prevalence and epidemic trends of TB within metropolitan Beijing. Here, we genotyped 1565 strains collected from 18 districts and counties in Beijing. As shown in Table 1 and Table 5, 1300 (82%) of these strains from the Beijing district belonged to the Beijing family, a finding very similar to results reported by from the Beijing district other labs [21].

The prevalence of the Beijing family in Beijing (82%) is higher than in Hong Kong (70%) [22], Taiwan (44.4%) [23], Vietnam (54%) [24], Thailand (44%) [25]and Russia (44.5%) [26].

Although the Beijing family was the dominant genotypic family in the Beijing district, genotypic polymorphisms were also evident. In this study, non-Beijing family strains from the T1, T2, T3, T4, MANU2 and U families. Although modern lineages were the most epidemiological strains in the Beijing metropolitan area, but ancient lineages were also been found. It was interesting, that U and CAS1-Delhi types were detected in Beijing. It has previously been reported [13,27] that the Middle East, Central, and Southern Asia are 'high incidence' areas for these strains. Hence, frequent movement of ethnic groups from the Xinjiang autonomous region may be a major factor explaining the increasing prevalence of these genotypes in Beijing. The T family genotype, one of the prevalent genotypes in Africa, Central, South America and Europe, was found in this study [28]. This result showed movement of groups from these regions may increase prevalence of these genotypes in Beijing. In addition, novel spoligotypes were identified in this study. It showed that the complexity of the Beijing metropolitan area strains source, which need more deep study.

The population structure of TB strains in Beijing consists of East Asian (Beijing), Euro-American, Indo-Oceanic, *Mycobacterium bovis* BCG and strains of unknown origin. The *Mycobacterium bovis* BCG strains deserve particular mentioned. The exact source of these strains is unknown. It is unclear if patients were really been infected by *Mycobacterium bovis* BCG or this was only a case

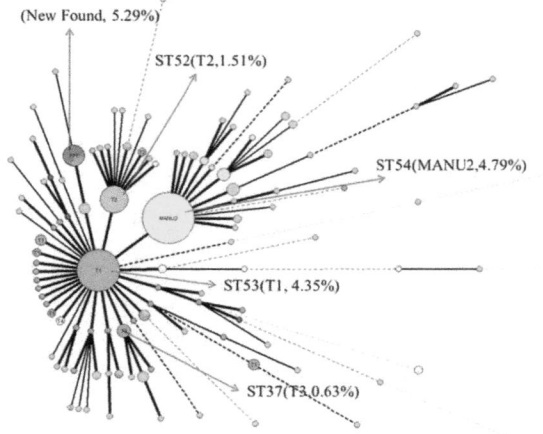
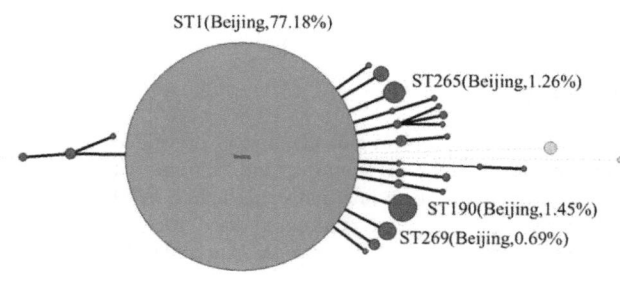

Figure 2. Minimum spanning tree showing the clustering by Spoligotyping of 1585 *M. tuberculosis* strains of Beijing. Each nodal point represents a particular spoligotype, and the size of nodal point is relative to the number of strains with that spoligotype. The percentage represents the proportion of each spoligotype. The annotations in the figure were the 8 most frequent spoligotypes.

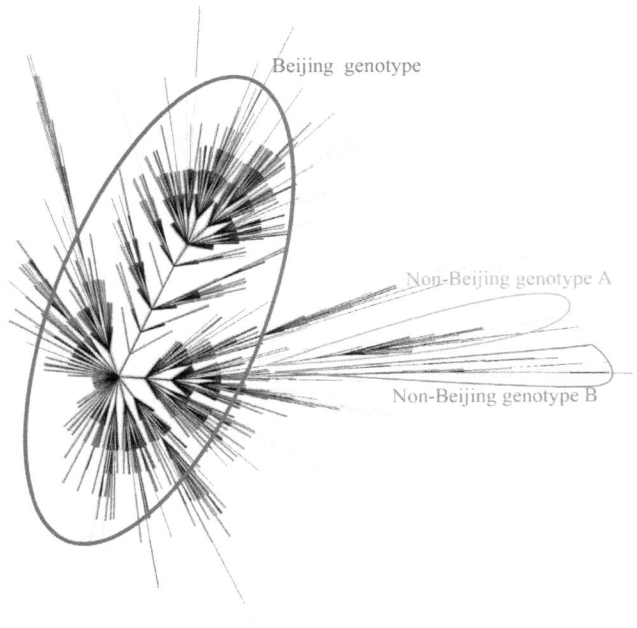

Figure 3. Minimum spanning tree showing the clustering by 12-locus (BJ) VNTR of 1053 *M. tuberculosis* strains of Beijing. Different clusters are shown in different colors; the largest light pink group represents Beijing family isolates, whereas orange, green and blue represent the non-Beijing family groups.

of laboratory contamination. We will carry out a full investigation in the future.

We also found that there were statistically significant differences between the number of male and female patients infected by Beijing genotype strains, with the ratio of infected males being significantly higher than that for infected female. However, there were no statistically significant associations between the Beijing strain and age or household registration. These results differ slightly from other areas [29], but these data should give an accurate picture of Beijing TB strains. Our results illustrated the fact that research on strains in the Beijing area must be based on a large-scale study in order to obtain real and accurate results.

In particular, because PCR products of some strains were not obtained despite repeated attempts when performing VNTR typing, the number of strains analyzed simultaneously by spoligotyping and VNTR methods was 1053. The ratio of the Beijing genotype obtained for Beijing was still 83.8% (n = 1053) (Table 5) very similar to 82% (n = 1585). These results indicate that our data does represent the TB situation in Beijing.

We believe that choice of appropriate VNTR loci in this study was critical for identifying the prevalent cluster in Beijing. In order to facilitate efficient genotyping of large numbers of strains from Beijing, we first selected a limited panel of typical VNTR loci that would correctly cluster the bacteria into the major clades and genotypic families. 15-locus and 24-locus VNTR genotype methods were commonly being considered to be optimized and were regarded as classical methods [10]. However, recent studies have shown that these loci were not always suited to studies of East Asian strains, especially in China where there is both a high incidence of TB and a high prevalence of the Beijing family strain [30]. Many studies have shown that the discriminatory power of

the 15-locus VNTR method was low when applied to the analysis of strains collected from the Chinese mainland [31], and 24-locus VNTR methods are too complex when used for the analyze multiple strains acquired from various regions of the Chinese mainland [32].

The same situation has been encountered in Japan, which also has a high prevalence of Beijing family strains. Studies there have suggested that 15 and 24-locus VNTR methods were poorly optimized for genotyping strains encountered in Japan. In 2008, Murase et al. [15] optimized JATA-12 VNTR loci for the analysis of Japanese strains. They found that these loci had the same discriminatory power as the 24-locus VNTR and were better adapted to East Asian countries which possess a high percentage of Beijing family strains. Gao et al. [16] described several methods in 2006. Their purpose was to identify the simplest and most accurate typing methods, and they found that the 16-locus VNTR method was the most suitable, especially for Beijing family strains. Recently, VNTR-15 and 8 single nucleotide polymorphisms (SNPs) methods have been standardized in Shanghai for the genotyping of Beijing strains [31]. Additional typing of three hypervariable loci (VNTR3820, VNTR4120, and VNTR3232) has also been used; Gao and colleagues have recommended combining the standard VNTR-15 and SNPs as first-line typing methods, and using hypervariable loci for second-line typing of clustered strains in molecular epidemiology studies of homogenous TB populations. However, the above reports have the same shortcoming: the number of strains was too small (325 in Japan, 289 in Shanghai), and thus the results were likely not representative of the broader situation; although they may be suitable for Japan or Shanghai, they may not be appropriate for Beijing. Our purpose here was to identify the most appropriate loci specific to the Beijing metropolitan area. The discriminative power of some loci was too low when apply to Beijing genotype strains. Other loci were excluded because PCR products of these loci were not consistently obtained despite repeated attempts. Other more loci needed more an in-depth study to determine when applies to Beijing genotype strains in different areas.

Which loci were the most suitable for analyzing Beijing district strains? Recently, precise answers are still not been known. If the discriminatory power of loci is too low, unrelated strains will be clustered together. However, if the discriminatory power of loci is too high, related strains will not be clustered. Our study was motivated by the need to select appropriate loci which not only discern the diversity of unrelated strains but also connect stains that can be clustered. Therefore, in this study, we identified 12 VNTR loci, QUB-11b, MIRU10, Mtub21, MIRU23, MIRU39, MIRU16, MIRU40, MIRU31, Mtub24, and Mtub04, MIRU20, and QUB-4156c, as useful genotyping loci; these loci were selected to describe the polymorphisms of these strains. Our newly established 12-locus VNTR typing method had almost equivalent discriminatory power for TB genotyping to that of the 24-locus VNTR (Table 8), especially when applied to Beijing family strains, even though it only contained 12 loci. Our results indicate that this 12-locus (BJ) VNTR method gives appropriate results and HGDI discriminatory power. Thus, this method may be a reference for other typing methods and genotype studies.

The results obtained on the basis of 12-locus (BJ) VNTR typing were slightly different with those of spoligotyping, but this may be due to the presence of two different isolates in some samples. We acquired the cluster result from the minimum spanning tree of spoligotype and VNTR data by BioNumerics software. These clustered strains by VNTR strongly suggest the possibility of the recent TB transmission, which need we investigate the background information of these patients carried clustered strains in the following study.

Our results present a comprehensive picture of the prevalence of different TB genotypes in Beijing and their distribution. Although the major epidemiological cluster identified was the Beijing genotype, considerable genetic diversity was present among the TB strains collected in Beijing. The Beijing family, which is distributed extensively around the world, represents a high proportion of TB strains responsible for ongoing transmission of TB in China, Japan, Korea, Russia, and Central Asia [33–35]. This work represents an important study on TB in the Beijing metropolitan area, and our hope is that it will facilitate new government policies and methods for controlling TB, especially in Beijing city.

To the best of our knowledge, this was the first extended and detailed genotyping study of the genetic diversity of TB in a metropolitan area of China using classical two genotyping approaches. This study has important public health implications for Beijing as well as other highly-populated regions of the world. We describe not only the genotypic structure of TB, including the presence of different genotype families dispersed in different districts of the city. Genotyping data can provide a broader understanding of TB epidemiology and outbreaks, information which can be applied to future TB control efforts. This study will add to the reference database of TB strains circulating in Beijing, and may facilitate the implementation of a more efficient anti-TB control program.

Materials and Methods

Collection and identification of clinical strains isolates from Beijing

A total of 1585 clinical strains, collected in Beijing, China, during 2009, were obtained from the Beijing Research Institute for Tuberculosis Control. All *M. tuberculosis* strains isolates were from new pulmonary TB patients, who were sputum smear-positive and/or culture-positive for *M. tuberculosis*. The isolates were recovered from −80°C stocks and were subcultured on solid Lowenstein–Jensen medium for 3–4 weeks at 37°C. The 1585 isolates were first assessed by spoligotyping, and then 1056 of these isolates were used to further optimize the highly polymorphic VNTR typing method by set to calculating the Hunter-Gaston discriminatory index (HGDI) values of the VNTR loci.

Ethics statement

These patients were able to access the study after they signed an Informed Consent form. The protocols applied in this study were approved by the Ethics Committee of the Beijing Research Institute for Tuberculosis Control. The retrospective study was approved by the Institutional Review Boards of Beijing Tuberculosis & Thoracic Tumor Research Institute.

Spoligotyping

Spoligotyping was performed according to a standard protocol [36]. Spoligotyping was used to identify the genotype of TB strain in the direct repeat (DR) locus as described previously [37]. Genomic DNA was extracted from freshly cultured bacteria. Bacteria were resuspended in 400 μL TE buffer (pH 8. 0), then heated in 100 degrees Celsius bath for 30 min. After incubating on ice for 2 min, samples were centrifuged at 10,000 g for 2 min. Genomic DNA in the supernatant was used as a PCR template. A commercially available Isogen Spoligotyping kit (Isogen Bioscience BV, the Netherlands) was used according to the manufacturer's instructions. All genomic DNA was amplified with primers DRa

(5′-GGTTTTGGGTCTGACGAC-3′) (biotinylated 5′ end) and DRb (5′-CCGAGAGGGGACGGAAAC-3′). Amplified products were then hybridized with a membrane. Images were detected with a chemiluminescence system, including the ECL detection liquid (GE Healthcare, Life Sciences) and X-ray films (Kodak, Rochester, NY). Beijing genotype strains were defined as those which hybridized to all of the last nine spacer oligonucleotides (spacers 35 to 43), while Beijing-like genotype strains were ones that hybridized to only some of the last nine spacers. The results were compared to the SpolDB4 database (an international spoligotype database at the Institute Pasteur de Guadeloupe. http://www.pasteur-guadeloupe.fr:8081/SITVITDemo/) [7].

VNTR typing

Sequences of primers used for amplification of the nine MIRU loci (MIRU-10, 16, 20,23,24,27,31,39 and 40), two QUB loci (QUB-11b, QUB-4156C) and one Mtub locus (Mtub21) selected and applied to these strains isolates collected in 2009 were listed in Table 4.

For each of the VNTR loci, we used a total PCR reaction volume of 20 μL. Each PCR reaction contained the following: 1 μL of DNA template; a 0.4 μM of each primer (Sbs Biotech, Beijing, China), 2×Taq PCR Master Mix (Tiangen Biotech, Beijing, China), and double-distilled H_2O to bring the total volume to 20 μL. Then PCR reaction conditions were as follows: initial denaturation at 95°C for 5 min, and then 35 cycles of denaturation at 95°C, annealing for 30 s at a temperature range 58–62°C and extension at 72°C for 45 s, followed by a final extension at 72°C for 7 min. PCR products were analyzed by electrophoresis in 2% agarose gels, using a 50 bp DNA ladder (TaKaRa-Bio Inc., Dalian, China) as the reference marker. H37Rv and double-distilled H_2O were used as positive and negative controls, respectively. The copy number of repeats was calculated using the following formula: (length of the PCR product minus length of the flanking regions)/length of one repeat copy unit.

Data analysis

Spoligotyping data, expressed in binary and octal formats, and VNTR data expressed in decimal format were analyzed using BioNumerics software (Version 5.0, Applied Maths, Belgium). Cluster analysis was performed and a dendrogram was generated in Bionumerics using the Dice similarity coefficient and UPGMA coefficient. Genotype analyses were performed in NatureEdge software (own).

Statistical analysis

The discriminatory power (the Hunter-Gaston discriminatory index (HGDI)) of each typing method was calculated according to a previously published method [38]:

$$HGDI = 1 - \left[\frac{1}{N(N-1)} \sum_{j=1}^{s} nj(nj-1) \right]$$

where N is the total number of isolates in the typing method, s is the number of distinct patterns discriminated by VNTR, and n_j is the number of isolates belonging to the jth pattern. The percentage clustering was calculated with the following formula: $(n_c - C)/N$, where N is the total number of isolates, C is the number of clusters, and n_c is the total number of clustered isolates [39].

Supporting Information

Figure S1   Map of Beijing showing the distribution of strains collected in various districts from Beijing.

Figure S2   The VNTR cluster of isolate strains collected in various districts from Beijing 2009 by BioNumerics.

Table S1   Description of most popular shared-types (SITs; n = 1565 isolates) and new found orphan isolates with corresponding spoligotyping defined lineagessublineages from M.tuberculosis.

Table S2   The comparison between Beijing family genotype and Non-Beijing family genotype in sex, age, and household by Spoligotyping result (n = 1585). There was a statistically significant difference in sex, but not found statistically significant associations with age and household registration.

Table S3   The clad structure of M. tuberculosis isolates strains from Beijing in 2009 year. The population clad structure mainly based on major M. tuberculosis complex clades.

Table S4   Spoligotype and VNTR original data of isolates strains from Beijing in 2009 year. The spoligotyping profiles of the 1565 isolates, and the 12-locus(BJ) VNTR repeats profiles of the 1053 isolates.

Acknowledgments

We are grateful to Dr. Chongguang Yang and Dr. Tao Luo for performing the BioNumerics analysis; Dr. Yu Pang for help with setting up the software. We also thank Dr. Changye Wang for depicting the map of Beijing city, and Dr. Dan Yu for help with data analysis. We thank Dr. Joy Fleming for her correction of the manuscript.

Author Contributions

Conceived and designed the experiments: YL SW CL. Performed the experiments: YL MT XW RW XJ WL YX XZ TW TM WW. Analyzed the data: YL XJ JZ. Contributed reagents/materials/analysis tools: QX ZZ HF YL WL. Contributed to the writing of the manuscript: YL CL.

References

1. WHO (2014) WHO Global tuberculosis report 2013. Global tuberculosis report Available: http://www.who.int/tb/publications/global_report/en/.
2. An Yansheng ZL, Tu D (2004) Impact of immigration on epidemic of tuberculosis in Beijing. The Journal of The Chinese Antituberculosis Association 26: 319–323.
3. Lixing Z (2003) Trends of tuberculosis incidence in Beijing. The Journal of The Chinese Antituberculosis Association 25: 204–208.
4. Liu Y, Li X, Wang W, Li Z, Hou M, et al. (2012) Investigation of space-time clusters and geospatial hot spots for the occurrence of tuberculosis in Beijing. Int J Tuberc Lung Dis 16: 486–491.
5. Jia ZW, Jia XW, Liu YX, Dye C, Chen F, et al. (2008) Spatial analysis of tuberculosis cases in migrants and permanent residents, Beijing, 2000–2006. Emerg Infect Dis 14: 1413–1419.
6. Wan K, Liu J, Hauck Y, Zhang Y, Zhao X, et al. (2011) Investigation on Mycobacterium tuberculosis diversity in China and the origin of the Beijing clade. PLoS One 6: e29190.
7. Brudey K, Driscoll JR, Rigouts L, Prodinger WM, Gori A, et al. (2006) Mycobacterium tuberculosis complex genetic diversity: mining the fourth international spoligotyping database (SpolDB4) for classification, population genetics and epidemiology. BMC Microbiol 6: 23.
8. Mazars E, Lesjean S, Banuls AL, Gilbert M, Vincent V, et al. (2001) High-resolution minisatellite-based typing as a portable approach to global analysis of Mycobacterium tuberculosis molecular epidemiology. Proc Natl Acad Sci U S A 98: 1901–1906.
9. Supply P, Allix C, Lesjean S, Cardoso-Oelemann M, Rusch-Gerdes S, et al. (2006) Proposal for standardization of optimized mycobacterial interspersed repetitive unit-variable-number tandem repeat typing of Mycobacterium tuberculosis. J Clin Microbiol 44: 4498–4510.
10. Barnes PF, Cave MD (2003) Molecular epidemiology of tuberculosis. N Engl J Med 349: 1149–1156.
11. Jiao WW, Mokrousov I, Sun GZ, Guo YJ, Vyazovaya A, et al. (2008) Evaluation of new variable-number tandem-repeat systems for typing Mycobacterium tuberculosis with Beijing genotype isolates from Beijing, China. J Clin Microbiol 46: 1045–1049.
12. Gutierrez MC, Ahmed N, Willery E, Narayanan S, Hasnain SE, et al. (2006) Predominance of ancestral lineages of Mycobacterium tuberculosis in India. Emerg Infect Dis 12: 1367–1374.
13. Thomas SK, Iravatham CC, Moni BH, Kumar A, Archana BV, et al. (2011) Modern and ancestral genotypes of Mycobacterium tuberculosis from Andhra Pradesh, India. PLoS One 6: e27584.
14. Comas I, Homolka S, Niemann S, Gagneux S (2009) Genotyping of genetically monomorphic bacteria: DNA sequencing in Mycobacterium tuberculosis highlights the limitations of current methodologies. PLoS One 4: e7815.
15. Murase Y, Mitarai S, Sugawara I, Kato S, Maeda S (2008) Promising loci of variable numbers of tandem repeats for typing Beijing family Mycobacterium tuberculosis. J Med Microbiol 57: 873–880.
16. Zhang L, Chen J, Shen X, Gui X, Mei J, et al. (2008) Highly polymorphic variable-number tandem repeats loci for differentiating Beijing genotype strains of Mycobacterium tuberculosis in Shanghai, China. FEMS Microbiol Lett 282: 22–31.
17. Pang Y, Zhou Y, Zhao B, Liu G, Jiang G, et al. (2012) Spoligotyping and drug resistance analysis of Mycobacterium tuberculosis strains from national survey in China. PLoS One 7: e32976.
18. Guo YL, Liu Y, Wang SM, Li CY, Jiang GL, et al. (2011) Genotyping and drug resistance patterns of Mycobacterium tuberculosis strains in five provinces of China. Int J Tuberc Lung Dis 15: 789–794.
19. Gori A, Bandera A, Marchetti G, Degli Esposti A, Catozzi L, et al. (2005) Spoligotyping and Mycobacterium tuberculosis. Emerg Infect Dis 11: 1242–1248.
20. Hill V, Zozio T, Sadikalay S, Viegas S, Streit E, et al. (2012) MLVA based classification of Mycobacterium tuberculosis complex lineages for a robust phylogeographic snapshot of its worldwide molecular diversity. PLoS One 7: e41991.
21. Lu B, Zhao P, Liu B, Dong H, Yu Q, et al. (2012) Genetic diversity of Mycobacterium tuberculosis isolates from Beijing, China assessed by Spoligotyping, LSPs and VNTR profiles. BMC Infect Dis 12: 372.
22. Chan MY, Borgdorff M, Yip CW, de Haas PE, Wong WS, et al. (2001) Seventy percent of the Mycobacterium tuberculosis isolates in Hong Kong represent the Beijing genotype. Epidemiol Infect 127: 169–171.
23. Jou R, Chiang CY, Huang WL (2005) Distribution of the Beijing family genotypes of Mycobacterium tuberculosis in Taiwan. J Clin Microbiol 43: 95–100.
24. Anh DD, Borgdorff MW, Van LN, Lan NT, van Gorkom T, et al. (2000) Mycobacterium tuberculosis Beijing genotype emerging in Vietnam. Emerg Infect Dis 6: 302–305.
25. Prodinger WM, Bunyaratvej P, Prachaktam R, Pavlic M (2001) Mycobacterium tuberculosis isolates of Beijing genotype in Thailand. Emerg Infect Dis 7: 483–484.
26. Toungoussova OS, Sandven P, Mariandyshev AO, Nizovtseva NI, Bjune G, et al. (2002) Spread of drug-resistant Mycobacterium tuberculosis strains of the Beijing genotype in the Archangel Oblast, Russia. J Clin Microbiol 40: 1930–1937.
27. Pannell CW (2011) China Gazes West: Xinjiang's Growing Rendezvous with Central Asia. Eurasian Geography and Economics 52: 105–118.
28. Weniger T, Krawczyk J, Supply P, Niemann S, Harmsen D (2010) MIRU-VNTRplus: a web tool for polyphasic genotyping of Mycobacterium tuberculosis complex bacteria. Nucleic Acids Res 38: W326–331.
29. Liu J, Tong C, Jiang Y, Zhao X, Zhang Y, et al. (2014) First Insight into the Genotypic Diversity of Clinical Mycobacterium tuberculosis Isolates from Gansu Province, China. PLoS One 9: e99357.
30. Iwamoto T, Yoshida S, Suzuki K, Tomita M, Fujiyama R, et al. (2007) Hypervariable loci that enhance the discriminatory ability of newly proposed 15-loci and 24-loci variable-number tandem repeat typing method on Mycobacterium tuberculosis strains predominated by the Beijing family. FEMS Microbiol Lett 270: 67–74.
31. Luo T, Yang C, Gagneux S, Gicquel B, Mei J, et al. (2012) Combination of single nucleotide polymorphism and variable-number tandem repeats for genotyping a homogenous population of Mycobacterium tuberculosis Beijing strains in China. J Clin Microbiol 50: 633–639.

32. Luo T, Yang C, Pang Y, Zhao Y, Mei J, et al. (2014) Development of a hierarchical variable-number tandem repeat typing scheme for Mycobacterium tuberculosis in China. PLoS One 9: e89726.

33. Jiang Y, Liu HC, Zheng HJ, Tang B, Dou XF, et al. (2012) Evaluation of four candidate VNTR Loci for genotyping 225 Chinese clinical Mycobacterium tuberculosis complex strains. Biomed Environ Sci 25: 82–90.

34. Wada T, Iwamoto T, Maeda S (2009) Genetic diversity of the Mycobacterium tuberculosis Beijing family in East Asia revealed through refined population structure analysis. FEMS Microbiol Lett 291: 35–43.

35. Mokrousov I, Ly HM, Otten T, Lan NN, Vyshnevskyi B, et al. (2005) Origin and primary dispersal of the Mycobacterium tuberculosis Beijing genotype: clues from human phylogeography. Genome Res 15: 1357–1364.

36. Kamerbeek J, Schouls L, Kolk A, van Agterveld M, van Soolingen D, et al. (1997) Simultaneous detection and strain differentiation of Mycobacterium tuberculosis for diagnosis and epidemiology. J Clin Microbiol 35: 907–914.

37. Lillebaek T, Andersen AB, Dirksen A, Glynn JR, Kremer K (2003) Mycobacterium tuberculosis Beijing genotype. Emerg Infect Dis 9: 1553–1557.

38. Hunter PR, Gaston MA (1988) Numerical index of the discriminatory ability of typing systems: an application of Simpson's index of diversity. J Clin Microbiol 26: 2465–2466.

39. Small PM, Hopewell PC, Singh SP, Paz A, Parsonnet J, et al. (1994) The epidemiology of tuberculosis in San Francisco. A population-based study using conventional and molecular methods. N Engl J Med 330: 1703–1709.

40. Wang J, Liu Y, Zhang CL, Ji BY, Zhang LZ, et al. (2011) Genotypes and characteristics of clustering and drug susceptibility of Mycobacterium tuberculosis isolates collected in Heilongjiang Province, China. J Clin Microbiol 49: 1354–1362.

41. Kremer K, Au BK, Yip PC, Skuce R, Supply P, et al. (2005) Use of variable-number tandem-repeat typing to differentiate Mycobacterium tuberculosis Beijing family isolates from Hong Kong and comparison with IS6110 restriction fragment length polymorphism typing and spoligotyping. J Clin Microbiol 43: 314–320.

42. Mokrousov I, Narvskaya O, Vyazovaya A, Millet J, Otten T, et al. (2008) Mycobacterium tuberculosis Beijing genotype in Russia: in search of informative variable-number tandem-repeat loci. J Clin Microbiol 46: 3576–3584.

3

Taxonomic Status, Phylogenetic Affinities and Genetic Diversity of a Presumed Extinct Genus, *Paraisometrum* W.T. Wang (Gesneriaceae) from the Karst Regions of Southwest China

Wen-Hong Chen[1,5], Yu-Min Shui[1]*, Jun-Bo Yang[2], Hong Wang[1]*, Kanae Nishii[3], Fang Wen[4], Zhi-Rong Zhang[2], Michael Möller[3]*

1 Key Laboratory for Plant Diversity and Biogeography of East Asia, Kunming Institute of Botany, Chinese Academy of Sciences, Kunming, Yunnan, China, 2 Plant Germplasm and Genomics Center, Germplasm Bank of Wild Species, Kunming Institute of Botany, Chinese Academy of Sciences, Kunming, Yunnan, China, 3 Science Division, Royal Botanic Garden Edinburgh, Edinburgh, Scotland, United Kingdom, 4 Guangxi Institute of Botany, Guangxi Zhuang Autonomous Region and Chinese Academy of Sciences, Guilin, Guangxi, China, 5 University of the Chinese Academy of Sciences, Beijing, China

Abstract

Background: The karst regions in South China have an abundance of endemic plants that face high extinction risks. The Chinese Gesneriaceae endemic *Paraisometrum mileense* (= *Oreocharis mileensis*), was presumed extinct for 100 years. After its re-discovery, the species has become one of five key plants selected by the Chinese forestry government to establish a new conservation category for plants with extremely small populations. For conservation purposes, we studied the phylogenetic and population genetic status of *P. mileense* at the three only known localities in Guangxi, Guizhou and Yunnan.

Methodology/Principal Findings: We collected 64 samples (52 species) of *Oreocharis* and 8 samples from three provinces of *P. mileense* and generated molecular phylogenies, and inferred that *P. mileense* represents a relatively isolated and derived taxonomic unit within *Oreocharis*. Phylogeographic results of 104 samples of 12 populations of *P. mileense* indicated that the populations in Yunnan have derived from those in Guangxi and Guizhou. Based on AFLP data, the populations were found to harbor low levels of genetic diversity ($He = 0.118$), with no apparent gradient across the species' range, a restricted gene flow and significant isolation-by-distance with limited genetic differentiation among the populations across the three provinces ($F_{ST} = 0.207$, $P<0.001$). The 10 populations in Yunnan were found to represent two distinct lineages residing at different altitudes and distances from villages.

Conclusion/Significance: The low levels of genetic diversity found in *P. mileense* are perhaps a consequence of severe bottlenecks in the recent past. The distribution of the genetic diversity suggests that all populations are significant for conservation. Current *in situ* and *ex situ* measures are discussed. Further conservation actions are apparently needed to fully safeguard this conservation flagship species. Our work provides a model of an integrated study for the numerous endemic species in the karst regions with extremely small populations.

Editor: Igor Mokrousov, St. Petersburg Pasteur Institute, Russian Federation

Funding: This work was supported by the National Natural Science Foundation of China (grant no. 31000258, 31470306), the National Key Basic Research Program of China (no. 2014CB954100), the Open Project of the Key Laboratory for Plant Diversity and Biogeography of East Asia, Kunming Institute of Botany (grant no. KLBB201304) and the project of Integrated Scientific Surveys to Daweishan National Park in China. The funders had no role in study design, data collection and analysis, decision to publish, or preparation of the manuscript.

Competing Interests: The authors have declared that no competing interests exist.

* Email: ymshui@mail.kib.ac.cn (YMS); wanghong@mail.kib.ac.cn (HW); m.moeller@rbge.ac.uk (MM)

Introduction

Paraisometrum mileense W.T.Wang was until relatively recently regarded as extinct in the wild because it had not been recollected for more than 100 years since its type specimen collection in 1906 [1,2]. In 2006, the plant was rediscovered in Shilin county, next to Mile county, Yunnan, where the type specimen was collected [1,3]. It was described as critically endangered (CR), possessing only 101–1000 individuals in a single population [4]. Interestingly, already in 2009, a second locality had been discovered in Longlin county in Guangxi, the province neighbouring Yunnan to the East [5], and soon afterwards, a third location, in Xingyi, in Guizhou province to the North of Guangxi, had been found [6] (Fig. 1), tripling the number of occurrence points. The total number of plants was estimated at >30,000 individuals [7], but up to 2010, detailed fieldwork by some of the authors (YMS and WHC) estimated the number to be significantly lower, with 630 mature plants in Yunnan, 150 in Guangxi, and 60 in Guizhou (Table 1). Incidentally, approximately 70–80% of the seedlings and young individuals died, and 50–60% of mature individuals sustained damage due to an extreme drought that occurred from 2010 to 2011 in the karst region in Yunnan where *P. mileense* grows (Fig. 2). This might partially explain the discrepancy in reported plant numbers, and illustrates the great vulnerability of the species to even short term climate fluctuations. Until recently, only the occurrence point in Guangxi was located in a provincial nature reserve (established in 2005), while the other two were outside protected areas. In 2011, however, the Chinese government set up a small reserve in Shilin, Yunnan, to protect the species at this primary rediscovery point (Fig. 2). The species is also one of five selected key plants to be used to establish a new category of protected areas in China for plant species with extremely small populations (PSESP) [4].

Until 2011, *Paraisometrum* W.T.Wang was regarded as a monotypic genus, but was then included in an enlarged genus *Oreocharis* Benth. as *O. mileensis* (W.T.Wang) Mich.Möller & A.Weber [8]. Irrespective of this recent inevitable taxonomic change, the species represents a highly threatened taxon. Furthermore, under its old name, it has received considerable attention as a strong flagship species for plant conservation in China, and we therefore use its original name here in this context, for consistency with current conservation initiatives, such as the establishment of the above new conservation category in China [4].

To be able to devise meaningful conservation strategies, knowledge of the taxonomic status, closest congeners and the level and distribution of genetic diversity within a taxon is essential [9–11]. Even though *P. mileense* has been included in phylogenetic analyses previously, its exact phylogenetic affinities in *Oreocharis* are still unclear. While it was clearly shown in the phylogenetic analyses that the species has evolved from within the enlarged *Oreocharis*, it fell on a polytomy with species of the *hitherto* genera *Ancylostemon* Craib, *Briggsia* Craib, *Isometrum* Craib, *Opithandra* B.L.Burtt and *Tremacron* Craib [8,12,13]. The generic characters to establish *Paraisometrum* are the presence of four upper corolla lobes, and one lower lobe of the pentamerous flowers [2]. Since this characteristic also occurs in other species of the newly defined *Oreocharis* (e.g. *O. saxatilis* (Hemsl.) Mich. Möller & A.Weber = *Ancylostemon saxatilis* Hemsl.) [14–17], the present work will address the species delineation of *P. mileense*. Furthermore, the plants in Guangxi and Guizhou appear to possess some floral features different from those in Yunnan (pers. obs. YMS and WHC). However, only two samples of *P. mileense*, one from Yunnan and one from Guangxi, were included in the most comprehensive study of *Oreocharis* to date [12]. This is not enough to test the taxon coherence in the light of modern approaches such as DNA barcoding [18–20]. Clearly, more molecular work was needed to include samples from all three occurrence points and from diverse species within the enlarged *Oreocharis*.

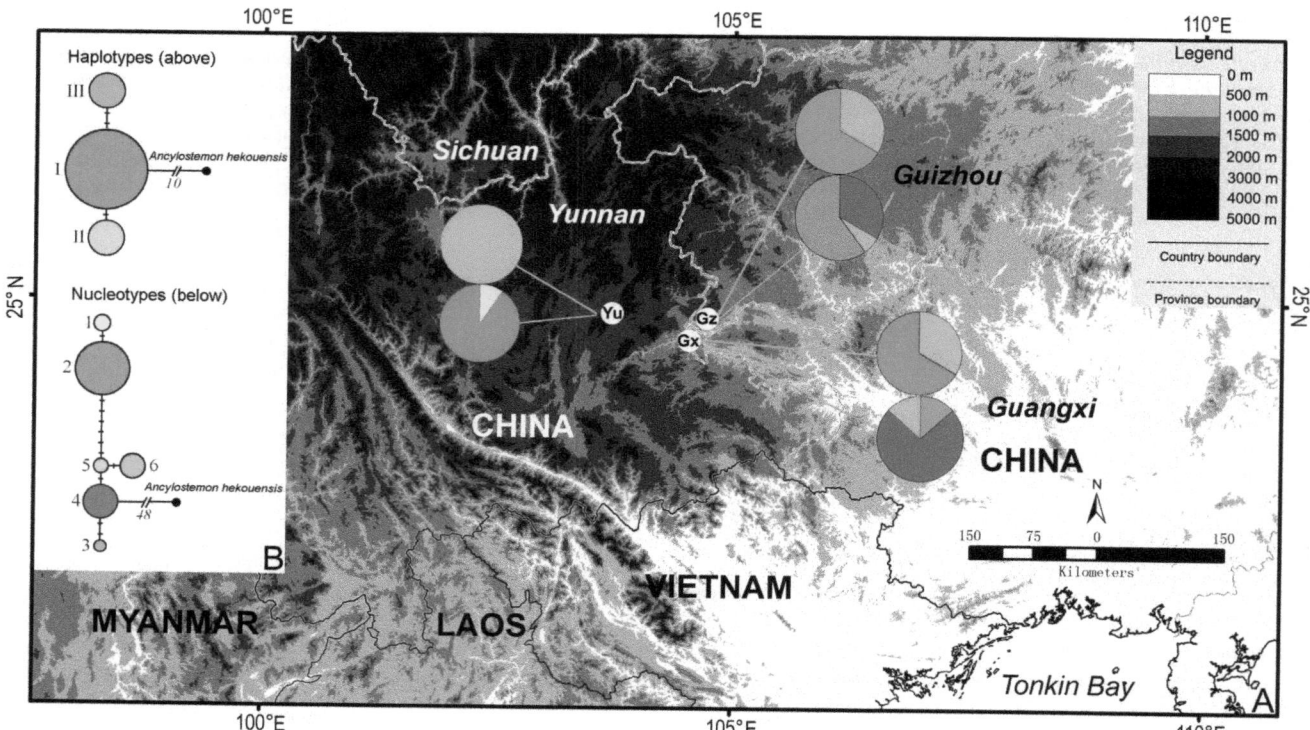

Figure 1. Map of localities of *Paraisometrum mileense* **in Yunnan, Guangxi and Guizhou.** A. Pie charts for haplotypes (above) and nucleotypes (below) are given for the three main geographical distribution areas. B. Median-joining networks of chloroplast haplotypes based on *trn*LF and *mat*K sequences (above) and of ITS nucleotypes (below) for *P. mileense* samples rooted on its closest relative *Ancylostemon hekouensis*.

Table 1. Detailed locality information for the 12 populations of *Paraisometrum mileense* used for AFLP analysis.

Population code	Locality name	Forest type	Position on slope	Altitude (m)	Distance (m) from village	Population size*	Area size (ha)	Sample number	Individual sample no
Yu1	Yunnan, Shilin, Guishan	secondary	lower	2001	25	46	0.40	5	Yu1a-e
Yu2	Yunnan, Shilin, Guishan	secondary	lower	2083	40	50	0.20	5	Yu2a-e
Yu3	Yunnan, Shilin, Guishan	secondary	lower	2112	20	52	0.25	5	Yu3a-e
Yu4	Yunnan, Shilin, Guishan	secondary	lower	1970	60	100	0.65	15	Yu4a-n
Yu5	Yunnan, Shilin, Guishan	primary	lower-middle	2149	800	76	0.25	8	Yu5a-e
Yu6	Yunnan, Shilin, Guishan	primary	lower-middle	2018	40	84	0.26	5	Yu6a-f
Yu7	Yunnan, Shilin, Guishan	primary	lower-middle	2077	600	120	1.20	5	Yu7a-e
Yu8	Yunnan, Shilin, Guishan	primary	middle-upper	2349	1000	32	0.20	5	Yu8a-e
Yu9	Yunnan, Shilin, Guishan	primary	upper	2412	1200	36	0.15	5	Yu9a-e
Yu10	Yunnan, Shilin, Guishan	primary	upper	2496	1500	34	0.16	10	Yu10a-i
Gx11	Guangxi, Longlin, Yacha	primary	upper	1183	1500	150	0.80	19	Gx11a-s
Gz12	Guizhou, Xingyi, Jingnan	primary	middle-upper	1405	1000	60	0.60	17	Gz12a-q

* - approximate number of mature plants per population based on data from August 2010.

There are many Gesneriaceae species endemic to the karst region in China, typically with small populations [7], such as *Primulina tabacum* Hance. This species occurs in only four populations distributed in Guangdong and Hunan with less than 1,000 plants in each [21]. Despite the relatively low plant number, a surprisingly high genetic diversity was found within the populations. Additionally, because of the long inter-population distances, a disruption of gene flow resulted in high population differentiation [10]. From a conservation perspective, the situation in *P. tabacum* was seen as 'a window of opportunity' to preserve a high level of extant genetic variation in the species. To determine whether a similar situation is present in *P. mileense*, we generated population genetic data using amplified fragment length polymorphisms (AFLPs) for individuals from the three localities in Guangxi, Guizhou and Yunnan. AFLP is a powerful tool for generating data from multiple loci for the detection of genetic variation without the need for pre-existing knowledge of genomic sequences [22,23], and have been successfully applied to small and relict populations (e.g. [10,24–27]), and a range of suitable analytical packages are available for these dominant markers (e.g. [27]).

Thus, our main aims were fourfold, to phylogenetically analyse sequence data, from the chloroplast intron–spacer sequences of *trn*LF, and the nuclear ribosomal internal transcribed spacer regions (nrITS), to a) test the taxonomic status of the species, b) to determine the phylogenetic position and relationships of *P. mileense* within the enlarged *Oreocharis*, c) to reconstruct the phylogeographic history of the *P. mileense* populations using *trn*LF, *mat*K, and ITS sequence data for individuals from 12 populations (or sub-populations) from the only three localities in Guangxi, Guizhou and Yunnan, and to acquire and analyse AFLP data for population samples to d) determine the levels of genetic diversity and differentiation of this species to assess its conservation requirements.

Materials and Methods

Ethics Statement

Of all sampled species, the target species *Paraisometrum mileense* is listed as one of five selected key plants to be used to establish a new category of protected areas in China for plant species with extremely small populations (PSESP). Approvals and permission for field studies were obtained from the Yunnan Forestry Bureau, China (permit no. [2011]–115). The GPS data are sensitive and are not provided for the protection of the plants according to the Yunnan forestry administrative request in the permit. However, we gave their approximate location in Fig. 1.

Study Materials

For the phylogenetic study, we expanded the ingroup sampling from 38 *Oreocharis* species plus one variety (41 samples) [12] to 52 species plus 2 varieties (73 samples), among which 11 species covered at least two population samples each. Of *P. mileense*, we included three samples each from Guangxi and Guizhou and two samples from Yunnan (Table S1). Thus, the *Oreocharis* ingroup sampling covered about 2/3 of the genus (53 out of about 80 species) [12]. The outgroup included 14 samples across six genera of straight-fruited advanced Asiatic and Malesian Gesneriaceae of subtribe Didymocarpinae (*sensu* [28]), with two *Didymocarpus* Wall. samples to root the trees [8,12].

The *Paraisometrum* materials for the phylogeographic analyses included 43 individuals from Yunnan (from 10 populations or sub-populations), 15 from the Guangxi population and 15 from the Guizhou population, and for the AFLP analysis 68 individuals

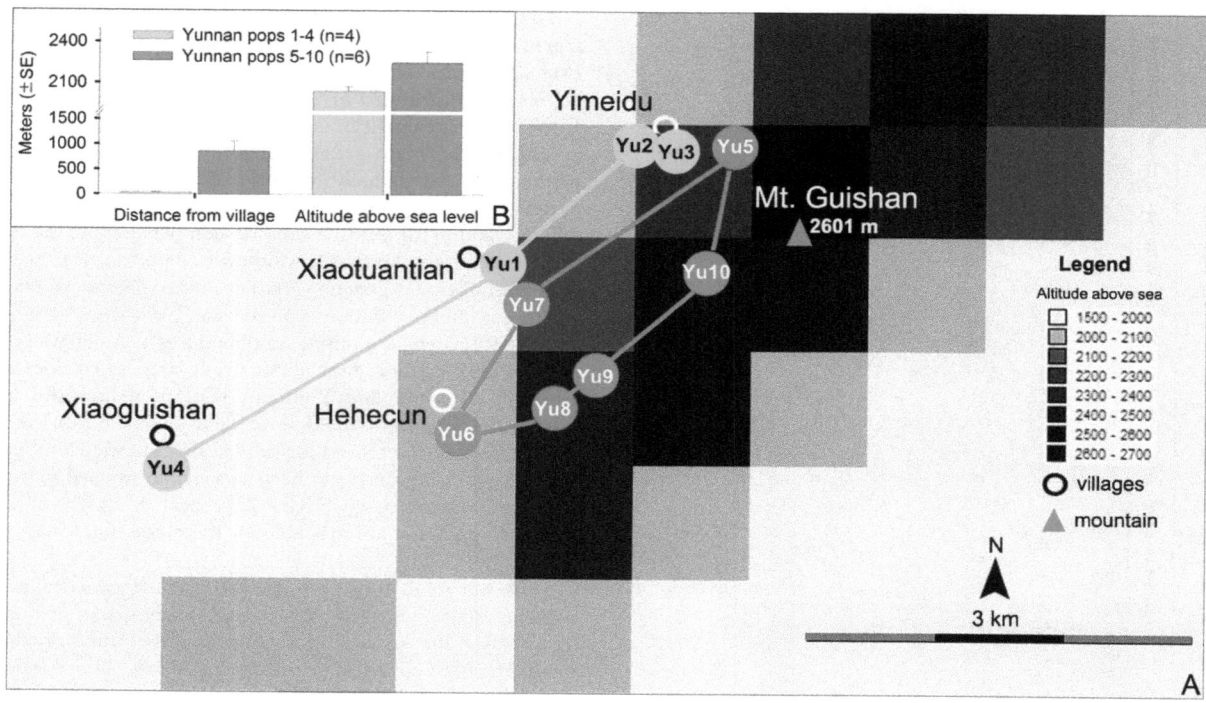

Figure 2. Populations of *Paraisometrum mileense* in Yunnan. A. Distribution of the ten populations analysed. B. Graphs of distance from village (left) and the average altitude (right) of populations 1–4 (light blue) and 5–10 (red).

from Yunnan (from 10 populations, 5–15 samples per population), 19 from the Guangxi population and 17 from the Guizhou population (Fig. 1; Table 1).

Molecular Methods

Genomic DNA was extracted from silica-gel-dried leaves collected in the field following a modified CTAB method [29,30]. The PCR primers for ITS were 'ITS5' and 'ITS4' [31], for *trn*LF 'C' and 'F' [32], and for *mat*K '3F' and '1R' [30]. PCR was performed in reactions containing 30–50 ng genomic DNA, 0.3 μl of each primer (5 μM/μl), 10 μl 2×*Taq* PCR MasterMix (Tiangen Biotech Co., Ltd, Beijing China: 0.1 U *Taq* polymerase/ μl, 0.5 nM of each dNTP, 20 mM Tris-HCl (pH 8.3), 100 mM KCl, 3 mM MgCl₂) and ddH₂O to make up 20 μl. PCR amplifications were conducted under the following profile: 95°C for 3 min followed by 35 cycles at 94°C for 30 s, at the annealing temperature specific for each primer pair for 30 s (ITS: 55°C; *trn*LF and *mat*K: 52°C), 72°C for 1 min, and a final extension step at 72°C for 5 min. After PCR amplicon purification, Sanger sequencing was carried out in 6 μl reactions containing 0.5 μl PCR product, 0.3 μl primer (5 μM/μl), 1.05 μl SeqBuffer, 0.3 μl BigDye Terminator Mix (Applied Biosystems, Foster City, USA) and 3.85 μl distilled H₂O. Sequencing reactions were cycled under the following conditions: 32 cycles at 96°C for 10 s, 50°C for 5 s, and 60°C for 3 min, and the products analysed on an ABI 3730×l sequencer (Applied Biosystems, Foster City, USA). The newly acquired sequences have been submitted to GenBank.

AFLP was performed based on Vos et al. [23]. Restriction endonuclease enzyme digestion and link reactions were performed in 20 μl reactions containing 4 μl template DNA (50 ng/μl), 1 μl Adaptor, 2 μl *Pst*I/*Mse*I, 2.5 μl 10× reaction buffer, 2.5 μl ATP (10 mM), 1 μl T4 Ligase, 7 μl distilled H₂O. The solution was centrifuged for a few minutes after stirring, and incubated at 37°C

for 5 h, then at 8°C for 4 h and then at 4°C overnight. Pre-amplification reactions were performed containing 2 μl template DNA, 1 μl Pre-ampmix, 0.5 μl dNTPs (Table 2), 2.5 μl 10× PCR buffer, 0.5 μl *Taq* polymerase, and 18.5 μl ddH₂O. PCR amplifications were conducted with the profile: 94°C for 2 min, followed by 30 cycles at 94°C for 30 s, 56°C for 30 s, 72°C for 80 s, with a final extension at 72°C for 5 min, and terminated at 4°C. Eight selective amplification primers were used for each *Pst*I and *Mse*I (Table 2). The selective amplification reactions were performed in 25 μl reactions containing 2 μl 1:20 diluted preamplification product, 2.5 μl 10× PCR buffer, 0.5 μl dNTP, 1 μl *Pst*I primer, 1 μl *Mse*I primer, 0.5 μl *Taq* polymerase, and 17.5 μl ddH₂O. PCR amplifications were conducted as follows: 94°C for 30 s, 65°C for 30 s, 72°C for 80 s for the first cycle, followed by 12 cycles with progressively decreasing annealing temperature by 0.7°C each cycle starting from 65°C to 56.6°C, which was followed by 23 cycles of 94°C for 30 s, 55°C for 30 s, 72°C for 80 s, with a final extension step at 72°C for 5 min, and then cooled to 4°C. The AFLP primers, *Taq* polymerase and dNTPs for PCR were purchased from Dingguochangsheng Biotechnology Co. Ltd. (Beijing, China). The amplified fragments were separated and detected with an ABI Prism 377 sequencer (Applied Biosystems, Foster City, USA). Due to ambiguous banding patterns or obvious PCR failures, 19 additional samples (GX11a-s) collected exclusively from Guangxi province were included (Table 1). The scoring error rate was about 1.3%, determined by replicating AFLP runs for 16 individuals (ca. 15% of all samples analysed) (*cf.* [33]).

Phylogenetic Analyses

The phylogenetic analyses were conducted on a matrix containing 86 ITS and *trn*LF sequences. ITS and *trn*LF sequences for 37 samples were newly acquired (Table S1), and for 49 samples

Table 2. Name and DNA sequences of primers and adaptors used in the AFLP experiments on *Paraisometrum mileense* samples.

	Primer name	Primer sequence
Adaptors	PstI 1	5′ -CTC GTA GAC TGC GTA CAT GCA
	PstI 2	5′ -TGT ACG CAG TCT AC
	MseI 1	5′ -GAC GAT GAG TCC TGA G
	MseI 2	5′ -TAC TCA GGA CTC AT
Pre-amplification primers	PstI	5′ -GAC TGC GTA CAT GCA G
	MseI	5′ -GAT GAG TCC TGA GTA A C
Selective amplification primers		
PstI primers (5 ng/μl)	PstI-1	5′ -GAC TGC GTA CAT GCA GAA
	PstI-2	5′ -GAC TGC GTA CAT GCA GAC
	PstI-3	5′ -GAC TGC GTA CAT GCA GAG
	PstI-4	5′ -GAC TGC GTA CAT GCA GAT
	PstI-5	5′ -GAC TGC GTA CAT GCA GTA
	PstI-6	5′ -GAC TGC GTA CAT GCA GTC
	PstI-7	5′ -GAC TGC GTA CAT GCA GTG
	PstI-8	5′ -GAC TGC GTA CAT GCA GTT
MseI primers (30 ng/μl)	MseI-1	5′ -GAT GAG TCC TGA GTA ACA A
(marked FAM)	MseI-2	5′ -GAT GAG TCC TGA GTA ACA C
	MseI-3	5′ -GAT GAG TCC TGA GTA ACA G
	MseI-4	5′ -GAT GAG TCC TGA GTA ACA T
	MseI-5	5′ -GAT GAG TCC TGA GTA ACT A
	MseI-6	5′ -GAT GAG TCC TGA GTA ACT C
	MseI-7	5′ -GAT GAG TCC TGA GTA ACT G
	MseI-8	5′ -GAT GAG TCC TGA GTA ACT T

retrieved from GenBank. The newly acquired DNA sequences were assembled and trimmed in Sequencher 4.1.4 (Gene Codes, Ann Arbor, MI, USA), and added to, and manually aligned with, the existing matrices. Maximum parsimony (MP) and Bayesian inference (BI) analyses followed Möller et al. [8,12] and Weber et al. [34], using PAUP* 4.0b10 [35] and MrBayes 3.2.2 [36,37], respectively.

The partition-homogeneity test in PAUP* (incongruence length difference test [38,39], on 1,000 replicates of repartitioning with tree bisection-reconnection (TBR) indicated no significant incongruence between the two datasets ($P = 0.28$) and the ITS and *trn*LF data sets were analysed combined. Parsimony was implemented on unordered and unweighted characters and through heuristic tree searches on 10,000 random starting trees with both TBR swapping and MulTrees on. Clade support was obtained as bootstrap indices with 10,000 heuristic replicates of random additions with TBR on and MulTrees off.

The most suitable substitution model for the BI analysis was obtained under the AIC criterion in MrModeltest [40], and was GTR+G for both, *trn*LF and the ITS spacers, and SYM+G for the 5.8S gene. Two independent runs of four MCMC chains of two million generations were run, sampled each 1,000th generation. The first 200 trees (10%) were discarded as burn-in (generations prior to stationarity of likelihood values), and the posterior probabilities (PP) obtained from 50% majority rule consensus trees obtained using the 'sumt' command in MrBayes. A low value of the average standard deviation of split frequencies (0.009025), the high correlation of the PP support values and the parallel distribution of variance values found between the two parallel runs

of the Bayesian analysis, and the level runs of the split posteriors indicated a good convergence of the MCMC runs (Table S2; Figs. S1–S3).

Phylogeographic Analyses

Chloroplast haplotypes were determined using combined chloroplast *trn*LF and *mat*K sequence data, and nucleotypes using ITS (GenBank accession numbers *trn*LF: KM062935–KM062942, KM06301–KM063145; *mat*K: KM063008–KM063080; ITS: KM062943–KM063007; KM063175–KM063182). Phylogeographic networks were reconstructed with NETWORK 4.6.1.1 (available at http://www.fluxus-engineering.com). *Ancylostemon hekouensis* Y.M.Shui & W.H.Chen (=*Oreocharis hekouensis* (Y.M.Shui & W.H.Chen) Mich.Möller & A.Weber) was used to root the networks, based on finding of the phylogenetic analyses here.

Population Genetic Analyses

The ABI trace files were analysed in GeneScan 3.1 (Applied Biosystems), and only intensive bands between 70 and 500 base pairs in size converted to a binary matrix (1 = band presence, 0 = band absence) (Table S3). Genetic diversity parameters were obtained in GenAlEx 6.5 [41], and included percentage of polymorphic loci (P), number of different alleles (Na), number of effective alleles (Ne), Shannon's Information index (SI), expected heterozygosity (He) and unbiased expected heterozygosity (uHe). To investigate for patterns among the *Paraisometrum* populations, we used STRUCTURE 2.3.4. [42] to assign individuals to genotypically distinct groups. We used the admixture model and the option for correlated allele frequencies as recommended by Falush et al. [43]. The program was run 14 times for each cluster (K) from K1 to K10. Each run of 100,000 iterations was preceded by 10,000 iterations as burn-in when convergence was achieved. We plotted the mean likelihood for each cluster L(K) against the cluster number (K). To establish the optimal number of clusters, the relationship between K and ΔK, the second order rate of change of the likelihoods, was plotted [44]. The distribution of genetic variation within and between populations and between regions was analysed in a hierarchical AMOVA in GenAlEx.

To test an isolation-by-distance scenario, a Mantel test was performed on the geographic and *ln* geographic distance *versus* Nei's genetic distance (*D*) in GenAlEx. For a 3D illustration of spatial relationships of the populations, a principal coordinate analysis (PCoA) was performed using the Jaccard distance in R-pack Le Progiciel R.4.0d10 [45]. To illustrate the genetic relationships between the populations, an unrooted Neighbor Joining (NJ) tree was reconstructed using Nei and Li's restriction site distances in PAUP*. Branch support was obtained from 1000 NJ bootstrap replicates in PAUP*.

Results

Phylogeny

The MP analysis recovered 12 most parsimonious trees of 1538 steps length with a CI of 0.5566 and RI of 0.6982. The genus *Oreocharis* in its new definition formed a strongly supported clade in the strict consensus tree (BS = 100%). *Paraisometrum* was nested deeply within this clade in a derived position with the *P. mileense* samples forming a strongly supported clade (BS = 100%) (Fig. S4). Sister to this clade were two samples of *A. hekouensis*, though with no branch support.

The BI trees showed the same phylogenetic positions with view to the *Paraisometrum* and *A. hekouensis* samples (Fig. 3), the sister-relationship of the two clades received only low support (PP = 0.63). The BI tree was more resolved both within the enlarged *Oreocharis* and within the *Paraisometrum* clade. *Paraisometrum* was nested deeply in a derived position within the clade dominated by species with yellow tetrandrous flowers (Fig. 3). The populations from Guizhou and Guangxi appeared intermixed in two groups not reflecting their origins (PP = 0.62), while the samples from Yunnan were most similar to each other (PP = 1.0), and somewhat distant to those from the other provinces.

Phylogeography

Among the 1676 bases in the combined *trn*LF and *mat*K matrix, three positions were variable among the *Paraisometrum* samples, resulting in three chloroplast haplotypes (Fig. 1; Table 3). Haplotype I was the only one in Yunnan and present in the other two provinces, while haplotype II was unique to Guangxi, and haplotype III was a private haplotype in the population from Guizhou (Fig. 1A). The median-joining network, rooted on *A. hekouensis*, placed haplotype I in a central position with haplotypes II and III as peripherals (Fig. 1B).

Among the 643 bases of the ITS region, eleven positions were variable, giving six different ITS nucleotypes among the samples with types 1 and 2 exclusive to Yunnan, type 4 and 5 shared among the Guangxi and Guizhou populations, while type 3 was private in Guangxi, and type 6 private in Guizhou (Fig. 1A; Table 3). The median-joined network placed type 4 present in Guangxi and Guizhou in the centre, from which first types only present in these two provinces have evolved; the ITS types 1 and 2 present in Yunnan were in a derived position (Fig. 1B).

Genetic Diversity

At the population level, the AFLP data suggested a low level of genetic diversity of the populations, particularly in Yunnan (Table 4). This is likely a result of the low number of individuals included per populations. At the region level, the data suggested that the Yunnan populations harboured an overall higher level of diversity. For example, the percentage of polymorphic loci was 90.88% in Yunnan, compared to 61.18% in Guangxi and 66.99% in Guizhou. Other genetic diversity indices were also slightly higher, such as the Shannon's Information index (Yunnan: 0.247, Guangxi: 0.219, Guizhou: 0.235), though the number of effective alleles and heterozygosity indices were very similar in the three provinces.

Genetic Structure

The AMOVA indicated that most genetic diversity resided within the populations of *P. mileense* (79%) and significant genetic differences existed among populations (21%, $F_{ST} = 0.207$, $P < 0.001$) (Table 5). When structured for regions, a similar genetic differentiation was observed between the three provinces (12%, $F_{CT} = 0.120$, $P < 0.001$) and among populations (12%, $F_{SC} = 0.144$, $P < 0.001$) (Table 6).

The optimal number of clusters for the STRUCTURE analysis was determined as K = 4 (Fig. 4 A and B). The analysis revealed a strong separation between the Yunnan populations 1–4 and 5–10 (Fig. 4C). The Guangxi population (Pop 11, Gx) was very homogeneous and distinct. The Guizhou population (Pop 12, Gz) was a mix of two types, one shared with the Guangxi population and the other individuals similar to a great degree to samples of the Yunnan populations 8 and 9. With K = 3, the patterns did not change greatly, except that the latter genotypes were clustering with the populations 1–4 from Yunnan. With K = 5, no further change in clustering was observed, the fifth genetic cluster scattering in small proportions among all populations. The Mantel test was significant using the geographic distance (r = 0.398; P = 0.01) and *ln* geographic distance (r = 0.383; P = 0.03).

The PCoA (1st axis 12% variance; 2nd axis = 7.7%; 3rd axis = 6.6%) separated the samples of the three regions strongly in the first axis, including the Yunnan populations 1–4 and 5–10 (Fig. 5A). Samples from Guangxi clustered very tightly together, while those from Guizhou and Yunnan were widely scattered. The samples from Guizhou were also scattered and almost overlapped with individuals from Guangxi at one end of their distribution, and with those from Yunnan, populations 5–10, at the other (Fig. 5A).

In the unrooted NJ tree, the samples clustered into four groups according to their province of origin including the split between Yunnan populations 1–4 and 5–10 (Fig. 5B). The Guangxi and Yunnan (populations 1–4) clusters received a high bootstrap support of 93%. These clusters were also characterized by short and uniform branch lengths. The same was the case for the cluster of Yunnan populations 5–10, except for three individuals, Yu8e, Yu9c and Yu9e. These had extremely long branches and were the same samples that seemed to be categorized as belonging to the Guizhou population in the STRUCTURE analysis (see Fig. 4C). The Guizhou population cluster was characterized by a loose clustering and relatively uniformly long branches (Fig. 5B).

Discussion

Taxonomic Status and Phylogenetic Affinities

The present study aimed to address several important issues surrounding this enigmatic, and until recently, monotypic genus that was thought to be extinct for 100 years. It was recently

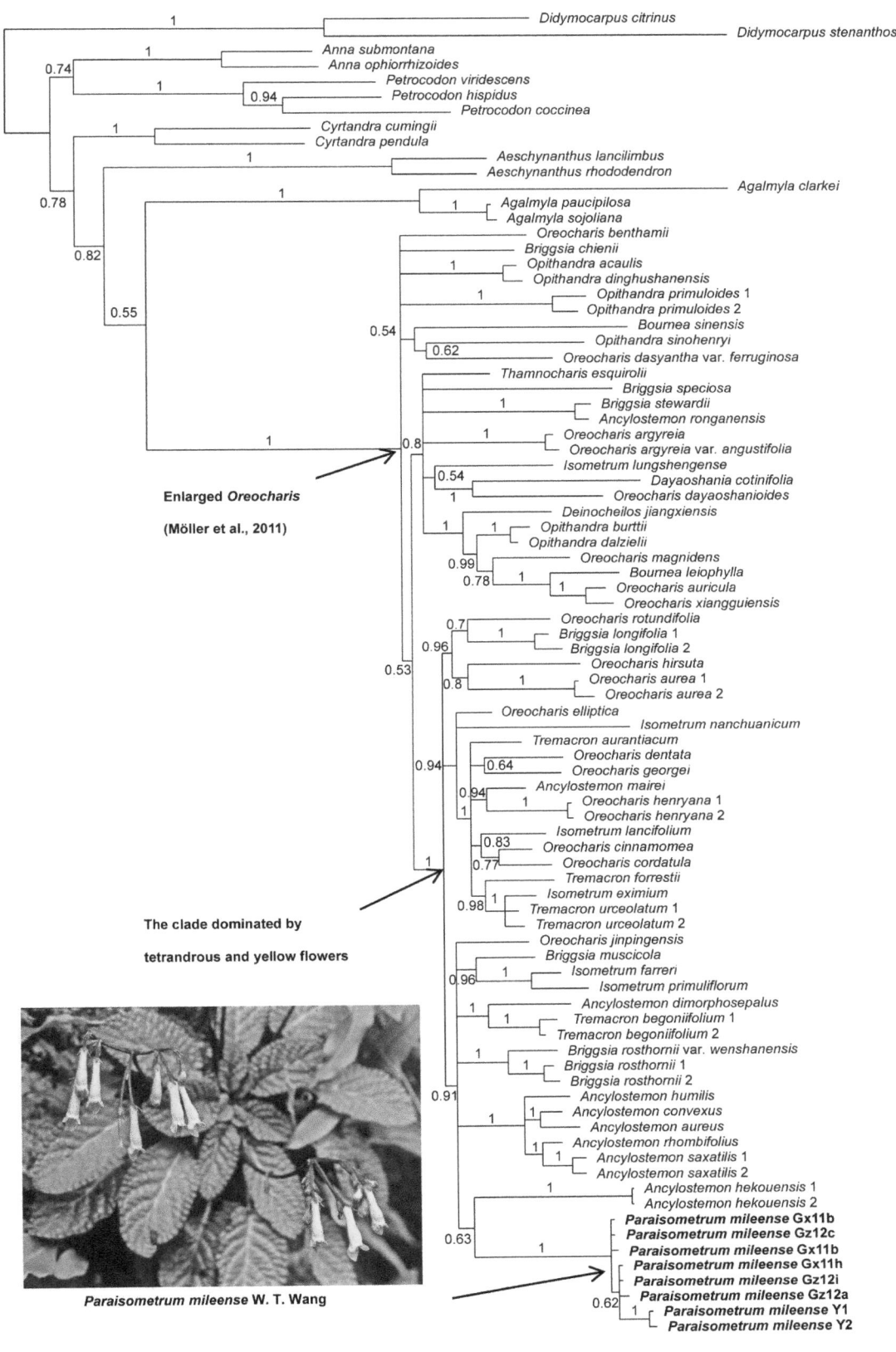

Enlarged *Oreocharis*

(Möller et al., 2011)

The clade dominated by

tetrandrous and yellow flowers

Paraisometrum mileense W. T. Wang

— 0.01 changes

Figure 3. Bayesian inference analysis placing *Paraisometrum mileense* in *Oreocharis*. Tree based on combined ITS and *trn*LF sequence data with average branch lengths and posterior probabilities. Photo of *P. mileense* by Yu-Min Shui.

Table 3. Haplotypes and nucleotypes found among the samples of *Paraisometrum mileense*.

Regions	N	Haplotypes			Nucleotypes					
		I	II	III	1	2	3	4	5	6
Yunnan (Pops1–10)	43	43	-	-	4	39	-	-	-	-
[Yunnan (Pops1–4)]	22	22	-	-	2	20	-	-	-	-
[Yunnan (Pops5–10)]	21	21	-	-	2	19	-	-	-	-
Guangxi (Pop11)	15	5	10	-	-	-	2	11	2	-
Guizhou (Pop12)	15	5	-	10	-	-	-	5	1	9
Total	73	-	-	-	-	-	-	-	-	-

reported to occur only in one population in Shilin, Yunnan [4], but we were able to sample from two further populations, from two different provinces. Since all *Paraisometrum* samples in the phylogenetic study fell in a single strongly supported clade separated from other *Oreocharis* taxa by a long branch, they can be regarded as a single taxonomic unit (Figs. 3, 4). Unlike previous phylogenetic analysis where relationships of this genus to other congeners were unresolved [8,12,13], we found indications that the closest species to *Paraisometrum mileense* is *Ancylostemon hekouensis*, a recently described species endemic to the karst region in Southwest China [46], with a similar leaf and flower morphology to *Paraisometrum*, but without the distinct 4-lobed upper corolla lip. The species with the most similar corolla shape to *P. mileense*, *Ancylostemon saxatilis* Hemsl., was found not closely related to *P. mileense* (Figs. 3, 4). This morphological homoplasy, is in line with previous findings of high levels of parallelism in the evolution of floral morphology in *Oreocharis* [12].

Phylogeographic History

Both, haplotype and nucleotype distribution, suggest a scenario of a close relationship between the Guangxi and Guizhou populations (shared nucleotypes), and a derived Yunnan population (lower haplo- and nucleotype diversity, derived nucleotypes) (Fig. 1, Table 3), and suggests a migration of *Paraisometrum* westward into Yunnan [9].

The significance in the Mantel test on the AFLP data indicates an isolation-by-distance scenario [47], and suggests a limited dispersal across the landscape. This may be a consequence of *Paraisometrum* being pollinated by bees, which have a relatively limited flying distance (e.g. [48,49]), and the seeds of *Paraisometrum*, that, though small (ca. 0.6 mm long), are not dust-like, as in orchids [50,51], to be dispersed far by wind, and do not have special dispersal aids, e.g. hooks or appendages (e.g. [52]) for long distance dispersal [53]. The significant genetic structure detected among the populations and regions (Tables 5, 6), further indicates a breakdown of their genetic connectivity. The ITS data may be used to give an indication of the divergence time between the *Paraisometrum* localities. Taking the average rate of evolution of the ITS spacers of 11 herbaceous plants (4.13×10^{-9} substitutions per site per year) [54,55], the Yunnan populations might have separated from the other populations around 1.6 million years ago (± 0.26 SE), at the beginning of the Pleistocene. This is a period that would cover repeated glacial-interglacial cycles during which limited secondary contact may have occurred (see below). This could explain the higher levels of within-population genetic variation observed here.

Our study provides a first insight into the history of a plant considered extinct in the wild. To fully address the evolution of the species, and to untangle historic from contemporary events, an increased sampling is required, additional markers employed such as microsatellites, combined with population demographic analyses (e.g. [56,57]), and, with view to conservation, ecological niche modelling (e.g. [58]).

Genetic Diversity and Differentiation

The STRUCTURE analysis indicated several noteworthy aspects of the *Paraisometrum* populations; firstly, that the Yunnan populations represented two distinct lineages with at least three gene pools and can be divided into two main populations, populations 1–4, and populations 5–10 with some admixture (Fig. 4). This bipartition of the Yunnan populations was also seen in the PCoA clustering analysis (Figs. 5A, S5) and the NJ tree (Fig. 5B). Both groups have the same haplotypes which indicates

Table 4. Genetic diversity indices based on AFLP data among the 12 populations of *Paraisometrum mileense* analysed.

Population	N	P	Na	Ne	SI	He	uHe
Yu1	5	33.37	0.744 (0.023)	1.160 (0.007)	0.155 (0.006)	0.100 (0.004)	0.111 (0.004)
Yu2	5	36.37	0.762 (0.024)	1.172 (0.007)	0.170 (0.006)	0.109 (0.004)	0.121 (0.004)
Yu3	5	33.01	0.712 (0.023)	1.161 (0.007)	0.156 (0.006)	0.100 (0.004)	0.112 (0.004)
Yu4	15	55.85	1.130 (0.024)	1.195 (0.007)	0.200 (0.006)	0.124 (0.004)	0.128 (0.004)
Yu5	8	44.34	0.938 (0.024)	1.202 (0.008)	0.195 (0.006)	0.125 (0.004)	0.133 (0.005)
Yu6	5	28.60	0.660 (0.022)	1.154 (0.007)	0.142 (0.006)	0.093 (0.004)	0.103 (0.004)
Yu7	5	32.95	0.727 (0.023)	1.174 (0.007)	0.162 (0.006)	0.106 (0.004)	0.118 (0.005)
Yu8	5	36.01	0.760 (0.024)	1.194 (0.008)	0.179 (0.006)	0.117 (0.004)	0.130 (0.005)
Yu9	5	43.66	0.891 (0.024)	1.197 (0.007)	0.199 (0.006)	0.126 (0.004)	0.140 (0.004)
Yu10	10	49.17	1.017 (0.024)	1.224 (0.008)	0.214 (0.006)	0.137 (0.004)	0.144 (0.005)
Gx	19	57.99	1.190 (0.024)	1.213 (0.008)	0.212 (0.006)	0.133 (0.004)	0.137 (0.004)
Gz	17	66.99	1.344 (0.023)	1.220 (0.007)	0.235 (0.006)	0.143 (0.004)	0.148 (0.004)
Total	8.7(0.036)	43.19 (3.48)	0.906 (0.007)	1.189 (0.002)	0.185 (0.002)	0.118 (0.001)	0.127 (0.001)

Mean over Loci for each region

	N	P	Na	Ne	SI	He	uHe
Yunnan	68	90.88	1.818 (0.014)	1.231 (0.007)	0.247 (0.006)	0.149 (0.004)	0.150 (0.004)
Guangxi	19	60.75	1.240 (0.024)	1.219 (0.008)	0.219 (0.006)	0.137 (0.004)	0.141 (0.004)
Guizhou	17	67.42	1.352 (0.023)	1.227 (0.007)	0.238 (0.006)	0.146 (0.004)	0.151 (0.004)

Grand Mean over Loci and pops

	N	P	Na	Ne	SI	He	uHe
Total	34.667(0.337)	73.01(9.14)	1.470(0.013)	1.226(0.004)	0.235(0.003)	0.144(0.002)	0.147(0.002)

Values are means (and SE).
N = no. of samples; P = percentage of polymorphic loci; Na = no. of different alleles; Ne = no. of effective alleles; SI = Shannon's Information index; He = expected heterozygosity; uHe = unbiased expected heterozygosity.

Table 5. Results of an unstructured hierarchical AMOVA on AFLP data of 12 populations of *Paraisometrum mileense*.

Source	df	SS	MS	Est. Var.	Var. (%)	F statistics	P
Among Pops	11	4914.428	446.766	36.513	21	$F_{ST}=0.207$	<0.001
Within Pops	92	12907.486	140.299	140.299	79	–	–
Total	103	17821.913	–	176.812	100	–	–

Table 6. Results of a structured hierarchical AMOVA on AFLP data of 12 populations of *Paraisometrum mileense*, with three regions, Yunnan (10 pops), Guangxi (1 pop), Guizhou (1 pop).

Source	df	SS	MS	Est. Var.	Var. (%)	F statistics	P
Among Regions	2	2243.796	1121.898	22.405	12	$F_{CT}=0.120$	<0.001
Among Pops	9	2670.632	296.737	23.581	13	$F_{SC}=0.144$	<0.001
Within Pops	92	12907.486	140.299	140.299	75	$F_{ST}=0.247$	<0.001
Total	103	17821.913	–	186.286	100	–	–

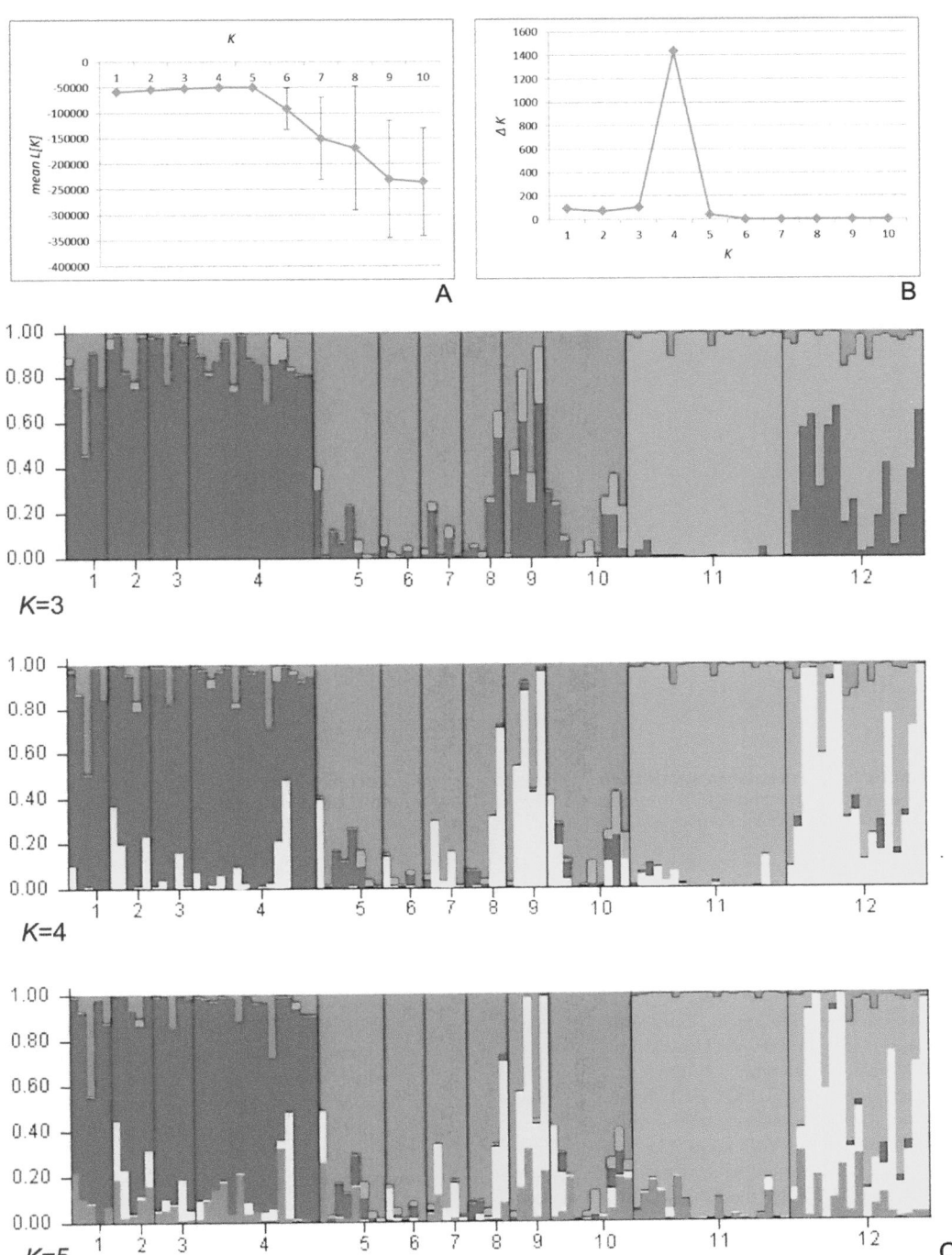

Figure 4. Results of the Bayesian inference STRUCTURE analysis on AFLP data of 12 *Paraisometrum mileense* populations. A. Plot of *K*-clusters *versus* mean (±SD) likelihoods (L[K]). B. *K* plotted against the second order rate of change of the likelihoods (ΔK). C. STRUCTURE clustering results for *K* = 3 to 5 as suggested in B. Numbers refer to populations in Table 1. 1–10 = Yunnan, 11 = Guangxi, 12 = Guizhou.

their common ancestry, and the same composition of ITS nucleotypes (Table 3), which suggests fragmentation of a previously more continuously distributed population. Populations 1–4 occur near villages, in the foothills of Mt. Guishan in secondary open forests with frequent disturbance due to human activities, while populations 5–10 grow mostly undisturbed further away from villages, above the foothills almost to the summit of Mt.

Guishan, and in more dense and primary forests (Fig. 2B, Table 1). Whether the bipartition is linked to the condition of the habitats surrounding the population groups, being disconnected from each other by disturbed forest, or has an older origin, would require further research. Secondly, the population in Guangxi was distinct and very homogeneous, also seen in the tight clustering in the PCoA (Fig. 5A) and the uniformly short branches

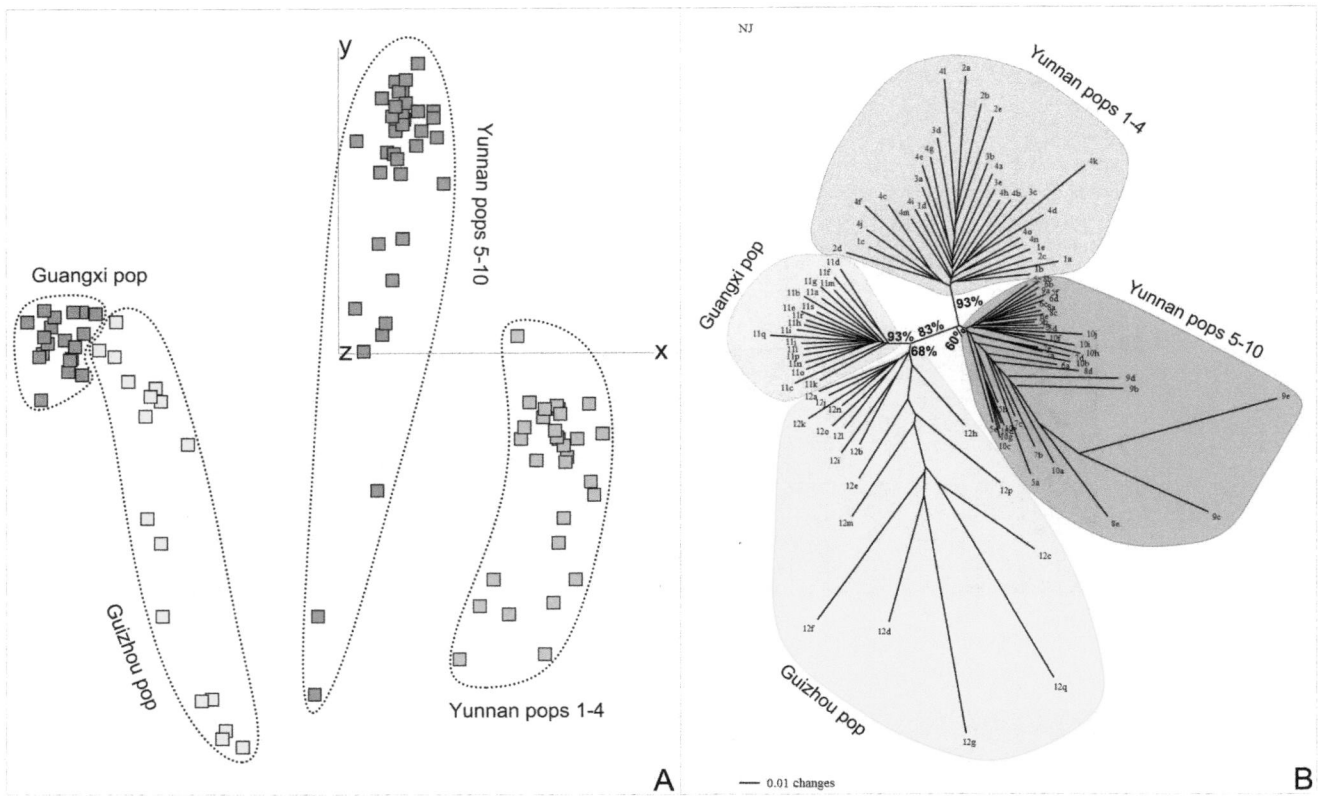

Figure 5. Four genetic lineages exist in *Paraisometrum mileense*. A. PCoA scatter plot based on AFLP data using the Jaccard distance on 12 populations of *P. mileense*. Axis 1(x) and 2(y). B. Unrooted NJ-tree based on Nei and Li's restriction site distances. Bootstrap values only given for internal branches (1000 Neighbor-Joining replicates).

in NJ tree (Fig. 5B). This might indicate that while the plants contain similar genetic diversity levels as other regions, the plants were closer related to each other possibly due to consanguineous matings, perhaps as a side effect of their small population sizes and limited area of distribution of less than a hectare (Table 1), and long distance to the other distribution localities with no apparent intervening populations known (Fig. 1). Thirdly, on the contrary, the internal and terminal branches of the population from Guizhou were long, and the STRUCTURE analysis suggested that some plants had a genetic makeup greatly similar to plants of the Yunnan populations 8 (plant Yu8e) and 9 in particular (plants Yu9c, Yu9e). This was not reflected in a mixed clustering of the respective Yunnan samples among the Guizhou samples, though their branches were unusually long (Fig. 5B). This might reflect genetic links and allele exchanges between these populations, perhaps due to habitat expansion during glacial-interglacial cycles that brought the populations in closer proximity (e.g. [58]). They are currently only some 120 km apart (Fig. 1).

The genetic diversity found at the population level was very low, although this might have been a consequence of the low number of individuals (often 5) included for analysis of the Yunnan populations. However, even when calculated within the three regions, the AFLP diversity levels were relatively low, with no marked difference between the regions (Table 4). Even at the species level, the genetic diversity of *Paraisometrum* was comparatively low (*He* = 0.144), and much lower compared to *Primulina tabacum* (*He* = 0.339), another herbaceous Gesneriaceae species from China [11]. *Primulina tabacum* and *P. mileense* have similar distribution sizes though in different provinces (*P. tabacum* in

Guangdong and Hunan, *Paraisometrum* in Guangxi, Guizhou and Yunnan), but both occur in limestone karst areas, though at different altitudes (*P. tabacum* below 300 m, *Paraisometrum* between 1,180–2,500 m) and in different aspects (*P. tabacum* grows around entrances of limestone caves, *Paraisometrum* in limestone forests). The high genetic diversity levels found in *P. tabacum* populations were explained by their refugial history and/ or breeding system [11]. The more restricted habitats in cave entrances of *P. tabacum*, as opposed to the more open forest habitats of *Paraisometrum* could be a factor that may have allowed a more continuous distribution and genetic connectivity of *Paraisometrum* populations. However, it is more likely that the greater population sizes of *P. tabacum* (1,000 individuals per population, as opposed to 30–150 individuals in *Paraisometrum* populations) has allowed *P. tabacum* to retain higher diversity levels. The similarly low levels of AFLP diversity compared to *Paraisometrum* found in other herbaceous perennials, such as *Trollius europaeus* L. (*He* = 0.158–0.229) [24], *Silene otitis* St.-Lag. (*He* = 0.167–0.240) [25], and *Draba aizoides* L. (*He* = 0.07–0.15) [26], were linked to effects of population fragmentation. Thus, the low genetic diversity in *Paraisometrum* is likely a result of a combination of fragmentation and small population size. Intriguingly though, in *Draba* L., the low genetic diversity levels were not correlated with a limited reproductive fitness, as suggested by their high germination rates. This may have important conservation implications and could be a field of study in *Paraisometrum* in the future.

Implications for Conservation

Our work provides an example of an integrated study for endemic species in the karst regions in South China with extremely small populations. It is well known that the karst region is characterized by a limestone topography with an abundance of endemic plants with extremely small populations [7,59,60]. In these regions, localised endemics occur often restricted to one or few limestone hills, which are usually isolated by non-limestone topography and the plants effectively occur on islands with high risks of extinction [7,14]. The three distribution points of *Paraisometrum* are isolated in different limestone forests, and the occurrence nearby villages is not unusual for endemic plants of the karst region [59,61].

The villages are commonly closely associated with characteristic forests on the limestone hills, the Feng Shui Forests. These are important for the water supply of the villages, but are recently influenced by activities of local people, such as free range poultry and goat keeping [62], which negatively affect the forest habitats.

With a view to conservation of *Paraisometrum*, the only three distribution points known to date appear to contain a significant amount of diversity within their populations with only a limited level of differentiation. Though, analysing a larger number of individuals per population might show some more resolution. Furthermore, the STRUCTURE analysis shows that they seem to have experienced some gene flow between Yunnan and Guizhou populations and are thus not genetically greatly isolated (Table 1). However, the calculated levels of genetic diversity were quite low, even when estimated across the regions including Yunnan (and not based on the small 'sub-populations' there). This is likely linked to the relatively small population sizes. The low genetic diversity might hinder adaptation to rapid climatic changes, as shown in the exceptionally dry season of 2011 during which they contracted significantly, due to the death of many immature individuals, and hardly any flower was produced in the Yunnan populations, their inflorescences withered (SYM, pers. observation). The relatively uniform distribution of genetic diversity in the three provinces of *Paraisometrum* does not immediately allow a prioritization of regions or populations for conservation efforts. Even though little differences in the genetic make-up were detected between the two Yunnan lineages, the one including populations 1–4, may deserve stronger protection due to their close proximity to human habitation and the negative effects these can bring (Fig. 2). For all of the reasons given above, conservation efforts should thus include all populations and include *in situ* as well as *ex situ* measures to safeguard this enigmatic species.

In situ measures relevant to *Paraisometrum* have already been implemented in Yunnan on a small scale at Shilin populations and involved the removal of *Ageratina adenophora* (Spreng.) R. M. King & H. Robinson, an invasive plant alien to China [63], the exclusion of goats through fencing in the plants, and by involving the local villagers to safeguard these plants that have a high potential for eco-tourism. Future *in situ* measures may also involve forest restoration [64], particularly for populations near villages. However, these may not be sufficient and *ex situ* conservation measures are required alongside *in situ* measures in the light of the observed strong detrimental effects of recent rainfall variation on the populations. This may, for the moment preclude transfer of plants from threatened habitats to new localities in the wild. Though, the plants are easily cultivated and propagated vegetatively and through seeds. Currently, about 56 plants are cultivated at the Kunming Botanical Garden, Yunnan, China, and seeds of three populations have been deposited at the Germplasm Bank of Wild Species, Plant Germplasm and Genomics Center, Kunming Institute of Botany, Yunnan, China. Thus, the immediate and medium term survival of the species seems ensured. However, seeds in germplasm banks are prone to genetic erosion during seed rejuvenation cycles even of inbreeding species [65]. Currently research is underway studying the pollinators to elucidate the reproductive strategy and breeding system to better devise a tailored *ex situ* strategy for *Paraisometrum mileense*.

Conclusions

The erstwhile monotypic genus *Paraisometrum mileense* (now *Oreocharis mileensis*) represents an isolated and independent taxonomic unit, and appears closest related to the erstwhile *Ancylostemon hekouensis* (now *Oreocharis hekouensis*). The populations in Yunnan seem to have derived from those in Guangxi and Guizhou with some evidence of limited gene transfer between Yunnan and Guizhou. The occurrence points in Yunnan were divided into two quite distinct lineages, while the population in Guangxi is relatively homogeneous. Overall, the populations of *P. mileense* contain relatively low levels of genetic diversity with no apparent gradient across the species' range. Several conservation measures are currently implemented, but additional actions are needed. From a conservation genetic perspective, all populations seem equally important for *in situ* and *ex situ* protection.

Supporting Information

Figure S1   Comparison of posterior probability values between run 1 versus run 2 (10% burn-in) of the Bayesian inference analysis.

Figure S2   Posterior probabilities of splits at selected increments over MCMC run1 (A) and 2 (B) of the Bayesian inference analysis.

Figure S3   Comparisons of topological differences within and among MCMC runs of the Bayesian inference analysis.

Figure S4   Maximum parsimony strict consensus tree of 12 parsimonious trees of 1538 steps, based on combined ITS and *trn*LF sequence data (CI = 0.5566; RI = 0.6982). Numbers above branches are bootstrap values (x denotes branches receiving <50% support).

Figure S5   PCoA scatter plot based on AFLP data using the Jaccard distance on 12 populations of Paraisometrum mileense. Axes 1 (x) and 3 (z).

Table S1   Collection details and GenBank numbers of *Oreocharis* (including *Paraisometrum mileense*) and outgroup samples used for the phylogenetic analyses.

Table S2   Characteristics of the Bayesian inference analysis of 86 samples of combined ITS and *trn*LF data.

Table S3   Binary AFLP data for 12 populations of Paraisometrum mileense from Yunnan (Yu), Guizhou (Gz) and Guangxi (GX).

Acknowledgments

We thankfully acknowledge Prof. Yi-Gang Wei, Dr. Qi Gao, Mr. Wen-Bin Xu, Guangxi Institute of Botany, Mr. Xiao-Qing Luo, Institute of Subtropical Crops of Guizhou Province and Mr. Bo Xiao, Malipo Forestry Bureau, and Dr. Zong-Xin Ren, Kunming Institute of Botany, Chinese Academy Sciences, for providing support in the field. We also like to thank two anonymous reviewers for their constructive comments on the manuscript. The Royal Botanic Garden Edinburgh is supported by the Rural and Environment Science and Analytical Services division (RESAS) in the Scottish Government.

Author Contributions

Conceived and designed the experiments: WHC HW MM. Performed the experiments: WHC JBY KN ZRZ. Analyzed the data: MM HW YMS. Contributed reagents/materials/analysis tools: YMS MM FW. Contributed to the writing of the manuscript: WHC MM YMS.

References

1. Smith L (2007) 'Extinct' plant found flowering in China. The Times website. Available: http://www.thetimes.co.uk/tto/news/world/article1973186.ece. Accessed 2014 Sep 1.
2. Weitzman AL, Skog LE, Wang WT, Pan KY, Li ZY (1997) New taxa, new combinations, and notes on Chinese Gesneriaceae. Novon 7: 423–435.
3. Shui YM, Cai J (2007) "100-years-lost" plant reemerges in Yunnan, China. Samara 12: 3.
4. Ma YP, Chen G, Grumbine RE, Dao ZL, Sun WB, et al. (2013) Conserving plant species with extremely small populations (PSESP) in China. Biodivers Conserv 22: 803–809.
5. Xu WB, Pan B, Huang YS, Ye XX, Liu Y (2009) *Paraisometrum* W.T.Wang, a newly recorded genus of Gesneriaceae from Guangxi, China. Guihaia 29: 581–583.
6. Gao Q, Xu WB (2011) *Paraisometrum* W. T. Wang, a newly recorded genus of Gesneriaceae from Guizhou, China. Acta Bot Boreal-Occid Sin 31 858–860.
7. Wei YG, Wen F, Möller M, Monro A, Zhang Q, et al. (2010) Gesneriaceae of South China. Nanning: Guangxi Science and Technology Publishing House. 777 p.
8. Möller M, Forrest A, Wei YG, Weber A (2011) A molecular phylogenetic assessment of the advanced Asiatic and Malesian didymocarpoid Gesneriaceae with focus on non-monophyletic and monotypic genera. Plant Syst Evol 292(3–4): 223–248.
9. Lowe A, Harris S, Ashton P (2007) Chapter 3. Genetic diversity and differentiation. In: Ecological Genetics: Design, Analysis, and Application. Hongkong: Blackwell Publishing. pp. 52–100.
10. Ni XW, Huang YL, Wu L, Zhou RC, Deng SL, et al. (2006) Genetic diversity of the endangered Chinese endemic herb *Primulina tabacum* (Gesneriaceae) revealed by amplified fragment length polymorphism (AFLP). Genetica 127: 177–183.
11. Wang CN, Möller M, Cronk QCB (2004) Population genetic structure of *Titanotrichum oldhamii* (Gesneriaceae), a subtropical bulbiliferous plant with mixed sexual and asexual reproduction. Ann Bot 93: 201–209.
12. Möller M, Middleton DJ, Nishii K, Wei YG, Sontag S, et al. (2011) A new delineation for *Oreocharis* incorporating an additional ten genera of Chinese Gesneriaceae. Phytotaxa 23: 1–36.
13. Tan Y, Wang Z, Sui XY, Hu GW, Motley T, et al. (2011) The systematic placement of the monotypic genus *Paraisometrum* (Gesneriaceae) based on molecular and cytological data. Pl Diversity Resources 33: 465–476.
14. Li ZY, Wang YZ (2004) Plants of Gesneriaceae in China. Zhengzhou: Henan Science and Technology Publishing House. 14–79 p.
15. Pan KY (1987) Taxonomy of the genus *Oreocharis* (Gesneriaceae). Acta Phytotax Sin 25: 164–293.
16. Wang WT, Pan KY, Li ZY (1990) Gesneriaceae. In: Wang WT, editor. Fl. Reipubl. Popularis Sin. 69: 190–203. Science Press, Beijing.
17. Wang WT, Pan KY, Li ZY, Weitzman AL, Skog LE (1998) Gesneriaceae. In: Wu ZY, Raven PH (eds) Flora of China. 18: 268–272. Beijing: Science Press & St Louis: Missouri Botanical Garden Press.
18. Jiao LJ, Shui YM (2013) Evaluating candidate DNA barcodes among Chinese *Begonia* (Begoniaceae) species. Pl Diversity Resources 35: 715–724.
19. Liu J, Möller M, Gao LM, Zhang DQ, Li DZ (2011) DNA barcoding for the discrimination of Eurasian yews (*Taxus* L., Taxaceae) and the discovery of cryptic species. Mol Ecol Res 11: 89–100.
20. Liu ML, Yu WB, Wang H (2013) Rapid identification of plant species and iflora: application of DNA barcoding in a large temperate genus *Pedicularis* (Orobanchaceae). Pl Diversity Resources 35: 707–714.
21. Ren H, Peng SL, Zhang JX, Jian SG, Wei Q, et al. (2003) The ecological and biological characteristics of an endangered plant, *Primulina tabacum* Hance. Acta Ecol Sin 23: 1012–1017.
22. Rafalski JA, Vogel JM, Morgante M, Powell W, Andre C, et al. (1997) Generating and using DNA makers in plants. In: Birren B, Lai E, editors. Non-mammalian Genome Analysis: A practical Guide. New York: Academic Press, pp. 75–134.
23. Vos P, Hogers R, Bleeker M, Reijans M, van de Lee T, et al. (1995) AFLP: a new technique for DNA fingerprinting. Nucleic Acids Res 23: 4407–4414.
24. Despres L, Loriot S, Gaudeul M (2002) Geographic pattern of genetic variation in the European globeflower *Trollius europaeus* L. (Ranunculaceae) inferred from amplified fragment length polymorphism markers. Mol Ecol 11: 2337–2347.
25. Lauterbach D, Ristow M, Gemeinholzer B (2012) Population genetics and fitness in fragmented populations of the dioecious and endangered *Silene otites* (Caryophyllaceae) Plant Syst Evol 298: 155–164.
26. Vogler F, Reisch C (2013) Vital survivors: low genetic variation but high germination in glacial relict populations of the typical rock plant *Draba aizoides*. Biodivers Conserv 22: 1301–1316.
27. Bonin A, Ehrich D, Manel S (2007) Statistical analysis of amplified fragment length polymorphism data: a toolbox for molecular ecologists and evolutionists. Mol Ecol 16: 3737–3758.
28. Weber A, Clark JL, Möller M (2013) A new formal classification of Gesneriaceae. Selbyana 31(2): 68–94.
29. Doyle JJ, Doyle JL (1987) A rapid DNA isolation procedure for small quantities of fresh leaf tissue. Phytochem Bull 19: 11–15.
30. Doyle JJ, Doyle JL (1990) Isolation of plant DNA from fresh tissue. Focus 12: 13–15.
31. White TL, Bruns T, Lee S, Taylor J (1990) Amplification and direct sequencing of fungal ribosomal RNA genes for phylogenetics. In: Innis MA, Gelfand D, Sninsky JJ, White TJ, editors. PCR protocols: a guide to methods and applications. San Diego: Academic Press, pp. 315–322.
32. Taberlet P, Gielly L, Pautou G, Bouvet J (1991) Universal primers for amplification of three non-coding regions of chloroplast DNA. Plant Mol Biol 17: 1105–1109.
33. Bonin A, Bellemain E, Bronken Eidesen P, Pompanoni F, Brochmann C, et al. (2004) How to track and assess genotyping errors in population genetics studies. Mol Ecol 13: 3261–3273.
34. Weber A, Wei YG, Puglisi C, Wen F, Mayer V, et al. (2011) A new definition of the genus *Petrocodon* (Gesneriaceae). Phytotaxa 23: 49–67.
35. Swofford DL (2002) PAUP*: phylogenetic analysis using parsimony (*and other methods), version 4. Sinauer, Sunderland, Massachusetts, USA.
36. Huelsenbeck JP, Ronquist F (2001) MRBAYES: Bayesian inference of phylogenetic trees. Bioinformatics 17: 754–755.
37. Ronquist F, Teslenko M, van der Mark P, Ayres D, Darling A, et al. (2011) MrBayes 3.2: Efficient Bayesian phylogenetic inference and model choice across a large model space. Syst Biol 61: 539–542.
38. Farris J, Källersjö SM, Kluge AG, Bult C (1994) Constructing a significance test for incongruence. Syst Biol 43: 570–572.
39. Farris J, Källersjö SM, Kluge AG, Bult C (1995) Testing significance of incongruence. Cladistics 10: 315–319.
40. Nylander JA, Wilgenbusch JC, Warren DL, Swofford DL (2008) AWTY (are we there yet?): a system for graphical exploration of MCMC convergence in Bayesian phylogenetics. Bioinformatics 24(4): 581–583.
41. Peakall R, Smouse PE (2006) GenAlEx 6.5: genetic analysis in Excel. Population genetic software for teaching and research-an update. Mol Ecol Notes 6: 288–295.
42. Pritchard JK, Stephens M, Donnelly P (2000) Inference of population structure using multilocus genotype data. Genetics 155: 945–959.
43. Falush D, Stephens M, Pritchard JK (2003) Inference of population structure using multilocus genotype data: linked loci and correlated allele frequencies. Genetics 164: 1567–1587.
44. Evanno G, Regnaut S, Goudet J (2005) Detecting the number of clusters of individuals using the software STRUCTURE: a simulation study. Mol Ecol 14: 2611–2620.
45. Casgrain P, Legendre P, Vaudor A (2005) The royal package for multivariate and spatial analysis, Version 4.0 (development release 10). Available: http://adn.biol.umontreal.ca/~numericalecology/old/R/index.html. Accessed 2014 Sep 1.
46. Chen WH, Shui YM (2006) *Ancylostemon hekouensis* (Gesneriaceae), a new species from Yunnan, China. Ann Bot Fenn 43: 448–450.
47. Wright S (1943) Isolation by distance. Genetics 28: 114–138.
48. Pasquet RS, Peltier A, Hufford MB, Oudin E, Saulnier J, et al. (2008) Long-distance pollen flow assessment through evaluation of pollinator foraging range suggests transgene escape distances. P Natl Acad Sci USA 105(36): 13456–13461.
49. Eckert JE (1933) The flight range of the honeybee. J Agric Res 47: 257–285.
50. Dressler RL (1993) Phylogeny and classification of the orchid family. Cambridge: Cambridge University Press, 314 p.
51. Murren CJ, Ellison AM (1998) Seed dispersal characteristics of *Brassavola nodosa* (Orchidaceae). Am J Bot 85(5): 675–680.
52. Kokubugata G, Hirayama Y, Peng CI, Yokota M, Möller M (2011) Phytogeographic aspects of *Lysionotus pauciflorus* sensu lato (Gesneriaceae) in

the China, Japan and Taiwan regions: phylogenetic and morphological relationships and taxonomic consequences. Plant Syst Evol 292(3–4): 177–188.

53. Chen WH (2013) Reproductive Biology of *Paraisometrum* W. T. Wang, an endemic to SW China– with special emphasis on its phylogenetic position and genetic diversity. Ph.D. Thesis, Kunming Institute of Botany, Chinese Academy of Sciences, Kunming, China.

54. Kay KM, Whittall JB, Hodges SA (2006) A survey of nuclear ribosomal internal transcribed spacer substitution rates across angiosperms: an appropriate molecular clock with life history effects. BMC Evol Biol 6: 36.

55. Puglisi C, Wei YG, Nishii K, Möller M (2011) *Oreocharis × heterandra* (Gesneriaceae): a natural hybrid from the Shengtangshan Mountains, Guangxi, China. Phytotaxa 38: 1–18.

56. Ribeiro RA, Lemos-Filho JP, Ramos ACS, Lovato MB (2011) Phylogeography of the endangered rosewood *Dalbergia nigra* (Fabaceae): insights into the evolutionary history and conservation of the Brazilian Atlantic Forest. Heredity 106: 46–57.

57. Liu J, Möller M, Provan J, Gao LM, Poudel RC, et al. (2013) Geological and ecological factors drive cryptic speciation of yews in a biodiversity hotspot. New Phytol 199: 1093–1108.

58. Poudel RC, Möller M, Gao LM, Ahrends A, Baral SR, et al. (2012) Using morphological, molecular and climatic data to delimitate yews along the Hindu Kush-Himalaya and adjacent regions. PLoS ONE 7(10): e46873.

59. Fang RZ, Bai PY, Huang GB, Wei YG (1995) A floristic study on the seed plants from tropics and subtropics of Dian-Qian-Gui. Acta Bot Yunnan Suppl. 5: 111–150.

60. Sun H (2013) Phytogeographical regions of China. In: Hong DY, Blackmore S, editors. Plants of China: A companion to the Flora of China. Beijing: Sciences Press. pp. 176–204.

61. Chen WH, Möller M, Shui YM, Wang H, Wen F, et al. (2014) Three new species of *Petrocodon* (Gesneriaceae), endemic to the limestone areas of Southwest China, and preliminary insights into the diversification patterns of the genus. Syst Bot 39(1): 316–330.

62. Xu HG, Zhou F (2011) Biodiversity in the karst area of Southwest Guangxi. Beijing: Encyclopedia of China Publishing House. 143 p.

63. Zhu L, Sun OJ, Sang WG, Li ZY, Ma KP (2007) Predicting the spatial distribution of an invasive plant species (*Eupatorium adenophorum*) in China. Landscape Ecol 22(8): 1143–1154.

64. Lamb D, Erskine PD, Parrotta JA (2005) Restoration of degraded tropical forest landscapes. Science 310 (5754): 1628–1632.

65. Parzies HK, Spoor W, Ennos RA (2000) Genetic diversity of barley landrace accessions (*Hordeum vulgare* ssp. *vulgare*) conserved for different lengths of time in *ex situ* gene banks. Heredity 84: 476–486.

4

Biogeography of Human Infectious Diseases: A Global Historical Analysis

Elizabeth Cashdan*

Department of Anthropology, University of Utah, Salt Lake City, Utah, United States of America

Abstract

Objectives: Human pathogen richness and prevalence vary widely across the globe, yet we know little about whether global patterns found in other taxa also predict diversity in this important group of organisms. This study (a) assesses the relative importance of temperature, precipitation, habitat diversity, and population density on the global distributions of human pathogens and (b) evaluates the species-area predictions of island biogeography for human pathogen distributions on oceanic islands.

Methods: Historical data were used in order to minimize the influence of differential access to modern health care on pathogen prevalence. The database includes coded data (pathogen, environmental and cultural) for a worldwide sample of 186 non-industrial cultures, including 37 on islands. Prevalence levels for 10 pathogens were combined into a pathogen prevalence index, and OLS regression was used to model the environmental determinants of the prevalence index and number of pathogens.

Results: Pathogens (number and prevalence index) showed the expected latitudinal gradient, but predictors varied by latitude. Pathogens increased with temperature in high-latitude zones, while mean annual precipitation was a more important predictor in low-latitude zones. Other environmental factors associated with more pathogens included seasonal dry extremes, frost-free climates, and human population density outside the tropics. Islands showed the expected species-area relationship for all but the smallest islands, and the relationship was not mediated by habitat diversity. Although geographic distributions of free-living and parasitic taxa typically have different determinants, these data show that variables that influence the distribution of free-living organisms also shape the global distribution of human pathogens. Understanding the cause of these distributions is potentially important, since geographical variation in human pathogens has an important influence on global disparities in human welfare.

Editor: Linda Anne Selvey, Curtin University, Australia

Funding: The author has no funding or support to report.

Competing Interests: The author has declared that no competing interests exist.

* Email: cashdan@anthro.utah.edu

Introduction

Geographic variation in infectious disease has played a major role in determining history's political and demographic winners and losers [1,2], and remains a significant factor shaping differential welfare across the world today. We know a great deal about the ecological conditions that influence the distribution of particular pathogens in particular parts of the world, but there have been comparatively few analyses of global pathogen distributions and their determinants. On the other hand, theory in geographical ecology has addressed global patterning in species distributions across a wide range of taxa. The aim of this paper is to evaluate some of those arguments in the context of human pathogens, by assessing the relative influence of environmental variables that have been found to shape species diversity in other taxa. Among the factors considered are climate (temperature and precipitation), island size and isolation, and human factors that enhance disease transmission (population density, sedentism, and roads).

The dataset is unusual in being historical and in taking as units of observation the local pathogen and environmental conditions prevailing at 186 mostly small-scale non-industrial societies around the globe (the Standard Cross-Cultural Sample, or SCCS). The data are specific to these locations, which are for the most part not near major population centers and transportation hubs. While the use of historical pathogen data poses obvious limitations in accuracy and precision, it has the potential to give a clearer picture of the role of the physical environment, since influential moderators (global travel, modern medicine and public health) played a smaller role than they do today. The more that global disease patterns rest on differential access to vaccines and antibiotics, good sanitation, and clean water, the more difficult it becomes to isolate the effect of climate and other biogeographical variables in a global analysis. The dataset also has the advantage, when compared to national data such as GIDEON, of being spatially focused and on a consistent scale. Finally, a number of relevant cultural variables have been coded for the SCCS, including several that are likely to affect pathogen abundance and diversity. The present study, therefore, complements global biogeographical pathogen analyses that have used modern datasets [3,4]. The analysis considers effects of latitude, climate, island size and area, and population density and mobility.

Species richness is greater at lower latitudes across a wide range of taxa, and there are reasons why we might expect this to hold for human pathogens also: parasite richness is strongly correlated with host species richness in area-based studies [5], and we know that host species are typically more diverse near the equator. Furthermore, parasite-associated host mortality is greater at lower latitudes [6]. However, data on latitudinal gradients in parasites are conflicting. A recent meta-analysis across a wide range of host species found no overall relationship between latitude and parasite species richness per host species [7], and among carnivores the opposite pattern was found, with more parasite diversity on hosts living far from the equator [8]. It is likely that the influence of latitude in such studies is obscured by differences among host species that affect parasite richness (host body size, density, geographical range), a problem that would be avoided by studying pathogen diversity on a globally-distributed host such as Homo sapiens [7,9]. Global studies of species richness in human pathogens have found such latitudinal gradients [3,4].

The reason for latitudinal gradients in species richness remains a subject of debate [10,11]. Energy and water availability affect organism abundance because they are central to metabolism, but it is less clear why more energy or water would lead to greater number of species; it is likely that there are several mechanisms, and that they vary by taxa [12]. Empirical studies have shown that temperature (used as a proxy for energy availability) is often correlated with species richness, but other studies have shown similar patterning with precipitation and habitat diversity. This study assesses the relative importance of these variables as predictors of historical pathogen number and prevalence. In addition to developing a global model, the study tests the hypothesis that temperature is more important in areas where it is limiting (i.e., areas far from the equator), while water [13] and habitat diversity [14] are more important in areas of energy abundance.

Species richness on islands is also shaped by island size and isolation. The MacArthur & Wilson [15] equilibrium model of island biogeography explained this relationship as a consequence of immigration and extinction rates: smaller islands have fewer species due to higher extinction rates and fewer habitats, and more isolated islands have fewer species due to lower colonization rates. Larger islands also attract more immigrants (target effect) and less isolated islands receive repeated immigration and so are less vulnerable to extinction (rescue effect). While the assumption of equilibrium is problematic and new dynamic theories have been developed [16,17], the influence of island area and isolation remain important. A separate analysis of 37 islands in the sample was therefore conducted to see whether the size and isolation of islands shape pathogen number and prevalence, and, if so, whether greater habitat diversity on larger islands could explain the relationship.

Finally, the SCCS also allows us to include in the models aspects of human demography and culture likely to affect pathogen growth and transmission. Host population density is a strong predictor of parasite species richness across a wide range of host taxa [7], including non-human primates [18], and the same is likely to be the case for humans. Skeletal and other evidence suggests that the neolithic transition to settled farming and husbandry was often accompanied by an increase in infectious disease; proposed reasons include the larger pool of susceptible hosts and wider contacts arising from larger, denser, and more permanent settlements, as well as exposure to new zoonoses and vectors associated with food production [19–21]. Similar factors are likely to lead to variation in pathogen exposure among the nonindustrial societies of the SCCS. These factors are evaluated here by modeling the effects of population density, sedentism, and road quality.

The analyses begin by looking at the environmental variables that affect pathogen diversity and prevalence globally. The sample is then divided into tropical and non-tropical regions, and the relative importance of these environmental factors in the two regions is compared. Finally, a set of analyses was performed on the island locations only, in order to evaluate predictions about the effects of island area, isolation, and habitat diversity.

Methods

Data sources

The analyses use data from the Standard cross-cultural sample (SCCS) of 186 non-industrial cultures (see Figure 1). Each SCCS society is pinpointed to the time and location of a key ethnographic description [22], with most dating to the early part of the twentieth century (interquartile range 1880–1939). Many sociocultural and environmental variables have been coded for this sample, with the open access electronic journal World Cultures (http://www.worldcultures.org) functioning as a repository. This paper uses both existing coded data for the SCCS and a newly-developed set of SCCS pathogen codes.

The new pathogen codes are described briefly below and in Cashdan & Steele [23], and in more detail in the supplementary materials. Datafile S1 includes the code and information to guide its use and interpretation, while Datafile S2 contains the coded data.

Pathogen data. The pathogen data for the new codes were derived from historical sources, chiefly global maps published in the mid-twentieth century. The codes reflect the prevalence levels of 8 pathogens: malaria, dengue, filariae, typhus, trypanosomes, leishmanias, schistosomes, and plague. Most of these pathogens include several related species, due to limitations of the source material. Prevalence levels were taken primarily from isolines on the epidemiological maps, and coded as 1 = absent, 2 = rare, 3 = sporadic or moderate prevalence, and 4 = epidemic or high prevalence. The prevalence levels of the different pathogens were combined, as described below, to form a pathogen prevalence index.

The coding procedure followed that used by Murray & Schaller [24] in their historical cross-national pathogen codes, but was made specific to local conditions by recording, for each of the 8 pathogens, the highest pathogen level (1–4) within a 100 km radius of each SCCS society. The main sources were the three volume series of maps in Rodenwaldt & Bader [25] and the maps and data in Simmons et al. [26], supplemented by data in Faust & Russell [27]. Low [28,29] developed a 7-pathogen index for the SCCS using different historical sources. The two codes are highly correlated, but Low's includes two pathogens (leprosy and spirochetes) not in the Cashdan-Steele dataset. A combined index was therefore created by converting Low's three-point scale for leprosy and spirochetes and the Cashdan-Steele four-point scale for the other eight pathogens to z-scores, and using the mean of the 10 z-scores as an index of pathogen prevalence (see Datafile S1). A high score on the index, therefore, indicates both more types of pathogens and more severe exposure. In order to get a measure that more closely reflects species richness, a second index was created in which pathogens were dichotomized as either present or absent. The score here is the number of pathogens out of a total possible of 10. All analyses were done with both the pathogen prevalence index and with number of pathogens.

Because of limitations in the source material, both codes are biased toward pathogens that are transmitted through arthropod

Figure 1. Standard cross-cultural sample locations.

and other vectors. A few of these also have non-human hosts. The prevalence of such diseases is likely to be strongly shaped by the geographic distribution of the vectors that transmit them and the species that host them. This bias has the disadvantage that a number of important diseases (e.g., measles and cholera) are omitted. It also has the advantage that the geographic patterning of this sample of diseases will be less affected by international travel and by socioeconomic and public health measures than are diseases spread via droplet and oral-fecal transmission.

Another limitation is that the historical data do not contain information on variation in sampling effort, and less was known about pathogens in remote areas like tropical Africa than in more economically developed parts of the world. Since the sources nonetheless indicate more pathogens in these tropical regions, particularly in central Africa, the effect of this bias is likely to be conservative. Sampling bias is likely to be most problematic in studying the influence of island area, since pathogens on very small islands might have been estimated from better-known larger islands in the vicinity. The implications of this potential bias is discussed in the results section on island analyses.

Environmental and island data. Energy measures included in this study were mean annual temperature, number of frost-free months, and within-year measures of temperature extremes [30]. Water availability was measured by yearly mean precipitation over a 20-year period [31], and within-year measures of wet and dry extremes, including lowest precipitation in dryest month and highest in wettest month [30]. All data were taken from weather station records closest in time and place to the focus of each SCCS society. Habitat diversity was coded as the number of vegetation types within a radius of 100 through 250 miles [31,32], based on world maps published in the 1960s [33]. Many sociocultural factors affect pathogen spread, directly or indirectly, and three are used in these analyses: population density [34], road quality [35], and sedentism [34]. These are ordinal variables, as described below. The environmental and cultural data analyzed here come from the 2003 World Cultures 14(1) data disks, although the original published sources were consulted for full variable definitions and coding procedures.

Island area and various measures of isolation were obtained from the UNEP (United Nations Environment Programme) Island Directory at http://islands.unep.ch, supplemented in a few cases by other sources. A few islands were so small that the 100 km radius used to calculate pathogens extended beyond the island

border. In these cases, if there was another island within that radius, the area of that island was added to the focal island.

Analysis

There are two parts to the analysis. The first uses the full sample of 186 locations to build a global model of significant environmental predictors of pathogen prevalence (using the prevalence index) and richness (using number of pathogens). The global model was built incrementally, beginning with a model of physical environmental variables (island vs. mainland, temperature, and precipitation) followed by a separate model of three related cultural environmental variables (density, sedentism, road quality). The significant predictors from the two models were then combined into a single global model. In each case, analysis began with single-factor regressions followed by multivariate models, and variables that were individually significant but did not contribute independently to the multivariate models were dropped. The model was then applied separately to tropical and non-tropical regions, because the strength of these predictors was hypothesized to differ by latitude. Because the aim of the global model was to compare the relative effects of the different predictors, multiple regressions report standardized (beta) coefficients.

The second part of the analysis uses only the subset of 37 island locations in order to test the specific hypotheses that pathogen number and prevalence are associated with island size and habitat diversity.

In conducting the regressions, island area and some climate variables were transformed with a natural log transform prior to regression in order to make relationships linear and improve residual distributions. Where necessary, a constant was added before the log transform in order to make the minimum value 1.0. Mean annual temperature was negatively skewed, so those data were also reflected about zero before the log transform and then reflected back so as to restore the original order.

The ordinal variables were handled in different ways, depending on the nature of the variable. Road quality was dichotomized into societies where only footpaths were present, originally coded 1 ($n=124$), and societies with roads of varying quality, originally coded 2-4 ($n=57$). Sedentism was dichotomized into the 117 societies that maintain permanent camps (5-6 in the original scale) and the 69 that move during the year (1-4 in the original scale). Unlike road quality and sedentism, which were defined by qualitative descriptors, the 7 levels of population density correspond nonlinearly to persons per square mile: (1) less than 1 person per 5 sq. mi, (2) 1 person per sq. mi – 1 person per 5 sq. mi, (3) 1.1 – 5 persons per sq. mi, (4) 5.1–25, (5) 26–100, (6) 101–500, and (7) more than 500 persons per sq. mi. Population density was analyzed in multivariate regressions as an interval variable, although the underlying density in persons per square mile cannot be directly inferred from the data. Prevalence levels of individual pathogens were used only in bivariate correlations with individual environmental variables, using Spearman's rank order correlations.

Validity checks. Two additional analyses were done to validate the global model. The first was to run it separately against the two codes from which the combined index was derived. This was done both as a check on coding accuracy (since the codes used different historical sources) and as a way to evaluate how vulnerable the model was to the particular pathogens chosen (since the codes differed somewhat in the diseases coded). Another check was done to see whether the dependent and independent variables were associated only because they varied similarly across space, which would be indicated if there was spatial autocorrelation in the residuals. For each pair of points, the (squared) difference between the residuals and the actual geographic distance was calculated, to see whether the two values were correlated. This was done at various scales of distance down to 200 km. The societies in the sample are geographically dispersed (stratified both by geographic region and language group) so spatial autocorrelation at smaller scales cannot be assessed.

SAS was used for all analyses.

Results

The first part of the analysis builds a global model using the full dataset, first by considering the physical environment, then the cultural environment, and finally both together in a single model. The sample is then divided into high and low latitude zones, to see how the relative importance of these variables differ by latitude. The final analyses are restricted to the island locations, in order to test specific predictions from island biogeography.

Global Analyses

Latitudinal gradients. The upper graph in Figure 2 shows that the pathogen prevalence index is negatively correlated with distance from the equator, particularly when island locations are excluded, and that island locations have lower pathogen scores than those on the mainland. Because the pathogen prevalence index conflates number of species and abundance, the lower graph uses an index based solely on pathogen presence or absence; it shows a similar picture, with islands having fewer pathogens than expected given their latitude. This result is consistent with the broader literature on island biogeography, which finds species richness to be reduced on islands, and will be discussed further in a later section that considers island area. First we turn to the climatic factors that might be influencing the latitudinal gradient. In this dataset, mean annual temperature and precipitation are both correlated with distance from the equator (mean annual temperature: $r = -.80$, $p < .0001$, $n = 180$; mean annual precipitation: $r = -.50$, $p < .0001$, $n = 186$), so the first question is which variable is more important in shaping pathogen distributions, and to what extent associated variables (climate extremes and variation) also play a role.

The physical environment (temperature, frost, precipitation). Figure 3 shows pathogens as a function of log mean annual temperature, subset in two ways to illustrate additional effects on the relationship. The upper graph shows

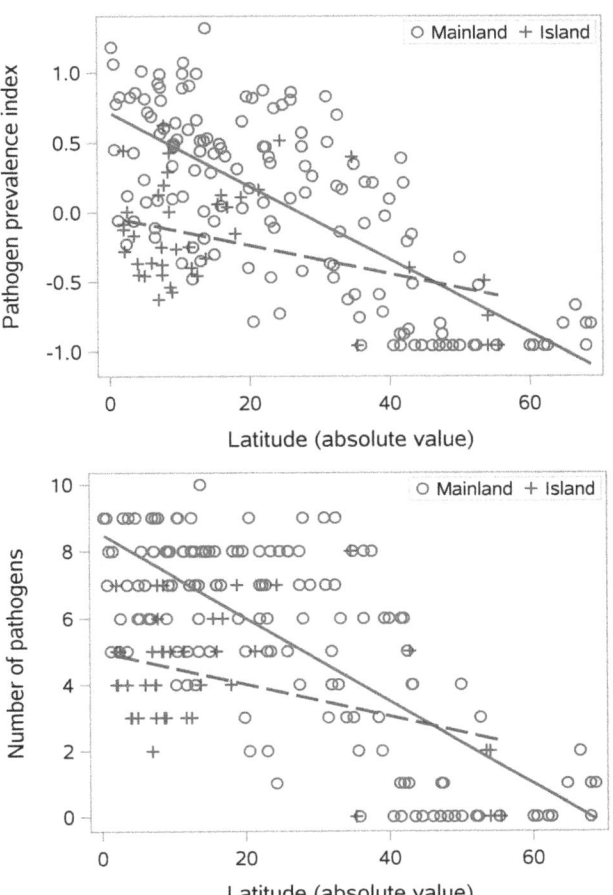

Figure 2. Pathogens by latitude. Separate regression lines for mainland and island locations.

that islands have lower pathogen scores than would be expected from their temperature, the same pattern seen with latitude. The lower graph, which excludes islands, shows that a year-round frost-free climate predisposes to more pathogens than would be expected from the climate's average temperature. Other measures of within-year temperature extremes were also analyzed, but were too highly correlated with mean annual temperature to be included in regressions. Temperature, frost, and islands have independent effects when included together in a multiple regression model: log mean annual temperature, frost months (dummy coded as some vs. none), and islands (dummy coded as island vs. mainland) together explain 39% of the variance in the pathogen prevalence index and 40% of the variance in number of pathogens, with the pathogen prevalence index being higher on the mainland ($\beta = .33$, $p < .0001$), in areas with high mean annual temperature ($\beta = .46$, $p < .0001$), and in frost-free climates ($\beta = -.26$, $p = .0002$).

Precipitation shows a more complicated relationship to pathogens, because two variables have independent effects: (a) mean annual precipitation and (b) the amount of precipitation in the driest part of the year. As will be shown below, these two precipitation variables affect different kinds of pathogens. Mean annual precipitation showed a modest ($R^2 = .13$) curvilinear relationship with pathogens best approximated with a third-order polynomial (see Figure 4; one influential precipitation value was removed from this graph and from the analyses). Extreme dryness

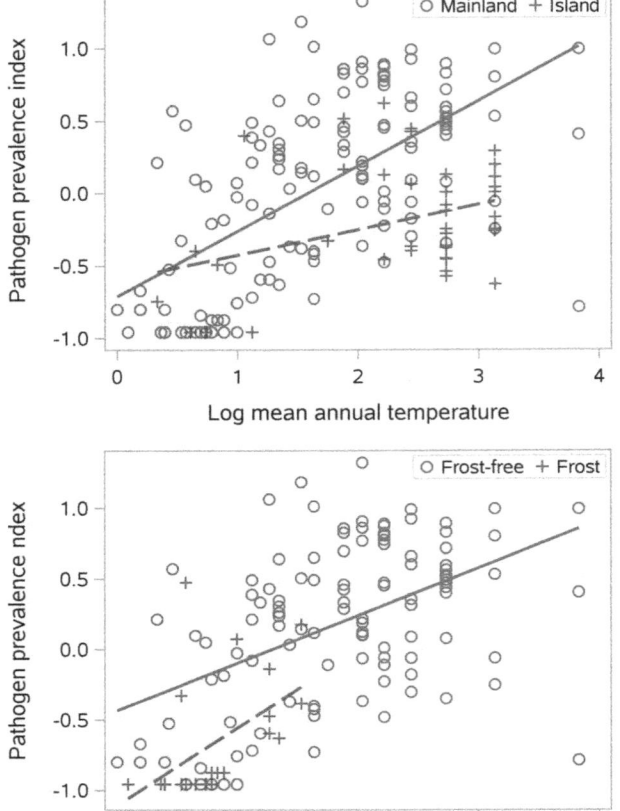

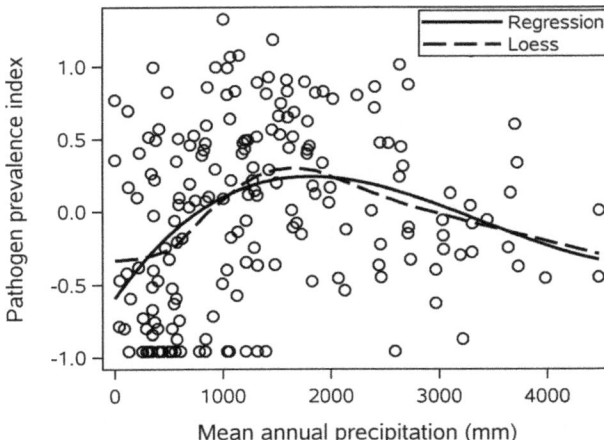

Figure 4. Pathogen prevalence index by mean annual precipitation. The regression line is a third-order polynomial.

Figure 3. Pathogens by natural log of mean annual temperature (c). Regression lines in the upper graph are subset by mainland vs. island locations. Regression lines in the lower graph are subset by locations with frost-free climates vs. those with one or more frost months.

climate variables are included in the model, while lowest precipitation in dryest month (one component of precipitation range) remains significant.

The effects of temperature, mean precipitation, and dryness differ for the different pathogens, and the patterning appears to reflect the ecology of the vector more than the type of pathogen. The mosquito-borne pathogens are a variable lot, including malaria (protozoans), dengue (virus), and filariae (nematodes), but all were worse in hot wet climates. Typhus (rickettsia) leishmanias (protozoans), and schistosomes (flukes) were all worse in areas with dry months, perhaps because of greater aggregation of vectors and hosts during drought. Bivariate correlations between the various pathogen groups and environmental predictors are summarized in Table 1.

The cultural environment (population density, mobility, roads). Pathogen distributions are affected by cultural as well as physical environmental factors. This section examines the effect of three cultural variables (population density, residential mobility, and roads) on pathogen distributions.

during part of the year (measured as log lowest precipitation during driest month) also increases the pathogen prevalence index (see Figure 5). Adding mean annual precipitation and seasonal dry extremes to the previous model increases the variance explained to 50% for both the pathogen prevalence index and number of pathogens.

This final climate model, showing the effects of the physical environmental variables (island vs mainland, temperature, frost, mean rainfall and seasonal dry extremes) has an adjusted $R^2 = .48$, $F(7, 171) = 24.52$, $p < .0001$ for the pathogen prevalence index and adjusted $R^2 = .48$, $F(7, 171) = 24.73$ for number of pathogens.

It has been suggested [36] that greater climate variation leads to lower diversity because organisms in such climates have evolved to be generalists, broadly tolerant of a wide range of climates. Guernier et al. [4] found the opposite to be the case for six groups of human pathogens: greater seasonal range in precipitation was associated with greater species diversity. In the present analysis, also, greater precipitation range (measured as maximum precipitation in wettest month minus lowest precipitation in dryest month) was associated with a higher pathogen prevalence index. However, dry extremes seem to be driving this relationship; precipitation range was not a significant predictor when the other

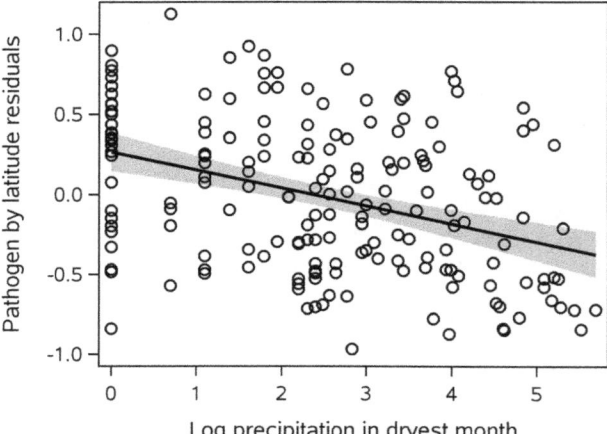

Figure 5. Pathogens by seasonal dry extremes, controlling for latitude. Dry extremes are measured by the lowest precipitation in the dryest month, logged. Regression with 95% confidence limits.

Table 1. Pathogen-specific correlations.

	Mean annual Temperature	Yrly Precip (linear)	Yrly Precip (polynomial)	Precipitation Dryest month	Number of Frost months	Population Density
Malaria	.53***	.42***	.41	−.07	−.41***	.36***
Dengue	.56***	.48***	.43	.06	−.38***	.48***
Filaria	.53***	.42***	.41	−.02	−.39***	.46***
Typhus	−.07	−.01	.22	−.39***	−.16*	.35***
Trypanosomes	.06	.18*	.20	−.03	−.12	.05
Leishmanias	.21**	.08	.14	−.22**	−.33***	.11
Schistosomes	.26***	.17*	.19	−.24***	−.24**	.18*
Plague	.07	.27**	.17	.01	−.12	.19*
Spirochetes	.35***	.26**	.20	−.13	−.35***	.38***
Leprosy	.38***	.35***	.23	−.19*	−.31 ***	.32***

Pathogen prevalence level by predictors in the final model of Table 2. Spearman's rank-order correlations were used for all variables other than annual precipitation. Two correlations are given for mean annual precipitation ("Yrly Precip"): the polynomial model uses the full sample, but without significance values, which are probably unreliable for the individual pathogen scores. The linear model is limited to the 78% of the sample where pathogens increase with precipitation (i.e., up to 2000 mm). Other variables were transformed as indicated in the text, except that frost was not dichotomized. Sample sizes are 180 for log temperature, log precipitation in dryest month, and number of frost months; sample sizes for the polynomial and linear correlations with mean annual precipitation are 185 and 143 respectively.
$*<.05, **<.01, ***<.001$

Better roads can be expected to broaden the geographic reach of pathogens by facilitating the movement of people, and of insect vectors transported inadvertently in the goods they carry. The mean pathogen prevalence index was higher in societies with roads (.29) as opposed to footpaths (−.14), $t(139) = −5.17, p < .0001$. Pathogens were also higher in more sedentary groups: societies with permanent camps had a mean pathogen prevalence index of .22 as compared with −.37 for more mobile groups ($t(184) = −7.7, p < .0001$). Population density was also, as expected, positively correlated with the pathogen prevalence index: $r = .46$ ($r_s = .47$), $p < .0001, n = 184$.

These variables are correlated, since all are associated with greater social complexity. Figure 6 shows that increases in population density are accompanied by a trend toward increased sedentism, although mobile populations have lower pathogens at the same degree of density. A check for collinearity supports keeping both sedentism and density in the model (variance inflation factor = 1.84), although with these variables in the model, road quality is no longer a significant predictor. The cultural environmental model, with just density and sedentism, explains 26% of the variance in the pathogen prevalence index and number of pathogens. For the pathogen prevalence index, the standardized coefficients were $\beta = .28, p = .001$ for density, $\beta = .27, p = .002$ for sedentism.

Examination of outliers in the climate analysis underscores the importance of considering the cultural as well as physical environment. For example, there was a highly influential point in the temperature and frost model. This point represents the Teda, a nomadic group in Chad with an unusually low pathogen score, given their local temperature and rainfall. None of the physical environmental factors in the dataset explain the discrepancy adequately, but their comparatively low pathogens are consistent with their very low density and high mobility at the time and place of their SCCS ethnographic description.

A combined global model (physical and cultural environmental variables). The effects of the physical and cultural environment on disease are not independent, and so the final global analysis considers the variables in a single model. In a combined model with the physical environmental variables,

residential mobility is no longer a significant predictor and is dropped from the final model. However, the earlier result suggests that the effect of density on pathogens in this model may be due both to its direct effects and to indirect effects resulting from associated decreased mobility.

Taken together, the results indicate that there are more pathogens and pathogen types on the mainland than on islands, and that pathogens increase with mean temperature, population density, and a frost-free climate. The relationship with precipitation is more complex, peaking at intermediate levels of mean

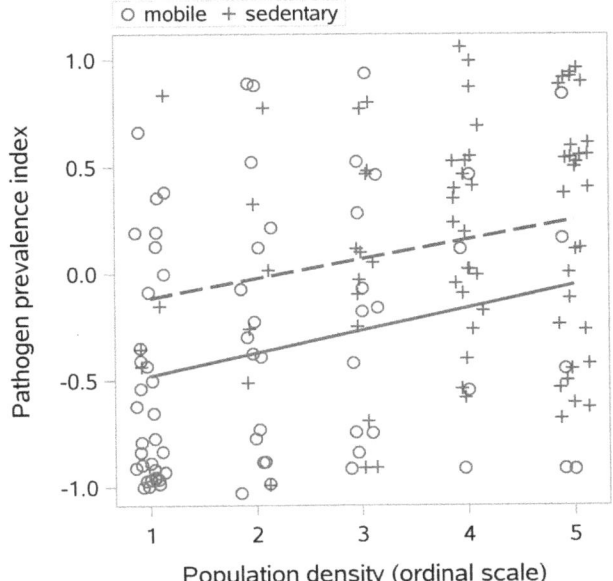

Figure 6. Pathogen prevalence index by density and residential mobility. Points have been jittered to avoid overlap. Population density is a nonlinear ordinal scale based on persons per sq. mi; see the methods section for density and mobility codes.

annual precipitation but also increasing in seasonally dry climates. The combined model explains 58% of the variance in both number of pathogens and the pathogen prevalence index. The regression statistics of this model are in Table 2.

Validity checks. As a check, the final model was run against each of the two databases from which the combined pathogen score was derived. Low [28,29] coded data on 7 pathogens using different historical sources, only two of which were used in the combined index. Using Low's 7-pathogen index as the dependent variable with this model produces an identical $R^2 = .58$, although the coefficient for density was smaller and that for temperature was larger. The greater influence of temperature using Low's data is probably because typhus and plague, which are unrelated to temperature in Table 1, were not included in that dataset. The other coefficients were similar to those of the combined index. A similar summation of the 8 pathogens in our new codes yields an $R^2 = .60$ with coefficients very similar to those of the combined index.

Another check was done to see whether the relationship between the independent and dependent variables in this model was due to spatial autocorrelation (e.g., whether independent and dependent variables were associated only because they vary similarly across space). Where this is the case, there will be spatial autocorrelation in the residuals. In order to evaluate this, the squared difference between the residuals of each pair of points was plotted against their great circle geographic distance, using both the full sample and subsets at increasingly smaller scales (points less than 3000, 1000, 500, and 200 km apart). Visual inspection indicated that the relationship was flat (r averaged $-.02$) at all scales of distance, none were statistically significant, and there were no trends with distance over this range.

Differences between high and low latitude regions

The model above is the best fit for global pathogen distributions, but recent literature suggests that more specific models may be appropriate at high and low latitudes. Energy availability appears to have a greater effect on species richness farther from the equator, whereas water [13] has been proposed as more important where energy is abundant. Habitat diversity [14] may also be more important at low latitudes. The sample was divided into tropical (low latitude) and non-tropical (high latitude) zones, and the results supported these expectations. Bivariate correlations by latitude zone are shown in Table 3.

High latitudes. As expected, temperature was a significant predictor only at high latitudes. In this region, the relationship was strongly linear ($r = .63$ for number of pathogens, $r = .66$ for the pathogen prevalence index). An unanticipated result was that the same is true for population density: it is a strong predictor of pathogens at high latitudes only. A multivariate model using only those two variables (log mean annual temperature and population density) explains 56% of the variance in the pathogen prevalence index and 61% in pathogen number at high latitudes. No other variables add significantly when those are in the model.

Low latitudes. The pattern in the tropical locations, in contrast, is shaped more by precipitation than by temperature. The relationship between precipitation and pathogens in the tropics is similar in shape to that shown in Figure 4 for the full sample, but the relationship is much tighter, the peak is at somewhat lower precipitation, and the pathogen decline at higher precipitation is more apparent. The best multivariate model of the pathogen prevalence index in the tropics includes mean annual precipitation as a third-degree polynomial together with population density (notwithstanding its weak bivariate relationship) and the dummy-coded island vs. continent (most of the islands in the sample are in the tropics). This model explains 44% of the variance in the pathogen prevalence index and 50% in pathogen number.

Table 3 shows that habitat diversity (measured as number of vegetation zones in a given radius) is also a factor in shaping pathogen diversity in the tropics. Habitat diversity remains significant when added to the other variables (precipitation, islands, density) in the tropical model. However, the overall R^2 is reduced, perhaps because of reduced sample size when that variable is included. Habitat diversity, alone among the variables

Table 2. Final regression model for predictors of the pathogen prevalence index and number of pathogens.

Variable	Prevalence Index		Number of Pathogens	
	β	p	β	p
Temperature	0.30	<.001	0.31	<.001
Frost months	−0.19	.001	−0.20	<.001
Island	0.31	<.001	0.31	<.001
Driest month	−0.18	.009	−0.19	.006
Pop Density	0.31	<.001	0.30	<.001
Precipitation (P)	1.78	<.001	1.62	<.001
P^2	−2.73	.001	−2.45	
P^3	1.15	.024	1.03	.006
	$F(8,168) = 29.57$		$F(8,168) = 29.23$	
	$R^2 = .58$		$R^2 = .58$	

Variable definitions:
Temperature: Log mean annual temperature (c).
Frost months: 1 = presence 0 = absence.
Island: 1 = mainland 0 = island.
Driest month: Log lowest precipitation in driest month (mm).
Pop Density: Population density (ordinal scale, 1–7).
Precipitation: Mean annual precipitation (mm).

Table 3. Bivariate correlations between independent variables and pathogen number and prevalence index, by latitude zone.

	Low latitudes		High latitudes	
	Number	**Prevalence**	**Number**	**Prevalence**
Mean temperature	−.20*	−.18	**.59*****	**.61*****
Mean precipitation	**.44*****	**.40****	.30	.35
Low precip dryest month	**−.33*****	**−.31 ****	**−.33****	**−.31****
Population density	.14	.17	**.67*****	**.69*****
Habitat diversity	**.36*****	**.24*****	.05	−.01

Note. Table shows Spearman's rank order correlation coefficients except for mean annual precipitation, where the correlation is based on the adjusted R^2 of a third-degree polynomial regression (no significance values are given for high latitudes because of poor fit diagnostics). Sample sizes for low/high latitudes: temperature 107/73, precipitation 111/75, precipitation dryest month 107/73, habitat diversity 97/75, density 109/75. Habitat diversity calculated at 150 miles radius; the correlation was slightly less at 100 miles.
$* < .05, ** < .01, *** < .001$

considered in this study, is a stronger predictor of number of pathogens than it is of the prevalence index. Habitat diversity presumably facilitates pathogen species richness via niche differentiation, whereas temperature, precipitation and population density also have direct effects on pathogen prevalence by enhancing pathogen growth and transmission.

The island model

We now turn to the subset of the sample consisting of island locations, in order to test the prediction from classical island biogeography that small islands will have fewer species than large islands. The analysis is based on 37 islands (New Guinea was excluded because it is home to four societies in the sample). The prediction is supported: controlling for distance from the equator, the partial correlation of log island area with pathogens is $r = .63, p < .0001$ for the pathogen prevalence index and $r = .61, p < .0001$ for number of pathogens. Figure 7 shows the relationship for islands in the tropics; the non-tropical islands are included in the statistics but are not shown in the figure because they span a wide latitudinal range.

As Figure 7 indicates, the linear relationship breaks down for the smallest islands. This is often the case in small islands, where the effect of area on species richness is overshadowed by stochastic factors [37,38]. In such cases, species richness typically plateaus at the lowest level, which may not be the case in these data. The leveling off with small islands could be an artifact of the poor resolution of historical pathogen data, reflecting extrapolation from better-known larger islands to poorly-sampled small islands nearby. A regression without the four smallest islands probably presents a more accurate picture of the relationship between pathogens and island area; in this model the largest island (Borneo) is also best removed as it is a highly influential point. Within this intermediate range of values, the relationship is linear with the log of island area, and the regression of log area and latitude on pathogen number provides a better fit: $F(2, 29) = 23.51, p < .0001, R^2 = .62$. The unstandardized coefficient for log island area on number of kinds of pathogens is 0.64.

Theory predicts that pathogens will also decrease with distance from the mainland. The relationship is weak in this dataset, and its independent effect is hard to evaluate since the smallest islands are also farthest from the mainland. Getting a good measure of isolation is difficult, since it involves not just distance to the nearest continent but to nearby islands that could be links to sources of greater diversity. Various distance measures were used to try to capture this, but none showed more than a weak correlation with

pathogens, or remained significant when island area was also included in the model. However, this could reflect measurement difficulties rather than relative importance. It is also possible that some of the area effect reflects the greater isolation of many of the smaller islands.

Discussion and Conclusions

Summary

Many of the variables that influence the distribution of free-living taxa also predict the number and prevalence index of

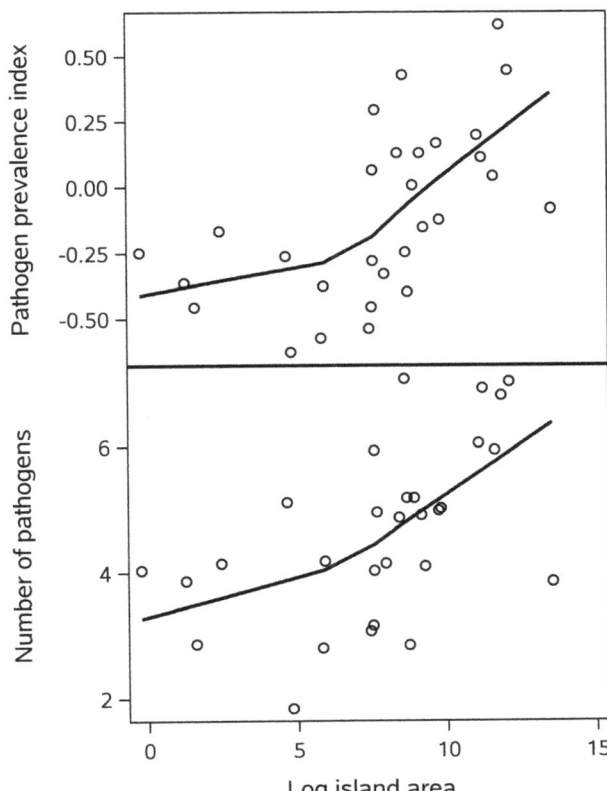

Figure 7. Pathogens by Island Area (tropical islands only), with loess curves.

human pathogens in this dataset. Pathogens increased with mean annual temperature and, controlling for mean temperature, in climates that remained free of frost throughout the year. The effect of temperature was highly significant, but only outside the tropics. Within the tropics, mean annual precipitation was a more important predictor, and was associated most strongly with mosquito-borne diseases (malaria, dengue, filariasis). Extreme seasonal dryness was also associated with more pathogens, especially typhus, leishmaniasis, and schistosomiasis. Finally, pathogens were worse in areas with high population density. Pathogens were also more numerous on the mainland than on islands, and on large as opposed to small islands. Most of the predictions derived from species diversity patterns in other taxa were supported, and are discussed in turn below, beginning with the island results.

Island biogeography

The classic model of island biogeography [15] predicts that there will be fewer species on islands that are small and isolated. The model uses simplifying assumptions [16], and the assumption of equilibrium is particularly problematic when studying the distribution of human infectious diseases. Nonetheless, and notwithstanding the heterogeneity of the 37 islands in this dataset, the classic predictions of island biogeography were upheld: islands have fewer kinds of pathogens than expected given their climate, and fewer in smaller than in larger islands. Island isolation did not have an independent effect, but may have had an indirect influence since many of the smaller islands in this dataset were also more isolated. Smaller islands also offer fewer types of habitat, but this appears not to be driving variation in pathogen richness in this dataset: although habitat diversity and area of islands were correlated, only island area had any relationship to pathogens.

The lower pathogen load on islands is consistent with Curtin's [39] meticulous accounting of historical troop mortality, which found that early in the 19th century some Pacific islands had a far lower pathogen load than would be expected by their climate, particularly Tahiti "which gave French troops a 100% mortality improvement over France" in the 1840s and continued to give benefits into the early 20th century ([39], p. 12). A similar pattern existed for Hawaii and New Zealand. (This ecological protection later made the islanders vulnerable to novel European diseases, which caused huge mortality throughout the Pacific).

Most of the pathogens in this dataset are vector-borne, and this is likely to enhance the influence of island area and isolation since two species, pathogen and vector, need to be present at the same time [40]. Some pathogens also require a minimum size of host population in order to remain endemic. This triple challenge can be expected to amplify the effect of isolation and extinction on pathogenic species on small islands, and among small isolated populations generally [41].

Latitudinal gradients

Both the pathogen prevalence index and number of pathogens show a strong latitudinal gradient in these data, with more pathogens closer to the equator. This could reflect vector ecology, since malaria (protozoans), dengue (virus) and filariae (nematodes) are taxonomically diverse, yet all are transmitted by mosquitos and all are most prevalent in hot wet climates where mosquitos are abundant. Typhus, leishmanias, and schistosomes, on the other hand, are transmitted by other vectors and intermediates (lice and fleas, sandflies, and freshwater snails, respectively) and had different environmental correlates. The latitudinal gradient in human pathogens may also reflect the ecology of alternate hosts. Rodents, an important reservoir for many human diseases, host

more viral parasites at lower latitudes, perhaps because of latitudinal gradients in their own viral vectors [42]. This complexity may contribute to the inconsistent findings regarding latitudinal gradients in parasite species richness across host taxa [7,9].

Environmental predictors

Although a latitudinal gradient in species richness exists across a wide range of free-living taxa, the cause of this pattern remains a topic of debate [10,11] and its relevance to parasitic taxa remains unclear. A number of mediators have been discussed in the literature; those discussed below include temperature (often used as a proxy for energy availability), frost, and precipitation.

Pathogen prevalence levels increased with temperature in 7 of the 10 pathogen groups, including the mosquito-borne pathogens, which are known to be highly temperature sensitive [43]. Although growth and survival of insect vectors, hence pathogen transmission, decline in extreme heat [44], the relationship with log temperature was linear in these data. There were also more kinds of pathogens as temperature increased. Locations with a year-round frost-free climate had more pathogens than would be expected from their mean annual temperature, probably by enabling pathogens and their vectors to overwinter.

Precipitation had a more complex relationship with the pathogen prevalence index, because mean yearly precipitation and extreme seasonal dryness had different effects and affected different pathogens. Mean annual precipitation had a curvilinear relationship with both the prevalence index and pathogen number, peaking at intermediate values and declining in very high precipitation areas. The association between pathogens and mean annual precipitation was especially strong in the tropics and for mosquito-borne pathogens; perhaps the decline in very wet areas is associated with mosquito larvae being washed out due to heavy rains and flooding.

Areas with little or no precipitation during the driest part of the year (lowest precipitation in driest month) also had more pathogens. Seasonal dryness affected a different group of pathogens than did mean temperature and precipitation. The greatest effect was on typhus. Typhus is transmitted by fleas and lice and can become worse in crowded conditions with poor sanitation, and when drought causes rodents (and the fleas they carry) to move near human habitation in search of water. An analysis of tree-rings in pre-industrial Central Mexico found that a significant drought occurred during the first year of all 22 large typhus outbreaks studied [45]. A similar effect via aggregation of people and the sand fly vector during dry periods has been associated with temporal changes in leishmaniasis in Brazil [46].

Guernier et al. [4] found precipitation range to be the single best predictor of species richness across six categories of human pathogens. In the present dataset, precipation range (highest precipitation in wettest month minus lowest precipitation in driest month) was also associated with significantly higher pathogen number, and with the prevalence index. However, range was not a significant predictor when other climate variables were included in the model, while seasonal dryness (one component of range) remained significant. In these data, therefore, the more influential aspect of precipitation range on pathogen distributions appears to be seasonal dryness.

Differences at low and high latitude

The relative importance of temperature and water as predictors of species richness has been shown to vary with latitude, with temperature being more important at high latitudes and availability of water more important at low latitudes, where energy is already abundant [13,47]. These patterns were also found in the pathogen data. In high latitude areas, mean annual temperature was the strongest climatic predictor of pathogen number and the prevalence index, whereas in tropical areas mean annual precipitation was the key climate variable. Surprisingly, population density was also a much stronger predictor in high latitude areas; one plausible reason for this finding is that there are more alternate animal hosts in the tropics, so that zoonotic pathogens may remain endemic there even when people are at low density.

Historical data: Advantages and limitations

Cross-national differences in pathogen prevalence today have been shown to reflect differential access to disease prevention measures more than environmental variables, although pathogen richness still shows the latitudinal gradient found with other taxa [3]. Historical data on remote populations, as used here, reduces the influence of public health and modern medicine, allowing for a clearer picture of the way environmental variables shape geographic patterning in pathogen prevalence. A second advantage of this dataset for environmental analysis is that the data describe local conditions at a consistent scale, rather than being based on national averages. In the present study, the pathogen prevalence index and number of kinds of pathogens show similar patterns, and are strongly environmentally determined.

Use of historical data has limitations as well as advantages, due primarily to lack of precision in the historical source material: pathogen distributions were for the most part not available at the species level, and prevalence was assessed by an index based on ordinal scales for each pathogen, rather than by direct counts of infected individuals. Limitations of the source material also bias the pathogen sample toward vector-borne pathogens, which are

less global than other pathogens [48] and are likely to show a stronger environmental signature. For this reason, prevalence and richness are more likely to be correlated with each other in this group of pathogens, even in modern datasets [49], and latitudinal gradients and climatic correlates are likely to reflect vector as well as pathogen ecology [50].

The results of this study show strong support for several theoretical and empirical findings in geographical ecology, and show that they explain human pathogen distributions on a global level. The results offer insights into past and present patterns of infectious disease, and provide information relevant to the likely effects of global warming on pathogens sensitive to temperature, frost, and seasonal dry extremes.

Acknowledgments

I am grateful to Adrian Bell, Doug Jones, Damian Murray, Bobbi Low, Alan Rogers, and Randy Thornhill for comments, suggestions, and support. Alan Rogers also helped with the spatial autocorrelation analysis.

Author Contributions

Conceived and designed the experiments: EC. Analyzed the data: EC. Wrote the paper: EC.

References

1. McNeill WH (1976) Plagues and Peoples. Garden City, NY: Anchor Press.
2. Crosby AW (1986) Ecological Imperialism: The Biological Expansion of Europe, 900–1900. Cambridge University Press.
3. Dunn RR, Davies TJ, Harris NC, Gavin MC (2010) Global drivers of human pathogen richness and prevalence. Proceedings of the Royal Society B: Biological Sciences 277: 2587–2595.
4. Guernier V, Hochberg ME, Guégan JF (2004) Ecology drives the worldwide distribution of human diseases. PLoS Biology 2: e141.
5. Kamiya T, ODwyer K, Nakagawa S, Poulin R (2014) Host diversity drives parasite diversity: meta-analytical insights into patterns and causal mechanisms. Ecography 37: 689–697.
6. Robar N, Burness G, Murray DL (2010) Tropics, trophics and taxonomy: the determinants of parasite-associated host mortality. Oikos 119: 1273–1280.
7. Kamiya T, O'Dwyer K, Nakagawa S, Poulin R (2014) What determines species richness of parasitic organisms? A meta-analysis across animal, plant and fungal hosts. Biological Reviews 89: 123–134.
8. Lindenfors P, Nunn CL, Jones KE, Cunningham AA, Sechrest W, et al. (2007) Parasite species richness in carnivores: Effects of host body mass, latitude, geographical range and population density. Global Ecology and Biogeography 16: 496–509.
9. Poulin R (2014) Parasite biodiversity revisited: frontiers and constraints. International journal for parasitology 44: 581–589.
10. Rosenzweig ML (1995) Species Diversity in Space and Time. Cambridge University Press.
11. Willig MR, Kaufman DM, Stevens RD (2003) Latitudinal gradients of biodiversity: Pattern, process, scale, and synthesis. Annual Review of Ecology, Evolution, and Systematics 34: 273–309.
12. Clarke A, Gaston KJ (2006) Climate, energy and diversity. Proceedings of the Royal Society B: Biological Sciences 273: 2257–2266.
13. Hawkins BA, Field R, Cornell HV, Currie DJ, Guégan JF, et al. (2003) Energy, water, and broad-scale geographic patterns of species richness. Ecology 84: 3105–3117.
14. Kerr JT, Packer L (1997) Habitat heterogeneity as a determinant of mammal species richness in high-energy regions. Nature 385: 252–254.
15. MacArthur RH, Wilson EO (1967) The Theory of Island Biogeography. Princeton University Press.
16. Lomolino MV (2000) A call for a new paradigm of island biogeography. Global Ecology and Biogeography 9: 1–6.

17. Whittaker RJ, Triantis KA, Ladle RJ (2008) A general dynamic theory of oceanic island biogeography. Journal of Biogeography 35: 977–994.
18. Nunn CL, Altizer S, Jones KE, Sechrest W (2003) Comparative tests of parasite species richness in primates. The American Naturalist 162: 597–614.
19. Barrett R, Kuzawa CW, McDade T, Armelagos GJ (1998) Emerging and re-emerging infectious diseases: The third epidemiologic transition. Annual Review of Anthropology: 247–271.
20. Eshed V, Gopher A, Pinhasi R, Hershkovitz I (2010) Paleopathology and the origin of agriculture in the Levant. American Journal of Physical Anthropology 143: 121–133.
21. Cohen MN, Armelagos GJ (1984) Paleopathology at the Origins of Agriculture. Academic Press.
22. Murdock G, White D (1969) Standard cross-cultural sample. Ethnology 8: 329–369.
23. Cashdan E, Steele M (2013) Pathogen prevalence, group bias, and collectivism in the standard cross-cultural sample. Human Nature 24: 59–75.
24. Murray D, Schaller M (2010) Historical prevalence of infectious diseases within 230 geopolitical regions: A tool for investigating origins of culture. Journal of Cross-Cultural Psychology 41: 99–108.
25. Rodenwaldt E, Bader RE (1961) World-Atlas of Epidemic Diseases (1952–1961). Hamburg, Germany: Falk-Verlag.
26. Simmons JS, Whayne TF, Anderson GW, Horack HM (1944) Global Epidemiology: A Geography of Disease and Sanitation. Philadelphia: J. B. Lippincott.
27. Faust E, Russell P (1964) Craig and Faust's Clinical Parasitology. Philadelphia: Lea and Febiger.
28. Low B (1990) Marriage systems and pathogen stress in human societies. American Zoologist 30: 325–340.
29. Low BS (1994) Pathogen intensity cross-culturally. World Cultures 8: 24–34.
30. Whiting JWM (1989) Climate data from weather stations. World Cultures 1: 179–199.
31. Cashdan E (2003) Ethnic diversity, habitat diversity and rainfall codes. World Cultures 13.
32. Cashdan E (2001) Ethnic diversity and its environmental determinants: Effects of climate, pathogens, and habitat diversity. American Anthropologist 103: 968–991.
33. Eyre SR (1968) Vegetation and Soils: A World Picture. E. Arnold Ltd., second edition.

34. Murdock GP, Wilson SF (1972) Settlement patterns and community organization: Cross cultural codes 3. Ethnology 11: 254–295.

35. Murdock GP, Morrow DO (1970) Subsistence economy and supportive practices: Cross-cultural codes 1. Ethnology 9: 302–330.

36. Stevens GC (1989) The latitudinal gradient in geographical range: How so many species coexist in the tropics. American Naturalist 133: 240–256.

37. Lomolino MV (2000) Ecology's most general, yet protean pattern: The species-area relationship. Journal of Biogeography 27: 17–26.

38. Lomolino M, Weiser M (2001) Towards a more general species-area relationship: Diversity on all islands, great and small. Journal of Biogeography 28: 431–445.

39. Curtin PD (1989) Death by Migration: Europe's Encounter with the Tropical World in the Nineteenth Century. Cambridge University Press.

40. Spurgin LG, Illera JC, Padilla DP, Richardson DS (2012) Biogeographical patterns and co-occurrence of pathogenic infection across island populations of Berthelots pipit (*Anthus berthelotii*). Oecologia 168: 691–701.

41. Black FL (1975) Infectious diseases in primitive societies. Science 187: 515–518.

42. Bordes F, Guégan JF, Morand S (2011) Microparasite species richness in rodents is higher at lower latitudes and is associated with reduced litter size. Oikos 120: 1889–1896.

43. Patz JA, Epstein PR, Burke TA, Balbus JM (1996) Global climate change and emerging infectious diseases. JAMA 275: 217–223.

44. Mordecai EA, Paaijmans KP, Johnson LR, Balzer C, Ben-Horin T, et al. (2013) Optimal temperature for malaria transmission is dramatically lower than previously predicted. Ecology Letters 16: 22–30.

45. Acuna-Soto R, Stahle D, Villanueva Diaz J, Therrell M (2007) Association of drought with typhus epidemics in Central Mexico. In: AGU Spring Meeting Abstracts. volume 1.

46. Thompson RA, de Oliveira Lima JW, Maguire JH, Braud DH, Scholl DT (2002) Climatic and demographic determinants of American visceral leishmaniasis in northeastern Brazil using remote sensing technology for environmental categorization of rain and region influences on leishmaniasis. The American Journal of Tropical Medicine and Hygiene 67: 648–655.

47. Whittaker RJ, Nogués-Bravo D, Araújo MB (2007) Geographical gradients of species richness: A test of the water-energy conjecture of Hawkins *et al.* (2003) using European data for five taxa. Global Ecology and Biogeography 16: 76–89.

48. Smith KF, Sax DF, Gaines SD, Guernier V, Guégan JF (2007) Globalization of human infectious disease. Ecology 88: 1903–1910.

49. Fincher C, Thornhill R (2008) Assortative sociality, limited dispersal, infectious disease and the genesis of the global pattern of religion diversity. Proceedings of the Royal Society B: Biological Sciences 275: 2587–2594.

50. Nunn CL, Altizer SM, Sechrest W, Cunningham AA (2005) Latitudinal gradients of parasite species richness in primates. Diversity and Distributions 11: 249–256.

5

Weak Population Structure in European Roe Deer (*Capreolus capreolus*) and Evidence of Introgressive Hybridization with Siberian Roe Deer (*C. pygargus*) in Northeastern Poland

Juanita Olano-Marin[1]*[¶]**, Kamila Plis**[1][¶]**, Leif Sönnichsen**[1,2]**, Tomasz Borowik**[1]**,
Magdalena Niedziałkowska[1]**, Bogumiła Jędrzejewska**[1]

1 Mammal Research Institute, Polish Academy of Sciences, Białowieża, Poland, **2** Leibniz Institute for Zoo and Wildlife Research, Berlin, Germany

Abstract

We investigated contemporary and historical influences on the pattern of genetic diversity of European roe deer (*Capreolus capreolus*). The study was conducted in northeastern Poland, a zone where vast areas of primeval forests are conserved and where the European roe deer was never driven to extinction. A total of 319 unique samples collected in three sampling areas were genotyped at 16 microsatellites and one fragment (610 bp) of mitochondrial DNA (mtDNA) control region. Genetic diversity was high, and a low degree of genetic differentiation among sampling areas was observed with both microsatellites and mtDNA. No evidence of genetic differentiation between roe deer inhabiting open fields and forested areas was found, indicating that the ability of the species to exploit these contrasting environments might be the result of its phenotypic plasticity. Half of the studied individuals carried an mtDNA haplotype that did not belong to *C. capreolus*, but to a related species that does not occur naturally in the area, the Siberian roe deer (*C. pygargus*). No differentiation between individuals with Siberian and European mtDNA haplotypes was detected at microsatellite loci. Introgression of mtDNA of Siberian roe deer into the genome of European roe deer has recently been detected in eastern Europe. Such introgression might be caused by human-mediated translocations of Siberian roe deer within the range of European roe deer or by natural hybridization between these species in the past.

Editor: Roscoe Stanyon, University of Florence, Italy

Funding: The project was financed by the Polish Ministry of Sciences and Higher Education (grant no. N N304 172536), and the budget of the Mammal Research Institute PAS in Białowieża and of the Leibniz Institute for Zoo and Wildlife Research in Berlin. JOM and LS were supported by the project Research Potential in Conservation and Sustainable Management of Biodiversity–BIOCONSUS in the 7th Framework Programme (contract no. 245737). The funders had no role in study design, data collection and analysis, decision to publish, or preparation of the manuscript.

Competing Interests: The authors have declared that no competing interests exist.

* Email: jolano@ibs.bialowieza.pl

¶ These authors are co-first authors on this work.

Introduction

Historical and recent events, shaped by both natural and anthropogenic factors, play an important role in the current patterns of genetic variation within the species. Climatic changes during the Quaternary, for example, defined the major genetic subdivisions of different taxa around the globe [1]. In more recent times, human practices such as agriculture, deforestation, development of infrastructure, hunting, and introduction of alien species, among others, have greatly affected the dynamics of natural populations, with consequences in the levels and distribution of genetic diversity within the species [2–4]. The microevolutionary consequences of human practices might have profound effects not only on threatened species living in small and isolated populations, but also on common and widespread species subject to strong management practices (e.g. ungulates) [5,6].

The European roe deer (*Capreolus capreolus*), one of the most common ungulates in Europe and an important game species, is distributed across the European continent from the Mediterranean to Scandinavia. Major genetic subdivisions within the European roe deer are probably the result of historical vicariant events in southern glacial refugia [7]. More recently, the European roe deer experienced considerable reductions in population numbers or even local exterminations at several locations within its continuous range, mainly as a consequence of deforestation and over-hunting during the late 19th and early 20th centuries, [8–10]. Reintroduction and re-stocking programs (often using non-indigenous animals) were carried out in a number of places in the world, both with hunting and conservational purposes [8,11]. Clearly, non-indigenous sources of individuals and local exterminations may have a tremendous impact on the genetic diversity of local populations of roe deer. In fact, genetic signatures (e.g. admixture, genetic drift) of such anthropogenic disturbances have been reported in several locations [7,12]. Additionally, recent studies have documented the introgression of Siberian roe deer (*C. pygargus*) mitochondrial DNA (mtDNA) genes into European roe deer populations in Poland, Lithuania and Russia [13,14]. Hybridization between these two closely related species might have been caused by natural processes which took place in the past or might be, at least partly, an effect of human-mediated introductions of Siberian roe deer within the range of European roe deer. Many events of such introductions took place in the European part of Russia and other countries in eastern Europe [8].

Human practices have not been completely detrimental for the European roe deer and, in fact, population expansions during the last two centuries have been attributed, among other factors, to the extension of cultivated fields providing a suitable and abundant source of food [9,15]. The ability of the European roe deer to thrive in modern human-modified landscapes might be related to its considerable morphological, behavioral and ecological variability, and to its ability to exploit a great variety of habitats (e.g. broadleaved, coniferous and mixed forests, agricultural landscapes, ecotonal strips, lowlands and highlands) [9]. Morphological, behavioral and ecological differences among animals living in areas with contrasting levels of forest cover have been described [16–18]; accordingly, distinct field and forest ecotypes of the species have been recognized, even though there is no consensus about the validity of this distinction [19]. Moreover, genetic analyses at allozyme loci did not support the distinction of field and forest ecotypes within the species [20], and more powerful genetic markers (i.e. microsatellites) have not yet been used to investigate the differentiation between animals living in environments with contrasting levels of forest cover. The European roe deer has also served as a model species for investigating the genetic effects of fragmentation and human disturbance. It has been shown, for example, that the combination of several landscape features (i.e. highways, rivers, canals) may lead to population genetic differentiation [21], and that genetic discontinuities correlate with transportation infrastructure [22].

In this study we investigated the influence of contemporary (ecological) and past events on the patterns of genetic diversity and population differentiation of European roe deer in northeastern Poland. Unlike in most of west Europe, our study site comprises vast areas of conserved primeval forests where the European roe deer was never driven to extinction. As such, it provides an interesting comparison with previous genetic studies in central and western Europe, where the effects of recent drastic habitat fragmentation and reductions in population numbers are likely to have profound influences on the patterns of genetic diversity of the species. On the other hand, our study site is located within an area where introgression of mtDNA of Siberian roe deer into the genome of European roe deer has recently been described [13,14]. Interestingly, local reports from the late 19th and early 20th centuries document the introduction of non-indigenous Siberian roe deer within our study site, with the purpose of increasing the size and quality of hunting trophies [23]. We analyzed molecular genetic variation at both mitochondrial and nuclear DNA across three sampling areas. The pattern of genetic structure was investigated with respect to the amount of forest cover, in order to establish whether modern and powerful genetic markers support the distinction between field and forest roe deer. The pattern of mtDNA diversity was analyzed in the context of previous phylogeographical studies of roe deer and historical literature.

Materials and Methods

Ethics statement

Government approval or licenses were not required for the collection of tissue samples (i.e. skin or muscles) from legally hunted animals, which were obtained through hunters and hunting associations. Hunted animals were shot with rifle during the hunting season, following the rules of the Polish hunting law. No animals were killed specifically for this study. Permissions for sampling of live animals were obtained from the Polish Ministry of Environment (Permit No. DLOPik-L-gl-6713/86b/07/ab) and the Local Ethical Commission in Białystok (Resolution No. 46/2008).

Study site and sampling of European roe deer

The study area consisted of three sampling sites (Białowieża, Knyszyn and Augustów, ca. 5340 km^2), distributed latitudinally in northeastern Poland (22°33′ –22°53′E, 52°26′ –54°17′N; max. span in distance: N–S 200 km, E–W 107 km; Fig. 1). The landscape in all sampling sites can be divided into three distinctive categories: open (arable lands covered by crop plantations and meadows), closed (coniferous, deciduous, and mixed forests), and mosaics of the arable land, meadows and forests. The northern-most sampling site, Augustów, covers ca. 2750 km^2 (with 58% of open and 33% of forested areas, with most of the latter belonging to Augustów Forest). The second sampling site, Knyszyn (740 km^2, with 71% of forests mainly belonging to Knyszyn Forest, and 28% of open areas), is located ca. 92 km south of Augustów. The southernmost sampling site, Białowieża (1850 km^2, with 44% of forests, mainly belonging to Białowieża Primeval Forest, and 53% of open land), is located ca. 56 km from Knyszyn.

We used a total of 328 roe deer tissue samples collected between 2004–2011:234 in Białowieża, 22 in Knyszyn, and 72 in Augustów (Fig. 1). With the exception of 33 live-captured individuals used for telemetry studies [24] and 4 lynx preys, samples were obtained through hunters and hunting associations, and consisted of parts of skin or muscles from legally hunted animals. The geographic coordinates of the samples were assigned according to the information provided by the hunters and were defined as the geometric central point of a hunting district where a particular animal was hunted. In case of the individuals followed by telemetry, the geographic coordinates corresponded to the place of capture. All samples were stored in 96% ethanol at −20°C prior to DNA extraction.

Definition of ecologically-relevant groups of roe deer

In order to investigate the influence of the amount of forest cover on the pattern of genetic structure, we defined groups of European roe deer inhabiting contrasting environments within our three sampling sites. We visually inspected the distribution of forest cover within the sampling sites with ArcMap 9.3.1 (ESRI Inc. 2009) (Fig. 1); individuals within areas of continuous forest cover were grouped as forest roe deer, while individuals within areas of open or mixed habitats were grouped as field roe deer. Given the potential role of road infrastructure on the genetic structuring of roe deer [22], field roe deer from the Białowieża area were further subdivided in two subgroups, one north and one south of the main road in that region. Groups were delimited by complex polygons and the percentage of forest cover within these polygons was calculated. Four samples could not be assigned to groups due to their relatively large geographic isolation with respect to the other samples.

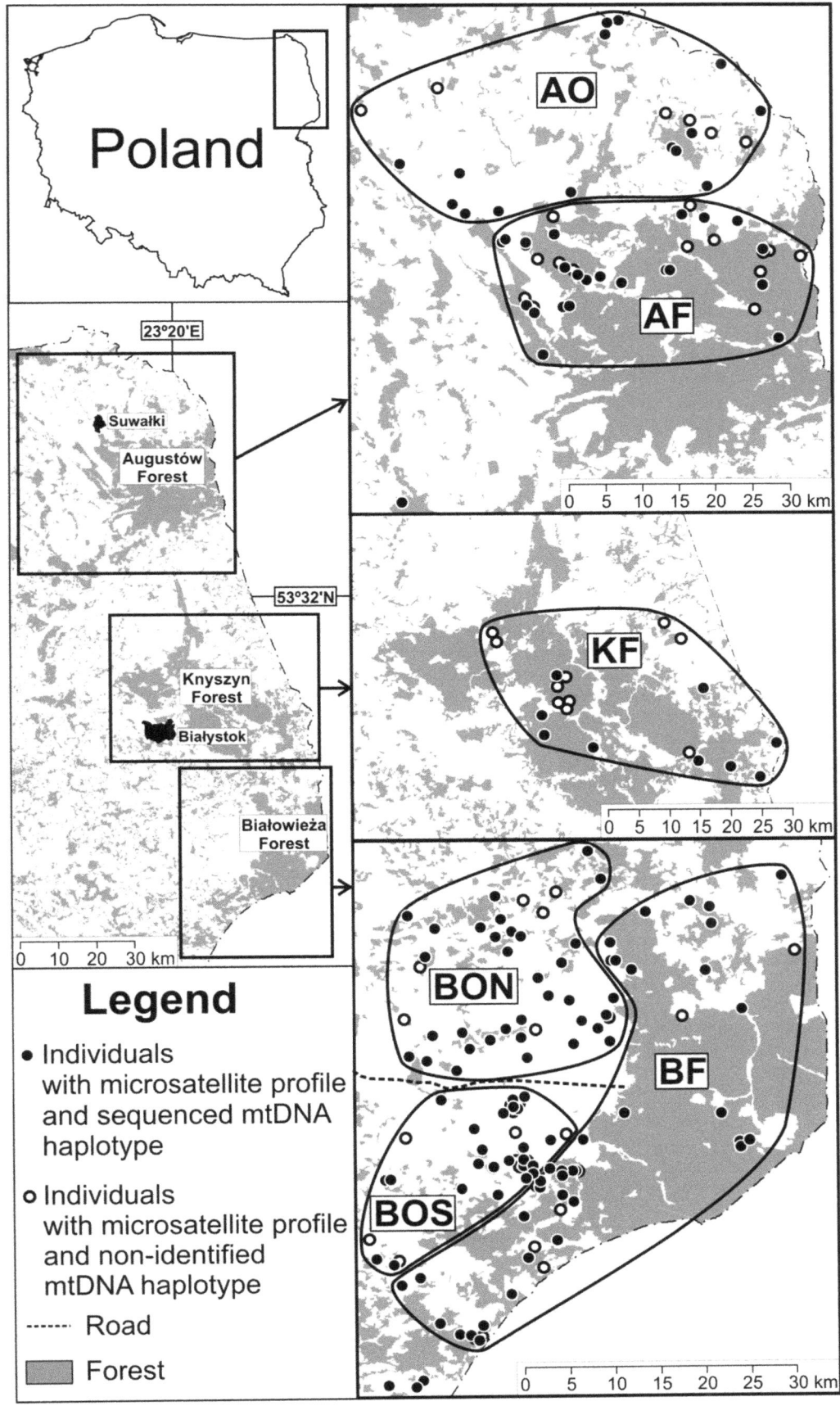

Figure 1. Study area and roe deer sampling in northeastern Poland. Samples were collected in and around three lowland forests: Augustów (A), Knyszyn (K) and Białowieża (B). Filled symbols indicate samples that were genotyped at both microsatellites and mtDNA control region, whereas open symbols indicate samples that were only typed with microsatellites. Polygons were drawn around regions with contrasting degrees of forest cover within each sampling site, which were used to define field (O - open habitat) and forest (F) groups of roe deer; a further subdivision in Białowieża separates open habitats in the north (N) and south (S) of a main road.

Genotyping

Total DNA from all samples was extracted with the Qiagen DNeasy Blood and Tissue Kit following the manufacturer's protocols. We genotyped each individual with 16 microsatellite markers that were reported as polymorphic for roe deer (Table S1). Products were separated in an ABI 3130 xl Genetic Analyzer with the GeneScan 400HD ROX Size Standard (Applied Biosystems). Genotypes were read with the software GeneMarker (Softgenetics). A fragment of mtDNA control region was amplified by PCR with the primers L-Pro and H-Phe [25]. Cycling conditions were 95°C for 15 min; 35 cycles of 94°C for 15 s, 56°C for 15 s, and 72°C for 1 min; and 72°C for 10 min. PCR products were purified using Clean Up (A&A Biotechnology, Gdańsk, Poland). Sequencing reactions were carried out in a 10 μl volume using the Big Dye sequencing kit v.3.1 (Applied Biosystems) with the forward primer. Products were purified with the Exterminator kit (A&A Biotechnology) and separated on an ABI 3130 xl Genetic Analyzer (Applied Biosystems). Sequencing results were analyzed with the ABI DNA Sequencing Analysis software and aligned in BioEdit v.7.0.9 [26].

Analyses of microsatellite data

We checked for duplicated samples with the package *allelematch* [27] for R [28]. From the 328 genotyped samples, 9 were excluded from further analysis because they showed identical or almost identical (>93% of similarity) profiles to other samples (i.e. some individuals were sampled more than once), or due to failures in amplification at 4 or more microsatellite loci. Allele frequencies, observed and expected heterozygosity, deviations from Hardy–Weinberg equilibrium (HWE), and F_{IS} were calculated with GENEPOP 4.1.4 [29] for each sampling area. Allelic richness, a measure that accounts for differences in samples sizes in estimates of the number of alleles, was calculated with HP-RARE 1.0 [30]. Null allele frequencies and genotyping errors were estimated with the program Micro-Checker [31]. False Discovery Rate corrections [32] were performed in R [28] to account for multiple testing.

In order to account for the possibility of genetic changes occurring over time, we performed an analysis of molecular variance (AMOVA) in Arlequin v.3.5.1.3 [33] with the samples grouped into year classes within each sampling area. We investigated the occurrence of population structure and the pattern of genetic differentiation among field and forest roe deer with different methods. First, we used the clustering program STRUCTURE [34]. We chose the admixture model and the option of correlated allele frequencies between populations, with and without sample location and the grouping of field and forest roe deer as prior information [35]; we let the parameter alpha (the degree of admixture between subpopulations) be inferred from the data and set lambda (the allele frequency prior) to 1. We conducted 10 independent replicate analyses for values of K (number of genetic clusters) between 1 and 10 with a burn-in period of 10 000 iterations and 100 000 Markov chain Monte Carlo (MCMC) cycles. We used the STRUCTURE HARVEST-ER [36] to compile and visualize the results from the STRUC-TURE runs and to calculate ΔK [37]. Second, we used the spatial model of clustering implemented in the program GENELAND

[38]. Uncertainty in the coordinates was set to 8×10^{-3} to account for ca. 1 km of uncertainty in the location assigned to the samples [21], and the maximum number of populations was set to 10. We used the correlated allele frequency model as described in Guillot [38], for runs with 1×10^6 MCMC iterations. Ten independent runs were made to look for convergence in the number of estimated K. Third, we further investigated the genetic substructure, isolation by distance, and differentiation groups of field and forest roe deer by means of multivariate methods implemented in the R-package *adegenet* [39]. An advantage of multivariate methods over other clustering algorithms is that the former do not rely on Hardy-Weinberg and linkage equilibrium for summarizing genetic variability. We used spatial Principal Component Analysis (sPCA) [40] in order to investigate spatial genetic patterns within the study area; the connection network between individuals was defined with the Delaunay triangulation [41], and global and local tests (with 9999 permutations) were performed as an aid for selecting the structures to be interpreted. Discriminant Analysis of Principal Components (DAPC) [42], a method designed to identify and describe clusters of genetically similar individuals, was used to visualize the genetic relatedness/differentiation between the previously defined groups of field and forest roe deer. For this DAPC, forty principal components of PCA and all (5) discriminant functions were retained. The function *find.clusters* was used to identify the number of clusters (K) in our data and to compare with the prior groups; for this, we covered values of K between 1 and 10, and followed the procedure outlined in Jombart *et al.* [42]. Fourth, we performed an analysis of molecular variance (AMOVA) in Arlequin v.3.5.1.3 [33] with the field and forest groups nested within each sampling area, for a hierarchical partition of the genetic variance within and among sampling sites and groups. Finally, we calculated D_{Jost} [43] as a measure of differentiation between field and forest groups and sampling areas with the R-package *diveRsity* v.1.3.2 [44]. We investigated the pattern of nuclear genetic differentiation among the mtDNA clades found, as well as the influence of individuals with Siberian haplotypes (see Results) on the genetic structure of roe deer in the study area by performing additional STRUC-TURE analyses and DAPC; for this, prior groups were defined according to the mtDNA clades or samples with Siberian haplotypes were excluded from the analyses.

Analyses of mtDNA data

We obtained good-quality mitochondrial control region sequences (ca. 610 bp) for 241 out of the 319 unique samples. For the remaining 78 samples, repeated amplification attempts with different PCR and sequencing conditions did not result in readable sequences. Mitochondrial control region sequences were aligned against a reference sequence of European roe deer (GenBank accession number AY625869.1) [7] and manually edited in BioEdit v.7.0.5.3 [26]. Measures of genetic diversity were estimated with Arlequin v.3.5.1.3 [33]. We determined the position of our roe deer haplotypes within the phylogenetic clades of the species described by Randi *et al.* [7], using published European and Siberian roe deer mtDNA sequences [7,13,14,45–51]. As the recently published mtDNA sequences by Matosiuk *et al.* [14] were shorter than ours, we performed the phylogenetic reconstruction both with our full-length sequences (610 bp) and with shortened ones (510 bp). Phylogenetic trees were reconstructed using MEGA 5 [52], with the neighbour-joining procedure and Tamura and Nei's TN93 genetic distance model, as described by Randi et al. [7]. Support for the internodes was assessed after 10 000 bootstrap resampling steps. We inferred the haplotype genealogy of our samples with a mtDNA network constructed with the median-joining procedure in Network 4.6 (http://www.fluxus-engineering.com/). Genetic differentiation among sampling sites and prior groups was estimated with AMOVA, F_{ST} [33] and spatial analysis of molecular variance (SAMOVA) [53], with field and forest groups of roe deer nested within each sampling area. In order to investigate the influence of individuals with Siberian haplotypes (see Results) on the pattern of genetic diversity/differentiation, we repeated all the analyses with samples carrying European haplotypes only.

Results

Genetic diversity of European roe deer in northeastern Poland

The genetic diversity of roe deer at the three sampling sites in northeastern Poland was high, both at microsatellite and mitochondrial markers (Table 1). One microsatellite locus, BMS119, was monomorphic and therefore excluded from further analyses. The number of alleles at the remaining 15 microsatellite loci ranged between 2–14 (8.67 on average), and the mean observed heterozygosity was 0.60. None of the loci showed evidence of scoring errors due to large allele drop-out or stutter peaks. Significant deviations from HWE were detected in loci NVHRT21 and Roe1 (in Białowieża), NVHRT73 (in Białowieża and Augustów), and NVHRT71 and ETH225 (in all sampling sites); with the exception of locus Roe1, all the deviations from HWE were due to heterozygote deficit. The frequency of null alleles at the loci with significant deficit of heterozygotes ranged between 0.11–0.41, with the highest values for locus NVHRT71. F_{IS} measured across loci was positive in all sampling areas, although low (≤0.07). Due to the strong influence of locus NVHRT71 on results of the DAPC and estimates of population differentiation, and to its high frequency of null alleles, we excluded it and present the results of all analyses done with the remaining 14 loci.

A total of 13 different mtDNA control region haplotypes were defined by 39 polymorphic sites (38 substitutions, 1 deletion). Complete sequences have been deposited in GenBank with accession numbers KM068160–KM068172. The genetic diversity was high (Table 1). The phylogenetic and network analyses (Fig. 2, Fig. S1 and S2) revealed three haplotypes (H1, H8 and H11) that were highly divergent and grouped with Siberian roe deer. Surprisingly, these haplotypes were carried by more than half (50.6%) of the individuals sampled. Siberian haplotypes clearly increased the measures of mtDNA diversity in all sampling areas (Table 1). The remaining haplotypes corresponded to European roe deer; most of them (H2, H4, H5, H6, H7, H10, H12, H13) grouped within the Central clade described by Randi et al. [7], one (H9) grouped within the East clade, and one considerably divergent haplotype (H3) was included within the Central or Western clades of the species (Fig. 2, Fig. S1 and S2). The main topology of the trees including full-length and shortened sequences was similar (Fig. 2, Fig. S1 and S2). Numbers of the different haplotype-clades varied considerably across the study area. The proportion of individuals carrying haplotypes of Siberian roe deer declined northwards, from 55% in Białowieża and Knyszyn (pooled) to 34% in Augustów; the opposite trend was revealed in the frequencies of the European roe deer haplotype belonging to the clade East (2% in Białowieża-Knyszyn, 21% in Augustów) and haplotype H3 (0.5 and 8.5%, respectively).

Genetic population structure

The visual inspection of the distribution and amount of forest cover within the study site resulted in the definition of 6 groups of roe deer (Fig. 1): 3 in forested areas (one in each sampling site, with 69 samples in Białowieża, 20 in Knyszyn, and 39 in Augustów) and 3 in more open fields (two in Białowieża: with 103 and 55 samples at the north and south of the main road, respectively; and one in Augustów with 29 individuals); 4 samples were not classified due to their geographic isolation. Field and forest groups clearly differed in their mean forest cover (17% vs. 70% for open and forested areas, respectively).

We found no evidence of genetic changes in the roe deer population over the study years (only 0.05% of the genetic variance at microsatellite loci occurred among years, $F_{SC} = 0.0005$, p = 0.3216). The clustering analyses in STRUCTURE (both with and without spatial prior information) showed a peak in the mean posterior probability (Ln P(D)) and in ΔK for K = 2. The pattern of assignment of the individuals to the two inferred clusters divided the study area along the north-south axis (Fig. 3 left panel): individuals from Białowieża (in the south) were mostly assigned to one cluster, whereas individuals from Augustów (in the north) were mostly assigned to the second cluster; individuals from Knyszyn (located between Białowieża and Augustów), on the other hand, showed a mixed pattern of assignment to the two inferred clusters. The GENELAND analysis (which employs a spatially-explicit clustering approach) also returned K = 2 as the most likely number of genetic clusters in six out of ten independent runs. The results matched exactly the pattern observed with STRUCTURE: all individuals from Białowieża and Augustów were assigned to either one or the other cluster, whereas the individuals from Knyszyn were assigned to both (Fig. 3 right panel). Neither the STRUCTURE nor the GENELAND analyses suggested a further population subdivision that could reflect genetic differences between individuals assigned to field and forest groups.

Similar results were obtained with the multivariate analyses. The first positive eigenvalue or global score of the sPCA was retained based on its spatial and variance components. A global test confirmed the existence of a global structure in our data (i.e. positive spatial autocorrelation, max(t) = 0.0104, $p = 0.0001$), whereas the local test did not detect any local pattern (i.e. negative spatial autocorrelation, max(t) = 0.0045, non-significant). The first global score differentiated Białowieża from the other two sampling areas, which appeared similar to each other (not shown). The slight pattern of isolation by distance, on the other hand, was not significant. The first principal component of the DAPC separated the individuals from Białowieża and Augustów in the extremes, sharing the space with the individuals from Knyszyn, which were placed in the middle between the other two sampling sites; a high degree of overlap between field and forest groups within sampling areas, and between animals on either side of the road in Białowieża was evident (Fig. 4). The function find.clusters identified 3–5 groups in the data; these groups did not match the prior groups defined by sampling areas and forest cover, and their members did not show any spatial clustering. The AMOVA with forest and field roe deer grouped within sampling sites revealed that 98.5% of the genetic variance occurred within the groups, whereas only 0.3% of the variance occurred among groups within sampling areas, and 1% among the three sampling areas. The pairwise measures of differentiation (D_{Jost}) between groups of

Table 1. Microsatellite and mtDNA diversity of roe deer *Capreolus capreolus* at three sampling regions in northeastern Poland.

Parameter	Sampling site			Total
	Białowieża	Knyszyn	Augustów	
Microsatellites				
Sample size	230	20	69	319
No. of alleles/locus	8.33	5.87	7.07	8.67
Allelic richness	6.07	5.87	5.98	-
Private allelic richness	0.54	0.40	0.60	-
Observed heterozygosity	0.62*	0.58*	0.61*	0.61*
Expected heterozygosity	0.66	0.63	0.65	0.66
Fis	0.07	0.07	0.06	0.07
mtDNA (all sampled roe deer)				
Sample size	184	10	47	241
No. haplotypes	13	6	6	13
No. polymorphic sites	39 (6.4%)	34 (5.6%)	38 (6.2%)	39 (6.4%)
(% of sequence length)				
Haplotype diversity h (SD)	0.714 (0.029)	0.844 (0.103)	0.794 (0.030)	0.756 (0.025)
Nucleotide diversity π (SD)	0.024 (0.012)	0.026 (0.014)	0.025 (0.013)	0.025 (0.012)
Pairwise divergence k (SD)	14.457 (6.503)	15.844 (7.672)	15.547 (7.055)	14.994 (6.726)
mtDNA (European roe deer only)				
Sample size	84	4	31	122
No. haplotypes	10	3	5	10
No. polymorphic sites	24 (3.9%)	12 (2.0%)	20 (3.3%)	24 (3.9%)
(% of sequence length)				
Haplotype diversity h (SD)	0.766 (0.040)	0.833 (0.222)	0.778 (0.035)	0.848 (0.019)
Nucleotide diversity π (SD)	0.006 (0.003)	0.011 (0.007)	0.012 (0.006)	0.009 (0.005)
Pairwise divergence k (SD)	3.750 (1.910)	6.500 (3.817)	7.307 (3.506)	5.266 (2.561)

Microsatellite diversity is based on 15 polymorphic loci. MtDNA diversity is based on 610 bp control region sequence. SD: standard deviation; * significant deviation (< 0.05) from Hardy-Weinberg equilibrium.

samples were extremely small (<0.028, in a 0–1 scale where 0 indicates no differentiation and 1 reflects complete differentiation, Table 2). The pattern of nuclear genetic structure did not change when only individuals with European mtDNA haplotypes were considered.

The mtDNA data revealed a similar pattern of genetic structure. Białowieża and Augustów showed some degree of differentiation reflected in relatively large and significant F_{ST} values (Table 2); F_{ST} values among roe deer ecotypes within the sampling areas, on the other hand, were mostly small and non-significant (Table 2). The AMOVA performed with mtDNA data, with forest and field roe deer grouped within sampling sites, showed that 93.2% of the genetic variance occurred within groups, 2.4% among groups within sampling areas, and 4.3% among the sampling sites. According to the SAMOVA (Fig. 5), the roe deer samples could be divided into 4 regional groups: two separating forest and field areas in Augustów, one grouping Knyszyn forest and field areas of Białowieża, and the last one corresponding to Białowieża forest; this subdivision, however, was not further supported by F_{ST} measures of genetic differentiation (Table 2).

Nuclear genetic differentiation among individuals with Siberian and European mtDNA haplotypes

The measures of microsatellite diversity did not differ among groups of animals with European, Siberian and non-identified

mtDNA haplotypes (not shown). The STRUCTURE analysis and DAPC with groups defined according to the mtDNA clade of the individuals (i.e. Siberian, clade Central, clade East, H3 or non-identified) suggested a genetic homogeneity in the nuclear genetic composition among most clades. In the DAPC, individuals with the haplotype H3 grouped apart from all the other individuals with different haplotypes (Fig. 6).

Discussion

Genetic diversity of roe deer in northeastern Poland in historical context

The relatively high levels of variability and the slight heterozygote deficiency at microsatellite loci of roe deer in northeastern Poland are consistent with findings from previous studies conducted across Europe (e.g., [54,55]). The high frequency of null alleles at loci with significant deficit of heterozygotes indicates a plausible cause for the observed departures from HWE. Other causes of heterozygote deficit (i.e. sex-biased dispersal, yearly shifts in allele frequencies, inbreeding), although cannot be completely ruled out, are not supported by other roe deer studies [56] and our own observations.

The values of mtDNA diversity indicate a high effective population size of roe deer in northeastern Poland, and are likely to reflect the complex history of the species in this region. Almost

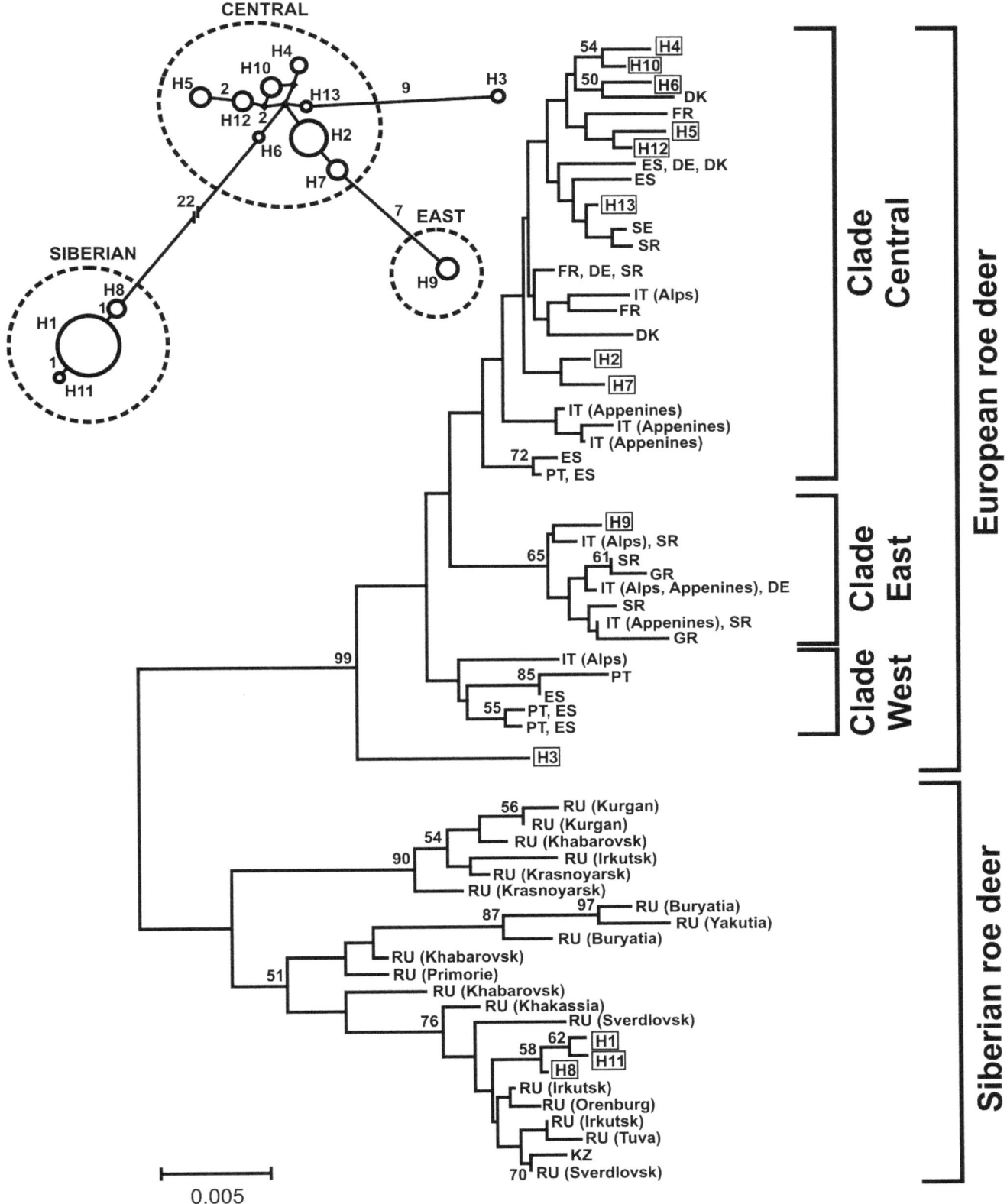

Figure 2. Phylogenetic relationship between the mtDNA haplotypes found in this study (H1–H13) and other published mtDNA control region sequences of European [7] and Siberian [45] roe deer. Geographic locations of published sequences of European roe deer (DE = Germany; DK = Denmark; ES = Spain; FR = France; GR = Greece; IT = Italy; PT = Portugal; SE = Sweden; SR = Serbia, Montenegro, and Kosovo) and Siberian roe deer (KZ = Kazakhstan; RU = Russia) are indicated in the phylogenetic tree. Numbers at nodes show support (≥50%) from 10 000 bootstrap replicates. European roe deer clades are defined according to Randi et al. [7]. In the median-joining network of the mtDNA haplotypes

found in this study (upper-left), clades are grouped within punctuated circles. The size of the solid circles is proportional to the number of individuals with a given haplotype in the whole study area. Numbers above the branches indicate the mutation steps between two haplotypes.

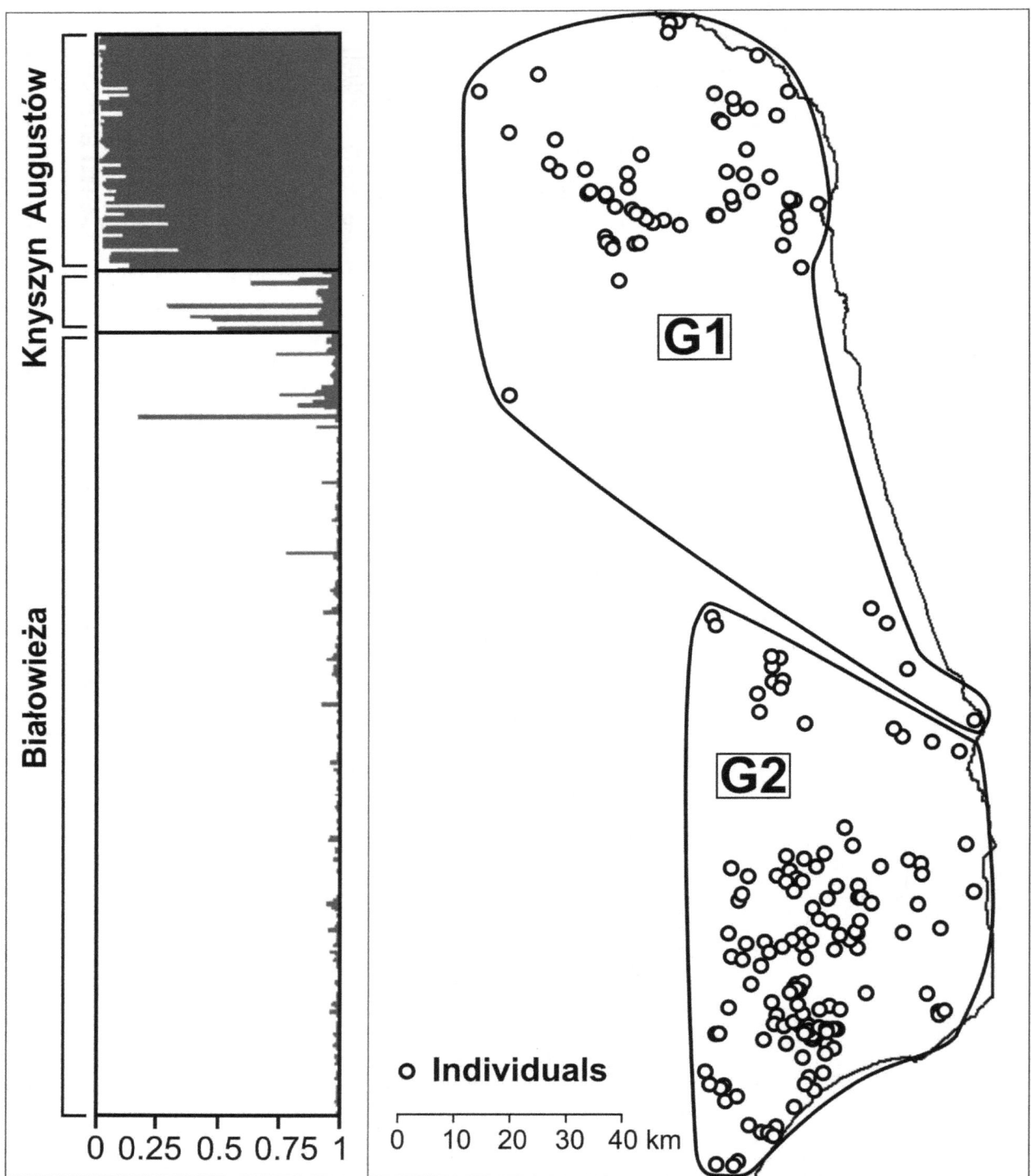

Figure 3. Population structure of roe deer in northeastern Poland according to microsatellite markers. Left panel: Results from the STRUCTURE analysis. The best supported number of genetic clusters was 2. Each vertical line represents one individual partitioned into 2 colored segments; each colored segment represents the estimated membership coefficient of the individuals to the two inferred clusters. Right panel: Map of estimated membership from the GENELAND analysis for K = 2. Each dot represents one individual and the polygons encircle individuals belonging to the two inferred genetic groups (G1, G2).

Table 2. Pairwise measures of differentiation (D$_{Jost}$) and Fst between sampling sites and groups of roe deer inhabiting areas with different degree of forest cover (denotations of groups as in Fig. 1).

Group	BOS	BON	BF	KF	AO	AF
BOS	–	0.0002	0.0000	0.0051	0.0133	0.0180
BON	0.0037	–	0.0042	0.0053	0.0234	0.0133
BF	−0.0017	0.0604*	–	0.0032	0.0278	0.0179
KF	−0.0471	−0.0473	0.0016	–	0.0028	0.0036
AO	0.0996*	0.1981*	0.0507	0.1075	–	0.0011
AF	0.0127	0.0790*	0.0091	−0.0018	0.0136	–

Above diagonal: D$_{Jost}$ calculated with microsatellite data; below diagonal: Fst calculated with mtDNA sequence data. The 95% confidence interval of all D$_{Jost}$ estimates ranged between 0.0001–0.1171. * Significant (<0.05) Fst values.

half of the individuals carried European roe deer mtDNA haplotypes. From these, eight were closely related and belonged to the Central clade [7], which, according to the mtDNA network and given that the roe deer was never extinct in northeastern Poland [57], are likely to be native in this region. The two remaining European haplotypes (H9 and H3), on the other hand, were highly divergent and were represented by few individuals mostly found in the north of the study area. Haplotype H3 was previously described by Zvychainaya et al. [45] in a European roe deer population in western Russia. In our analyses, this haplotype

clustered within the Western or the Central European roe deer clades and, therefore, its origin cannot be clearly determined. Four of the European roe deer haplotypes found in our study (H4, H5, H6 and H12) have not been previously described by other authors [7,13,14,45–51].

More than half of all the individuals studied carried Siberian roe deer mtDNA haplotypes, which is in concordance with the results of recently published studies of Lorenzini et al. [13] and Matosiuk et al. [14], who conducted phylogeographic analyses of roe deer populations including samples from Poland and Lithuania. One

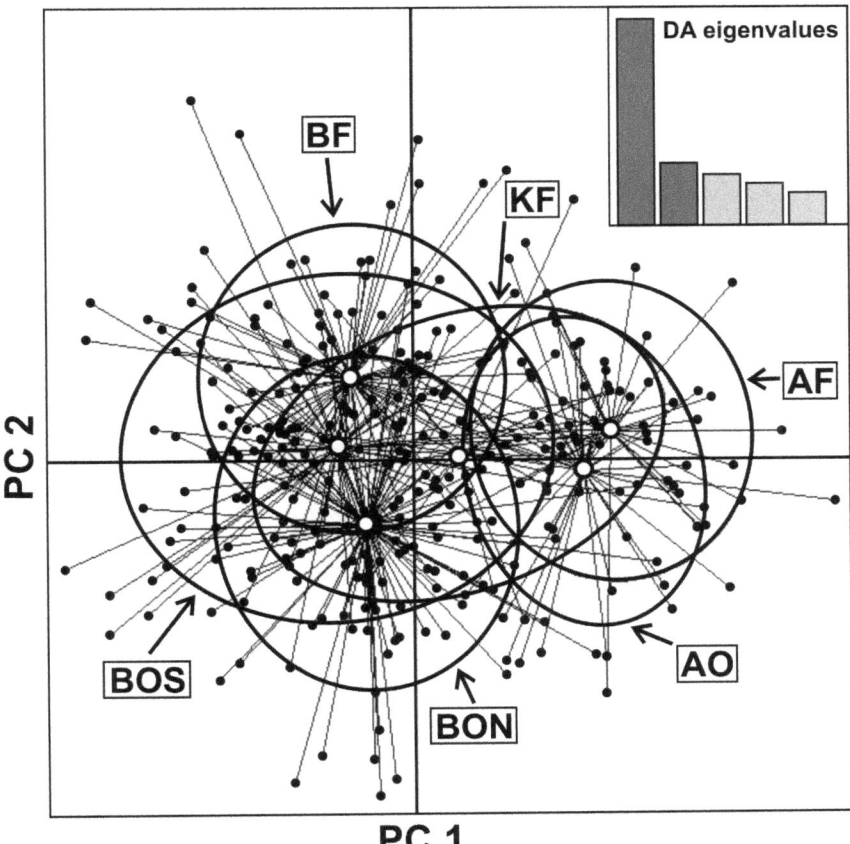

Figure 4. Nuclear genetic differentiation of roe deer groups inhabiting fields and forests according to the Discriminant Analysis of Principal Component (DAPC). The first two principal components are shown. Denotations of groups of roe deer as in Fig. 1.

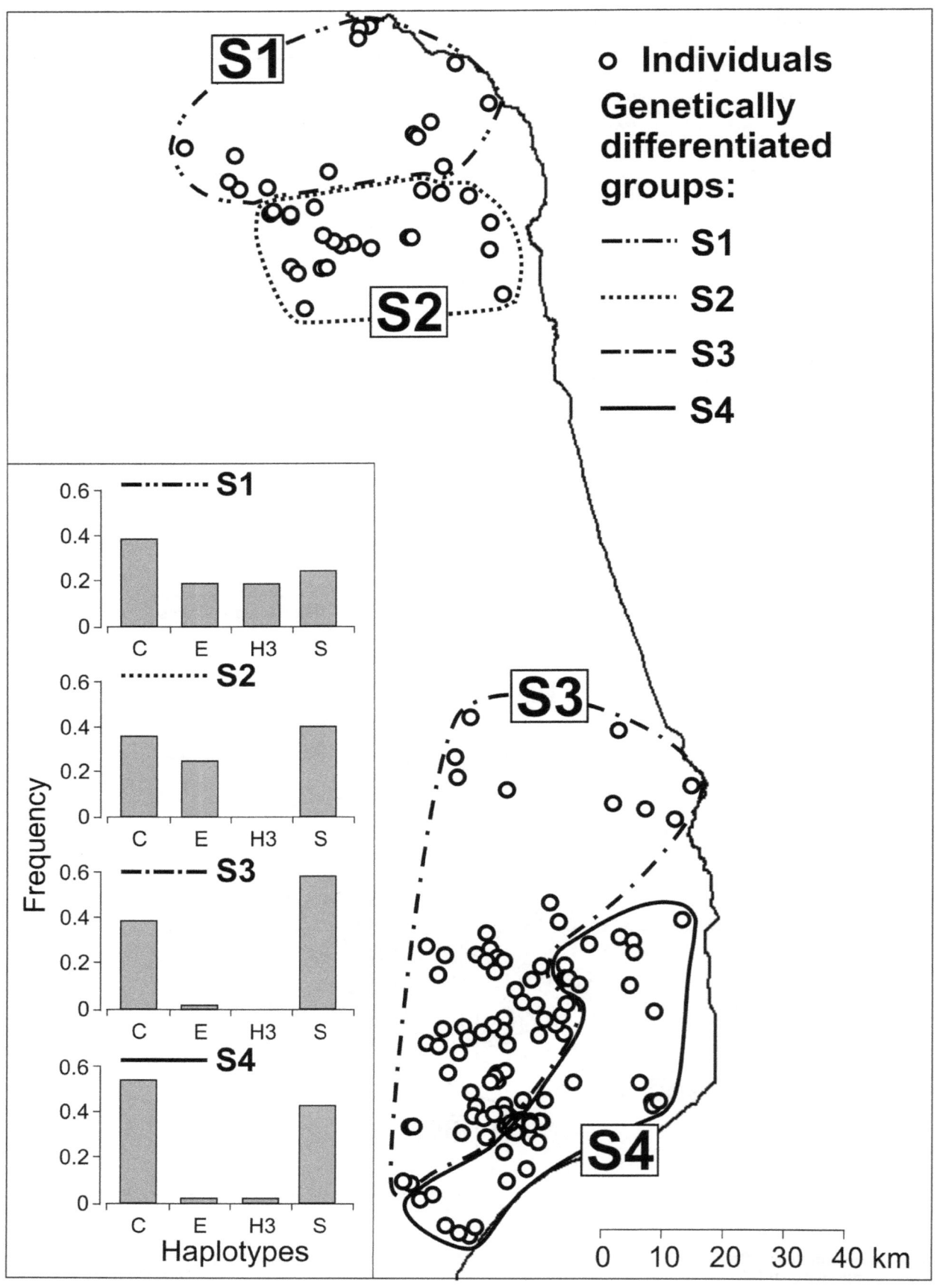

Figure 5. Groups of roe deer (S1–S4) based on mtDNA data according to SAMOVA, and frequencies of haplotype clades (C = Central, E = East, H3, S = Siberian) in each of the defined groups.

out of the three Siberian mtDNA haplotypes found in our study areas (H11) was not described before by other authors [7,13,14,45–51]. In our study, individuals possessing Siberian roe deer haplotypes did not show any remarkable phenotypic feature that would put in doubt their identity as European roe deer (hunting associations, pers. comm., and our own observations), nor did they present difficulties for amplification of nuclear markers. Moreover, no nuclear genetic differentiation (including sex-specific markers) (see [14]) among animals carrying divergent mtDNA haplotypes was found.

Phylogeographic studies of the genus *Capreolus* estimate a divergence time between European and Siberian roe deer of about 2–3 million years [25]. Nowadays, their natural distribution is allopatric, with European roe deer occupying most of western Europe, and Siberian roe deer naturally found across the temperate zone of eastern Europe and Asia. A narrow contact zone between the two species occurs at the westernmost limit of the Siberian roe deer distribution, at the Khoper and Don rivers in the European part of Russia [58]. European and Siberian roe deer differ in characters such as body size, morphometric traits and karyotype [8]. Despite a large degree of reproductive isolation between the two species, hybridization in captivity has been demonstrated and it probably also occurs in natural conditions but is difficult to document [8]. Successful production of hybrids is

more likely to occur in crosses between Siberian females and European males, as the smaller European roe deer females usually die while giving birth to large hybrid fetuses or give birth to dead young [8,58]. Introgressive hybridization of Siberian mtDNA into the European roe deer gene pool is, therefore, not unlikely. However, it has not been reported in phylogeographic studies of European roe deer at central and western parts of its geographical distribution [7,12,54,59] and was only recorded in the eastern part of the species range (this study, [13,14]), reaching up to 78% of hybrids in a roe deer population in the Moscow region, Russia [46].

Given the fact that our study area lies far beyond the actual range of Siberian roe deer [8], the finding of a high proportion (but low numbers) of Siberian haplotypes in otherwise European-looking roe deer in northeastern Poland is particularly interesting. Both anthropogenic factors and natural processes have an impact on the present distribution of roe deer in Europe [8]. On one hand, our observations might be caused by human-mediated introduction(s) of Siberian roe deer [8,23] and the posterior introgressive hybridization with the local European roe deer; on the other hand, they might be an effect of natural processes which took place in eastern Europe in the past (as postulated by [13] and [14]). In fact, the structure of the mtDNA network of the Siberian haplotypes found in this study (with one very common haplotype

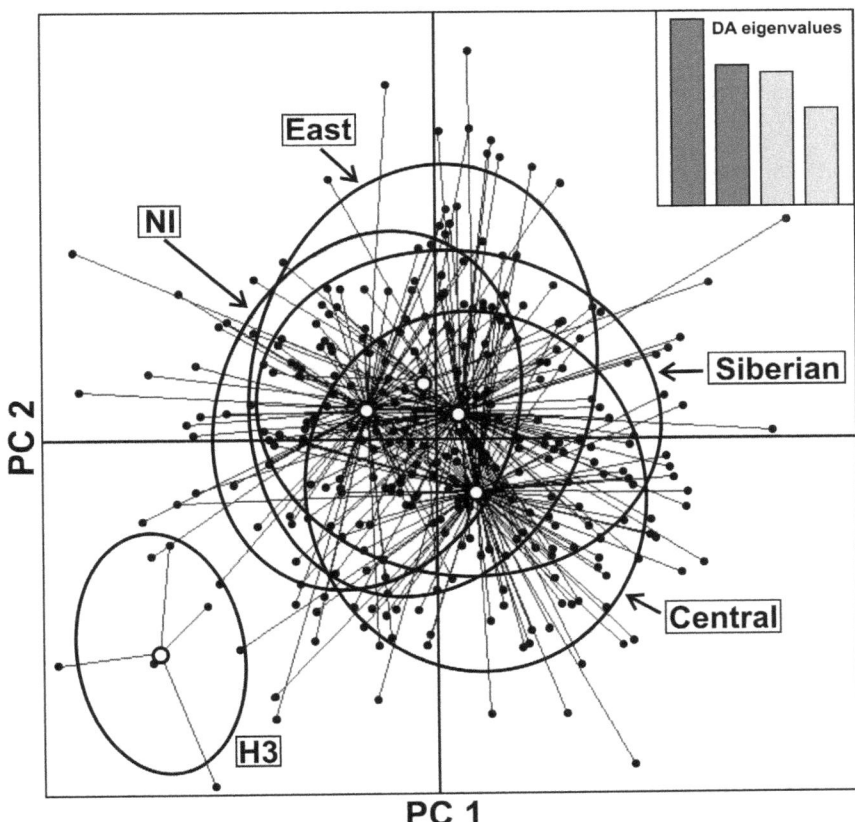

Figure 6. Scatterplot of the DAPC showing the nuclear genetic differentiation (according to microsatellite markers) between groups of individuals with European (Central, East and H3), Siberian and non-identified (NI) mtDNA haplotypes. The first two principal components are shown.

and two minor ones) might indicate a bottleneck caused by a low number of founders. Historical evidence from the Białowieża region supports the hypothesis of human-mediated introductions, with a first recorded translocation and posterior release of eight Siberian roe deer (probably from the Ural mountains) and their progeny to the Białowieża Primeval Forest (BPF) in 1891 [23]. At that time, the population of European roe deer in BPF was estimated at about 600 individuals, and since then its numbers fluctuated between 300 and 6100 animals [57]. Posterior translocations from BPF to other areas in Poland were also described by Karcov [23]. Unfortunately, not all translocations were documented in written sources, but many of them might have taken place in eastern Poland (personal communication with members of the Polish Hunting Association). The fact that the number of individuals carrying haplotypes of Siberian roe deer declined northwards from Białowieża (where the documented introductions took place) to Augustów Forests, also supports the hypothesis of human-mediated translocations.

Alternatively, the presence of mtDNA of Siberian roe deer in European roe deer populations in eastern Europe might be caused by sympatric distribution of the two species in the past and natural introgression of Siberian roe deer mtDNA into the European roe deer genome [13,14]. According to Danilkin [8], the historical range of Siberian roe deer spread further west than the present distribution of the species. Lorenzini et al. [13] and Matosiuk et al. [14] suggest that, after the Last Glacial Maximum, the range of Siberian and European roe deer overlapped in central and eastern Europe, and natural hybridization occurred at that time. Interestingly, Siberian haplotypes found both in our study and by Matosiuk et al. [14] grouped in the phylogenetic tree with samples collected in Kazakhstan, the Irkutsk region and Khakassia, which might suggest that founders of European roe deer populations in northeastern Poland originated from this area. Lorenzini et al. [13] claims that the lack of divergence in nuclear DNA between individuals with Siberian and European mtDNA can be a proof that the mtDNA introgression is much older than 200 years ago, when the translocation by humans took place. Given the lack of genetic data from roe deer populations in large areas of eastern Europe and western Asia, it is not possible, in our opinion, to definitively establish whether the introgression of Siberian roe deer genes into the local populations of European roe deer has natural or human-mediated causes. The genetic analyses of ancient samples of roe deer from eastern Europe could help to resolve this issue.

Introgression of Siberian roe deer mtDNA into local populations of European roe deer may have evolutionary implications. Mitochondrial DNA can be responsible for adaptation of organisms to changing environmental conditions (e.g. [60,61]). In our case, hybrids possessing Siberian mtDNA could be better adapted, for example, to severe winters, an important agent of roe deer mortality in eastern Poland [62]. A similar conclusion was drawn by Matosiuk et al. [14], who compared the distribution of Siberian roe deer haplotypes in the population of European roe deer in Poland with environmental factors. However, the hypothesis of adaptive advantage of Siberian roe deer gene introgression into European populations calls for further studies.

Population structure and landscape features

All of the microsatellite and mtDNA analyses showed a pattern of slight north-south differentiation that corresponded to the geographic origin of the roe deer samples. The continuous distribution of roe deer across the whole study area, the lack of significant barriers for their movement, the low differentiation between sampling sites, and the absence of a significant pattern of

isolation by distance suggests that populations from Augustów and Białowieża should be seen as extremes in the continuous genetic variation of roe deer within the study site, rather than as discrete differentiated groups. Given that the mean dispersal distances of roe deer are small (less than 2 km) [9], and although the furthest recorded distance from the place of capture in our study area was relatively large (22.1 km), it is not surprising to find a certain amount of differentiation between the furthermost areas (separated by ca. 140 km) in this study. These low levels of structure among sampling areas were not only caused by individuals carrying Siberian haplotypes, as analyses excluding these individuals revealed the same general pattern found with all samples. The lack of genetic differentiation among groups of animals found on different sides of the main road in the Białowieża area, suggests that this type of non-fenced transportation infrastructure does not act as a barrier for roe deer dispersal. Fenced highways, in contrast to non-fenced roads, railways and other linear landscape elements, seem to have an impact on the movement of roe deer and cause some degree of population differentiation [22]. The pattern of nuclear genetic differentiation in the studied roe deer population did not considerably differ from the one found with mtDNA, suggesting that both males and females have a similar contribution to the observed pattern. Female philopatry in the roe deer might result in a stronger signal of genetic differentiation at mtDNA (relative to nuclear markers) at relatively large spatial scales [63]. At smaller spatial scales, however, the lack of differences in fine-scale genetic structuring between males and females [64] might explain the concordant pattern of mitochondrial and nuclear genetic structure found here.

We did not find a clear genetic support for the subdivision of roe deer into field and forest ecotypes. The genetic differentiation between groups of individuals inhabiting areas with different levels of forest cover within the same sampling site was very small; this was true at all three sampling areas and with both nuclear and mitochondrial markers. Our results with presumably neutral genetic markers support earlier findings of a lack of differentiation between roe deer ecotypes using allozymes [20]. These results are not surprising; genetic differentiation between roe deer ecotypes may only occur through a mechanism (e.g. spatial, behavioral, morphological or ecological) that could generate some kind of reproductive isolation between them. In all three sampling regions, there were no evident barriers for dispersal between areas with different degrees of forest cover. In fact, with telemetry data, we were able to document the dispersal of individuals between habitats with different degrees of forest cover (L. Sönnichsen, unpublished data). The slight morphological, physiological and behavioral differences between roe deer inhabiting forests and fields that have been used to support the definition of ecotypes [16–18], are thus likely to reflect the phenotypic plasticity of the species in response to differences in forest cover, predation pressure and food availability [9].

Supporting Information

Figure S1 Phylogenetic relationships between the mtDNA haplotypes found in this study (H1–H13; marked with red points) and other published mtDNA control region sequences (N = 243) of European and Siberian roe deer [5–14] with length of 610 bp. Numbers at nodes show support (≥50%) from 10.000 bootstrap replicates. European roe deer clades are defined according to Randi et al. [5]. Each clade is marked with a different color.

Figure S2 Phylogenetic relationship between the mtDNA haplotypes found in this study (H1–H13; marked

with red points) and other published mtDNA control region sequences (N = 215) of European and Siberian roe deer [5–13] with length of 510 bp. Numbers at nodes show support (≥50%) from 10.000 bootstrap replicates. European roe deer clades are defined according to Randi *et al.* [5]. Each clade is marked with a different color. Due to the shortening of our sequences, there are no differences between haplotypes H2 and H13.

Table S1 Microsatellite diversity of roe deer at three sampling areas in northeastern Poland. N = number of samples, A = number of alleles, He = expected heterozygosity, Ho = observed heterozygosity.

Acknowledgments

We thank the personnel of the State Forests, hunters and colleagues from the MRI PAS involved in the collection of roe deer samples, M. Górny for help with GIS analyses, H. Zalewska and B. Marczuk for help in the lab, and the group of BIOCONSUS fellows for the regular discussions. We are grateful to the editor and the three anonymous reviewers for their constructive criticism that allowed us to improve our manuscript.

Author Contributions

Conceived and designed the experiments: LS BJ. Performed the experiments: JOM KP. Analyzed the data: JOM KP LS TB MN BJ. Contributed reagents/materials/analysis tools: TB KP LS. Contributed to the writing of the manuscript: JOM KP MN BJ. Critical revision of the manuscript: JOM KP LS TB MN BJ.

References

1. Hewitt G (2000) The genetic legacy of the Quaternary ice ages. Nature 405: 907–913.
2. Ledig FT (1992) Human impacts on genetic diversity in forest ecosystems. Oikos 63: 87.
3. Harris R, Wall W, Allendorf F (2002) Genetic consequences of hunting: what do we know and what should we do? Wildl Soc Bull 30: 634–643.
4. Balkenhol N, Waits LP (2009) Molecular road ecology: exploring the potential of genetics for investigating transportation impacts on wildlife. Mol Ecol 18: 4151–4164.
5. Linnell JDC, Zachos FE (2011) Status and distribution patterns of European ungulates: genetics, population history and conservation. Putman R, Apollonio M, Andersen R (eds.): Ungulate management in Europe: problems and practices. Cambridge: Cambridge University Press.
6. Niedziałkowska M, Jędrzejewska B, Wójcik JM, Goodman SJ (2012) Genetic structure of red deer population in northeastern Poland in relation to the history of human interventions. J Wildl Manage 76: 1264–1276.
7. Randi E, Alves P, Carranza J, Milosevic-Zlatanovic S, Sfougaris A, et al. (2004) Phylogeography of roe deer (*Capreolus capreolus*) populations: the effects of historical genetic subdivisions and recent nonequilibrium dynamics. Mol Ecol 13: 3071–3083.
8. Danilkin AA (1996) Behavioural Ecology of Siberian and European Roe Deer. London: Chapman & Hall. 277 p.
9. Andersen R, Duncan P, Linnell JDC (1998) The European roe deer: the biology of success. Oslo: Scandinavian University Press. 384 p.
10. Randi E (2005) Management of wild ungulate populations in Italy: Captive-breeding, hybridisation and genetic consequences of translocations. Vet Res Commun 29: 71–75.
11. Whitehead G (1993) The Whitehead encyclopedia of deer. Shrewsburg, England: Swan Hill Press. 597 p.
12. Vernesi C, Pecchioli E, Caramelli D, Tiedemann R, Randi E, et al. (2002) The genetic structure of natural and reintroduced roe deer (*Capreolus capreolus*) populations in the Alps and central Italy, with reference to the mitochondrial DNA phylogeography of Europe. Mol Ecol 11: 1285–1297.
13. Lorenzini R, Garofalo L, Qin X, Voloshina I, Lovari S (2014) Global phylogeography of the genus *Capreolus* (Artiodactyla: Cervidae), a Palaearctic meso-mammal. Zool J Linn Soc 170: 209–221.
14. Matosiuk M, Borkowska A, Świslocka M, Mirski P, Borowski Z, et al. (2014) Unexpected population genetic structure of European roe deer in Poland: an invasion of the mtDNA genome from Siberian roe deer. Mol Ecol 23: 2559–2572.
15. Hewison AJM, Vincent JP, Joachim J, Angibault JM, Cargnelutti B, et al. (2001) The effects of woodland fragmentation and human activity on roe deer distribution in agricultural landscapes. Can J Zool 79: 679–689.
16. Kałuziński J (1974) The occurrence and distribution of field ecotype of roe deer in Poland. Acta Theriol 19: 291–300.
17. Fruziński B, Kałuziński J, Baksalary J (1982) Weight and body measurements of forest and field roe deer. Acta Theriol 27: 479–488.
18. Majewska B, Pielowski Z, Łabudzki L (1982) The level of some energy-metabolism indexes in forest and field populations of roe deer. Acta Theriol 27: 471–477.
19. Hofmann R, Saber A, Pielowski Z, Fruziński B (1988) Comparative morphological investigations of forest and field ecotypes of roe deer in Poland. Acta Theriol 33: 103–114.
20. Kurt F, Hartl G, Volk F (1993) Breeding strategies and genetic variation in European roe deer *Capreolus capreolus* populations. Acta Theriol 38: 187–194.
21. Coulon A, Guillot G, Cosson J, Angibault J, Aulagnier S, et al. (2006) Genetic structure is influenced by landscape features: empirical evidence from a roe deer population. Mol Ecol 15: 1669–1679.
22. Hepenstrick D, Thiel D, Holderegger R, Gugerli F (2012) Genetic discontinuities in roe deer (*Capreolus capreolus*) coincide with fenced transportation infrastructure. Basic Appl Ecol 13: 631–638.
23. Karcov G (1903) Belovezhskaya Pushcha. Ee istoricheskii ocherk, sovremennoe okhotniche khozaistvo i vysochaishe okhoty v Pushche. Sankt Petersburg, Russia: A. Marks.
24. Sönnichsen L (2012) Under pressure - time and space management of European roe deer [Doctoral Thesis]. Berlin, Germany: Freie Universität Berlin.
25. Randi E, Pierpaoli M, Danilkin AA (1998) Mitochondrial DNA polymorphism in populations of Siberian and European roe deer (*Capreolus pygargus* and *C. capreolus*). Heredity 80: 429–437.
26. Hall T (1999) BioEdit: a user-friendly biological sequence alignment editor and analysis program for Windows 95/98/NT. Nucleic Acids Symp Ser 41: 95–98.
27. Galpern P, Manseau M, Hettinga P, Smith K, Wilson P (2012) Allelematch: an R package for identifying unique multilocus genotypes where genotyping error and missing data may be present. Mol Ecol Res 12: 771–778.
28. R Core Team (2012) R: a language and environment for statistical computing. Vienna, Austria: R Foundation for Statistical Computing.
29. Rousset F (2008) GENEPOP'007: a complete re-implementation of the GENEPOP software for Windows and Linux. Mol Ecol Res 8: 103–106.
30. Kalinowski ST (2005) hp-rare 1.0: a computer program for performing rarefaction on measures of allelic richness. Mol Ecol Notes 5: 187–189.
31. Van Oosterhout C, Hutchinson WF, Wills DPM, Shipley P (2004) MICRO-CHECKER: software for identifying and correcting genotyping errors in microsatellite data. Mol Ecol Notes 4: 535–538.
32. Benjamini Y, Hochberg Y (1995) Controlling the false discovery rate: a practical and powerful approach. J R Stat Soc Ser B Methodol 57: 289–300.
33. Excoffier L, Lischer HEL (2010) Arlequin suite ver 3.5: a new series of programs to perform population genetics analyses under Linux and Windows. Mol Ecol Res 10: 564–567.
34. Pritchard JK, Stephens M, Donnelly P (2000) Inference of population structure using multilocus genotype data. Genetics 155: 945–959.
35. Hubisz MJ, Falush D, Stephens M, Pritchard JK (2009) Inferring weak population structure with the assistance of sample group information. Mol Ecol Res 9: 1322–1332.
36. Earl DA, vonHoldt BM (2012) STRUCTURE HARVESTER: a website and program for visualizing STRUCTURE output and implementing the Evanno method. Conserv Genet Res 4: 359–361.
37. Evanno G, Regnaut S, Goudet J (2005) Detecting the number of clusters of individuals using the software STRUCTURE: a simulation study. Mol Ecol 14: 2611–2620.
38. Guillot G (2008) Inference of structure in subdivided populations at low levels of genetic differentiation - the correlated allele frequencies model revisited. Bioinformatics 24: 2222–2228.
39. Jombart T (2008) adegenet: a R package for the multivariate analysis of genetic markers. Bioinformatics 24: 1403–1405.
40. Jombart T, Devillard S, Dufour A-B, Pontier D (2008) Revealing cryptic spatial patterns in genetic variability by a new multivariate method. Heredity 101: 92–103.
41. Upton G, Fingleton B (1985) Spatial data analysis by example: point pattern and quantitative data. New York, USA: Wiley.
42. Jombart T, Devillard S, Balloux F (2010) Discriminant analysis of principal components: a new method for the analysis of genetically structured populations. BMC Genetics 11: 94.
43. Jost L (2008) G(ST) and its relatives do not measure differentiation. Mol Ecol 17: 4015–4026.

44. Keenan K, McGinnity P, Cross TF, Crozier WW, Prodöhl PA (2013) diveRsity: An R package for the estimation and exploration of population genetics parameters and their associated errors. Methods Ecol Evol 4: 782–788.

45. Zvychainaya EY, Danilkin AA, Kholodova MV, Sipko TP, Berber AP (2011) Analysis of the variability of the control region and cytochrome b gene of mtDNA of *Capreolus pygargus* Pall. Biol Bull 38: 434–439.

46. Zvychaynaya EY, Kiryakulov VM, Kholodova MV, Danilkin AA (2011) Roe deer (*Capreolus*) from Moscow area: analysis of mitochondrial control region polymorphism. Vestn Okhotovedeniya 8: 168–172.

47. Vorobieva NV, Sherbakov DY, Druzhkova AS, Stanyon R, Tsybankov AA, et al. (2011) Genotyping of *Capreolus pygargus* fossil DNA from Denisova cave reveals phylogenetic relationships between ancient and modern populations. PLoS One 6(8): e24045.

48. Xiao CT, Zhang MH, Fu Y, Koh HS (2007) Mitochondrial DNA distinction of northeastern China roe deer, Siberian roe deer, and European roe deer, to clarify the taxonomic status of northeastern China roe deer. Biochem Genet 45: 93–102.

49. Koh HS, Bayarlkhagva D, Jang KH, Han ED, Jo JE, et al. (2013) Genetic divergence of the Siberian roe deer from Korean Jeju Island (*Capreolus pygargus ochraceus*), reexamined from nuclear IRBP and mitochondrial cytochrome b and control region sequences of *C. pygargus*. J Biol Res 19: 46–55.

50. Baker KH, Hoelzel AR (2012) Evolution of population genetic structure of the British roe deer by natural and anthropogenic processes (*Capreolus capreolus*). Ecol Evol 3: 89–102.

51. Gentile G, Vernesi C, Vicario S, Pecchioli E, Caccone A, et al. (2009) Mitochondrial DNA variation in roe deer (*Capreolus capreolus*) from Italy: Evidence of admixture in one of the last *C. c. italicus* pure populations from central-southern Italy. Ital J Zool 76: 16–27.

52. Tamura K, Peterson D, Peterson N, Stecher G, Nei M, et al. (2011) MEGA5: molecular evolutionary genetics analysis using maximum likelihood, evolutionary distance, and maximum parsimony methods. Mol Biol Evol 28: 2731–2739.

53. Dupanloup I, Schneider S, Excoffier L (2002) A simulated annealing approach to define the genetic structure of populations. Mol Ecol 11: 2571–2581.

54. Lorenzini R, Lovari S (2006) Genetic diversity and phylogeography of the European roe deer: the refuge area theory revisited. Biol J Linn Soc 88: 85–100.

55. Zachos F, Hmwe S, Hartl G (2006) Biochemical and DNA markers yield strikingly different results regarding variability and differentiation of roe deer (*Capreolus capreolus*, Artiodactyla: Cervidae) populations from northern Germany. J Zool Syst Evol Res 44: 167–174.

56. Coulon A, Cosson J, Morellet N, Angibault J, Cargnelutti B, et al. (2006) Dispersal is not female biased in a resource-defence mating ungulate, the European roe deer. Proc R Soc B-Biol Sci 273: 341–348.

57. Jędrzejewska B, Jędrzejewski W, Bunevich AN, Miłkowski L, Krasiński ZA (1997) Factors shaping population densities and increase rates of ungulates in Białowieża Primeval Forest (Poland and Belarus) in the 19th and 20th centuries. Acta Theriol 42: 399–451.

58. Danilkin AA (1995) *Capreolus pygargus*. Mamm Species 512: 1–7.

59. Wiehler J, Tiedemann R (1998) Phylogeography of the European roe deer *Capreolus capreolus* as revealed by sequence analysis of the mitochondrial control region. Acta Theriol Suppl. 5: 187–197.

60. Lane N (2009) Biodiversity: On the origin of bar codes. Nature 462: 272–274.

61. Toews DPL, Brelsford A (2012) The biogeography of mitochondrial and nuclear discordance in animals. Mol Ecol 21: 3907–3930.

62. Okarma H, Jędrzejewska B, Jędrzejewski W, Krasiński ZA, Miłkowski L (1995) The roles of predation, snow cover, acorn crop, and man-related factors on ungulate mortality in Białowieża Primeval Forest, Poland. Acta Theriol 40: 197–217.

63. Nies G, Zachos F, Hartl G (2005) The impact of female philopatry on population differentiation in the European roe deer (*Capreolus capreolus*) as revealed by mitochondrial, DNA and allozymes. Mamm Biol 70: 130–134.

64. Bonnot N, Gaillard J-M, Coulon A, Galan M, Cosson J-F, et al. (2010) No difference between the sexes in fine-scale spatial genetic structure of roe deer. PLoS One 5(12): e14436.

6

Genetic Diversity of and Differentiation among Five Populations of Blunt Snout Bream (*Megalobrama amblycephala*) Revealed by SRAP Markers: Implications for Conservation and Management

Wei Ji[1,2✌], **Gui-Rong Zhang**[1,2✌], **Wei Ran**[1], **Jonathan P. A. Gardner**[1,2,3], **Kai-Jian Wei**[1,2]*, **Wei-Min Wang**[1], **Gui-Wei Zou**[4]

1 Key Laboratory of Freshwater Animal Breeding, Ministry of Agriculture, College of Fisheries, Huazhong Agricultural University, Wuhan, P. R. China, **2** Freshwater Aquaculture Collaborative Innovation Centre of Hubei Province, Wuhan, P. R. China, **3** School of Biological Sciences, Victoria University of Wellington, Wellington, New Zealand, **4** Key Laboratory of Freshwater Biodiversity Conservation, Ministry of Agriculture, Yangtze River Fisheries Research Institute, Chinese Academy of Fishery Sciences, Wuhan, P. R. China

Abstract

The blunt snout bream (*Megalobrama amblycephala*) is an important freshwater aquaculture fish throughout China. Because of widespread introductions of this species to many regions, the genetic diversity of wild and natural populations is now threatened. In the present study, SRAP (sequence-related amplified polymorphism) markers were used to assess genetic diversity of blunt snout bream. Three natural populations (Liangzi Lake, Poyang Lake and Yuni Lake, one cultured population (Nanxian) and one genetic strain ('Pujiang No. 1') of blunt snout bream were screened with 88 SRAP primer combinations, of which 13 primer pairs produced stable and reproducible amplification patterns. In total, 172 bands were produced, of which 132 bands were polymorphic. Nei's gene diversity (*h*) and Shannon's information index (*I*) values provided evidence of differences in genetic diversity among the five populations (Poyang Lake>Liangzi Lake>Nanxian> 'Pujiang No. 1'>Yuni Lake). Based on cluster analysis conducted on genetic distance values, the five blunt snout bream populations were divided into three groups, Poyang Lake and Liangzi Lake (natural populations), Nanxian and 'Pujiang No. 1' (cultured population and genetically selected strain), and Yuni Lake (natural population). Significant genetic differentiation was found among the five populations using analysis of molecular variance (AMOVA), with more genetic divergence existing among populations (55.49%), than within populations (44.51%). This molecular marker technique is a simple and efficient method to quantify genetic diversity within and among fish populations, and is employed here to help manage and conserve germplasm variability of blunt snout bream and to support the ongoing selective breeding programme for this fish.

Editor: Hanping Wang, The Ohio State University, United States of America

Funding: This work was supported by the Modern Agriculture Industry Technology System Construction Projects of China entitled "Staple Freshwater Fishes Industry Technology System" (Grant No. CARS-46-05), the National R&D Infrastructure and Facility Development Program of China (Grant No. 2006DKA30470-002-03), and the Major Science and Technology Program for Water Pollution Control and Treatment (Grant No. 2012ZX07202-004-02). The funders had no role in study design, data collection and analysis, decision to publish, or preparation of the manuscript.

Competing Interests: The authors have declared that no competing interests exist.

* Email: kjwei@mail.hzau.edu.cn

✌ These authors contributed equally to this work.

Introduction

Freshwater fishes are very diverse given the relatively small global size of the freshwater habitat, but are among the most endangered organisms globally [1]. Many species are now considered to be under threat in terms of reductions in their natural distributional ranges, and also reductions in numbers and sizes of populations as a consequence of factors such as pollution, replacement by invasive or deliberately introduced species, over-fishing, the aquarium pet trade, river flow modification and habitat change. A large body of literature points to the effects of such threats in terms of both localised and global extinctions of many freshwater fish species on all human-inhabited continents

[1–7]. It has been noted that freshwater habitats in general (i.e., as a broad ecosystem type) are particularly susceptible to climate change impacts, both direct and indirect, and that consideration of even the most optimistic climate change scenarios points to the likelihood of *ex-situ* management of many species for their survival [8]. At a lesser scale of threat, the deliberate human-mediated movement of fishes between previously isolated lakes or rivers as a fisheries management activity contributes to the eroding of genetic differences between intra-specific populations, and thereby threatens the genetic integrity and genetic adaptation of local fish populations [9,10]. As a consequence, there is now an increasing awareness of the need to balance the protection and genetic integrity of freshwater fish populations against the need to exploit many such populations as an important source of protein for human populations [11].

The magnitude of the threat faced by freshwater fishes of the world is highlighted by two review papers [1,3]. Both papers point out that freshwater fishes and their associated fisheries are important sources of employment, recreation and culture, and in developing regions they provide a significant contribution to the provision of local human food sources. Both papers also make the point that freshwater systems, because of their landscape positions, are subject to a number of externalities (threats not directly associated with or a function of the water body itself) that exacerbate problems of local fish endemism and non-substitut-ability. Cooke *et al.* note that many different groups and organizations have a role to play in conserving freshwater fishes and conclude by noting that failure to engage with the public will hinder conservation efforts and outcomes [1], while Dudgeon *et al.* had previously noted that threats to freshwater systems may constitute the greatest conservation challenge yet faced, and may require a new paradigm for biodiversity protection [3].

The Yangtze River is the world's third longest river system, with more than 3000 tributaries and 4000 lakes [2]. The system is notable for its speciose fish fauna and for high levels of endemism [2,12]. However, it is also notable for the modifications that it has undergone and for the threats now faced by its biota ([2,12] and references therein). The blunt snout bream (*Megalobrama amblycephala* Yih, 1955) (Teleostei: Cyprinidae) is an important endemic freshwater fish that has been widely favoured and is now cultured in China as a delicacy. It was originally distributed in the middle Yangtze River and a few accessory lakes, of which Yuni Lake, Liangzi Lake and Poyang Lake are the major sources [13]. Since having been recognized as a new species in the 1950s [14], this fish has become a principal species for freshwater aquaculture in China. This bream has been widely introduced all over the country because of its ease of culture, rapid growth rate, resistance to disease, high catchability and many other advantages [15]. In 2012, the total production of the bream reached 705,821 tonnes [16]. After domestication over a fifty year period, the aquaculture performance of many cultured populations has deteriorated as indicated by growth depression and the early onset of maturation. This deterioration is thought to be due to inbreeding and poor management of broodstocks [17]. Furthermore, as a consequence of environmental change and over-exploitation of resources, natural populations of this species have decreased substantially over the last few decades. As a consequence, the germplasm resources and gene pool diversity of this species in natural inland waters are now threatened by introductions of fish with unknown genetic histories, artificial propagation and poor stocking practices [13]. At the present time, because of these ongoing threats to the integrity of blunt snout bream populations, it is urgently required to clarify genetic diversity and population structure of this bream in order to effectively protect and utilize natural populations and their genetic resources.

Genetic diversity has been estimated in several natural and genetically selected populations of blunt snout bream using morphometrics and isozymes [13], RAPDs [18–20], mtDNA sequencing and RFLPs [21–23], and microsatellite markers [24–26]. For the first time however, we assess genetic diversity in natural populations of blunt snout bream and contrast this with genetic diversity estimates derived from a genetically selected strain and a cultured population. Sequence-related amplified polymorphism (SRAP) is a molecular marker system developed for selective amplification of open reading frames [27]. These polymorphisms result mainly from various promoters, introns and spacers among different species and individuals. SRAP is a

highly reproducible and highly informative technique for assessing genetic diversity in comparison with other PCR-based techniques [28,29]. SRAP markers have been successfully applied to the assessment of genetic diversity, strain identification and linkage map construction in a number of species, principally plants of commercial value [30–34]. More recently, SRAPs have also been applied to several aquatic animals [35–39] and have been shown to be highly informative and reliable markers for the evaluation of genetic diversity and population differentiation.

In this study, SRAP markers are used to investigate population genetic diversity and differentiation among three natural populations, a genetically selected strain and a cultured population of blunt snout bream in China. Our aim is to better understand the genetic diversity and integrity of natural populations based upon geography and the genetic affinities of selected strain and cultured populations based upon history. This work contributes to the critical need for germplasm conservation while at the same time supporting the ongoing selective breeding and improvement of blunt snout bream.

Materials and Methods

Ethics Statement

No specific permits were required for the field studies described here. The sampling lakes and ponds are not privately-owned or protected in any way, and the field sampling activities did not involve endangered or protected species.

Sample collection and DNA extraction

In total, 234 individuals from five blunt snout bream populations were included in this study. Samples were collected from three natural populations (Liangzi Lake, Poyang Lake and Yuni Lake), from one selected strain ('Pujiang No. 1' from Shanghai), and from one cultured population (Nanxian County in Hunan Province). Details of the sampled populations are provided in Table 1 (we use the term 'population' for ease of use, although we recognise that each group tested here may not be a biologically defined population) and the sampling sites are showed in Figure 1. As far as can be ascertained, the Liangzi Lake, Poyang Lake and Yuni Lake populations represent a good geographic spread of natural populations from the species' original range of distribution and have been least impacted by the translocation of fish (between these sites, from other regions, from selected strains, etc). Because of the long history of human-mediated movement of fish the identification of additional natural populations is now increasing difficult, so we have focussed on what we believe to be the best source of natural (wild) fish. For sampling, all individuals were captured from lakes or ponds with a haul seine and a piece of pectoral fin (~0.5 cm^2) was clipped from each individual without anesthesia. After sampling, the individuals of the selected and cultured populations were returned to their corresponding ponds, whereas the individuals of the three natural populations were transported to the Ezhou fish breeding base of Huazhong Agricultural University (HZAU) for further breeding programme. The clipped fin tissues were preserved in 95% ethanol. Total DNA was extracted from fin tissues using the phenol-chloroform method [40]. DNA concentration and quality were assessed by electro-phoresis on a 0.8% agarose gel. The animal research oversight committee (AROC) of HZAU had knowledge of the fish sampling plans prior to their approval of the present animal research protocol. This study was approved by the Institutional Animal Care and Use Committees (IACUC) of HZAU.

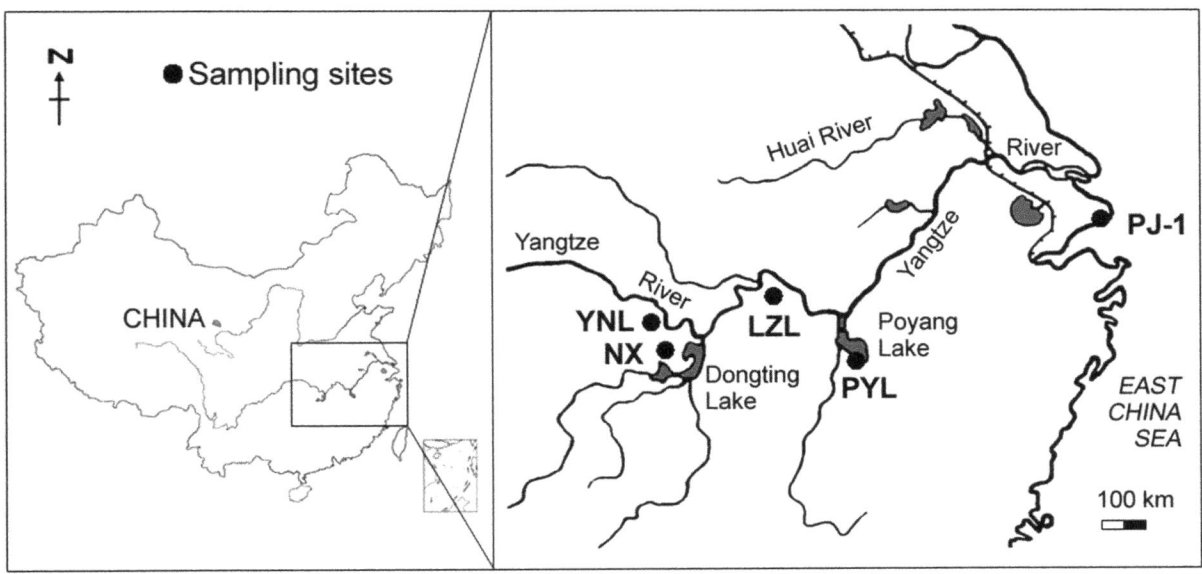

Figure 1. Sampling sites of *Megalobrama amblycephala* **in China.** Sample codes are given in Table 1.

SRAP reactions

A set of 19 SRAP primers (eight forward and eleven reverse, see Table 2) was obtained from Li and Quiros [27], giving a total of 88 primer pair combinations to be used to search for polymorphisms among all individuals. Thirteen polymorphic primer pairs were used for PCR amplification of fish from all five populations because, after preliminary testing, only these primers provided informative and repeatable results.

All SRAP PCR reactions were carried out in a final volume of 15 μL containing 1.5 μL 10×PCR buffer, 2.0 mM Mg²⁺, 0.25 mM dNTP, 0.6 U *Taq* DNA polymerase (Fermentas), 0.5 μM each of primers, and 20 ng template DNA. All PCRs were conducted in a PTC-200 Thermocycler under the following conditions: an initial denaturing step was performed at 94°C for 5 min, followed by five cycles of 1 min at 94°C, 1 min at 35°C, 1 min at 72°C, then 35 cycles at 94°C for 1 min, 50°C for 1 min, 72°C for 1 min, followed by a final extension step at 72°C for 10 min. PCR products were separated on a 6% non-denaturing

polyacrylamide gel via electrophoresis and silver staining was used to visualize the results.

Data analysis

Using a 100 bp DNA marker ladder as a size standard, SRAP size-dependent fragments were scored as absent (0) or present (1) to obtain a binary matrix. Only clear, strong and reproducible bands in the size range of 60 to 1124 bp were scored. POPGENE 1.32 was used to calculate number of polymorphic bands (N), percent of polymorphic bands (P), Nei's gene diversity (h), Shannon's information index (I), Nei's unbiased genetic distance (D) and genetic similarity (S) estimates [41]. Based on Nei's D, a NJ tree with bootstrap support values (10,000 replications) was constructed using POPTREE2 [42] to illustrate the relationship among populations. Genetic population structure was quantified by analysis of molecular variance (AMOVA) using ARLEQUIN 3.11 [43]. The pairwise fixation index (F_{ST}) was used to evaluate the level of genetic differentiation among populations. Sequential

Table 1. Sampling locations and sample sizes of natural, cultured and genetically selected populations of *Megalobrama amblycephala*.

Population	Code	Population type	Sampling location	Coordinates (lat., long.)	Sampling date	Sample size
Liangzi Lake	LZL	natural	Liangzi Lake, Hubei Province	30°16′N, 114°31′E	Jan 2009	47
Poyang Lake	PYL	natural	Poyang Lake, Jiangxi Province	28°36′N, 116°11′E	Jan 2009	48
Yuni Lake	YNL	natural	Yuni Lake, Hubei Province	29°50′N, 112°08′E	Jan 2009	48
'Pujiang No. 1'	PJ-1	genetically selected strain (F₇)	Songjiang fish breeding station, Shanghai	31°02′N, 121°05′E	July 2010	48
Nanxian	NX	cultured	Shuanghu fish farm, Nanxian county, Hunan Province	29°20′N, 112°18′E	July 2010	43

Table 2. Primer sequences of SRAP markers used in this study.

Forward Primer	Sequence (5'-3')	Reverse Primer	Sequence (5'-3')
Me1	TGAGTCCAAACCGGATA	Em1	GACTGCGTACGAATTAAT
Me2	TGAGTCCAAACCGGAGC	Em2	GACTGCGTACGAATTTGC
Me3	TGAGTCCAAACCGGAAT	Em3	GACTGCGTACGAATTGAC
Me4	TGAGTCCAAACCGGACC	Em4	GACTGCGTACGAATTTGA
Me5	TGAGTCCAAACCGGAAG	Em5	GACTGCGTACGAATTAAC
Me6	TGAGTCCAAACCGGTAA	Em6	GACTGCGTACGAATTGCA
Me7	TGAGTCCAAACCGGTCC	Em7	GACTGCGTACGAATTCAA
Me8	TGAGTCCAAACCGGTGC	Em8	GACTGCGTACGAATTCTC
		Em9	GACTGCGTACGAATTCGA
		Em10	GACTGCGTACGAATTCAG
		Em11	GACTGCGTACGAATTCCA

Bonferroni corrections [44] were used to correct for multiple testing.

To further evaluate the extent of genetic differentiation among the five populations we carried out a discriminant function analysis (DFA) to determine the population assignment success for each individual fish based on its genetic diversity. The SRAP markers had 140 variable bands, a number too large for the DFA to work with, so we summarised the genetic variation using principal components analysis (PCA). The first 40 eigen vectors, which explained 82% of the variation in the SRAP data set, were used in the DFA. The DFA was based on original group size (i.e., the number of fish per population that was actually sampled). The DFA and PCA were carried out using STATISTICA v7 software (StatSoft Ltd).

Results

SRAP analysis

Eighty-eight primer combinations were tested and of these, 13 primer pairs (Me1/Em3, Me1/Em4, Me1/Em6, Me1/Em7, Me4/Em1, Me4/Em2, Me4/Em8, Me5/Em1, Me5/Em2, Me5/Em4, Me5/Em7, Me8/Em8, and Me8/Em9) showed clear, reproducible and polymorphic banding patterns. The number of amplified bands for each primer pair ranged from 8 to 21 (average 13). In total, 172 bands were detected, of which 132 bands showed polymorphism, giving a polymorphic value of 76.7%.

Genetic diversity

For fish from the five populations (mean ± SD sample size of 46.8±2.17), the number of polymorphic bands (N) and the percentage of polymorphic bands (P) were highest in the PYL population ($N = 100$, $P = 58.14\%$), and lowest in the YNL population ($N = 50$, $P = 29.07\%$). Nei's gene diversity (h) across the five populations ranged from 0.090 (YNL) to 0.185 (PYL), and Shannon's information index (I) ranged from 0.137 (YNL) to 0.280 (PYL) (Table 3).

Genetic differentiation among populations

The SRAP markers revealed a high level of genetic differentiation among the five blunt snout bream populations. Pairwise F_{ST} values ranged from 0.351 between the geographically adjacent natural populations of Liangzi Lake (LZL) and Poyang Lake (PYL), to 0.685 between the Yuni Lake (YNL) population

and the genetically selected strain of 'Pujiang No. 1' (PJ-1). The differences in genotype frequencies between all pairs of populations were highly significant ($P<0.001$) (Table 4).

Analysis of molecular variance (AMOVA) of hierarchical gene diversity indicated that 44.51% of the genetic variation was explained by within-population variation, whereas 55.49% of the variation was explained by among-population variation, with a significant fixation index ($F_{ST} = 0.555$, $P<0.001$) (Table 5).

Nei's unbiased genetic distances (D) among all populations ranged from 0.087 (LZL-PYL) to 0.270 (YNL-NX), and Nei's unbiased genetic similarities (S) ranged from 0.763 (YNL-NX) to 0.917 (LZL-PYL) (Table 6).

The NJ dendrogram constructed using Nei's unbiased genetic distance (Figure 2) showed that the populations were grouped into three clusters. The LZL (Liangzi Lake) population clustered with its geographically close population of Poyang Lake (PYL). The genetically selected population of 'Pujiang No. 1' (PJ-1) and the cultured population of Nanxian (NX) clustered together but separated from LZL and PYL samples. The YNL (Yuni Lake) population formed a distinct branch, separated from all other populations. Bootstrap support values for these clusters were all very high.

Discriminant function analysis (DFA) of the first 40 eigen values from the PCA resulted in 100% correct assignment. That is, all fishes from all populations were correctly assigned to their populations of sampling. The DFA result is consistent with the F_{ST}, AMOVA and NJ analysis, all of which reveal a high level of genetic differentiation among fish from the five populations.

Discussion

China's freshwater fish diversity is high at 920 species in 302 genera [12]. Among the recorded species, most (73%) are Cypriniformes, and across the country's 9 freshwater regions, river discharge contributes most to explaining variation in species richness [12]. Within the Yangtze River basin itself, 361 species and subspecies have been recorded, of which 177 are endemic [2]. Twenty-five fish species have been identified as being endangered [45]. It has been suggested that hydrological alterations pose the greatest threat to the basin's fish diversity and that the most immediate restoration need is for reconnection between the Yangtze River and its lakes [2]. In concluding, Fu et al. noted that the cluster of lakes within the middle Yangtze River region requires protection for the preservation of migratory fishes and

Table 3. Genetic diversity results for five populations of *Megalobrama amblycephala* based on SRAP markers.

Parameter	Population				
	LZL	PYL	YNL	PJ-1	NX
Number of polymorphic bands (*N*)	88	100	50	56	64
Percent of polymorphic bands (*P*, %)	51.16	58.14	29.07	32.56	37.21
Nei's gene diversity (*h*) (mean ± SD)	0.152±0.183	0.185±0.195	0.090±0.166	0.107±0.176	0.126±0.183
Shannon's information index (*I*) (mean ± SD)	0.235±0.265	0.280±0.280	0.137±0.240	0.162±0.256	0.191±0.266

that there is a need to identify areas of high fish biodiversity to mitigate loss of fish biodiversity in the Yangtze River basin [2]. In our study, we focus on the genetic diversity of the blunt snout bream, a cyprinid, in the cluster of lakes in the middle Yangtze River region. The focus of our research is primarily around conservation, but we recognize that there may be tension between conservation planning and fishing/aquaculture activity, and as such our ultimate aim is to contribute new knowledge to both conservation and fisheries management.

Genetic diversity

All four indices (number of polymorphic bands *N*, percent of polymorphic bands *P*, Nei's gene diversity *h*, Shannon's information index *I*) showed the same order from largest to smallest for the five populations. Both *h* and *I* indicated that the highest genetic diversity level was observed in the PYL population, followed by the LZL population, and the lowest was in the YNL population, all three being natural populations of *Megalobrama amblycephala*. As suggested by Li *et al.* [13], the genetic diversity level of fish within Niushan Lake (a subsidiary lake of Liangzi Lake) is a little higher than that within Yuni Lake. Our SRAP-based results are however, not in agreement with previous estimates of genetic diversity based on RAPD analysis that had shown that the genetic similarity of blunt snout bream from Yuni Lake (0.962) is very similar to that of fish from Liangzi Lake (0.959) [18]. However, mtDNA control region sequence analysis by Li [21] revealed a low level of population diversity within Yuni Lake compared to Liangzi Lake and Poyang Lake, with the haplotype diversity and nucleotide diversity index values detected in fish from Yuni Lake being only half of those for fish from Liangzi and Poyang Lakes. These mtDNA-based findings are similar to the results reported by us based on SRAP markers which represent the most up to date estimates of genetic diversity. The highest SRAP-based variation was observed for fish from PYL, which is the lake least affected by human activity (e.g., transfer of fish). The values reported for fish

from this lake therefore give some indication of the natural level of genetic variability that may be expected to occur in wild fish, and also a measure of the impact (reduced genetic variability) associated with human-mediated activities such as breeding programmes, inbreeding, and transfer of fish from one lake to another.

For the cultured (Nanxian) and selectively bred ('Pujiang No. 1') populations, the levels of genetic diversity were much lower than within the native PYL and LZL populations. These results are in agreement with investigations on largemouth bass (*Micropterus salmoides*) and large yellow croaker (*Pseudosciaena crocea*) [46,47], and are consistent with the known impacts of hatchery-based breeding programmes in general ([9,10] and references therein). In summary, as a result of artificial selection, the chance of the effective number of individuals contributing to the population declines, with an associated decline in population genetic diversity and an increase in inbreeding depression.

An unexpected outcome of our research is that the genetic diversity level of fish from the cultivated and breeding populations was found to be greater than that of wild blunt snout bream from one native lake population (Yuni Lake, YNL). This surprising result may be due to the degradation of germplasm resources of YNL over a reasonably short period of time. Previously, Li *et al.* [13] and Zhang [18] reported that the genetic diversity of blunt snout bream from Yuni Lake was similar to that of fish from Liangzi Lake at the beginning of the early 21st century. Subsequently, Li [21] reported that the genetic diversity of fish from Yuni Lake was nearly 50% of the level of fish from Liangzi Lake, a sharp decline of genetic diversity of fish from Yuni Lake. Whilst different marker types (e.g., allozymes, mtDNA sequencing) have been used by the different studies, each study is internally consistent and points to a decrease in genetic diversity of fish from Yuni Lake over a short period of time. Our study has revealed that the genetic diversity of blunt snout bream from YNL is lower than that of a cultured population and a selectively bred population. We

Table 4. Pairwise fixation index values (F_{ST}) between pairs of blunt snout bream (*Megalobrama amblycephala*) populations.

Population	LZL	PYL	YNL	PJ-1	NX
LZL	——				
PYL	0.351***	——			
YNL	0.533***	0.573***	——		
PJ-1	0.573***	0.572***	0.685***	——	
NX	0.544***	0.526***	0.659***	0.434***	——

Note:
***Significant at $\alpha = 0.001$ after sequential Bonferroni correction for multiple testing.

Table 5. AMOVA analysis of genetic variation in five populations of blunt snout bream based on SRAP markers.

Source of variation	df	Sum of squares	Variance components	Percentage of variation (%)	F_{ST}-value
Among-populations	4	2376.04	12.48	55.49	0.555***
Within-populations	229	2293.57	10.02	44.51	
Total	233	4669.61	22.50		

Note:
***Significant at $\alpha = 0.001$ level.

suggest that the results obtained for fish from YNL are attributable to the rapid intensification of artificial culture, propagation, and over-fishing of blunt snout bream in Yuni Lake during the past 10 years, which have resulted in much reduced genetic diversity of bream germplasm at this location. Whilst further research is needed to determine if this is actually the case, our research highlights the potential negative impact of culture, selective breeding programmes and multiple transfer of fishes on native genetic diversity of this species of fish in lakes of the middle Yangtze River region.

Population differentiation

F_{ST} is one of the most widely used measures for evaluating the genetic differentiation between/among populations, which can provide important insights into the evolutionary processes that cause the genetic differentiation among populations [48]. Whilst the magnitude of F_{ST} values is to some extent species- or even group-specific, as a general rule F_{ST} values of 0–0.05 represent little differentiation, values of 0.05–0.25 indicate moderate differentiation, and values higher than 0.25 indicate very great differentiation among populations [49]. In the present study, all values of F_{ST} between the five populations were >0.25, indicative of a high level of genetic differentiation among these five populations. The highest F_{ST} value detected (0.685) was between PJ-1 and YNL, suggesting that pronounced genetic differentiation now exists between the selectively bred strain (PJ-1) and the natural YNL population. These findings were supported by our discriminant function analysis (DFA) which was able to assign each and every fish to its population of origin, are comparable to the results reported by [50], and indicate how rapidly genetic differentiation can be achieved between wild breeding and selectively bred fishes (the PJ-1 fish are from an F_7 strain). The five populations of blunt snout bream clustered into three groups, as revealed by NJ analysis. Two of the three wild populations, PYL and the geographically close population of LZL, clustered together, while the third wild population (YNL) was highly

differentiated from the other two wild populations (in the NJ analysis YNL is identified as a distinct branch). In addition, we tried to construct a NJ tree based on pairwise F_{ST} between populations at the same time. The NJ tree based on pairwise F_{ST} was consistent with the results above (data not shown). These results confirm earlier observations that obvious genetic divergence occurred between fish from YNL and LZL or PYL, and no divergence occurred between LZL and PYL [21]. Historically, PYL, LZL and YNL (that is, Poyang Lake, Liangzi Lake, and Yuni Lake, respectively) were linked to the Yangtze River. Yuni Lake was separated from the Yangtze River about forty years ago due to human activities [13], which isolated its fish and has prevented genetic exchange among populations that were previously connected. This isolation has resulted in a build-up of genetic differentiation between fish from Yuni Lake and the two other lakes (i.e., there has been a brief but substantial period of divergence between the fish populations). LZL and PYL were still connected via the Yangtze River and closer to each other geographically, so there was still gene flow between these two lakes, therefore a closer genetic relationship was found between the populations from these two lakes. Whilst fish from these two lakes exhibit the highest levels of pairwise similarities, they are still sufficiently genetically differentiated to be distinct. These findings support the contention of Fu et al. [2] that lakes within the central Yangtze River region are in need of further protection and that connections between them and the river itself should be re-established. Finally, the genetically selected population of PJ-1 and the cultured population of NX clustered together to form a third, separate group that lies between the two previously described wild lake groups. The PJ-1 fish were produced through long-term artificial selective breeding since 1986, based on original wild population of blunt snout bream from Yuni Lake [26]. This selectively-bred line now has a growth rate that is increased by 30% compared to wild fish [17,51]. The distinct genetic differences among the three groups as revealed by NJ analysis indicate that genetic divergence has occurred between the selective

Table 6. Nei's unbiased genetic distance (below diagonal) and genetic similarity (above diagonal) among five populations of blunt snout bream.

Population	LZL	PYL	YNL	PJ-1	NX
LZL	——	0.917	0.831	0.837	0.844
PYL	0.087	——	0.802	0.805	0.815
YNL	0.186	0.221	——	0.776	0.763
PJ-1	0.178	0.217	0.254	——	0.916
NX	0.169	0.204	0.270	0.088	——

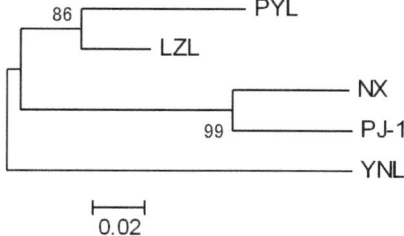

Figure 2. NJ dendrogram of relationships among blunt snout bream populations based on Nei's unbiased genetic distance. Percent bootstrap support values for 10,000 replications are indicated at nodes. See Table 1 for population abbreviation definitions.

breeding population (PJ-1), the cultured population (NX) and the wild populations (LZL, PYL and YNL). Similar results have also been reported for grass carp and mandarin fish [35,52]. This finding illustrates that restocking of lakes by fish from selectively bred programmes should be carried out with great care and a

References

complete understanding of the genetic diversity that may exist between fish strain and recipient lake, if extensive interbreeding between donor strain and wild lake fish is expected.

In conclusion, the SRAP system has been successfully employed to assess population genetic diversity among domesticated and wild genotypes of blunt snout bream from China. Analysis revealed that cultivated, genetically selected and wild bream germplasm are highly differentiated and that wild lake populations have high levels of genetic diversity. Comparison of the genetic diversity of wild fish to that of cultured and selectively bred fish illustrates the need for greater care managing this important resource so that the aquaculture programme to support this species is not disadvantaged in the future and so that wild populations of fish are protected for both conservation and fisheries purposes.

Author Contributions

Conceived and designed the experiments: KJW GRZ WJ. Performed the experiments: GRZ WJ WR. Analyzed the data: KJW WR JPAG. Contributed reagents/materials/analysis tools: WMW GWZ. Wrote the paper: WJ GRZ KJW JPAG.

1. Cooke SJ, Lapointe NWR, Martins EG, Thiem JD, Raby GD, et al. (2013) Failure to engage the public in issues related to inland fishes and fisheries: strategies for building public and political will to promote meaningful conservation. J Fish Biol 83: 997–1018.
2. Fu C, Wu J, Chen J, Wu Q, Lei G (2003) Freshwater fish biodiversity in the Yangtze River basin of China: patterns, threats and conservation. Biodivers and Conserv 12: 1649–1685.
3. Dudgeon D, Arthington AH, Gessner MO, Kawabata Z-I, Knowler DJ, et al. (2006) Freshwater biodiversity: importance, threats, status and conservation challenges. Biol Rev 81: 163–182.
4. Freitas CEC, Siqueira-Souza FK, Florentino AC, Hurd LE (2013) The importance of spatial scales to analysis of fish diversity in Amazonian floodplain lakes and implications for conservation. Ecol Freshw Fish DOI: 10.1111/eff.12099.
5. Lintermans M (2013) Recovering Australia's threatened freshwater fish. Mar Freshw Res, Special Issue 64 (9): 1–133.
6. Reid GM, MacBeath TC, Csatádi K (2013) Global challenges in freshwater-fish conservation related to public aquariums and the aquarium industry. International Zoo Yearbook 47: 6–45.
7. Reis RE (2013) Conserving the freshwater fishes of South America. International Zoo Yearbook 47: 65–70.
8. Pearce-Kelly P, Khela S, Ferri C, Field D (2013) Climate-change impact considerations for freshwater-fish conservation, with special reference to the aquarium and zoo community. International Zoo Yearbook 47: 81–92.
9. Ariki H, Schmid C (2010) Is hatchery stocking a help or harm? Evidence, limitations and future directions in ecological and genetic surveys. Aquaculture 308: S2–S11.
10. Hold N, Murray LG, Kaiser MJ, Hinz H, Beaumont AR, et al. (2013) Potential effects of stock enhancement with hatchery-reared seed on genetic diversity and effective population size. Can J Fish Aquat Sci 70: 330–338.
11. Bondad-Reantaso MG (2007) Assessment of freshwater fish seed resources for sustainable aquaculture. FAO Fisheries Technical Paper 501: 1–628.
12. Kang B, Deng J, Wu Y, Chen L, Zhang J, et al. (2013) Mapping China's freshwater fishes: diversity and biogeography. Fish and Fisheries DOI: 10.1111/faf.12011.
13. Li SF, Cai WQ, Zhou BY (1991) Morphology and biochemical genetic variations among populations of blunt snout bream (*Megalobrama amblycephala*). J Fish China 15: 204–211.
14. Yih PL (1955) Notes on *Megalobrama amblycephala*, sp. nov., a distinct species from *M. terminalis* (Richardson). Acta Hydrobiol Sin 2: 115–122.
15. Ko HW (1975) An excellent fresh-water food fish, *Megalobrama amblycephala*, and its propagating and culturing. Acta Hydrobiol Sin 5: 293–312.
16. Fisheries Bureau of the Agriculture Ministry of China (2013) China fishery statistical yearbook. Beijing: Chinese Agricultural Press. 31 p.
17. Li SF, Cai WQ (2003) Genetic improvement of the herbivorous blunt snout bream (*Megalobrama amblycephala*). Naga, WorldFish Center Quarterly 26(1): 20–23.
18. Zhang DC (2001) Study on genetic diversity of bluntnose black bream from Yunihu and Liangzihu lakes. J Chin Three Gorges Univ, Nat Sci 23: 282–284.
19. Zou SM, Li SF, Cai WQ (2005) SCAR transformation of a RAPD marker in blunt snout bream "Pujiang No. 1". J Fish China 29: 296–299.

20. Li SF, Yang HY, Zou SM (2005) Effects on genetic structure from quick inbreeding and estimation of inbreeding response in cultured populations of *Megalobrama amblycephala*. J Fish China 29: 161–165.
21. Li HH (2008) mtDNA sequence variation and genetic structure of the *Megalobrama amblycephala* from Yuni Lake, Liangzi Lake and Poyang Lake. Freshw Fisheries 38(4): 63–65.
22. Tang SJ, Li SF, Cai WQ (2008) Analysis of mitochondrial DNA in blunt snout bream (*Megalobrama amblycephala*) populations with different ploidy levels. J Fish Sci China 15: 222–229.
23. Bian CY, Dong S, Tan SZ (2007) PCR-RFLP analysis of mtDNA D-loop region of three populations of bluntnose blackbream *Megalobrama amblycephala*. J Dalian Fish Univ 22: 175–179.
24. Li SW, Chang YM, Liang LQ, Sun XW (2006) Rapid isolation of microsatellites from genome of Bluntnose black bream (*Megalobrama amblycephala*). J Fish Sci China 13: 187–192.
25. Tang SJ, Li SF (2007) Preliminary analysis of genetic variation of blunt snout bream (*Megalobrama amblycephala*) populations with different ploidy levels. J Shanghai Fish Univ 16: 97–102.
26. Tang SJ, Li SF, Cai WQ (2009) Development of microsatellite markers for blunt snout bream *Megalobrama amblycephala* using 5′-anchored PCR. Mol Ecol Resour 9: 971–974.
27. Li G, Quiros CF (2001) Sequence-related amplified polymorphism (SRAP), a new marker system based on a simple PCR reaction: its application to mapping and gene tagging in *Brassica*. Theor Appl Genet, 103: 455–461.
28. Budak H, Shearman RC, Parmaksiz I, Dweikat I (2004a) Comparative analysis of seeded and vegetative biotypes buffalograsses based on phylogenetic relationship using ISSRs, SSRs, RAPDs and SRAPs. Theor Appl Genet 109: 280–288.
29. Jones N, Ougham H, Thomas H, Pašakinskienė I (2009) Markers and mapping revisited: finding your gene. New Phytol 183: 935–966.
30. Ferriol M, Picó B, Nuez F (2003) Genetic diversity of a germplasm collection of *Cucurbita pepo* using SRAP and AFLP markers. Theor Appl Genet 107: 271–282.
31. Budak HR, Shearman C, Parmaksiz I, Gaussoin RE, Riorsdan TP, et al. (2004b) Molecular characterization of buffalograss germplasm using sequence-related amplified polymorphism markers. Theor Appl Genet 108: 328–334.
32. Sun SJ, Gao W, Lin SQ, Zhu J, Xie BG, et al. (2006) Analysis of genetic diversity in *Ganoderma* population with a novel molecular marker SRAP. Appl Microbiol Biotechnol 72: 537–543.
33. Qiao LX, Liu HY, Guo BT, Weng ML, Dai JX, et al. (2007) Molecular identification of 16 *Porphyra* lines using sequence-related amplified polymorphism markers. Aquat Bot 87: 203–208.
34. Zhang F, Chen SM, Chen FD, Fang WM, Chen Y, et al. (2011) SRAP-based mapping and QTL detection for inflorescence-related traits in chrysanthemum (*Dendranthema morifolium*). Mol Breed 27: 11–23.
35. Zhang ZW, Han YP, Zhong XM, Zhang ZY, Cao ZM, et al. (2007) Genetic structure analyses of grass carp populations between wild and cultured ones. J Fish Sci China 14: 720–725.
36. Xin WT, Sun ZW, Yin HB, Sun Y (2009) Identification of sex-associated SRAP marker in *Pelteobagrus fulvidraco*. J Northeast Forestry Univ 37(5): 112–113.
37. Zhang HY, He MX (2009) Segregation pattern of SRAP maker in F1 generation of *Pinctada martensii* family. Mar Sci Bull 28(2): 50–56.

38. Ding WD, Cao ZM, Cao LP (2010) Molecular analysis of grass carp (*Ctenopharyngodon idella*) by SRAP and SCAR molecular markers. Aquacult Int 18: 575–587.

39. Hu ZH, Xu JZ, Wang YB, Chai XJ (2010) SRAP analysis of genetic variation between female and male swimming crab (*Portunus trituberculatus*). J Shanghai Ocean Univ 19: 734–738.

40. Sambrook J, Fritsch EF, Maniatis T (1989) Molecular cloning: a laboratory manual, 2nd edn. New York: Cold Spring Harbor Laboratory Press.

41. Yeh FC, Yang RC, Boyle T (1999) POPGENE version 1.32: microsoft window-based freeware for population genetic analysis. The Centre for International Forestry Research, University of Alberta, Alberta, Canada. Available: http://www.ualberta.ca/~fyeh/.

42. Takezaki N, Nei M, Tamura K (2010) POPTREE2: software for constructing population trees from allele frequency data and computing other population statistics with windows interface. Mol Biol Evol 27: 747–752.

43. Excoffier L, Laval G, Schneider S (2005) Arlequin ver. 3.0: an integrated software package for population genetics data analysis. Evol Bioinform Online 1: 47–50.

44. Rice WR (1989) Analyzing tables of statistical tests. Evolution 43: 223–225.

45. Yue P, Chen Y (1998) Pisces. In: Wang S editor. China red data book of endangered animals. Beijing: Science Press.

46. Liang SX, Sun XW, Bai JJ, Gao JS (2008) Genetic analysis for cultured largemouth bass (*Micropterus salmoides*) in China with microsatellites. Acta Hydrobiol Sin 32: 694–700.

47. Wang J, Su YQ, Quan CG, Ding SX, Zhang W (2001) Genetic diversity of the wild and reared *Pseudosciaena crocea*. Chin J Oceanol Limnol 19: 152–156.

48. Holsinger KE, Weir BS (2009) Genetics in geographically structured populations: defining, estimating and interpreting F_{ST}. Nat Rev Genet 10(9): 639–650.

49. Wright S (1978) Evolution and the genetics of populations, Vol. 4: variability within and among natural populations. Chicago: The University of Chicago Press.

50. Zhao Y, Li SF, Tang SJ (2009) Genetic variations among late selected strains and wild populations of blunt snout bream (*Megalobrama amblycephala*) by ISSR analysis. J Fish China 33: 893–900.

51. Li SF, Cai WQ (2000) Two-way selective response of *Megalobrama amblycephala*. J Fish China 24: 201–205.

52. Fang ZQ, Chen J, Zheng WB, Wu YY, Xiao Z (2005) RAPD analysis of wild population and cultivated population in *Siniperca chuatsi* Basilewsky. J Dalian Fish Univ 20: 16–19.

7

Investigating Population Genetic Structure in a Highly Mobile Marine Organism: The Minke Whale *Balaenoptera acutorostrata acutorostrata* in the North East Atlantic

María Quintela[1,2], Hans J. Skaug[1,3], Nils Øien[4], Tore Haug[5], Bjørghild B. Seliussen[1], Hiroko K. Solvang[4], Christophe Pampoulie[6], Naohisa Kanda[7], Luis A. Pastene[7], Kevin A. Glover[1,8]*

1 Dept. of Population Genetics, Institute of Marine Research, Bergen, Norway, 2 BIOCOST Research Group, Dept. of Animal Biology, Plant Biology and Ecology, University of A Coruña, A Coruña, Spain, 3 Department of Mathematics, University of Bergen, Bergen, Norway, 4 Dept. of Marine Mammals, Institute of Marine Research, Bergen, Norway, 5 Dept. of Marine Mammals, Institute of Marine Research, Tromsø, Norway, 6 Marine Research Institute of Iceland, Reykjavik, Iceland, 7 Institute of Cetacean Research, Tokyo, Japan, 8 Department of Informatics, Faculty of Mathematics and Natural Sciences, University of Bergen, Bergen, Norway

Abstract

Inferring the number of genetically distinct populations and their levels of connectivity is of key importance for the sustainable management and conservation of wildlife. This represents an extra challenge in the marine environment where there are few physical barriers to gene-flow, and populations may overlap in time and space. Several studies have investigated the population genetic structure within the North Atlantic minke whale with contrasting results. In order to address this issue, we analyzed ten microsatellite loci and 331 bp of the mitochondrial D-loop on 2990 whales sampled in the North East Atlantic in the period 2004 and 2007–2011. The primary findings were: (1) No spatial or temporal genetic differentiations were observed for either class of genetic marker. (2) mtDNA identified three distinct mitochondrial lineages without any underlying geographical pattern. (3) Nuclear markers showed evidence of a single panmictic population in the NE Atlantic according to STRUCTURE's highest average likelihood found at K = 1. (4) When K = 2 was accepted, based on the Evanno's test, whales were divided into two more or less equally sized groups that showed significant genetic differentiation between them but without any sign of underlying geographic pattern. However, mtDNA for these individuals did not corroborate the differentiation. (5) In order to further evaluate the potential for cryptic structuring, a set of 100 *in silico* generated panmictic populations was examined using the same procedures as above showing genetic differentiation between two artificially divided groups, similar to the aforementioned observations. This demonstrates that clustering methods may spuriously reveal cryptic genetic structure. Based upon these data, we find no evidence to support the existence of spatial or cryptic population genetic structure of minke whales within the NE Atlantic. However, in order to conclusively evaluate population structure within this highly mobile species, more markers will be required.

Editor: Valerio Ketmaier, Institute of Biochemistry and Biology, Germany

Funding: Funding was provided by Norwegian Ministry of Fisheries and Coastal Affairs. The funders had no role in study design, data collection and analysis, decision to publish, or preparation of the manuscript.

Competing Interests: The authors have declared that no competing interests exist.

* Email: kevin.glover@imr.no

Introduction

Anthropogenic activities are key factors affecting wildlife populations, including altering population structure and distribution patterns [1–4]. Overexploitation by the whaling industry led to serious declines in many of the world's populations of whales. Currently, the IUCN conservation status "least concern" is applicable to only ~20% of whale species and only 8% show increasing population trends [5]. Marine mammals are highly mobile and may travel large distances (*e.g.* Stevick *et al.* [6]). A number of factors are thought to play a role in shaping the genetic structuring of cetacean populations such as the complex social structure (*e.g.* matrilineal based groups), the resource specialization and the great capacity for learning [7–9].

Minke whales, the second smallest baleen whales (about 10 m in length), are currently considered as two species [10]: the cosmopolitan common minke whale (*Balaenoptera acutorostrata*, Lacepede, 1804) and the Antarctic minke whale (*B. bonaerensis*, Burmeister, 1867), which is confined to the Southern hemisphere with the exceptions of rare inter-oceans migration events [11,12]. The former is further divided into three sub-species: the North Atlantic (*B.a. acutorostrata*), the North Pacific (*B.a. scammoni*), and the dwarf common minke whale (*B.a.* unnamed sub-species), which is thought to be restricted to the Southern hemisphere.

B.a. acutorostrata occurs in the entire North Atlantic during the Northern hemisphere summer months, limited in the northern range by the ice [13]. Although their winter distribution and thus the location of breeding areas is unknown, they probably fit the general ecological pattern of large cetaceans in the Northern hemisphere and migrate to lower latitudes, inhabiting temperate and tropical waters where pairing and birth of calves takes place

[14]. Calves are born between November and March after a gestation period of ten months [15,16]. In the western North Pacific, two *B.a. scammoni* breeding populations on either side of Japan are known to mix on feeding grounds in the Okhotsk Sea [17].

The minke whale is still harvested in significant numbers, and the management of *B.a. acutorostrata* in the North Atlantic is regulated under the Revised Management Procedure (RMP) developed by the Scientific Committee of the International Whaling Commission which also regularly reviews the species status through Implementation Reviews, the last one completed in 2009 [18]. The RMP implements the concept of Management Areas, which are currently outlined by taking into account different factors including distribution, life history parameters, local conservation threats such as bycatch, pollution, direct human exploitation and competition with fisheries, as well as differences in national legislation [19]. Five Management Areas have been established in the Eastern North Atlantic (*i.e.* "IWC Small Areas" [20]). The main concern of this outline is that, for minke whales, which is a migratory species, the small Management Areas would not reflect the real population boundaries but instead temporary mixed assemblages [21]. Therefore, careful assessment of the genetic diversity and genetic structure of the populations is essential to enable any successful conservation strategy.

Distinct breeding populations have not been identified for North Atlantic minke whales. Hence, the assessment of genetic structuring of minke whales within the North Atlantic has been based upon samples collected in the feeding grounds and stranded individuals. The question of population genetic structure within this species remains unresolved with partially conflicting results. There seems to be a general agreement regarding the absence of any clear spatial genetic structuring at mitochondrial level [22–25] although the possibility of co-existence of two breeding populations of common minke whales in the North Atlantic was proposed by Palsbøll [26] after finding two main groups of genotypes when analyzing restriction fragment length polymorphism on mtDNA. Likewise, whereas some studies based on nuclear markers [24,27,28] failed to reveal any genetic differentiation between individuals from the central and north-eastern parts of the North Atlantic; some other insights based on stable isotopes and heavy metals [29], levels of radioactive caesium ^{137}Cs [30], persistent organochlorines [30], microsatellites [23] or isozymes [31–33] suggest a geographic substructuring across different areas of the North Atlantic. Recently, Anderwald *et al.* [24], using a set of ten microsatellites, reported the possible existence of two cryptic stocks across the North Atlantic. All these uncertainties regarding population identification and assessment have further increased the scientific and political controversy that whaling already poses [34] and therefore, the need to elucidate population genetic structure within this species.

Norway conducts a commercial harvest of minke whale, *B.a. acutorostrata* in the Northeast Atlantic, and each year, approximately 500 whales are captured across five IWC Management Areas (Fig. 1). In order to enforce domestic regulation and compliance within this harvest, an individual-based DNA register (NMDR) has been maintained since 1996 [12]. This register contains genetic data of ten microsatellites and mtDNA for approximately 8000 whales harvested during the period 1996–2011. In addition, the register includes biometric information together with the geographic position of captures, what provides a powerful database to investigate the potential genetic structure of this species in the NE Atlantic.

The main objective of this study was to investigate spatial and temporal genetic structure of *B. a. acutorostrata* harvested in the NE Atlantic IWC Management Areas during the period 2007–2011. Secondly, we examined the possible existence of cryptic populations distributed across the North Atlantic as proposed by Anderwald *et al.* [24]. Therefore, the present study also included a set of samples from 2004 to match their sampling time frame and hence to enable comparisons. To achieve this second objective, conventional genetic analyses as well as simulation studies were conducted.

Material and Methods

Sampling, genotyping, and mtDNA sequencing

The time frame of the present study circumscribes to the period 2007–2011 when the genotyping of the individuals was performed by the Institute of Marine Research in Bergen, following very strict procedures to ensure the data quality [35,36]. In addition, samples collected in 2004 have been included for comparative purposes with former studies. Thus, the present data consists of genetic data from 2990 whales (2156 females and 834 males) that were harvested in the period from April to September. The distribution of individuals per sex, year and Management Areas is shown in Table 1. No animals were killed to provide samples for the present study as all the samples analyzed existed prior to it and were included in the NMDR [12] from which all the information used in the present work has been obtained; *i.e.* the biometrics, the position of the catches, the microsatellite genotypes and the mtDNA sequences of each of the 2990 individuals that were analyzed. The analytical approaches used for nuclear and mitochondrial markers at the NMDR were the following: DNA was extracted twice from muscle stored in ethanol using Qiagen DNeasy Blood & Tissue Kit following manufacturer's instructions and DNA concentration was measured on a Nanodrop. Ten microsatellite loci: EV1*Pm*, EV037*Mn* [37]; GATA028, GATA098, GATA417 [38]; GT023, GT211, GT310, GT509, GT575 [39] were amplified in three multiplex reactions based on a 2 minute hot start at 94°C, denaturizing for 20 seconds at 94°C, annealing for 45 seconds, elongation at 72°C for 1 minute and a final hold at 4°C. Multiplex specific conditions are detailed in Glover *et al.* [12]. Individuals were sexed using specific primers for the ZFY/ZFX gene [40].

The D-loop region of mtDNA was amplified by performing forward sequencing of one DNA isolate and reverse sequencing of the second one. The first PCR reaction yielded a 1066 bp amplification product, which was forward strand sequenced. The second PCR reaction entailed the amplification of a 331 bp product that was sequenced in the reverse direction. PCR conditions for the two directions were identical, thus containing 0.5 units Go Taq polymerase (Promega), 1.5 mM MgCl$_2$, 0.2 mM dNTP and 0.2 μM of each primer. Forward product used primers MT4(M13F) and MT3(M13Rev) modified from Árnason *et al.* [41], whereas primers for the reverse product were: BP15851(M13F), modified from Larsen *et al.* [42] and MN312(M13R), modified from Palsbøll *et al.* [43]. PCR conditions were: hot start at 94°C for 2 min, followed by 30 cycles of denaturizing at 94°C for 50 seconds annealing at 53°C for 50 seconds and elongation at 72°C for 3 min 30 seconds, and finally a 10 min elongation at 72°C and a 4°C hold.

Genetic structure according to microsatellites

Total number of alleles, allelic richness and the inbreeding coefficient F_{IS} per population and per year were calculated with MSA [44], whereas observed (H_O) and unbiased expected heterozygosity (UH_E) were computed with GenAlEx [45]. The genotype distribution of each locus per year class and its direction

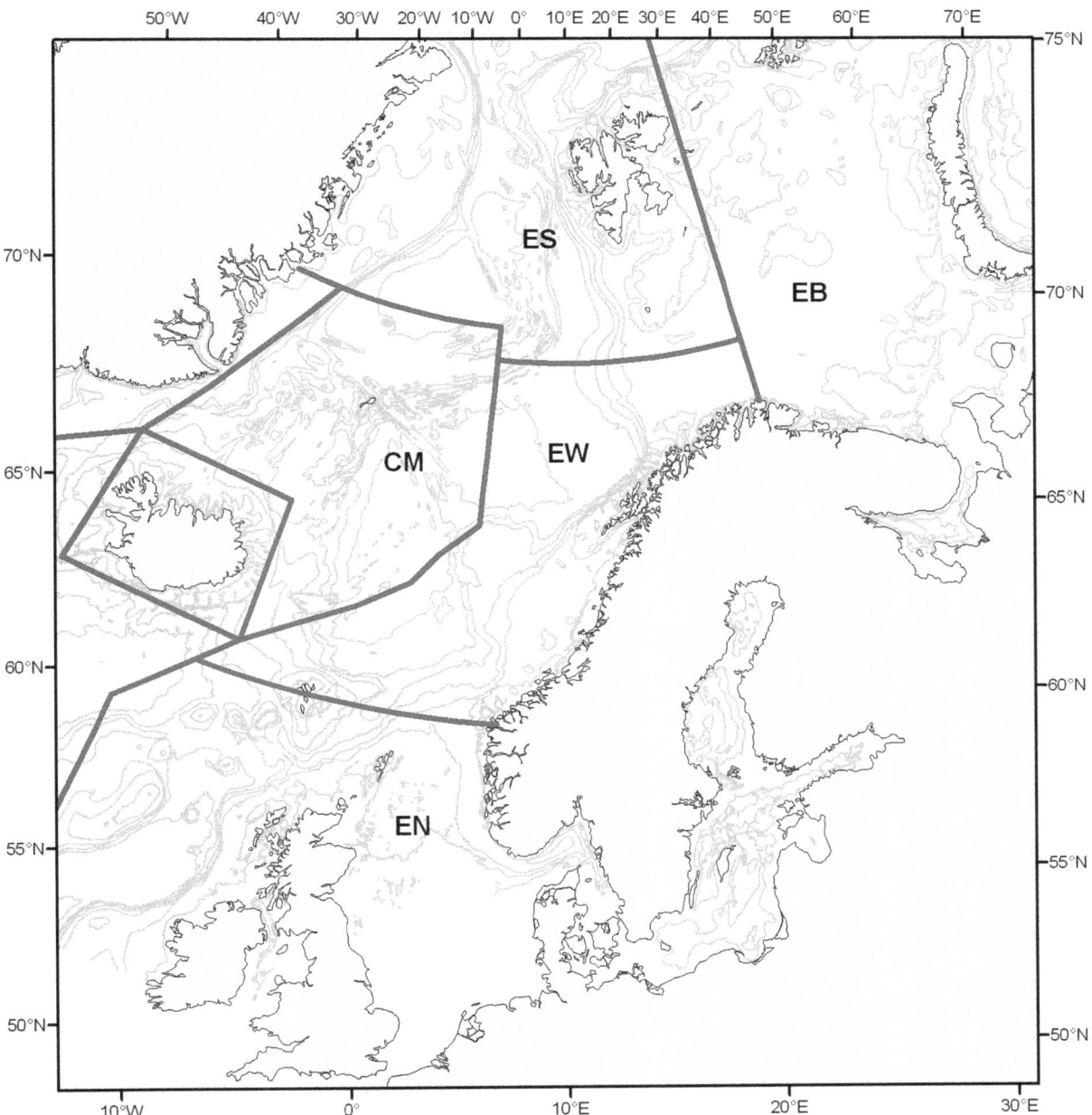

Figure 1. Geographic distribution of the five International Whaling Commission (IWC) Management Areas: ES (Svalbard-Bear Island area), EB (Eastern Barents Sea), EW (Norwegian Sea and coastal zones off North Norway, including the Lofoten area), EN (North Sea), and CM (Western Norwegian Sea-Jan Mayen area).

(heterozygote deficit or excess) was compared with the expected Hardy-Weinberg distribution using the program GENEPOP 7 [46] as was the linkage disequilibrium. Both were examined using the following Markov chain parameters: 10000 steps of dememorisation, 1000 batches and 10000 iterations per batch.

We used several methods to estimate population structure, including STRUCTURE [47], BAPS [48], and traditional F_{ST} [49] and R_{ST} analyses [50]. Slatkin's R_{ST} is an analogue of Wright's F_{ST} [51], adapted to microsatellite loci by assuming a high-rate stepwise mutation model instead of a low-rate K- or infinite-allele mutation model.

Both genetic differentiation among Management Areas per year class, and the level of temporal population genetic differentiation were tested using the Analysis of Molecular Variance (AMOVA) implemented in ARLEQUIN v.3.5.1.2 [52]. We also calculated the pairwise F_{ST} between populations from year class 2004 to 2011.

Both STRUCTURE [47,53,54] and BAPS [48,55] conduct a Bayesian analysis to identify hidden population structure, the former using allele frequency and linkage disequilibrium information from the data set directly, the latter identifying populations with different allele frequencies. Thus, BAPS first infers the most likely individual clusters in the sample population and then

Table 1. Distribution of females (F) and males (M) per Management Area (Fig. 1) on a per year class basis.

Year	Period	MANAGEMENT AREAS										Total
		EW		ES		EB		EN		CM		
		F	M	F	M	F	M	F	M	F	M	
2004	25th April – 23rd September	102	83	107	2	100	23	52	29	17	0	515
2007	22nd April – 22nd August	89	83	265	11	8	20	44	47	0	0	567
2008	30th April – 5th September	52	90	212	8	9	11	47	39	25	5	498
2009	11th April – 15th September	84	87	229	14	3	0	24	25	0	0	466
2010	29th April – 11th September	60	80	252	12	11	6	23	4	1	0	449
2011	1st May – 27th August	93	110	160	24	78	18	9	3	0	0	495
	Total	480	533	1225	71	209	78	199	147	43	5	2990

performs the most likely admixture of genotypes [55]; an approach that is more powerful in identifying hidden structure within populations [56].

We used the Bayesian model-based clustering algorithms implemented in STRUCTURE v. 2.3.4 to identify genetic clusters under a model assuming admixture and correlated allele frequencies without using population information. Ten runs with a burn-in period consisting of 100000 replications and a run length of 1000000 Markov chain Monte Carlo (MCMC) iterations were performed for a number of clusters ranging from $K = 1$ to $K = 5$. If applicable, we then used STRUCTURE Harvester [57] to calculate the Evanno et al. [58] ad hoc summary statistic ΔK, which is based on the rate of change of the 'estimated likelihood' between successive K values. The usual scenario where this approach is appropriate are those cases where once the real K is reached, L(K) at larger Ks plateaus or continues increasing slightly and the variance between runs increases. Hence, the estimated 'log probability of data' does not provide a correct estimation of the number of clusters and instead, ΔK accurately detects the uppermost hierarchical level of structure [58]. Runs were automatized with the program ParallelStructure [59] that controls the program STRUCTURE and distributes jobs between parallel processors in order to significantly speed up the analysis time. Afterwards, runs were averaged with CLUMPP version 1.1.1 [60] using the LargeKGreedy algorithm and the G' pairwise matrix similarity statistics. Averaged runs were graphically displayed using barplots on a per year class basis.

Secondly, we used BAPS 6.0 [48] for a number of clusters ranging between $K = 1$ and $K = 5$ (10 runs per K), and then we performed the most likely admixture of genotypes [55], again on a per year class basis.

Mitochondrial DNA

Estimates of genetic diversity were calculated with DnaSP [61] and consisted of number of segregating sites, average number of pairwise nucleotide differences, nucleotide diversity and haplotype diversity.

Demographic changes were examined using three different approaches: Tajima's D [62], Fu's F_S [63] and by comparing mismatch distributions of pairwise nucleotide differences between haplotypes to those expected under a sudden population expansion model [64–66]. The analyses were implemented in the program ARLEQUIN v.3.5.1.2, and P-values were generated using 10000 simulations.

We used Tajima's D and Fu's Fs to test for shift in the allele frequency spectrum compared to a neutral Wright-Fisher model consistent with population expansion under neutral evolution. The neutrality test Fs [63] has been shown to be a powerful test to detect population growth when large sample sizes are available [67]. Large and negative significant values of Fs indicate an excess of recent mutations (haplotypes at low frequency) compared to those expected for a stable population, which can be interpreted as a signature of recent population growth, genetic hitchhiking or population expansion following a bottleneck event [63]. Demographic changes were also investigated by calculating the raggedness index of the observed mismatch distribution for each of the populations according to the population expansion model. This measure quantifies the smoothness of the observed mismatch distribution. Small raggedness values represent a population which has experienced sudden expansion (possibly following a bottleneck) whereas higher values suggest stationary populations [68,69]. Unimodal distributions are expected for populations that recently expanded or experienced a bottleneck, as individuals within a population will present similar haplotype divergence (in terms of nucleotide differences) [64,66]. In contrast, a multimodal or 'ragged' distribution is expected for a stable or slowly declining population [64]. Statistical significance for the mismatch distributions was obtained using a goodness-of-fit test based on the sum of squared deviations between the observed and expected distributions [70] and the Harpending's raggedness index, rg [68] after 10000 simulations using the estimated parameters of the expected distribution for a population expansion.

The evolutionary relationships between haplotypes were examined with the software Network [71] using the median-joining algorithm to build an unrooted cladogram. Networks were built separately for every year class and also for the full data set ranging from 2004 to 2011. Singletons were removed, transitions weights were changed into 10 whereas tranversions and gaps were changed into 30; epsilon was set at 10, and the MP option [72] was enabled to delete redundant links and median vectors.

BAPS clustering was used to validate Network results and thus the program was run 100 times for the number of clusters reported by the median-joining tree.

Testing the hypothesis of cryptic stock clustering of North Atlantic minke whale

Anderwald et al. [24] identified genetic sub-structuring of North Atlantic minke whales and proposed the existence of two putative cryptic stocks. In their paper, the detection of genetic differentiation among minke whale individuals was enhanced by the use of an outgroup in STRUCTURE, and thus they included 30 individuals of B.a. scammoni from the Sea of Japan as an outgroup. We added this approach to our former STRUCTURE analyses using two different outgroups: firstly, 95 individuals of the subspecies Pacific minke whale (B. a. scammoni); secondly, 93 individuals of the Antarctic minke whale (B. bonaerensis) and, thirdly, both outgroups simultaneously. STRUCTURE and BAPS analyses together with the assessment of genetic differentiation between groups of individuals after clustering procedures are exhaustively detailed in the File S1 in Supporting Information.

In addition to the above analyses, and to test the alternative hypothesis of minke whales constituting a panmictic population, we created a set of 100 in silico generated panmictic populations based on the allele frequencies observed in our samples. Hence, at each of the ten loci, the allelic values (two per individual) were put in a pool, and then randomly re-assigned to individuals, thereby preserving the original allele frequencies. The resulting in silico simulated panmictic populations were analysed automatizing STRUCTURE with the program ParallelStructure under a model assuming admixture and correlated allele frequencies without using population information. Ten runs per K ranging from 1 to 5, a burn-in period of 100000 replications, and a run length of 1000000 MCMC iterations were followed by Evanno's test. For the sake of the comparison, ten of the populations yielding K = 2 after Evanno's test were averaged with CLUMPP and pairwise F_{ST} between resulting clusters was performed as above. In addition, BAPS analyses were also conducted for K ranging from 1 to 5.

Detection of sex-biased dispersal

The potential for sex-biased dispersal was investigated using the microsatellite data with the methods described by Goudet et al. [73] and implemented in GenAlEx [45]. The statistics used were: F_{IS}, F_{ST}, Ho, Hs (the within group gene diversity), the mean corrected assignment index (mAIc) and the variance around the assignment index (vAIc) [74,75]. When comparing allele frequencies between individuals of the dispersing sex and those of the more philopatric one, a greater similarity is expected among the

more dispersing sex. Likewise, expectations would be mAIc to be higher in the more philopatric sex, while vAIc should be lower [73]. Female philopatry and male dispersal are the expected patterns for mammalian species based on the expectation that partuating females will be more dependent on local resources [76]. Thus, a one-tailed Mann–Whitney U-test was used to test if dispersal was biased toward males, as in most marine mammals.

Results

Spatial and temporal genetic structure according to microsatellites

The sex distribution per Management Area across years was biased towards females in 73% of the cases (Table 1). Namely, in ES, females were 7–54 fold more abundant than males, whereas in CM no males were reported with the exception of 5 individuals in 2008. The spatial distribution showed that females and males overlapped in latitudes below 71°N but hardly any males occurred in the northernmost regions.

The microsatellite data set contained no missing data and both the number of alleles (116–124), and allelic richness (11.5–12.3) were stable across the six year classes analysed (Table 2). Observed heterozygosity ranged between 0.757 and 0.795, and unbiased expected heterozygosity, between 0.768 and 0.801. Analysis of HWE revealed that at the significance level of α 0.05, 4.8% of loci by sample combinations displayed significant deviations; whereas this number decreased to 0.4% at the significance level of α 0.001. LD was detected 68 times (5.6%) at α 0.05 and 9 (0.7%) at α 0.001.

The analysis of geographic genetic structuring among Management Areas revealed no differentiation over time. Thus, AMOVA performed separately for each of the year classes reported no significant F_{ST} in any case (Table 3). Likewise, all the pairwise comparisons between Management Areas were non-significant for all year classes analysed with the only exception of EB-EW in 2010 being marginally significant at $R_{ST} = 0.024$ ($P = 0.046$) but not after Bonferroni correction ($P = 0.0017$).

Similar to results of spatial genetic structure above, AMOVA reported high temporal genetic stability, and the pairwise comparison between the different year classes (Table 4) yielded only one significant albeit very weak pairwise F_{ST} between years 2007–2008 ($F_{ST} = 0.0004$, $P = 0.0270$); which was no longer significant after Bonferroni correction.

The individual analysis of every Management Area also showed temporal genetic stability as a general picture. Hence, the AMOVA performed separately in each of the five IWC zones yielded a non-significant F_{ST} that exhibited 0.003 as the highest value. Likewise, the pairwise comparisons between years within each area were also non-significant with the exception of area EN. The pairwise matrix for EN reached significance in three out of the total fifteen cases corresponding to the comparisons between year 2008 and years 2004, 2007 and 2009 respectively.

STRUCTURE showed that the highest average likelihood was found to be K = 1 in all year classes together with a decreasing trend of LnP(D) across consecutive values of K (Table A in File S1). In these situations, although there is no need to perform Evanno's test; we found that ΔK took its highest value for K = 2 in all the year classes with the exception of 2008 where K = 3. The common feature to all the sampling years were the low values reported for ΔK, which reached its maximum in 2011 ($\Delta K = 16.7$) but otherwise ranged between 3 and 8. Barplots for each year class are shown in Fig. A in File S1. For K≥2, every individual showed that the membership to each cluster was evenly distributed among groups producing flag-like barplots.

Table 2. Minke whale microsatellites.

Year	No alleles	Ar	No private alleles	Ho	uHe	F_{IS}
2004	119	11.7	2	0.757±0.015	0.768±0.010	0.0057±0.0243
2007	120	11.7	0	0.776±0.012	0.770±0.010	−0.0050±0.0185
2008	123	12.1	3	0.791±0.012	0.777±0.010	−0.0150±0.0205
2009	124	12.3	4	0.790±0.018	0.775±0.013	−0.0057±0.0183
2010	116	11.6	0	0.787±0.024	0.801±0.022	0.0093±0.0229
2011	116	11.5	1	0.795±0.017	0.778±0.010	−0.0057±0.0151

Summary statistics per year showing total number of alleles, allelic richness (based on minimum sample size of 449 diploid individuals), number of private alleles, observed heterozygosity (average ± SE), unbiased expected heterozygosity (average ± SE), and inbreeding coefficient (F_{IS}) (average ± SD).

BAPS showed that most likely K was 3 for year classes 2008, 2010 and 2011 and 4 for the remaining ones. No admixture was detected in any of the sampling years but in 2004 with one admixed individual.

Mitochondrial DNA

A total of 92 haplotypes were found in the complete dataset (*i.e.* year classes 2004 and 2007–2011), 25 of them unique (0.8% of the individuals). Six of the haplotypes were shared by 4–9% of the individuals whereas the most abundant one was present in 806 whales (27%). The number of haplotypes found per year class ranged mostly between 36 and 38 (Table 5), and took its maximum value in 2004 ($N_H = 62$). Both haplotype (H_D) and nucleotide diversity (π) showed high and stable values across the years (Table 5).

The distribution of the most common haplotypes was even across Management Areas and AMOVA revealed that no differentiation was observed among them in any of the sampling years (Table 3), with the exception of the significant pairwise comparison EB-ES ($F_{ST} = 0.008$, $P = 0.035$) in 2004. Similarly, high temporal stability (Table 4) was reported with one weak but marginally significant pairwise comparison: 2010–2011 ($F_{ST} = 0.0019$, $P = 0.043$), although not significant after Bonferroni correction.

A median-joining tree (Fig. 2) was built for the total data set (*i.e.* 2004 and 2007–2011 excluding singletons) given the temporal stability detected across year classes. A central ancestral haplotype was reported in 27% of the individuals, whereas none of the remaining ones exceeded a frequency of 9%. This ancestral haplotype did not show any phylogeographic structure, *i.e.* it was not linked to any of the Management Areas as it was present in relatively even proportions in each of them. The MJ-tree suggests the existence of three lineages: a central one that evolved through two episodes of expansion, with haplotypes connected between them via single mutational steps in the vast majority of cases. Two of the lineages gathered 85% of the haplotypes in a quite even distribution whereas the third one accounted for the 15% remaining. The haplotype composition per lineages revealed by Network perfectly matched BAPS clustering for K = 3, with 100% coincidence. The same individuals that showed significant differentiation in three lineages at mtDNA yielded $F_{ST} = 0.00018$ ($P = 0.8966$) when analysed for microsatellites.

Different insights (Table 6) invoke population size expansion such as: a) large negative and significant Fu's Fs, b) negative albeit non-significant Tajima's D values (except in 2010), c) small raggedness values, d) unimodal mismatch distributions (not shown), and e) the star-shape of haplotype network.

Examining potential cryptic population structure using clustering methods

Evanno's test revealed that K = 2 was the most likely scenario in all the clustering approaches performed per year class, either with or without outgroups, and even regardless of the number of outgroups included in the analysis (with the exception of year class 2010 without outgroup that showed K = 3). Thus, when using outgroups, NE Atlantic minke whales (*B.a. acutorostrata*) constituted a compact cluster whereas the outgroups (i.e *B.a. scammoni* or/and *B. bonaerensis*) constituted a second compact one (see Fig. B in File S1). Therefore, in such a case, we needed to explore K = 3 to have NE Atlantic minke whales divided into two groups

Table 3. Genetic differentiation into Management Areas per year class.

Year	Microsatellites		mtDNA	
	F_{ST}	R_{ST}	F_{ST} (Haplotype frequency)	F_{ST} (Tamura-Nei)
2004	0.0000 (0.6836)	0.0000 (0.6827)	0.0012 (0.2524)	0.0010 (0.4364)
2007	0.0000 (0.6964)	0.0000 (0.8940)	0.0000 (0.6157)	0.0000 (0.7972)
2008	0.0000 (0.9613)	0.0000 (0.9442)	0.0000 (0.8310)	0.0000 (0.5750)
2009	0.0004 (0.2406)	0.0000 (0.7125)	0.0000 (0.5240)	0.0006 (0.3684)
2010	0.0000 (0.9374)	0.0000 (0.4030)	0.0000 (0.4723)	N.C.[NOTE]
2011	0.0000 (0.9867)	0.0000 (0.6003)	0.0012 (0.2652)	0.0000 (0.4919)

Summary of AMOVA (F_{ST} and *P*-value) conducted with ARLEQUIN with 10000 permutations at microsatellites and mtDNA.
[NOTE]N.C. not calculated. Nucleotide composition too unbalanced for Tamura-Nei correction.

Table 4. Temporal genetic differentiation: Pairwise F_{ST} between year classes calculated with ARLEQUIN for microsatellites (lower diagonal) and mtDNA (upper diagonal).

	2004	2007	2008	2009	2010	2011
2004		0.0005	0.0012	0.0005	0.00000	0.0037
2007	0.00000		0.0002	0.00000	0.00000	0.0024
2008	0.00023	**0.00043***		0.00000	0.00000	0.00000
2009	0.00000	0.00000	0.00016		0.00000	0.0005
2010	0.00000	0.00000	0.00015	0.00000		**0.0019***
2011	0.00006	0.00000	0.00000	0.00000	0.00018	

Significance calculated after 10000 permutations. Values highlighted in boldface type as significant at P<0.05 (*) lost significance after Bonferroni correction.

(*e.g.* Fig. 3c–f, and Fig. A3 in File S1). The distribution of those individuals into clusters was conditioned to overcoming a threshold of membership of 0.50. Although an exhaustive report of all the results regarding cryptic clustering can be found in the File S1 in Supporting Information, the main findings of each assignment procedure were as follows:

a) **STRUCTURE without outgroup (K = 2).-** Although individuals showed very narrow ranges of membership (0.51–0.63, average 0.59) to clusters (Table B a,b in File S1); there was a significant albeit weak genetic differentiation between groups per year class demonstrated by pairwise F_{ST} (Table B a,b in File S1), Fisher's exact test ($\chi^2 =$ infinity, df = 20, *P*<0.0001) and Factorial Correspondence Analyses (despite a low percentage of total variation explained by the two first axes ranging between 3.97 and 4.22%). The same individuals genotyped at mtDNA did not produce any significant F_{ST} in any sampling year (Table B a,b in File S1). GeneClass corroborated clustering with an average percentage of correct assignment of 86% (ranging from 83.5 to 90.2%, Table J in File S1).

b) **STRUCTURE with outgroups (K = 3).**

  a. Pacific minke whale (*B.a. scammoni*) as an outgroup (Fig. C in File S1, left column).- The average membership to cluster was higher than when using STRUCTURE without outgroups (0.88). Individuals were divided into two clusters that showed weak albeit significant genetic differentiation (average F_{ST} between clusters was 0.0130). The same individuals genotyped at mtDNA did not show any genetic differentiation between clusters (Table D in File S1).

  b. Antarctic minke whale (*B. bonaerensis*) as an outgroup (Fig. C in File S1, right column).- The average membership to cluster was higher, 0.95, whereas the average F_{ST} between clusters was slightly lower, 0.0122. Again, the individuals from both clusters genotyped at mtDNA did not reveal any genetic differentiation (Table E in File S1).

  c. We used a conservative approach and divided the NE Atlantic minke whales into two groups after taking the consensus of the results of the analyses with Antarctic and Pacific outgroups together. This means that individuals were assigned to cluster 1 or 2 after comparing the assignment obtained after Antarctic and Pacific analyses. Likewise, a number of individuals was left unassigned and this comprised those that did

not reach the inferred ancestry 0.5 threshold plus the mismatches between both procedures (*e.g.* individuals that belonged to cluster 1 with Antarctic outgroup and to cluster 2 in the Pacific clustering). Again a weak but significant differentiation between groups per year class was shown by pairwise F_{ST} (Table F in File S1), Fisher's exact test ($\chi^2 =$ infinity, df = 20, *P*<0.0001) and Factorial Correspondence Analyses (albeit the low percentage of the total variation explained by the two first axes ranging from 3.9 to 4.2%). Once more, the same individuals genotyped at mtDNA did not show any evidence of genetic differentiation (Table F in File S1). GeneClass corroborated STRUCTURE consensus clustering with a high percentage of correct assignment (97–98.6%) across all year classes with the exception of 2008 that was slightly lower (85.6%), Table J in File S1.

c) **BAPS for K = 2.-** BAPS divided individuals of each sampling year class into two groups of even size for year classes 2008, 2009 and 2011 whereas for the remaining ones, the size ratio was around 1.4–1.5 (Table H in File S1). No admixed individuals were detected in any of the sampling years. A weak albeit significant F_{ST} (Table H in File S1) was found between groups per year class, a differentiation that was further confirmed by Fisher's exact test ($\chi^2 =$ infinity, df = 20, *P*<0.0001) and Factorial Correspondence Analyses (percentage of the total variation explained by the two first axes ranging between 3.73 and 4.10%). The same individuals genotyped at mtDNA only produced significant F_{ST} for year classes 2007 and 2008 based on Tamura-Nei distance and haplotype frequencies, respectively (Table H in File S1). GeneClass corroborated BAPS clustering with a percentage of correct assignment of 100% in all the cases (Table J in File S1).

The geographic distribution of individuals after both procedures of clustering was slightly different. Hence, STRUCTURE-clustered individuals were evenly distributed among Management Areas per year class whereas for the BAPS-clustered ones, this distribution was less homogeneous in some of the cases (Fig. D and Table K in File S1).

The analyses of the 100 *in silico* generated panmictic populations with STRUCTURE revealed that, again and like the real data: a) the highest average likelihood was detected at K = 1, and b) a decreasing trend of LnP(D) across consecutive values of K was found in all the cases. As formerly reported, even if Evanno's test is not applicable in this situation, we wanted to test

Table 5. Minke whale mtDNA.

Year	N	N_H	N_{UH}	S	k	$\Pi \times 10^2$	H_D
2004	515	62	29	21	3.071	0.969	0.908±0.008
2007	567	49	22	26	2.863	0.906	0.895±0.008
2008	498	38	14	21	2.854	0.900	0.886±0.010
2009	466	36	11	24	2.824	0.891	0.884±0.010
2010	449	38	14	18	2.940	0.931	0.897±0.009
2011	495	37	10	20	2.781	0.877	0.864±0.011

Summary of diversity statistics: Sample size (N), number of haplotypes (N_H), number of unique haplotypes (N_{UH}), number of segregating sites (S), average number of pairwise nucleotide differences (k), nucleotide diversity (π) and haplotype diversity (H_D: mean ± SD).

its outputs and thus we found that the most likely number of clusters showed different values: $K = 2$ in 58% of the cases, $K = 3$ in 33% and $K = 4$ in the remaining 9%. Similarly, low values of ΔK (ranging from 1 to 15) were also reported for the 100 panmictic populations. CLUMPP was performed on a set of ten randomly chosen simulated populations that showed $K = 2$ after Evanno's method, and individuals were distributed into clusters after overcoming a threshold of 0.50. In all cases, both clusters showed similar size (ratio 1–1.3) and 7–19% of individuals were left unassigned (Table M in File S1). Although the range of membership to cluster was very low (0.51–0.64), pairwise F_{ST} between groups exhibited low (0.012–0.020) but significant values ($P < 0.0001$). Importantly, these values were equal in magnitude to the observations based upon the real data reported above. BAPS analyses showed that, in spite of dealing with panmictic populations, in no case the most likely K was found to be 1. Instead, the number of putative populations took the following values: 3 (4% of the cases), 4 (38%) and 5 (58%) respectively.

Detection of sex-biased dispersal

According to the expectations that dispersal should be biased towards males, as in most of mammals, mAIc was lower in males than in females (-0.051 vs. 0.020) and vAIc was higher (2.90 vs. 2.35) Fig. G in File S1, whereas the rest of the statistics (F_{IS}, F_{ST}, Ho and Hs) took almost identical values in both sexes. However, the Mann–Whitney U-test proved to be non-significant ($P > 0.5$) therefore we were unable to detect sex-biased dispersal in North Atlantic minke whales. Furthermore, when performing a two-tailed U-test we found a non-significant result ($P = 0.953$) that would not support a higher female dispersal either.

Discussion

Overall, the total data set ($N = 2990$) consisted of 28% males and 72% females; proportions that exactly coincide with Anderwald et al. [24] and are very similar to the 21% males 79% females reported by Andersen et al. [23]. This uneven presence of sexes was also reflected in the sex composition across Management Areas (Table 1), which also agrees with Andersen et al. [23] and Anderwald et al. [24] and corresponds to the known segregational behaviour with respect to sex and age during summer as mature females tend to occur further north than males [77–79].

Microsatellite loci used here exhibited a range of variation of genetic diversity comparable to what has been formerly reported for the same species [23–25] as well as for other balaenopterids such as Bryde's whales, *B. brydei* [80]; fin whales, *B. physalus* [81,82]; sei whales, *B. borealis* [83]; bowhead whales, *B. mysticetus* [84,85] and gray whales, *Eschrichtius robustus* [86]. Likewise, a similar magnitude of mtDNA genetic diversity, measured either as nucleotide (average of 0.009) or haplotype diversity (average of 0.9), was formerly reported for *B.a. acutorostrata* within the same geographic area [22–25,43] and resembles what has been described for other whale species [83,86–91]. However, the nucleotide diversity reported for the Antarctic minke whale (*B. bonaerensis*) is higher (0.0159) and this was interpreted as the Antarctic species having larger long-term effective population size than *B.a. acutorostrata* [22]. It has been proposed that the current size of the Antarctic minke whale population is unusually high as an indirect result of the whaling that killed more than 2 million of large whales leading to competitive release for smaller krill-eating species [92].

The mismatch distribution analyses were consistent with exponential population expansion suggesting that populations of

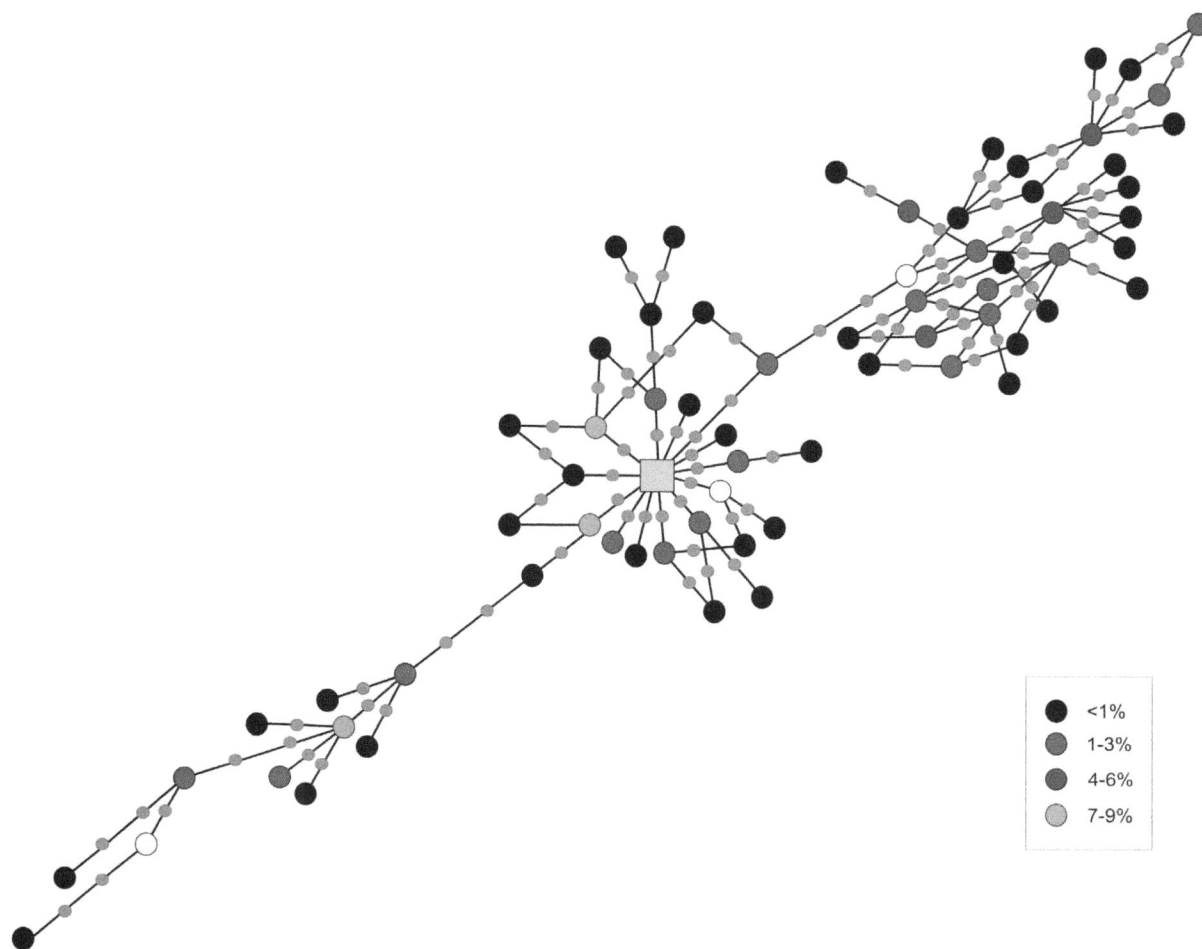

Figure 2. Median-joining network of mtDNA haplotypes corresponding to the period 2004 and 2007–2011. Haplotypes are represented as circles which area is not proportional to its relative frequency for simplicity. Instead, the frequency of haplotypes is depicted through the color code detailed in the legend. The green square represents the ancestral and more abundant haplotype (present in 27% of the individuals). The minimum number of steps connecting parsimoniously two haplotypes is indicated as a red dot, and the open circles represent extinct or missing haplotype that might have not been sampled (mv).

North Atlantic minke whale are not at equilibrium, something that had already been reported in the literature for this species in the same geographic area [24,25]. Earlier studies proposed that this expansion followed the last glacial maximum, as seen for various other cetacean species in the North Atlantic [24].

Spatial genetic structure

The present molecular markers (ten microsatellite loci and 331 bp of mtDNA D-loop) studied on 2990 individuals congruently failed to reveal any genetic differentiation among Management Areas during the period 2004 and 2007–2011. Only two

Table 6. Minke whale mtDNA.

Year	Fu's F_S	Tajima's D	SSD (P-value)	rg (P-value)
2004	**−24.916 (0.0002)**	−0.0062 (0.5859)	0.0098 (0.4413)	0.0153 (0.6398)
2007	**−23.861 (0.0000)**	−0.6057 (0.3004)	0.0105 (0.4586)	0.0159 (0.6831)
2008	**−13.279 (0.0119)**	−0.1924 (0.4987)	0.0115 (0.4099)	0.0158 (0.6806)
2009	**−10.710 (0.0279)**	−0.5333 (0.3516)	0.0128 (0.4439)	0.0193 (0.6339)
2010	**−13.711 (0.0104)**	0.2234 (0.6487)	0.0136 (0.3308)	0.0192 (0.5360)
2011	**−12.002 (0.0188)**	−0.1403 (0.5213)	0.0117 (0.4639)	0.0178 (0.6527)

Analyses of population stability (Tajima's D and Fu's F_S tests) and population expansion (sum of squared deviations, SSD and raggedness, rg mismatch distribution tests). Significant values are indicated with boldface type.

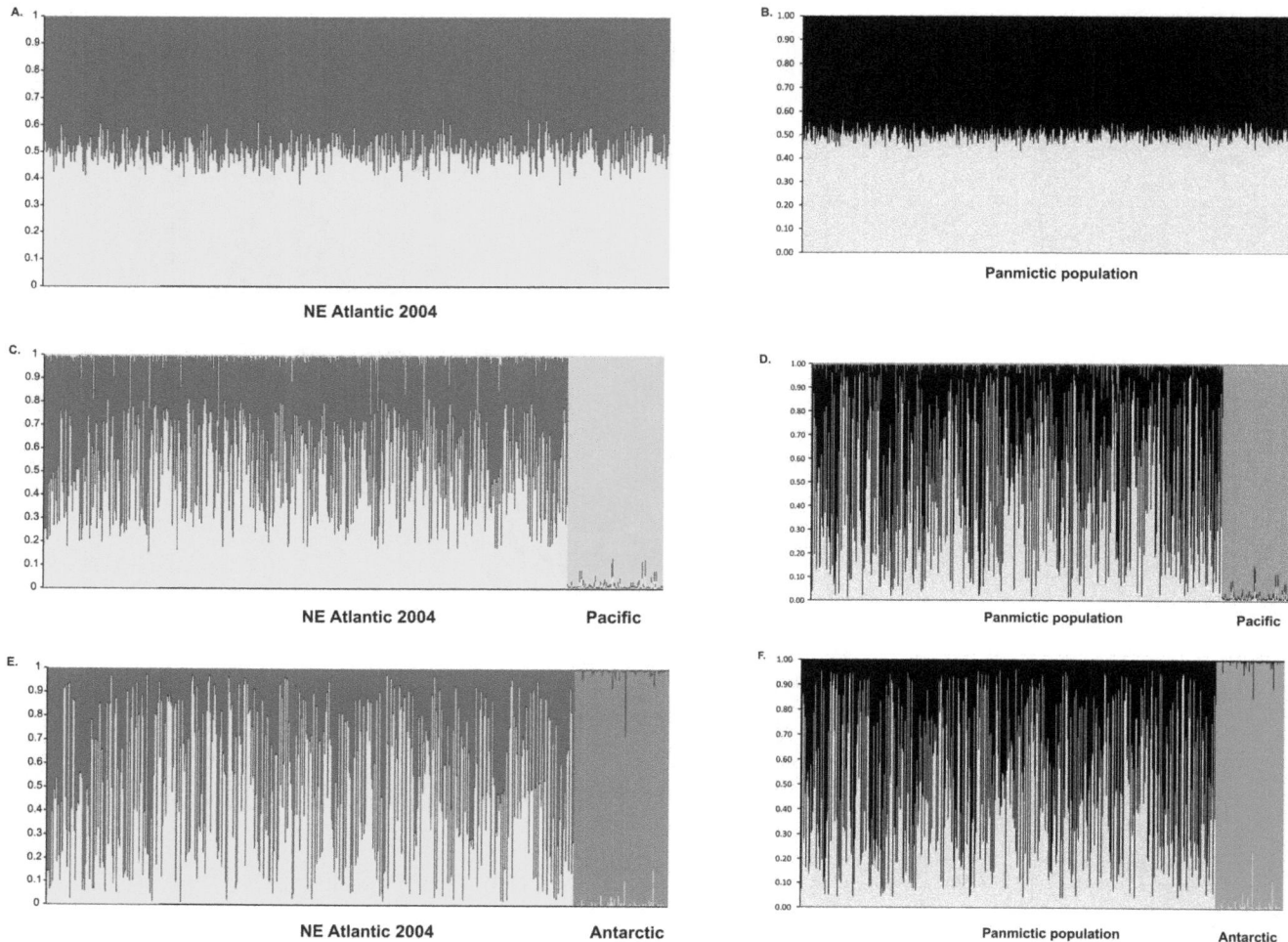

Figure 3. Example of comparison between real populations and the simulated panmictic ones. Bayesian clustering of North East Atlantic minke whale corresponding to year class 2004 (left column) and to a randomly chosen simulated panmictic population (right column). Inferred ancestry of individuals was calculated after averaging ten STRUCTURE runs with CLUMPP for K = 2 (barplots a,b) and K = 3 (barplots c–f). The outgroups were 95 individuals of the Pacific subspecies (*B. a. scammoni*) and 93 individuals of the Antarctic species (*B. bonaerensis*).

weak and marginally significant pairwise comparisons were recorded: EB-EW for microsatellites in 2010 ($R_{ST} = 0.024$, $P = 0.046$) and EB-ES for mtDNA in 2004 ($F_{ST} = 0.008$, $P = 0.035$). This translates to 1% of pairwise tests showing some spatial and temporal divergence, and neither were significant following Bonferroni correction for multiple testing. This lack of spatial genetic differentiation was the case when analyzing each of the six year classes separately, and also for all the specimens combined in the same AMOVA analysis, both for mitochondrial ($F_{ST} = 0.0005$, $P = 0.1134$) and nuclear ($F_{ST} = 0.0000$, $P = 0.9473$) DNA. Likewise, none of the resulting pairwise comparisons between areas were significant. The analyses using the joint data set seems legitimate given the temporal genetic stability found at both markers, stability that had already been reported within similar time frames [23,25,27].

The lack of geographic genetic differentiation as revealed in the present study is in agreement with some former studies of minke whales in the N Atlantic that were based on nuclear [24,25,27,28] and mitochondrial DNA markers [22–25]. However, the consensus about this issue is far from being commonplace, as the opposite scenario has also been reported for nuclear markers [23,31–33]. In particular, Andersen *et al.* [23] suggested the existence of four

genetically discrete subpopulations in the Atlantic (*i.e.* West Greenland, NE Atlantic, North Sea and Central North Atlantic) and indicated that this could be the result of the profound ecological differences between feeding areas (environmental conditions, prey availability) posing different selective pressures, coupled with a strong affiliation between mother and calf to the feeding site. Seasonal site fidelity that had been already reported for minke whales [93–95] as well as for other species such as humpback whales [96].

We also tested Tiedemann's [97] thesis that states that, for marine large mammals, the F_{ST} obtained for females would reflect the maximum spatial genetic differentiation of the species. Through the population structure observed in the maternally inherited mtDNA, Baker *et al.* [96] demonstrated that humpback whales show strong fidelity to migration destinations such as feeding grounds. Following a similar approach, we performed AMOVA across Management Areas by pooling all females sampled between 2007 and 2011 and, once more, we found no genetic differentiation for microsatellites ($F_{ST} = 0.00009$, $P = 0.326$) or mtDNA ($F_{ST} = -0.005$, $P > 0.05$). This result disagrees, again, with Andersen *et al.* [23] who reported a significant F_{ST} at both markers for females.

In conclusion, our data set of ten microsatellites and 331 bp of mtDNA control region failed to reveal any spatial genetic variation across 2990 individual whales harvested in the five management areas in the NE Atlantic for the years 2004 and 2007–2011.

Cryptic population genetic structure

The division of North Atlantic minke whale into two mtDNA lineages had already been reported [25,26], and Palsbøll [26] suggested the presence of two potential breeding populations coexisting at feeding grounds in the North Atlantic. The division of mtDNA showing a lack of concordance between haplotypes and geographic regions was first mentioned by Bakke et al. [22] who proposed the existence of two or more differentiated populations sharing the same feeding grounds. However, the pattern observed in mtDNA might also reflect a residual ancestral polymorphism or a "recent" isolation of two populations at breeding sites, which roam through large parts of the North Atlantic Ocean during the feeding migration, as proposed by Palsbøll [26], Bakke et al. [22] and Pampoulie et al. [25]. Our results also agree with the discordance between haplotypes and geographic areas; however, we support the division of mtDNA into three distinct lineages (Fig. 2), with a central group that evolved through two different expansion episodes. This possible expansion was further corroborated by large negative and significant Fu's Fs, negative Tajima's D (except in 2010), small raggedness values and unimodal mismatch distributions (Table 6). The lack of connections among lineages further suggested genetic differentiation. Importantly, microsatellites did not corroborate this result.

Nuclear markers provided no evidence to reject the hypothesis that North Atlantic minke whales constitute a single panmictic population. This is in spite of certain insights from STRUCTURE analyses conducted both in this study and in Anderwald et al. [24] that appeared to spuriously suggest the existence of cryptic subpopulations. First, LnP(D) obtained after the STRUCTURE analyses conducted here revealed that $K = 1$ was the most likely number of clusters, both for the real data distributed in six year classes and for the 100 simulated panmictic populations. In all cases, a clear decreasing trend of LnP(D) along consecutive values of K was recorded. The ad hoc statistic ΔK based on the rate of change in the log probability of data between successive K values obtained through Evanno's test can, unfortunately, never validate $K = 1$ [58]. Furthermore, this test is not even applicable in situations of decreasing pattern of LnP(D) [58]. However, when ignoring this limitation, Evanno's test showed that the highest ΔK was found at values ranging between $K = 2$ and $K = 4$. Thus, in the real data, 5 out of the 6 cases yielded $K = 2$ at the Evanno criterion, whereas year class 2008 reported $K = 3$. Likewise, the 100 in silico generated panmictic populations revealed $K = 2$ in a majority of cases (58%) whereas the remaining ones were distributed between $K = 3$ (33%) and $K = 4$ (9%). Hence, both LnP(D) pointed at 1, together with the fact that the highest ΔK was found mainly at $K = 2$ in the real and simulated panmictic populations, supports that North Atlantic minke whale constitutes one single panmictic population.

In addition, when Evanno's test is computed in non-pertinent situations and seems to reveal substructuring in the population ($K = 2$), there are multiple features that strongly suggest a false result. The first hint to be considered is the low values of ΔK, which is an indication that the strength of the signal detected by STRUCTURE is weak [58]. In our case, both the 100 simulated panmictic populations and the six real ones showed extremely low values of ΔK (ranging mainly from 1 to 10). In contrast, when the differentiation signal between two populations is strong, i.e. when the number of clusters is unequivocally two, ΔK exhibits

significantly higher values. Thus, for instance, when we conducted STRUCTURE analyses for Atlantic minke whales including Antarctic or Pacific whales as an outgroup (Fig. B in File S1), the value of ΔK at $K = 2$ was higher by three orders of magnitude than the one found when running STRUCTURE without outgroups. Secondly, when $K = 1$ but the model is forced for $K = 2$, most individuals will have a probability around 0.5 and 0.5 of belonging to cluster 1 and cluster 2 respectively. Our results also corroborated this extent as the inferred membership to clusters ranged from 0.51 to 0.64 in the real and the ten in silico generated panmictic populations. However, even in this situation, a weak albeit significant F_{ST} between clusters (average value of 0.010 in the real data and of 0.016 in the panmictic populations) can still be found (see Tables A2, A13 in File S1). The fact that values of F_{ST} are of similar magnitude in the real data and in the 100 panmictic populations sheds important doubts about the reliability of such genetic structure.

When running STRUCTURE with outgroups to enhance the genetic differentiation, the resulting barplots for $K = 3$ (Fig. C in File S1) showed that the subdivision of NE Atlantic minke whales revealed a higher inferred membership to cluster compared to when no outgroups were used. Furthermore, when the outgroup was the Pacific subspecies (B.a. scammoni), the averaged inferred membership was 0.89 but when the outgroup was the Antarctic species (B. bonaerensis), this value was even higher (average 0.95) and the percentage of non-assigned individuals was slightly lower. This higher inferred membership to cluster could be expected to result in a higher genetic differentiation. However, the resulting F_{ST} values between these clusters were virtually identical in the following cases: real data without outgroups, real data with outgroups, simulated panmictic populations without outgroups, one randomly chosen simulated panmictic population without outgroups (Tables A2, A4, A5, A6, A13 in File S1). Additionally, all of these values overlap with the level of genetic differentiation observed using a similar approach in Anderwald et al. 2011 [24]. The fact that both real and simulated panmictic populations showed the same patterns further increased the doubts upon the reliability of the clustering analyses upon which subdivision of North East Atlantic minke whales into cryptic populations has been suggested [24]. Furthermore, in most cases, the distribution of the individuals belonging to clusters 1 and 2 across Management Areas was surprisingly similar for the six year classes sampled (Table K in File S1), and in the in silico generated panmictic populations.

Anecdotally, North Atlantic minke whales have been suggested to follow an annual migration cycle between Arctic feeding grounds and Southern breeding grounds. The information on sightings of minke whales in the Southern North Atlantic is however very scarce [78] and one or more breeding grounds have so far not been demonstrated. Also, foetuses in different stages of development have been found in catches from the northern feeding grounds [78], indicating that mating may take place even there. The hypothesis of panmixia could therefore be well supported by these observations, also implying that separate breeding grounds may not exist.

As a general picture, the data analysed here show that while nuclear markers suggest panmixia, mitochondrial markers reveal the existence of three distinct lineages in North East Atlantic minke whales; which can be a reflection of the different time scales both type of markers represent. Besides, due to maternal inheritance, mitochondrial genes have lower effective migration rates than nuclear genes [98], and random drift is faster for the haploid, maternally inherited mt genome compared to a diploid, biparentally inherited nuclear locus [99]. Furthermore, an accelerated substitution rate of the mitochondrial genome contributes to faster differentiation [100]. Thus, the aforementioned discordance of higher population subdivision in mtDNA than in nuclear DNA is indicative of migration and breeding sex ratios not being biased [101]. Accordingly, and in agreement with Pampoulie et al. [25], we did not reach statistical support for the hypothesis of male-biased dispersal in this species. In contrast, male-biased dispersal has been reported for other whale species such as sperm whales [102,103] or gray whales [86].

Conclusions

The population structure of North Atlantic minke whale, B.a. acutorostrata, has been the subject of a long debate with contrasting results, partially driven by the fact that most previous studies have been limited by low numbers of samples, or genetic markers, or a combination of both. In order to shed further light on this topic, we conducted a spatial, temporal and cryptic population analysis of 2990 whales harvested in the North East Atlantic during the period 2004 and 2007–2011. This large data set, which has been genotyped according to strict protocols upon which the NMDR is based [12], and is thus of very high data quality [36], failed to reveal any indication of geographical or temporal population genetic structure within the NE Atlantic based upon the analysis of ten microsatellites and 331 bp of the mitochondrial D-loop. Furthermore, while three mtDNA lineages were revealed in the data, these did not show any underlying geographic pattern, and possibly represent an ancestral signal. In order to address the possibility of cryptic population structure as suggested by Anderwald et al. [24], we run STRUCTURE using a similar approach. However, while Evanno's test might seem to suggest the existence of two genetically differentiated clusters per year class, there were a number of facts strongly suggesting that these results were potentially an artefact. Firstly, as this approach can never validate K = 1, it shows K = 2 instead but with a very low value of ΔK, which is an indication of a very weak genetic signal. Furthermore, there was a lack of corroboration with mtDNA, in each case there was close to a 50/50 division between individuals into groups 1 and 2, and there was an absence of any clear geographic pattern underlying the clusters. The suspicion that these analyses would spuriously reveal population substructure was subsequently confirmed when it was possible to falsely create two cryptic populations in our in silico generated panmictic populations. These displayed more or less identical genetic characteristics both as in the real data in this study, and in the study by Anderwald et al. [24]. Therefore, we conclude that there is at present no or very little evidence to suggest that the minke whale displays spatial or cryptic population genetic structure throughout the North East Atlantic. However, it is also duly acknowledged that all studies conducted thus far have been limited by low numbers of genetic markers. Therefore, in order to conclusively evaluate the potential for spatial or cryptic population genetic structure within this highly mobile species, significantly larger numbers of markers will be required. Recent publication of the minke whale genome [104] will represent a major resource to identify the numbers of markers needed to address this issue in the future.

Supporting Information

File S1   Supporting Information. File S1 contains detailed information on the following issue: "Testing the hypothesis of cryptic stock clustering in North East Atlantic minke whales": including Material and Methods, and Results. This appendix also comprises eight figures (Fig. A–G) and thirteen tables (Table A–

M). **Figures. Fig. A1.** Bayesian clustering of North East Atlantic minke whales genotyped at 8 microsatellites for the six sampled year classes. Inferred ancestry of individuals was calculated after averaging ten STRUCTURE runs with CLUMPP after Evanno's test. **Fig. A2.** Bayesian clustering of North East Atlantic minke whales genotyped at 10 microsatellites for the six sampled year classes. Inferred ancestry of individuals was calculated after averaging ten STRUCTURE runs with CLUMPP after Evanno's test. **Fig. B.** Bayesian clustering of North East Atlantic minke whale year class 2004 with outgroups: a) 95 individuals of the subspecies Pacific minke whale (B. a. scammoni); b) 93 individuals of the Antarctic minke whale (B. bonaerensis), and c) both former outgroups together. The number of clusters that best fitted the data was K = 2 after Evanno's [58] test in each case. This scenario was consistent across year classes. **Fig. C.** Bayesian clustering of North East Atlantic minke whale with outgroups in each year class. In the column to the left, the outgroup are 95 individuals of the subspecies Pacific minke whale (B. a. scammoni) whereas in the column to the right, the outgroup are 93 individuals of the Antarctic minke whale (B. bonaerensis). The number of clusters that best fitted the data was distinctively K = 2 after Evanno's [58] test in each case. **Fig. D.** Geographic distribution of individuals after different clustering methods: BAPS and STRUCTURE for microsatellites. Pie charts represent the percentage of individuals belonging to clusters 1 (dark grey) and 2 (light grey) per Management Area taking year class 2008 as an example (the full data for all the year classes is available in Table K in File S1). **Fig. E.** Bayesian clustering of individuals of ten of the simulated panmictic populations that showed K = 2 after Evanno's test. Inferred ancestry of individuals was calculated after averaging ten STRUCTURE runs with CLUMPP. **Fig. F.** Distribution of pairwise F_{ST} after 10000 random clustering of North Atlantic minke whale individuals per year class into two groups. **Fig. G.** Frequency distributions of the corrected assignment index (AIc) for 2156 females (light grey bars above axis) and 834 males (dark grey bars below axis). AIc values differed among sexes, males having on average negative values (−0.051) and higher variance (2.90) and females positive values (0.020) with lower variance (2.35). However, Mann–Whitney U-test proved sex-biased dispersal to be non-significant (P>0.5). **Tables. Table A.** Summary result of STRUCTURE without outgroups: a) Data set of 8 microsatellites. b) Data set of 10 microsatellites. **Table B.** STRUCTURE without outgroups: Clustering of individuals per year class after Evanno's test (the two cases that showed the highest Evanno's ΔK at K = 3 are depicted in italics and analysed for K = 2 for comparison): Number of individuals per cluster and range of inferred membership to each of them (in brackets). Summary of the results of the AMOVA (F_{ST} and P-value) conducted with Arlequin with 10000 permutations. Analyses were performed for the same sets of individuals genotyped at mtDNA. Statistically significant values were highlighted in boldface type. Negative F_{ST} values found at mtDNA were transformed into 0. a) Data set of 8 microsatellites. b) Data set of 10 microsatellites. **Table C.** Summary statistics after STRUCTURE clustering showing total number of alleles, number of private alleles, observed heterozygosity (average ± SE), unbiased expected heterozygosity (average ± SE), and inbreeding coefficient (F_{IS}) (average ± SD). We show in italics the distribution of the individuals for K = 2 for the two year classes that showed the highest Evanno's ΔK at K = 3. a) Data set of 8 microsatellites. b) Data set of 10 microsatellites. **Table D.** STRUCTURE with the Pacific minke whale subspecies (B. a. scammoni) as an outgroup. Clustering of individuals per year class and one randomly chosen simulated panmictic population after Evanno's test and CLUMPP averaging: Number of individuals per cluster and range of inferred membership to each of them (in brackets). Summary of the results of the AMOVA (F_{ST} and P-value) conducted with Arlequin with 10000 permutations. Analyses were performed for the same sets of individuals genotyped at mtDNA. Statistically significant values were highlighted in boldface type. Negative F_{ST} values found at mtDNA were transformed into 0. **Table E.** STRUCTURE with Antarctic minke whale species (*B. bonaerensis*) as an outgroup. Clustering of individuals per year class and one randomly chosen simulated panmictic population after Evanno's test and CLUMPP averaging: Number of individuals per cluster and range of inferred membership to each of them (in brackets). Summary of the results of the AMOVA (F_{ST} and P-value) conducted with Arlequin with 10000 permutations. Analyses were performed for the same sets of individuals genotyped at mtDNA. Statistically significant values were highlighted in boldface type. Negative F_{ST} values found at mtDNA were transformed into 0. **Table F.** STRUCTURE consensus clustering of individuals (*i.e.* agreement between Antarctic and Pacific outgroup clustering) into two groups per year class. Summary of the results of the AMOVA (F_{ST} and P-value) conducted with Arlequin with 10000 permutations. Analyses were performed for the same sets of individuals at mtDNA. Statistically significant values are highlighted in boldface type. **Table G.** Summary statistics after STRUCTURE consensus clustering (*i.e.* consensus between Antarctic and Pacific outgroup clustering) showing total number of alleles, allelic richness (minimum sample size), number of private alleles, observed heterozygosity (average $\pm$ SE), unbiased expected heterozygosity (average $\pm$ SE), and inbreeding coefficient (F_{IS}) (average $\pm$ SD). **Table H.** BAPS clustering of individuals genotyped with microsatellites into two groups per year class. Summary of the results of the AMOVA (F_{ST} and P-value) conducted with ARLEQUIN with 10000 permutations. Analyses were performed for the same sets of individuals at mtDNA. Statistically significant values were highlighted in boldface type. **Table I.** Summary statistics after BAPS clustering showing total number of alleles, allelic richness (minimium sample size), number of private alleles, observed heterozygosity (average $\pm$ SE), unbiased expected heterozygosity (average $\pm$ SE), and inbreeding coefficient (F_{IS}) (average $\pm$ SD). **Table J.** GeneClass self-assignment: Percentage of individuals genotyped at microsatellites that were correctly assignment after clustering procedures. **Table K.** Number of individuals genotyped at microsatellites per Management Areas after clustering with BAPS and STRUCTURE (with and without outgroup). ND = No data. **Table L.** Matrix of numbers and percentage of coincident individuals when comparing the three clustering methods: BAPS, STRUCTURE without outgroup (STR), and STRUCTURE with outgroup (STR consensus). The percentage of coincident individuals was calculated by dividing the number of by the lowest number of individuals in the corresponding cluster. STRUCTURE analyses were performed with 8 microsatellites. **Table M.** STRUCTURE clustering of individuals in the 10 randomly selected simulated panmictic populations showing K = 2 after Evanno's test. Number of individuals per cluster and range of inferred membership to each of them (in brackets); number of non-assigned individuals (and % of the total). Summary of the results ofthe AMOVA (F_{ST} and P-value) conducted with Arlequin with 10000 permutations. Statistically significant values were highlighted in boldface type.

Acknowledgments

We thank François Besnier for help with ParallelStructure and simulations, and Lúa López and Rodolfo Barreiro for insightful discussions. We are grateful to Rus Hoelzel and Pia Anderwald for constructive criticism to earlier drafts of this manuscript.

Author Contributions

Conceived and designed the experiments: MQ HJS NØ TH KAG. Performed the experiments: BBS KAG. Analyzed the data: MQ HJS HKS. Contributed reagents/materials/analysis tools: HJS NØ TH NK LAP KAG. Wrote the paper: MQ HJS NØ TH BBS HKS CP NK LAP KAG.

References

1. Laikre L, Schwartz MK, Waples RS, Ryman N (2010) Compromising genetic diversity in the wild: unmonitored large-scale release of plants and animals. Trends in Ecology & Evolution 25: 520–529.
2. Allendorf FW, Luikart G (2006) Conservation and the Genetics of Populations: Blackwell Publishing.
3. Frankham R (2005) Stress and adaptation in conservation genetics. Journal of Evolutionary Biology 18: 750–755.
4. Glover KA, Quintela M, Wennevik V, Besnier F, Sørvik AGE, et al. (2012) Three decades of farmed escapees in the wild: A spatio-temporal analysis of Atlantic salmon population genetic structure throughout Norway. PLoS ONE 7: e43129.
5. IUCN (2013) International union for the conservation of nature and natural resources. In: species Irlot, editor: www.iucnredlist.org.
6. Stevick PT, McConnell BJ, Hammond PS (2002) Patterns of Movement. In: Hoelzel AR, editor. Marine Mammal Biology: An Evolutionary Approach. Oxford, U.K.: Blackwell Publishing. pp. 185–216.
7. Hoelzel A (1998) Genetic structure of cetacean populations in sympatry, parapatry, and mixed assemblages: implications for conservation policy. Journal of Heredity 89: 451–458.
8. Whitehead H, Dillon M, Dufault S, Weilgart L, Wright J (1998) Non-geographically based population structure of South Pacific sperm whales: dialects, fluke-markings and genetics. Journal of Animal Ecology 67: 253–262.
9. Hoelzel AR (2009) Evolution of population genetic structure in marine mammal species In: Bertorelle G, Bruford MW, Hauffe HC, Rizzoli A, Vernesi C, editors. Population Genetics for Animal Conservation. Cambridge: Cambridge University Press. pp. 410.
10. Rice DW (1998) Marine Mammals of the World. Systematics and Distribution. Society for Marine Mammalogy 4: 1–231.
11. Glover K, Kanda N, Haug T, Pastene L, Oien N, et al. (2013) Hybrids between common and Antarctic minke whales are fertile and can back-cross. BMC Genetics 14: 25.
12. Glover KA, Haug T, Øien N, Walløe L, Lindblom L, et al. (2012) The Norwegian minke whale DNA register: a data base monitoring commercial harvest and trade of whale products. Fish and Fisheries 13: 313–332.
13. Johnsgård Å (1966) The distribution of Balaenopteridae in the North Atlantic Ocean. In: Norris KS, editor. Whales, Dolphins and Porpoises. California: University of California Berkely Press. pp. 114.
14. Stewart BS, Leatherwood S (1985) Minke Whale, *Balaenoptera acutorostrata* Lacépède, 1804. In: Ridgway SH, Harrison SR, editors. Handbook of Marine Mammals, Volume 3: The Sirenians and Baleen Whales: Academic Press. pp. 91–136.
15. Jonsgård Å (1951) Studies on the little piked whale or minke whale (*Balaenoptera acutorostrata* Lacépède). Norsk Hvalfangsttid 40.
16. Sergeant DE (1963) Minke whales, *Balaenoptera acutorostrata* Lacépède, of the Western North Atlantic. Journal of the Fisheries Research Board of Canada 20: 1489–1504.
17. Wada S (1991) Genetic distinction between two minke whale stocks in the Okhotsk Sea coast of Japan.
18. IWC (2010) Report of the Scientific Committee. Journal of Cetacean Research and Management 11: 1–98.
19. Donovan GP (1991) A review of IWC stock boundaries. Cambridge, U.K. 39–68 p.
20. IWC (1992) Annex K. Report of the working group on North Atlantic minke trials. Rep Int Whal Comm 42: 246–251.
21. Hoelzel AR (1991) Whaling in the dark. Nature 352: 481–481.
22. Bakke I, Johansen S, Bakke Ø, El-Gewely MR (1996) Lack of population subdivision among the minke whales (*Balaenoptera acutorostrata*) from Icelandic and Norwegian waters based on mitochondrial DNA sequences. Marine Biology 125: 1–9.
23. Andersen LW, Born EW, Dietz R, Haug T, Øien N, et al. (2003) Genetic population structure of minke whales *Balaenoptera acutorostrata* from Greenland, the North East Atlantic and the North Sea probably reflects different ecological regions. Marine Ecology Progress Series 247: 263–280.

24. Anderwald P, Daníelsdóttir AK, Haug T, Larsen F, Lesage V, et al. (2011) Possible cryptic stock structure for minke whales in the North Atlantic: Implications for conservation and management. Biological Conservation 144: 2479–2489.

25. Pampoulie C, Daníelsdóttir AK, Víkingsson GA (2008) Genetic structure of the North Atlantic common minke whale (*Balaenoptera acutorostrata*) at feeding grounds: a microsatellite loci and mtDNA analysis. SC/F13/SP17 SC/60/PFI10 XXX: 1–17.

26. Palsbøll PJ (1989) Restriction fragment pattern analysis of mitochondrial DNA in minke whales, *Balaenoptera acutorostrata*, from the Davis Strait and the Northeast Atlantic: Copenhagen.

27. Martínez I, Pastene LA (1999) RAPD-typing of Central and Eastern North Atlantic and Western North Pacific minke whales, *Balaenoptera acutorostrata*. ICES Journal of Marine Science 56: 640–651.

28. Martínez I, Elvevoll EO, Haug T (1997) RAPD typing of north-east Atlantic minke whale (*Balaenoptera acutorostrata*). ICES Journal of Marine Science 54: 478–484.

29. Born EW, Outridge P, Riget FF, Hobson KA, Dietz R, et al. (2003) Population substructure of North Atlantic minke whales (*Balaenoptera acutorostrata*) inferred from regional variation of elemental and stable isotopic signatures in tissues. Journal of Marine Systems 43: 1–17.

30. Hobbs KE, Muir DCG, Born EW, Dietz R, Haug T, et al. (2003) Levels and patterns of persistent organochlorines in minke whale (*Balaenoptera acutorostrata*) stocks from the North Atlantic and European Arctic. Environmental Pollution 121: 239–252.

31. Daníelsdóttir AK, Sverrir DH, Sigfríður G, Alfreð Á (1995) Genetic variation in northeastern Atlantic minke whales (*Balaenoptera acutorostrata*). Developments in Marine Biology 4: 105–118.

32. Daníelsdóttir AK, Duke EJ, Árnason A (1992) Genetic variation at enzyme loci in North Atlantic minke whales, *Balaenoptera acutorostrata*. Biochemical Genetics 30: 189–202.

33. Árnason A (1995) Genetic markers and whale stocks in the North Atlantic ocean: a review. Developments in Marine Biology Volume 4: 91–103.

34. Skåre M (1994) Whaling: A sustainable use of natural resources or a violation of animal rights? Environment: Science and Policy for Sustainable Development 36: 12–31.

35. Skaug HJ, Bérubé M, Palsbøll PJ (2010) Detecting dyads of related individuals in large collections of DNA-profiles by controlling the false discovery rate. Molecular Ecology Resources 10: 693–700.

36. Haaland O, Glover K, Seliussen B, Skaug H (2011) Genotyping errors in a calibrated DNA register: implications for identification of individuals. BMC Genetics 12: 36.

37. Valsecchi E, Amos W (1996) Microsatellite markers for the study of cetacean populations. Molecular Ecology 5: 151–156.

38. Palsbøll PJ, Bérubé M, Larsen AH, Jørgensen H (1997) Primers for the amplification of tri- and tetramer microsatellite loci in baleen whales. Molecular Ecology 6: 893–895.

39. Bérubé M, Jørgensen H, McEwing R, Palsbøll PJ (2000) Polymorphic dinucleotide microsatellite loci isolated from the humpback whale, *Megaptera novaeangliae*. Molecular Ecology 9: 2181–2183.

40. Bérubé M, Palsbøll P (1996) Identification of sex in cetaceans by multiplexing with three ZFX and ZFY specific primers. Molecular Ecology 5: 283–287.

41. Árnason U, Gullberg A, Widegren B (1993) Cetacean mitochondrial DNA control region: sequences of all extant baleen whales and two sperm whale species. Molecular Biology and Evolution 10: 960–970.

42. Larsen AH, Sigurjonsson J, Øien N, Vikingsson G, Palsbøll P (1996) Populations genetic analysis of nuclear and mitochondrial loci in skin biopsies collected from Central and Northeastern North Atlantic humpback whales (*Megaptera novaeangliae*): Population identity and migratory destinations. Proceedings of the Royal Society of London Series B: Biological Sciences 263: 1611–1618.

43. Palsbøll PJ, Clapham PJ, Mattila DK, Larsen F, Sears R, et al. (1995) Distribution of mtDNA haplotypes in North-Atlantic humpback whales: The influence of behaviour on population structure. Marine Ecology Progress Series 116: 1–10.

44. Dieringer D, Schlötterer C (2003) MICROSATELLITE ANALYSER (MSA): A platform independent analysis tool for large microsatellite data sets. Molecular Ecology Notes 3: 167–169.

45. Peakall R, Smouse PE (2006) GenAlEx 6: genetic analysis in Excel. Population genetic software for teaching and research. Molecular Ecology Notes 6: 288–295.

46. Rousset F (2008) GENEPOP'007: a complete re-implementation of the genepop software for Windows and Linux. Molecular Ecology Resources 8: 103–106.

47. Pritchard JK, Stephens M, Donnelly P (2000) Inference of population structure using multilocus genotype data. Genetics 155: 945–959.

48. Corander J, Waldmann P, Marttinen P, Sillanpaa MJ (2004) BAPS 2: enhanced possibilities for the analysis of genetic population structure. Bioinformatics 20: 2363–2369.

49. Weir BS, Cockerham CC (1984) Estimating F-statistics for the analysis of population structure. Evolution 38: 1358–1370.

50. Slatkin M (1995) A measure of population subdivision based on microsatellite allele frequencies. Genetics 139: 457–462.

51. Wright S (1969) Evolution and the Genetics of Populations. Chicago: University of Chicago Press. 295 p.

52. Excoffier L, Laval G, Schneider S (2005) Arlequin ver. 3.0: An integrated software package for population genetics data analysis. Evolutionary Bioinformatics Online 1: 47–50.

53. Falush D, Stephens M, Pritchard JK (2003) Inference of population structure using multilocus genotype data: Linked loci and correlated allele frequencies. Genetics 164: 1567–1587.

54. Hubisz M, Falush D, Stephens M, Pritchard J (2009) Inferring weak population structure with the assistance of sample group information. Molecular Ecology Resources 9: 1322–1332.

55. Corander J, Waldmann P, Sillanpaa MJ (2003) Bayesian analysis of genetic differentiation between populations. Genetics 163: 367–374.

56. Corander J, Marttinen P (2006) Bayesian identification of admixture events using multilocus molecular markers. Molecular Ecology 15: 2833–2843.

57. Earl DA, von Holdt BM (2012) STRUCTURE HARVESTER: a website and program for visualizing STRUCTURE output and implementing the Evanno method. Conservation Genetics Resources 4: 359–361.

58. Evanno G, Regnaut S, Goudet J (2005) Detecting the number of clusters of individuals using the software STRUCTURE: a simulation study. Molecular Ecology 14: 2611–2620.

59. Besnier F, Glover KA (2013) ParallelStructure: a R package to distribute parallel runs of the population genetics program STRUCTURE on multi-core computers. PLoS ONE 8: e70651.

60. Jakobsson M, Rosenberg NA (2007) CLUMPP: a cluster matching and permutation program for dealing with label switching and multimodality in analysis of population structure. Bioinformatics 23: 1801–1806.

61. Rozas J, Sánchez-del Barrio JC, Messeguer X, Rozas R (2003) DnaSP, DNA polymorphism analyses by the coalescent and other methods. Bioinformatics 19: 2496–2497.

62. Tajima F (1989) Statistical Method for Testing the Neutral Mutation Hypothesis by DNA Polymorphism. Genetics 123: 585–595.

63. Fu YX (1997) Statistical tests of neutrality of mutations against population growth, hitchhiking and background selection. Genetics 147: 915–925.

64. Slatkin M, Hudson RR (1991) Pairwise comparisons of mitochondrial DNA sequences in stable and exponentially growing populations. Genetics 129: 555–562.

65. Rogers AR (1995) Genetic evidence for a Pleistocene population explosion. Evolution 49: 608–615.

66. Rogers AR, Harpending H (1992) Population growth makes waves in the distribution of pairwise genetic differences. Molecular Biology and Evolution 9: 552–569.

67. Ramos-Onsins SE, Rozas J (2002) Statistical properties of new neutrality tests against population growth. Molecular Biology and Evolution 19: 2092–2100.

68. Harpending HC (1994) Signature of ancient population growth in a low-resolution mitochondrial DNA mismatch distribution. Human Biology 66: 591–600.

69. Rogers AR, Harpending HC (1983) Population structure and quantitative characters. Genetics 105: 985–1002.

70. Schneider S, Excoffier L (1999) Estimation of past demographic parameters from the distribution of pairwise differences when the mutation rates vary among sites: Application to human mitochondrial DNA. Genetics 152: 1079–1089.

71. Bandelt HJ, Forster P, Rohl A (1999) Median-joining networks for inferring intraspecific phylogenies. Molecular Biology and Evolution 16: 37–48.

72. Polzin T, Vahdati Daneshmand S (2003) On Steiner trees and minimum spanning trees in hypergraphs. Operations Research Letters 31: 12–20.

73. Goudet J, Perrin N, Waser P (2002) Tests for sex-biased dispersal using bi-parentally inherited genetic markers. Molecular Ecology 11: 1103–1114.

74. Favre L, Balloux F, Goudet J, Perrin N (1997) Female-biased dispersal in the monogamous mammal *Crocidura russula*: Evidence from field data and microsatellite patterns. Proceedings of the Royal Society of London Series B: Biological Sciences 264: 127–132.

75. Mossman CA, Waser PM (1999) Genetic detection of sex-biased dispersal. Molecular Ecology 8: 1063–1067.

76. Greenwood PJ (1980) Mating systems, philopatry and dispersal in birds and mammals. Animal Behaviour 28: 1140–1162.

77. Øien N (1988) Length distributions in catches from the north-eastern Atlantic stock of minke whales. Rep Int Whal Comm 38: 289–295.

78. Horwood J (1990) Biology and exploitation of the minke whale. Boca Raton, Florida: CRC Press, Inc.

79. Laidre KL, Heagerty PJ, Heide-Jørgensen MP, Witting L, Simon M (2009) Sexual segregation of common minke whales (*Balaenoptera acutorostrata*) in Greenland, and the influence of sea temperature on the sex ratio of catches. ICES Journal of Marine Science: Journal du Conseil 66: 2253–2266.

80. Kanda N, Goto M, Kato H, McPhee MV, Pastene LA (2007) Population genetic structure of Bryde's whales (*Balaenoptera brydei*) at the inter-oceanic and trans-equatorial levels. Conservation Genetics 8: 853–864.

81. Bérubé M, Urbán J, Dizon AE, Brownell RL, Palsbøll PJ (2002) Genetic identification of a small and highly isolated population of fin whales (*Balaenoptera physalus*) in the Sea of Cortez, México. Conservation Genetics 3: 183–190.

82. Bérubé M, Aguilar A, Dendanto D, Larsen F, Notarbartolo Di Sciara G, et al. (1998) Population genetic structure of North Atlantic, Mediterranean Sea and

Sea of Cortez fin whales, *Balaenoptera physalus* (Linnaeus 1758): analysis of mitochondrial and nuclear loci. Molecular Ecology 7: 585–599.

83. Kanda N, Goto M, Yoshida H, Pastene LA (2009) Stock structure of sei whales in the North Pacific as revealed by microsatellite and mitochondrial DNA analyses.

84. Jorde PE, Schweder T, Bickham JW, Givens GH, Suydam R, et al. (2007) Detecting genetic structure in migrating bowhead whales off the coast of Barrow, Alaska. Molecular Ecology 16: 1993–2004.

85. Morin PA, Archer FI, Pease VL, Hancock-Hanser BL, Robertson KM, et al. (2012) Empirical comparison of single nucleotide polymorphisms and microsatellites for population and demographic analyses of bowhead whales. Endangered Species Research 19: 129–147.

86. Lang AR, Weller DW, Leduc RG, Burdin AM, Brownell RLj (2010) Genetic differentiation between Western and Eastern (*Eschrichtius robustus*) gray whale populations using microsatellite markers. University of Nebraska - Lincoln. 19 p.

87. Sremba AL, Hancock-Hanser B, Branch TA, LeDuc RL, Baker CS (2012) Circumpolar diversity and geographic differentiation of mtDNA in the critically endangered Antarctic blue whale (*Balaenoptera musculus intermedia*). PLoS ONE 7: e32579.

88. LeDuc RG, Weller DW, Hyde J, Burdin AM, Rosel PE, et al. (2002) Genetic differences between western and eastern gray whales (*Eschrichtius robustus*). Journal of Cetacean Research and Management 4: 1–5.

89. LeDuc RG, Dizon AE, Pastene LA, Kato H, Nishiwaki S, et al. (2007) Patterns of genetic variation in Southern Hemisphere blue whales and the use of assignment test to detect mixing on the feeding grounds. Journal of Cetacean Research and Management 9: 73–80.

90. Olavarría C, Baker CS, Garrigue C, Poole M, Hauser N, et al. (2007) Population structure of South Pacific humpback whales and the origin of the eastern Polynesian breeding grounds. Marine Ecology Progress Series 330: 257–268.

91. Patenaude NJ, Portway VA, Schaeff CM, Bannister JL, Best PB, et al. (2007) Mitochondrial DNA diversity and population structure among Southern right whales (*Eubalaena australis*). Journal of Heredity 98: 147–157.

92. Ruegg KC, Anderson EC, Scott Baker C, Vant M, Jackson JA, et al. (2010) Are Antarctic minke whales unusually abundant because of 20th century whaling? Molecular Ecology 19: 281–291.

93. Gill A, Fairbainrns RS (1995) Photo-identification of the minke whale *Balaenoptera acutorostrata* off the Isle of Mull, Scotland. In: Arnoldus Schytte

Blix LW, Øyvind U, editors. Developments in Marine Biology: Elsevier Science. pp. 129–132.

94. Dorsey EM, Stern SJ, Hoelzel AR, Jacobsen J (1990) Minke whales (*Balaenoptera acutorostrata*) from the west coast of North America: individual recognition and small-scale site fidelity. 357–368 p.

95. Dorsey EM (1983) Exclusive adjoining ranges in individually identified minke whales (*Balaenoptera acutorostrata*) in Washington state. Canadian Journal of Zoology 61: 174–181.

96. Baker CS, Slade RW, Bannister JL, Abernethy RB, Weinrich MT, et al. (1994) Hierarchical structure of mitochondrial DNA gene flow among humpback whales *Megaptera novaeangliae*, world-wide. Molecular Ecology 3: 313–327.

97. Tiedemann R, Hardy O, Vekemans X, Milinkovitch MC (2000) Higher impact of female than male migration on population structure in large mammals. Molecular Ecology 9: 1159–1163.

98. Birky CW, Maruyama T, Fuerst P (1983) An approach to population and evolutionary genetic theory for genes in mitochondria and chloroplasts, and some results. Genetics 103: 513–527.

99. Palumbi SR, Baker CS (1994) Contrasting population structure from nuclear intron sequences and mtDNA of humpback whales. Molecular Biology and Evolution 11: 426–435.

100. Avise JC, Ellis D (1986) Mitochondrial DNA and the evolutionary genetics of higher animals [and Discussion]. Philosophical Transactions of the Royal Society of London B, Biological Sciences 312: 325–342.

101. Birky CW, Fuerst P, Maruyama T (1989) Organelle gene diversity under migration, mutation, and drift: equilibrium expectations, approach to equilibrium, effects of heteroplasmic cells, and comparison to nuclear genes. Genetics 121: 613–627.

102. Engelhaupt D, Rus Hoelzel A, Nicholson C, Frantzis A, Mesnick S, et al. (2009) Female philopatry in coastal basins and male dispersion across the North Atlantic in a highly mobile marine species, the sperm whale (*Physeter macrocephalus*). Molecular Ecology 18: 4193–4205.

103. Lyrholm T, Leimar O, Johanneson B, Gyllensten U (1999) Sex-biased dispersal in sperm whales: contrasting mitochondrial and nuclear genetic structure of global populations. Proceedings of the Royal Society of London Series B: Biological Sciences 266: 347–354.

104. Yim H-S, Cho YS, Guang X, Kang SG, Jeong J-Y, et al. (2014) Minke whale genome and aquatic adaptation in cetaceans. Nat Genet 46: 88–92.

8

Fractured Genetic Connectivity Threatens a Southern California Puma (*Puma concolor*) Population

Holly B. Ernest[1,2*¤], **T. Winston Vickers**[1], **Scott A. Morrison**[3], **Michael R. Buchalski**[1,2], **Walter M. Boyce**[1]

1 Wildlife Health Center, School of Veterinary Medicine, University of California Davis, Davis, California, United States of America, 2 Wildlife and Ecology Unit, Veterinary Genetics Laboratory, School of Veterinary Medicine, University of California Davis, Davis, California, United States of America, 3 The Nature Conservancy, San Francisco, California, United States of America

Abstract

Pumas (*Puma concolor*; also known as mountain lions and cougars) in southern California live among a burgeoning human population of roughly 20 million people. Yet little is known of the consequences of attendant habitat loss and fragmentation, and human-caused puma mortality to puma population viability and genetic diversity. We examined genetic status of pumas in coastal mountains within the Peninsular Ranges south of Los Angeles, in San Diego, Riverside, and Orange counties. The Santa Ana Mountains are bounded by urbanization to the west, north, and east, and are separated from the eastern Peninsular Ranges to the southeast by a ten lane interstate highway (I-15). We analyzed DNA samples from 97 pumas sampled between 2001 and 2012. Genotypic data for forty-six microsatellite loci revealed that pumas sampled in the Santa Ana Mountains (n = 42) displayed lower genetic diversity than pumas from nearly every other region in California tested (n = 257), including those living in the Peninsular Ranges immediately to the east across I-15 (n = 55). Santa Ana Mountains pumas had high average pairwise relatedness, high individual internal relatedness, a low estimated effective population size, and strong evidence of a bottleneck and isolation from other populations in California. These and ecological findings provide clear evidence that Santa Ana Mountains pumas have been experiencing genetic impacts related to barriers to gene flow, and are a warning signal to wildlife managers and land use planners that mitigation efforts will be needed to stem further genetic and demographic decay in the Santa Ana Mountains puma population.

Editor: Adam Stow, Macquarie University, Australia

Funding: The work was supported by the following: California State Parks (WMB; http://www.parks.ca.gov/), California Department of Fish and Wildlife (HBE, WMB; https://www.dfg.ca.gov/), The Nature Conservancy (WMB, SAM; http://www.nature.org/); The McBeth Foundation (WMB; http://mcbethfoundation.com/); The Anza Borrego Foundation (WMB; http://theabf.org/), Nature Reserve of Orange County (WMB; http://www.naturereserveoc.org/); The National Science Foundation (WMB; http://www.nsf.gov/); and private donors (HBE, WMB). The Nature Conservancy (SAM) helped with sample and data collection and preparation of the manuscript. California Department of Fish and Wildlife, helped with sample and data collection. Otherwise, the funders had no role in study design, data collection and analysis, decision to publish, or preparation of the manuscript.

Competing Interests: The authors have declared that no competing interests exist.

* Email: hernest@uwyo.edu

¤ Current address: Department of Veterinary Sciences, University of Wyoming, Laramie, Wyoming, United States of America

Introduction

Genetic diversity, demography, and abundance – biological characteristics that influence population viability – can vary across a species' distribution. Species that are generally perceived as wide-ranging and abundant are sometimes relegated to status as "least conservation concern", in spite of indicators signaling concern and frequently, lack of data. Pumas (*Puma concolor*; also known as mountain lion, cougar, and in Florida, panther) epitomize this dilemma. Although pumas in California have not been subjected to hunting since 1972, and were designated as a Specially Protected Mammal in 1990 [1], there is minimal active management and little scientifically validated data on statewide or regional population numbers. Pumas in southern California have one of the lowest annual survival rates among any population in North America, on par with rates seen in hunted populations (unpublished data). They are under increasing threats from habitat loss and fragmentation, and mortality from vehicle strikes, depredation permits, poaching, public safety kills, wildfire, and poisoning [2,3]. Timely evaluation of potential threats to population viability is imperative in order to prioritize conservation activities to prevent collapse of some populations.

The human population of southern California is over 20 million [4] and expected to exceed 30 million by 2060 [5]. This increasing population will likely result in further loss, fragmentation, and degradation of natural habitats in the region. Habitat fragmentation south of greater Los Angeles has effectively turned the Santa Ana Mountain range in mostly Orange and Riverside counties into a 'mega-fragment' of habitat, surrounded to the west, north, and east by dense urban land uses. The only remaining montane and foothill habitat linkage connecting the Santa Ana Mountain range to other mountains of the Peninsular Range is a southeasterly swath of habitat bisected by a very heavily traveled 10-lane highway, Interstate 15 (I-15) (Figure 1).

Population viability of pumas in the Santa Ana Mountains (a geography henceforth referred to as distinct from the broader Peninsular Ranges to the east) has been of conservation concern

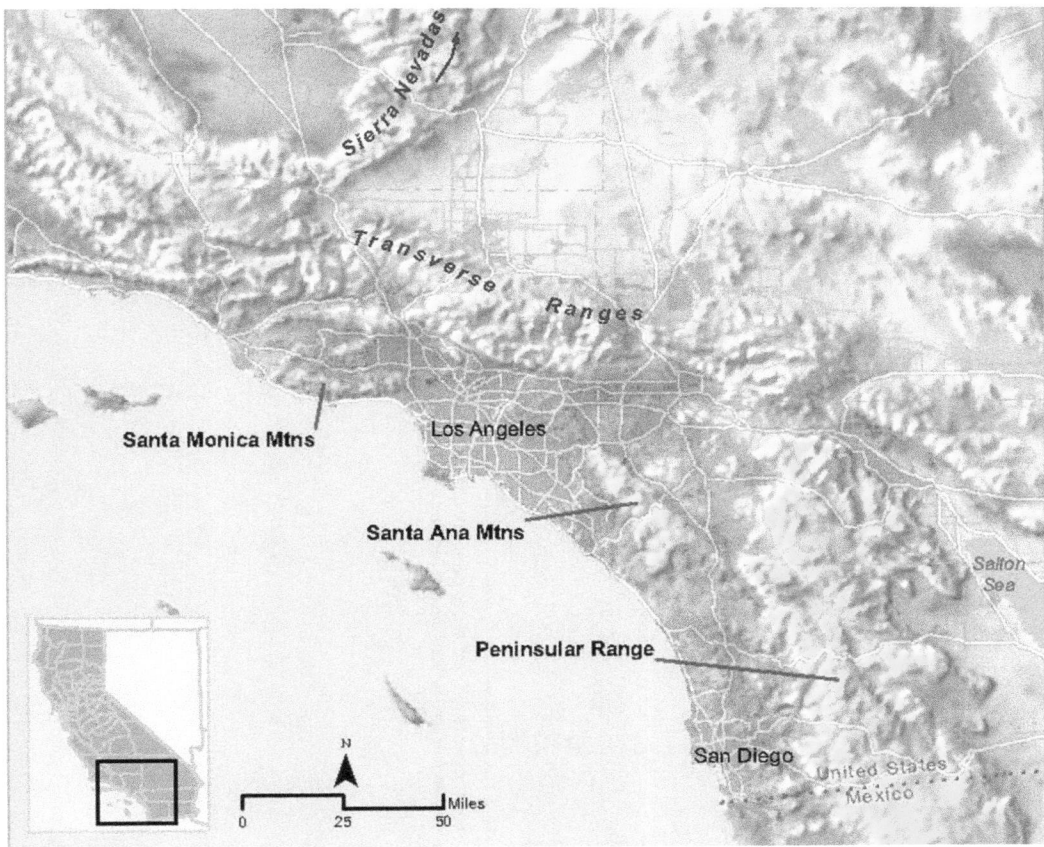

Figure 1. Topographic map depicting location of Santa Ana Mountains, eastern Peninsular Ranges in southern California, and adjacent regions. Inset shows location in the state of California.

for decades. Population monitoring and modeling in the 1980s highlighted that urbanization and highways were fragmenting puma habitat (e.g., [6]), and that in turn motivated efforts to protect habitat connectivity in the region (e.g., [7,8]). As part of a statewide assessment of puma genetic diversity and population structure, Ernest et al. [9] employed an 11-locus microsatellite panel and found that, for a limited sample size (n = 14) Santa Ana pumas had lower genetic diversity than other populations in California. Since 2001, pumas in the region have been the subject of an ongoing study by the Karen C. Drayer Wildlife Health Center of the University of California, Davis (UCD) School of Veterinary Medicine. Telemetry data from 74 pumas in the UCD study has confirmed that minimal connectivity (only one GPS-collared puma over ten years was documented to transit successfully; unpublished data) exists between the Santa Ana Mountains and the eastern Peninsular Ranges across I-15, confirming that previous connectivity concerns were warranted.

We conducted a detailed appraisal of the genetic diversity, relatedness, and population structure of southern California puma populations. Using 97 samples collected over 12 years as part of the UCD study, and a 46-locus microsatellite panel, we evaluated levels of genetic diversity, estimated effective population sizes and tested whether genetic data supported a hypothesis of recent bottleneck in the populations. We assessed whether genetics reflected our telemetry observations of infrequent puma crossings of I-15 between the Santa Ana Mountains and the Peninsular Ranges to the east. Additionally we explored inter-population gene flow at multiple time scales by employing methods that reflect

recent (a few generations) and more historical (tens or more generations). Finally, we tested our hypothesis that the Santa Ana population had lower genetic diversity than those sampled from other regions in California.

Materials and Methods

Samples
We obtained blood or tissue samples for analysis of nuclear DNA from pumas captured for telemetry studies, and from those found dead or killed by state authorities for livestock depredation or public safety in San Diego, Orange, Riverside, and San Bernardino counties of southern California (n = 97) during 2001–2012 (Figure 2). Pumas captured for telemetry were captured and sampled as detailed in [10]. Forty-two samples were collected to the west of I-15 in the Santa Ana Mountains, and 55 samples were collected in the Peninsular Ranges to the east of I-15. A small number of additional samples were collected from deceased animals in San Bernardino County just to the north of the Peninsular Range across Interstate Highway 10. For population genetic comparisons with pumas sampled elsewhere throughout California, a 257 sample subset of our statewide puma DNA data archive was employed (regions and sample sizes detailed in Table 1 and depicted in Figure 1 in [9])

Ethics Statement
Animal handling was carried out in strict accordance with the recommendations and approved Protocol 10950/PHS, Animal

Table 1. Genetic diversity summary statistics for southern California pumas (n = 97) relative to other populations in California (n = 257).

Sampling Region	Abbrev.		N	Na	AR	Ho	He	I	%P
North Coast	NC	Mean	29	3.6	2.0	0.41	0.44	0.80	98%
		SE		0.2	0.1	0.03	0.03	0.05	
Modoc Plateau & Eastern Sierra Nevada	MP-ESN	Mean	51	4.2	2.4	0.52	0.54	0.98	100%
		SE		0.3	0.1	0.03	0.03	0.05	
Western Sierra Nevada	WSN	Mean	47	4.2	2.4	0.47	0.51	0.95	98%
		SE		0.2	0.2	0.03	0.03	0.06	
Central Coast north	CC-N	Mean	83	3.2	1.9	0.41	0.41	0.70	98%
		SE		0.2	0.1	0.03	0.03	0.06	
Central Coast: central	CC-C	Mean	21	3.4	2.1	0.43	0.46	0.81	96%
		SE		0.2	0.1	0.03	0.03	0.05	
Santa Monica Mountains / Central Coast: South	CC-S	Mean	26	2.2	1.7	0.38	0.33	0.53	76%
		SE		0.1	0.1	0.04	0.03	0.05	
Peninsular Range-East	PR-E	Mean	55	3.1	2.0	0.43	0.41	0.74	87%
		SE		0.2	0.2	0.04	0.04	0.07	
Santa Ana Mountains	SAM	Mean	42	2.3	1.6	0.33	0.32	0.54	80%
		SE		0.2	0.1	0.03	0.03	0.05	

Abbrev. = region abbreviations used in Tables and Figures. Mean with standard error (SE). N = sample size. Na = average number of different alleles per locus. AR = allelic richness, standardized to sample size. Ho = observed heterozygosity. He = expected heterozygosity. I = Shannon's information index (Sherwin et al 2006). %P = percent of polymorphic loci. Regions are detailed further in text and generally follow California Bioregions designations. (http://biodiversity.ca.gov/bioregions.html).

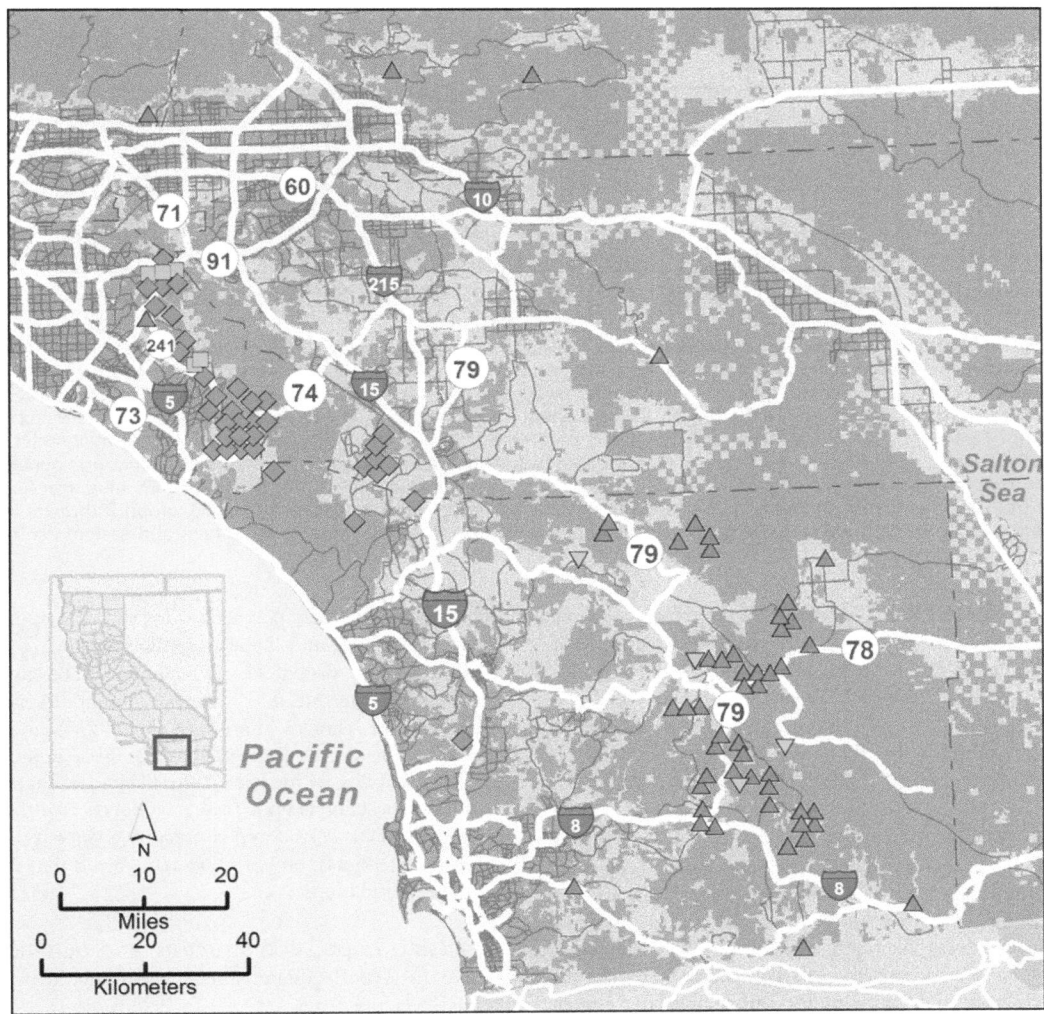

Figure 2. Map of puma capture locations in the Santa Ana Mountains and eastern Peninsular Ranges of southern California. Colors of symbols represent genetic group assignment inferred from Bayesian clustering analysis (STRUCTURE analysis, see Figure 4). Genetic group A-1 = green diamonds; A-2 = red triangles (apex at top). One male puma (M86) captured in the Santa Ana Mountains had predominant genetic assignment to the A-2 (red) genetic group. Five individuals (light green squares) captured in the Santa Ana Mountains had partial assignment to the A-2 group (M91, F92, M93, M97 and F102). Molecular kinship analysis showed that M86 and a female (F89) captured in the Santa Ana Mountains were parents of pumas M91, F92, and M93 (captured in the Santa Ana Mountains). Puma M97 assigned in parentage to M86 and F61, while F102 had unknown parentage (no parentage assignments; due possibly to her death early in project prior to collection of most of the samples). Three individuals (orange triangles, apex at bottom), had partial assignment (however, less than 20%) to A-1.

Welfare Assurance number A3433-01, with capture and sampling procedures approved by the Animal Care and Use Committee at the University of California, Davis (Protocol #17233), and Memoranda of Understanding and Scientific Collecting Permits from the California Department of Fish and Wildlife (CDFW). Permits and permissions for access to conserved lands at puma capture and sampling sites were obtained from CDFW, California Department of Parks and Recreation, The Nature Conservancy, United States (US) Fish and Wildlife Service, US Forest Service, US Bureau of Land Management, US Navy/Marine Corps, Orange County Parks Department, San Diego County Parks Department, San Diego State University, Vista Irrigation District, Rancho Mission Viejo/San Juan Company, Sweetwater Authority, California Department of Transportation (CalTrans), and the City of San Diego Water Department.

DNA Extraction and Microsatellite DNA data collection

Whole genomic DNA was extracted using the DNeasy Blood & Tissue Kit (QIAGEN, Valencia, CA, USA). Fifty microsatellite DNA primers were initially screened for this project. Forty-six loci that performed well in multiplex PCR (using the QIAGEN Multiplex PCR kit; QIAGEN) and conformed to expectations for Hardy-Weinberg and linkage equilibria were selected for ultimate analysis [11,12,13]. One sex-identification locus (Amelogenin) was used to confirm sex in samples from degraded puma carcasses [14].

PCR products were separated with an ABI PRISM 3730 DNA Analyzer (Applied Biosystems Inc., Foster City, CA, USA) with each capillary containing 1 μL of a 1:10 dilution of PCR product and deionized water, 0.05 μL GeneScan-500 LIZ Size Standard and 9.95 μL of HiDi formamide (both products Applied Biosystems Inc.) that was denatured at 95°C for 3 min. Products

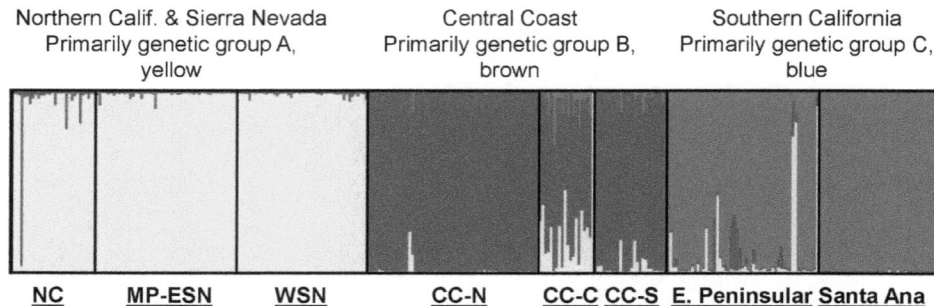

Figure 3. California puma population genetic structure. STRUCTURE bar plot displaying the genetic clustering relationship of southern California pumas relative to others in California. Three major genetic groups, A (blue, on right), B (brown, in center), and C (yellow, on left), are evident for analysis of 354 individuals sampled throughout California. Abbreviations: NC = North Coast, MP-ESN = Modoc Plateau & Eastern Sierra Nevada, WSN = Western Sierra Nevada, CC-N = Central Coast: north, CC-C = Central Coast: central, CC-S = Central Coast: South (Santa Monica Mountains), PR-E = Peninsular Range-East, SAM = Santa Ana Mountains. The plot is organized by grouping individuals in order of their geographic region sampling source. Proportional genetic assignment for each puma is represented by a vertical bar, most easily visualized for pumas that genetically assigned to a group different from most others sampled in its region (for example one individual with over 80% brown and 8% blue near far left of group A). Pumas primarily from the Sierra Nevada Range and northern California are represented by group A (yellow), group B (brown) includes primarily Central Coast pumas and group C (blue) represents primarily southern California pumas (Santa Ana Mountains and eastern Peninsular Ranges).

were visualized with STRand version 2.3.69 [15]. Negative controls (all reagents except DNA) and positive controls (well-characterized puma DNA) were included with each PCR run. Samples were run in PCR at each locus at least twice to assure accuracy of genotype reads and minimize risk of non-amplifying alleles. For >90% samples, loci that were heterozygous were run at least twice and homozygous loci were run at least three times.

Genetic diversity

The number of alleles (Na), allelic richness (AR; incorporates correction for sample size), observed heterozygosity (Ho), expected heterozygosity (He), Shannon's information index [16], and tests for deviations from Hardy-Weinberg equilibrium were calculated using software GenAlEx version 6.5 [17,18]. Shannon's information index provides an alternative method of quantifying genetic diversity and incorporates allele numbers and frequencies. Testing for deviations from expectations of linkage equilibrium was conducted using Genepop 4.2.1 [19], and we tested for the presence of null alleles using the program ML RELATE [20]. We assessed significance for calculations at alpha = 0.05 and used

sequential Bonferroni corrections for multiple tests [21] in tests for Hardy-Weinberg and linkage equilibria.

The average probability of identity (PID) was calculated two ways using GenAlEx: 1) assuming random mating (PID_{RM}) without close relatives in a population [22], and 2) assuming that siblings with similar genotypes occur in a population (PID_{SIBS}) [23]. Probability of identity is the likelihood that two individuals will have the same genetic profile (genotype) for the DNA markers used. PID_{SIBS} is considered conservative since it probably conveys a higher likelihood; however, we recognized that siblings occurred in these populations.

Assessing population structure and genetic isolation

We used a Bayesian genetic clustering algorithm (STRUC-TURE version 2.3.4 [24,25]) to determine the likely number of population groups (K; genetic clusters) and to probabilistically group individuals without using the known geographic location of sample collection. We used the population admixture model with a flat prior and assumed that allele frequencies were correlated among populations, and ran 50,000 Markov chain Monte Carlo repetitions following a burnin period of 10,000 repetitions. First,

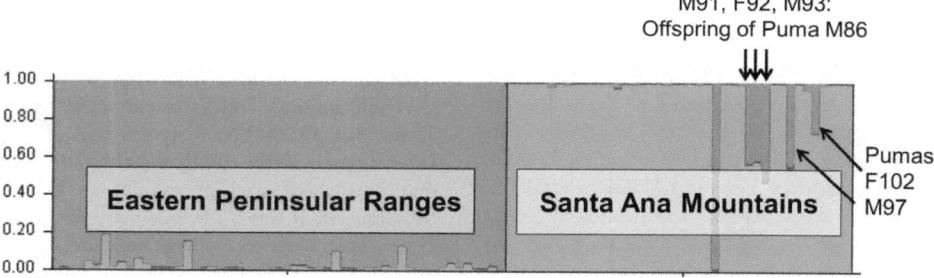

Figure 4. Southern California puma population genetic structure. Bar Plot displaying results of STRUCTURE analysis focused on genotypic data from 97 southern California pumas (the blue block from Figure 3). With removal of the strong genetic signal from northern California and Central Coast samples (see Figure 3), two distinct southern California groupings were inferred, C-1 (green, on right) and C-2 (red, on left). These reflect the two regions: Santa Ana Mountains to the west of I-15 (predominantly genetic group C-1) and eastern Peninsular Ranges to the east of I-15 (predominantly genetic group C-2). Genetic clustering is dependent on genetic variance among samples included in the analysis. One male puma (M86) captured in the Santa Ana Mountains has predominant genetic assignment to the C-2 (red) genetic group (the predominant genetic cluster for PR-E), and five others had partial assignment to the C-2 group (M91, F92, M93, M97 and F102). Molecular kinship analysis showed that M86 and a female (F89) assigning to the C-1 genetic group were parents of pumas M91, F92, and M93 (all were captured in the Santa Ana Mountains).

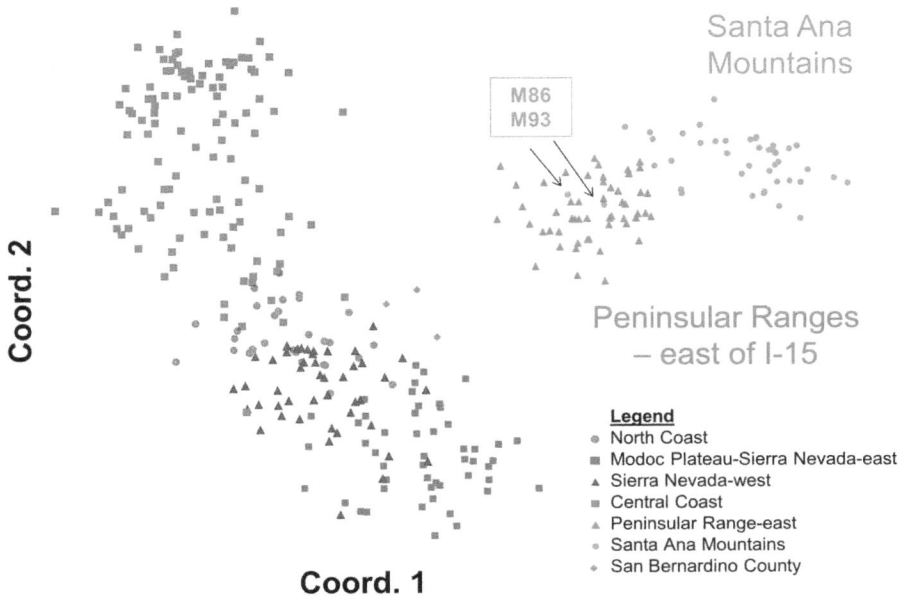

Figure 5. Principal Coordinates analyses (PCoA) constructed using genetic covariance matrices (GenAlEx) for 354 California puma genetic profiles including 97 from southern California. Patterns displayed for first two axes of variation within the genetic data set. Each point, color-coded to its sampling region, represents an individual puma. Note that colors in PCoA diagrams reflect geographic source of samples and not STRUCTURE genetic cluster assignment. Abbreviations and sample sizes per Table 1. Arrows denote pumas described in Figure 4.

an analysis including 354 statewide puma genotypes (97 from southern California and 257 from other regions) was run to estimate the probability of one through 10 genetic clusters (K), with each run iterated three times. Second, given the output of the statewide run, we ran an analysis using only the 97 southern California puma genotypes to estimate the probability of one through five K, with each run iterated three times. Employing STRUCTURE HARVESTER [26] we averaged log probability

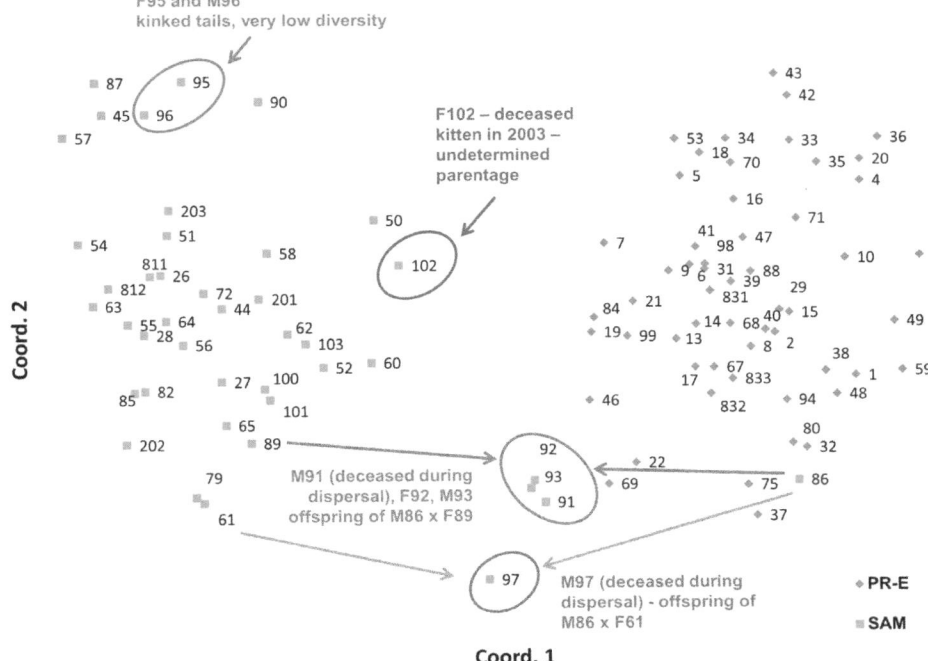

Figure 6. Principal Coordinates analyses (PCoA) via covariance matrices for 97 southern California puma genetic profiles as conducted in GenAlEx. Patterns displayed for first two axes of variation within the genetic data set. Each point represents an individual puma, and has sample identification number and color-coding to sampling region. Note that colors in PCoA diagrams reflect geographic source of samples and not STRUCTURE genetic cluster assignment. Abbreviations and sample sizes per Table 1.

Table 2. Wright's F_{ST} values indicate that southern California mountain lion populations are genetically distinct from other populations in California.

	NC	MP-ESN	WSN	CC-N	CC-C	CC-S	PR-E	SAM
North Coast (NC)	0							
Modoc Plateau & Eastern Sierra Nevada (MP-ESN)	0.09	0						
Western Sierra Nevada (WSN)	0.05	0.03	0					
Central Coast: north (CC-N)	0.13	0.14	0.12	0				
Central Coast: central (CC-C)	0.08	0.09	0.06	0.07	0			
Santa Monica Mountains (CC-S)	0.17	0.15	0.12	0.14	0.10	0		
Peninsular Range-East (PR-E)	0.13	0.08	0.09	0.12	0.09	0.16	0	
Santa Ana Mountains (SAM)	0.18	0.16	0.17	0.19	0.17	0.27	0.07	0

Note that one of the geographically closest puma populations, Santa Monica Mountains, has highest F_{ST} with the Santa Ana population, evidence of high genetic isolation for both regions. Probability, P(random $\geq$ data) based on 9999 permutations for all values are <0.001. Abbreviation definitions and sample sizes are included in Table 1.

of the data given K, log Pr(X|K), statistics across the multiple runs for each of the K estimates. In each case (statewide and southern California), we selected the K value of highest probability by identifying the set of values where the log Pr(X|K) value was maximized and subsequently selected the minimum value for K that did not sacrifice explanatory ability [27,28,29]. We defined membership to a cluster based upon the highest proportion of ancestry to each inferred cluster.

To further assess and visualize genetic relationships among regions and individuals, we performed principal coordinates analyses (PCoA) via covariance matrices with data standardization [30] using GenAlEx. This is a technique that allowed us to explore and plot the major patterns within the data sets. The PCoA process located major axes of variation within our multidimensional genotype data set. Because each successive axis explains proportionately less of the total genetic variation, the first two axes were used to reveal the major separation among individuals. Employing Genalex software, a pairwise, individual-by-individual genetic distance matrix was generated and then used to create the PCoA.

Wright's F-statistic, $F_{ST,}$ was calculated to appraise how genetic diversity was partitioned between populations. As implemented in GenAlEx, we used Nei's [31] formula, with statistical testing options offered through 9999 random permutations and bootstraps.

Detecting migrants

We used GENECLASS2 version 2.0.h [32] to identify first-generation migrants, i.e. individuals born in a population other than the one in which they were sampled. Genetic clusters identified during STRUCTURE analysis were treated as putative populations. GENECLASS2 provides different likelihood-based test statistics to identify migrant individuals, the efficacy of which depends on whether all potential source populations have been sampled. We first calculated the likelihood of finding a given individual in the population in which it was sampled, L_h, assuming all populations had not been sampled. We then calculated L_h/L_{max}, the ratio of L_h to the greatest likelihood among the populations [33], which has greater power when all potential source populations have been sampled. The critical value of the test statistic (L_h or L_h/L_{max}) was determined using the Bayesian approach of Rannala and Mountain [34] in combination with the resampling method of Paetkau et al. [33]; i.e., Monte Carlo simulations carried out on 10,000 individuals with the significance level set to 0.01.

Testing for bottlenecks and inferring effective population size

We tested for evidence of recent population size reductions in Santa Ana Mountains and eastern Peninsular Range regions with one-tailed Wilcoxon sign-rank tests for heterozygote excess in the program BOTTLENECK version 1.2.02 [35]. The program evaluates whether the reduction of allele numbers occurred at a rate faster than reduction of heterozygosity, a characteristic of populations which have experienced a recent reduction of their effective population size (Ne) [35,36]. This bottleneck genetic signature is detectable by this test for a finite time, estimated to be less than 4 times Ne generations [37]. These tests were performed using the two-phase (TPM, 70% step-wise mutation model and 30% IAM) model of microsatellite evolution and 10,000 iterations.

We then estimated contemporary Ne for each of the two regions based on gametic disequilibrium with sampling bias correction [38] using LDNE version 1.31 [39]. Ne is formally defined as the size of the ideal population that would experience the same

Table 3. Effective population size estimations and indications of recent genetic bottlenecks in southern California pumas.

	Mode	TPM	Ne (P-CI; JK-CI)
Santa Ana Mtns	Shifted mode	0.009	5.1 (3.3–6.7; 3.3–6.6)
Peninsular Range, East	Normal L	0.19	24.3 (21.7–27.3; 20.6–28.8)

Listed by column are p-values for population bottleneck tests (Wilcoxon sign-rank test; BOTTLENECK) assuming the two-phase (TPM) model of microsatellite evolution. Effective size (Ne) estimations (95% CI) based on data from 42 microsatellite loci. The Santa Ana Mountains population exhibited clear evidence of a population bottleneck. Effective population size estimate using the point estimate linkage disequilibrium method of (LDNE, Waples 2006) with 95% confidence intervals (CI) for both parametric (P) and jackknifed (JK) estimates.

amount of genetic drift as the observed population [40]. These analyses excluded alleles occurring at frequencies ≤0.05, and we used the jackknife method to determine 95% confidence intervals [38].

Relatedness analyses: pairwise coefficient and internal

Molecular kinship analysis was conducted using a number of software packages. Pairwise relatedness among individuals was evaluated using the algorithm of Lynch and Ritland [41], with reference allele frequencies calculated and relatedness values averaged within each southern California population, as implemented in GenAlEx. Partial molecular kinship reconstruction was conducted using a consensus of outputs from the GenAlEx pairwise relatedness calculator, ML Relate [20], CERVUS version 3.0.3 [42], and Colony version 2.0.3.1 [43,44]. Individual genetic diversity (also called internal relatedness) was assessed using Rhh [45] as implemented in R statistical software [46]. This is a measure of genetic diversity within each individual (an estimate of parental relatedness [47]), and we averaged over individuals for each of the two regions of southern California. Significance of differences between means was evaluated using t tests.

Results

Forty-two of the 46 loci that we employed were polymorphic in southern California and selected for the subsequent analyses. The average probabilities of identity with assumptions of either random mating (PID$_{RM}$) or mating among sibs (PID$_{SIBS}$) across the 42 loci for the eastern Peninsular Ranges were (PID$_{RM}$) 6.3×10^{-22} and (PID$_{SIBS}$) 3.1×10^{-10}, and for the Santa Ana Mountains were (PID$_{RM}$) 2.8×10^{-15} and (PID$_{SIBS}$) 1.1×10^{-7} respectively. These very small values indicate that the panel of genetic markers provided very high resolution to distinguish individuals. For

example, given this data the probability of seeing the same multi-locus genotype in more than one puma was less than one in nine million for Santa Ana Mountains pumas.

Genetic diversity

Measures of genetic variation including allelic diversity, heterozygosity, Shannon's information index, and polymorphism, were lower for Santa Ana pumas than most of those tested from other regions of California (Table 1). Such low genetic diversity indicators were approached only by pumas in the Santa Monica Mountains (Ventura and Los Angeles Counties), a neighboring remnant puma population in the north Los Angeles basin (Figure 1).

Population Structure

Bayesian clustering analysis (STRUCTURE; Figure 3 of statewide puma genetic profiles (n = 354), including 97 from southern California, also support genetic distinctiveness of Santa Ana Mountains and eastern Peninsular Range pumas from other populations in the state. Three main genetic groups (A, B, and C) were evident in the analysis (Figure 3) The 97 pumas sampled in southern California (right-hand set of bars in Figure 3, with samples from Santa Ana and eastern Peninsular Range pumas labeled) predominantly cluster within genetic group C. The Santa Ana pumas assign very tightly to group C (0.996 average probability assignment), while pumas of the eastern Peninsular Ranges showed more variable assignment (0.93 average probability assignment), with 9 individuals (16%) having less than 0.90 assignment. Pumas sampled in the Central Coast of California (which included Santa Monica Mountains pumas) make up the central set of bands, and those individuals predominantly assign to the genetic group B. Pumas sampled in the other regions of California (North Coast Ranges, Modoc Plateau, western Sierra

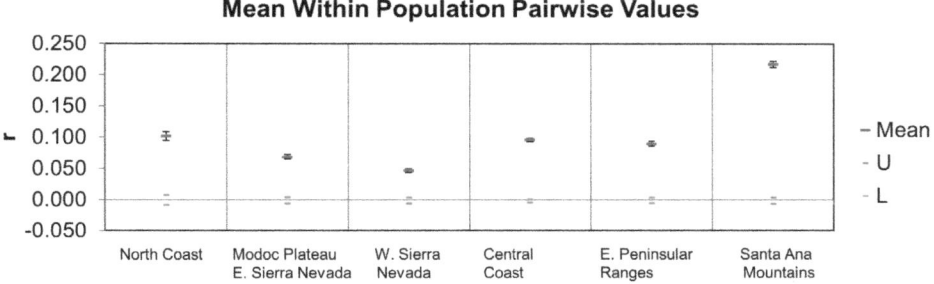

Figure 7. Average pairwise relatedness (r; blue bars with confidence intervals) for pumas sampled in southern California relative to other regions in California. Algorithm of Lynch and Ritland (1999) as implemented in GenAlEx. Expected range for "unrelated" is shown as red bars with confidence intervals. The average relatedness of Santa Ana Mountain pumas is higher than those sampled in Peninsular Ranges east of I-15 and for any other region tested in California. Relatedness in the Santa Ana Mountains pumas approaches second order family relationship (half sibs, niece-aunt, grandparent-grandchild, etc.). Abbreviations listed in Table 1.

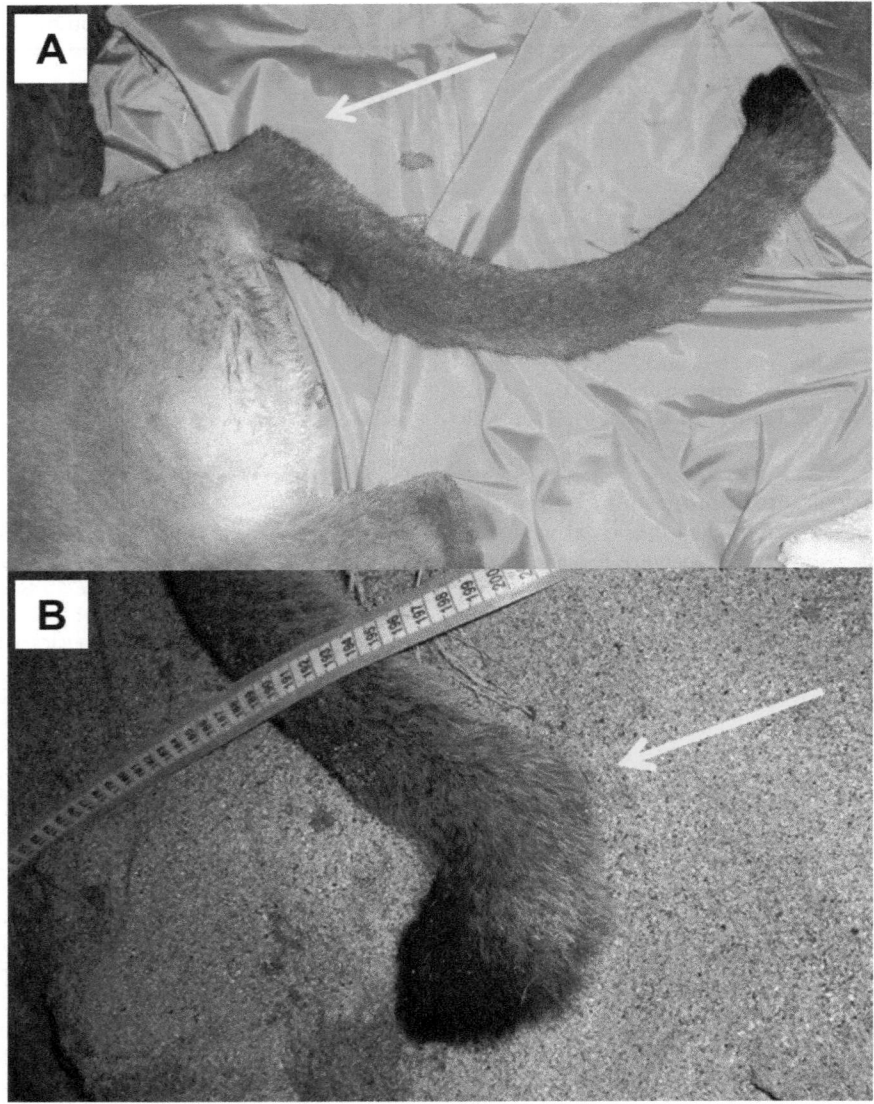

Figure 8. Photographs of kinked tails of pumas F95 (a) and M96 (b). Arrows indicate kink sites. Puma F95 had tail kink at base of tail and Puma M96 had tail kink near distal tip of tail. These two pumas had among the lowest genetic diversity measured in this study.

Nevada, and eastern Sierra Nevada) predominantly cluster with the genetic group A. Notably, there are individuals sampled in each geographic area which cluster with a genetic group that is not the dominant one in that area, suggesting dispersal events and/or genetic exchange that have occurred to varying degrees in each region.

A STRUCTURE analysis focused only on genetic data from the 97 southern California pumas indicated two distinct genetic groups (C-1 and C-2 shown in Figure 4). Pumas sampled in the eastern Peninsular Range region east of I-15 group primarily with C-2 and those of the Santa Ana Mountain region on the west side of I-15 group with C-1. An exception to the consistent genetic clustering was an adult male (M) puma (M86), that was captured in the Santa Ana Mountains but clustered with pumas from the eastern Peninsular Ranges (primarily genetic group C-2). Five other pumas captured in the Santa Ana Mountains had a 30–50% assignment to the C-2 group (M91, F92, M93, M97 and F102). Molecular kinship analysis showed that M86 and a female (F89)

captured in the Santa Ana Mountains and assigned to the C-1 genetic group were the likely parents of three of these pumas (M91, F92, and M93) (results of relatedness and kinship analyses). M86 also was the likely parent of another puma in the group (M97), an offspring of another female (F61) that was sampled in Santa Ana Mountains and clustered with the C-1 genetic group. F102 was a <1 year old female killed by a vehicle in 2003 prior to collection of the majority of samples from adults in the Santa Ana Mountains.

Principal coordinates analysis of statewide puma genetic profiles (n = 354) (PCoA; Figure 5) allowed graphical examination of the first two major axes of multivariate genetic variation, and confirmed and added detail to the genetic distinctiveness of southern California pumas relative to others in California. The PCoA also reinforced the distinctiveness of pumas sampled in the Santa Ana Mountains from those sampled in the eastern Peninsular Ranges. Most pumas sampled in the Santa Ana Mountains align in a cloud of data points distinct from the eastern

Peninsular Range pumas, and were the most genetically distant from all other pumas tested in California (Figure 5). The analysis also confirms the STRUCTURE findings that M86 who was sampled in the Santa Ana Mountains genetically aligns with the pumas sampled in the Peninsular Ranges, as does one of his offspring, M93 (see Figure 6 for additional detail). The PCoA position of data points for three pumas sampled in the San Bernardino Mountains north of Peninsular Ranges (pink diamonds in Figure 5) illustrates an intermediate genetic relationship between pumas from the rest of California and pumas sampled in the eastern Peninsular Ranges and Santa Ana Mountains, and suggests that they may represent transitional gene flow signature between southern California and regions to the north and east.

PCoA analysis of only the samples collected in the Santa Ana and Peninsular Ranges (Figure 6) confirms the findings from the STRUCTURE analysis indicating genetic distinctiveness of these two populations despite geographic proximity. Siblings M91, F92, and M93 (offspring of F89 and M86 according to our kinship reconstructions) as well as M97 (likely offspring of a female puma captured in the Santa Ana Mountains, F61, and M86, according to kinship reconstructions) are located graphically midway between their parents' PCoA locations.

Genetic isolation

Wright's F_{ST} calculations (Table 2) indicate that Santa Ana Mountains pumas are the most isolated of those tested throughout California (p = 0.0001). Despite the short distance (as short as the distance across the I-15 Freeway) between the Santa Ana Mountains and the eastern Peninsular Range region, F_{ST} was surprisingly high (0.07) given the very close proximity of the two regions (separated only by an interstate highway). The Santa Monica Mountains pumas and Santa Ana Mountains pumas had the highest F_{ST} (0.27; lowest gene flow) of all pairwise comparisons in the state, demonstrating a high level of genetic isolation between these regions. The Santa Monica Mountains and Santa Ana Mountains are less than 100 km direct distance apart, through the center of Los Angeles. However the more likely distance for puma travel between these two mountain ranges, avoiding urban areas and maximizing upland habitat, would likely exceed 300 km (estimated using coarse measurements on Google Earth, Google, Inc.).

Detection of migrants

GENECLASS2 identified four individuals as first-generation migrants (P<0.01), four with the L_h method (pumas F75, M80, M86, and M99), and one with the L_h/L_{max} ratio (M86, which was detected using both likelihood methods). Pumas F75, M80, and M99 were all captured from the San Bernardino Mountains (Figure 2) at the northern extent of the study region, yet clustered with individuals from the Eastern Peninsular Range during STRUCTURE analysis. Their migrant designation may suggest immigration from populations north of Los Angeles and/or a distinct genetic population within the San Bernardino region. Puma M86 was captured in the Santa Ana Mountains, but assigned strongly to the eastern Peninsular Range genetic cluster, indicating a seemingly clear population of origin. This individual assignment is in accord with the clustering results from STRUCTURE (Figure 4).

Evidence of genetic bottlenecks

The Santa Ana Mountains population exhibited clear evidence of a population bottleneck (Table 3; Wilcoxon sign-rank test for heterozygote excess, and detection of a shift in the allele frequency distribution mode [36]; BOTTLENECK software). The eastern

Peninsular Range mountain lions did not show a strong signature of a bottleneck.

Effective population size

Effective population size (Ne) estimations using the linkage disequilibrium method (LDNe program) were 5.1 for the Santa Ana Mountains population and 24.3 for mountain lions in the eastern Peninsular Ranges. Statistical confidence intervals for both regions, given the genetic data, were tight (Table 3).

Relatedness: pairwise coefficient and internal

The average pairwise coefficient of relatedness (r, Figure 7) was highest in Santa Ana Mountains pumas relative to all others tested in California (0.22; 95% confidence interval of 0.22–0.23), a level that approaches second order kinship relatedness (half-sibs, grantparent/grandchild, aunt-niece, etc). The value for the eastern Peninsular Ranges was 0.10 (confidence interval of 0.09–0.10), less than that of third order relatives (first cousins, great-grandparent/great grandchild). Other regions of California averaged similar or lower values to those of eastern Peninsular Ranges (Figure 7).

Among pumas sampled in the Santa Ana Mountains, the population average (0.14) for internal relatedness as implemented in rHH software was significantly higher (t test; p = 5.8×10^{-6}) than for those sampled in the eastern Peninsular Ranges (0.001). Of a group of six pumas which clustered near one another in PCoA (Figure 6), five have among the lowest individual genetic diversity measured in southern California (Puma ID [Internal Relatedness value]: F45 [0.37], F51 [0.37], M87 [0.28], F90 [0.21], F95 [0.38], and M96 [0.33]). Notably, pumas F95 and M96 (highest internal relatedness) were observed with kinked tails at capture in the Santa Ana Mountains (Figure 8).

Discussion

Pumas of the Santa Ana Mountains are genetically depauperate, isolated, and display signs of a recent and significant bottleneck. In general, coastal California puma populations have less genetic diversity and less gene flow from other populations than those farther inland [9] (Table 1). This study showed that two coastal populations (Santa Ana Mountains and Santa Monica Mountains) had particularly low genetic variation and gene flow from other regions. Lack of gene flow is likely due in part to natural barriers to puma movement: geography and habitat (Pacific Ocean to the west; less hospitable desert habitat bounding certain regions, etc.). However, our data suggest that anthropogenic developments on the landscape are playing a large role in genetic decay in the Santa Ana Mountains puma population. As large solitary carnivores with sizable habitat requirements, pumas are extremely sensitive to habitat loss and fragmentation [48,49].

The genetic bottleneck in the Santa Ana Mountains pumas is estimated at less than about 80 years, depending on definitions of effective population size (Ne) and puma generation time. Luikhart and Cornuet [37] state that the bottleneck signatures decay after "4 times Ne [here estimated to be 5.1] generations". Logan and Sweanor [50] estimated generation time for their New Mexico population of pumas to be 29 months (2.4 years) for females. If an allowance of 2.4–4.0 years is made for generation times (unknown) in the Santa Ana Mountains population, the maximum estimated time since a bottleneck would be about 40–80 years. This was a period of tremendous urban development and multi-lane highway construction in southern California, particularly I-15 [51]. It is likely that the potential for connectivity between the Santa Ana Mountains and the Peninsular Range-East region will continue to be eroded by ongoing increases in traffic volumes on I-15, and conversion of unconserved lands along the I-15 corridor by development and agriculture [8,48,52].

An isolated population of pumas in the Santa Monica Mountains to the north of the Santa Ana Mountains also exhibit low values relative to other western North American populations (see Table 2 in [53]. Santa Monica pumas are isolated by urbanization of a megacity and busy wide freeways (Ventura county, including greater Los Angeles region [53]. Multiple instances of intraspecific predation, multiple consanguineous matings (father to daughter, etc.), and lack of successful dispersal highlight a suite of anthropogenic processes also occurring in the Santa Ana Mountains. Our collective findings of kinked tails *and* very low genetic diversity in Santa Ana pumas F95 and M96 may portend manifestations of genetic inbreeding depression similar to those seen in Florida panthers [54,55]; however recognizing that kinked tails can have non-genetic etiologies.

Our analyses suggest that the Santa Ana Mountains puma population is highly challenged in terms of genetic connectivity and genetic diversity, a result hinted at in Ernest et al. [9] and now confirmed to be an ongoing negative process for this population. This compounds the demographic challenges of low survival rates and scant evidence of physical connectivity to the Peninsular Ranges east of I-15 (unpublished data). Beier [6] documented these same challenges during the 1990's, and data from the ongoing UCD study suggest the trends have accelerated. Substantial habitat loss and fragmentation has occurred and is continuing to occur; Burdett et al. [10] estimated that by 2030, approximately 17% of puma habitat that was still available in 1970 in southern California will have been lost to development, and fragmentation will have rendered the remainder more hazardous for pumas to utilize. Riley et al [53] document a natural "genetic rescue" event: the 2009 immigration and subsequent breeding success of a single male to the Santa Monica Mountains. This introduction of new genetic material into the population was paramount to raising the critically low level of genetic diversity, as also exemplified by the human-mediated genetic augmentation of Florida Panthers with Texas puma stock [56].

These findings raise concerns about the current status of the Santa Ana Mountains puma population, and the longer-term outlook for pumas across southern California. In particular, they highlight the urgency to maintain – and enhance – what connectivity remains for pumas (and presumably numerous other species) across I-15. Despite warnings [6,9] about potential serious impacts to the Santa Ana Mountains puma population if concerted conservation action was not taken, habitat connectivity to the Peninsular Ranges has continued to erode. We are hopeful that these new genetic results will motivate greater focus on connectivity conservation in this region. Indeed, the Santa Ana Mountains pumas may well serve as harbingers of potential consequences throughout California and the western United States if more attention is not paid to maintaining connectivity for wildlife as development progresses.

Acknowledgments

Samples, data, and expertise were provided by multiple people and agencies, including California Department of Fish and Wildlife (R. Botta, B Gonzales, M. Kenyon, P. Swift, S. Torres, D. Updike, and others), California State Parks, National Park Service, The Nature Conservancy, UC Davis Wildlife Health Center, UC Davis Veterinary Genetics Laboratory, and the US Geological Survey. We thank the following for their technical assistance: L. Dalbeck, T. Drazenovich, T. Gilliland, J. George, and M. Plancarte. GIS data management and project cartography was provided by B. Cohen and J. Sanchez. Field work assistance was provided by J. Bauer, C. Bell, P. Bryant, D. Dawn, M. Ehlbroch, D. Krucki, K. Logan, B. Martin, B. Millsap, M. Puzzo, D. Sforza, L. Sweanor, C. Wiley, E. York, and numerous volunteers. Thanks to E. Boydsen, K. Crooks, R. Fisher, and L. Lyren for assistance coordinating field projects and sample acquisition. We thank anonymous reviewers for their constructive comments.

Author Contributions

Conceived and designed the experiments: HBE TWV WMB. Performed the experiments: HBE TWV MRB WMB. Analyzed the data: HBE TWV MRB. Contributed reagents/materials/analysis tools: HBE TWV SAM WMB. Contributed to the writing of the manuscript: HBE TWV SAM MRB WMB.

References

1. Mansfield TM, Torres SG (1994) Trends in mountain lion depredation and public safety threats in California. Paper 33. Proceedings of the Sixteenth Vertebrate Pest Conference, Santa Clara, CA. Editors: WS Halverson and AC Crabb. University of California, Davis.

2. Riley SPD, Pollinger JP, Sauvajot RM, York EC, Bromley C, et al. (2006) FAST-TRACK: A southern California freeway is a physical and social barrier to gene flow in carnivores. Mol Ecol 15: 1733–1741.

3. Riley SPD, Bromley C, Poppenga RH, Uzal FA, White L, et al. (2007) Anticoagulant exposure and notoedric mange in Bobcats and Mountain Lions in urban Southern California. J Wildl Manage 71: 1874–1884.

4. U.S. Census Bureau (2012) 2010 Census of Population and Housing, Summary Population and Housing Characteristics, CPH-1-6, California, U.S. Government Printing Office, Washington, DC.

5. California Department of Finance (2013) New Population Projections: California to surpass 50 million in 2049. Press release. Available: http://www.dof.ca.gov/research/demographic/reports/projections/p-1/documents/Projections_Press_Release_2010-2060.pdf Accessed 2 December 2013.

6. Beier P (1995) Dispersal of juvenile cougars in fragmented habitat. J Wildl Manage 59: 228–237.

7. Luke C, Penrod K, Cabanero CR, Beier P, Spencer W, et al. (2004) A Linkage Design for the Santa Ana-Palomar Mountains Connection. Unpublished report. San Diego State University, Field Station Programs and South Coast Wildlands. Available at: http://www.scwildlands.org/reports/SCML_SantaAna_Palomar.pdf.

8. Morrison SA, Boyce WM (2009) Conserving connectivity: Some lessons from mountain lions in southern California. Conserv Biol 23: 275–285.

9. Ernest HB, Boyce WM, Bleich VC, May B, Stiver SJ, et al. (2003) Genetic structure of mountain lion (Puma concolor) populations in California. Conserv Genet 4: 353–366.

10. Burdett CL, Crooks KR, Theobald DM, Wilson KR, Boydston EE, et al. (2010) Interfacing models of wildlife habitat and human development to predict the future distribution of puma habitat. Ecosphere 1: 1.

11. Ernest HB, Penedo MCT, May BP, Syvanen M, Boyce WM (2000) Molecular tracking of mountain lions in the Yosemite Valley region in California: genetic analysis using microsatellites and faecal DNA. Mol Ecol 9: 433–441.

12. Kurushima JD, Collins JA, Well JA, Ernest HB (2006) Development of 21 microsatellite loci for puma (Puma concolor) ecology and forensics. Mol Ecol Notes 6: 1260–1262.

13. Menotti-Raymond M, David VA, Lyons LA, Schaffer AA, Tomlin JF, et al. (1999) A genetic linkage map of Microsatellites of the domestic cat (Felis catus). Genomics 57: 9–23.

14. Pilgrim KL, Mckelvey KS, Riddle AE, Schwartz MK (2005) Felid sex identification based on noninvasive genetic samples. Mol Ecol Notes 5: 60–61.

15. Toonen RJ, Hughes S (2001) Increased throughput for fragment analysis on an ABI Prism 377 automated sequencer using a membrane comb and STRAND software. BioTechniques 31: 1320–1324.

16. Sherwin WB, Jabot F, Rush R, Rossetto M (2006) Measurement of biological information with applications from genes to landscapes. Mol Ecol 15: 2857–2869.

17. Peakall R, Smouse PE (2006) GenAlEx 6: genetic analysis in Excel. Population genetic software for teaching and research. Mol Ecol Notes 6, 288–295.

18. Peakall R, Smouse PE (2012) GenAlEx 6.5: genetic analysis in Excel. Population genetic software for teaching and research – an update. Bioinformatics 28: 2537–2539.

19. Rousset F (2008) genepop'007: a complete re-implementation of the genepop software for Windows and Linux. Mol Ecol Resour 8: 1755–0998.

20. Kalinowski ST, Wagner AP, Taper ML (2006) ML-RELATE: a computer program for maximum likelihood estimation of relatedness and relationship. Mol Ecol 6: 576–579.

21. Rice WR 1989. Analysing tables of statistical tests. Evolution 43: 223–5.

22. Taberlet P and Luikart G (1999), Non-invasive genetic sampling and individual identification. Biol J Linn Soc 68: 41–55.

23. Waits LP, Luikart G, Taberlet P (2001) Estimating the probability of identity among genotypes in natural populations: cautions and guidelines. Mol Ecol 10: 249–56.

24. Pritchard JK, Stephens M, Donnelly P (2000). Inference of population structure using multilocus genotype data. G3 155: 945–959.

25. Hubisz MJ, Falush D, Stephens M, Pritchard J (2009) Inferring weak population structure with the assistance of sample group information. Mol Ecol Resour 9: 1322–1332.

26. Earl DA, vonHoldt BM (2012) STRUCTURE HARVESTER: a website and program for visualizing STRUCTURE output and implementing the Evanno method. Conserv Genet Resour 4: 359–361.

27. Pritchard JK, Wen W (2002) Documentation for structure version 2.3.4. As viewed June 1 2013: http://pritch.bsd.uchicago.edu.

28. Evanno G, Regnaut S, Goudet J (2005) Detecting the number of clusters of individuals using the software STRUCTURE: a simulation study. Mol Ecol 14: 2611–2620.

29. Waples RS, Gaggiotti O (2006) What is a population? An empirical evaluation of some genetic methods for identifying the number of gene pools and their degree of connectivity. Mol Ecol 15, 1419–1439.

30. Orloci L (1978) Multivariate analysis in vegetation research. 2nd ed. Dr W. Junk. The Hague. 451 p.

31. Nei M (1977) F-statistics and analysis of gene diversity in subdivided populations. Ann Hum Genet 41: 225–233.

32. Piry S, Alapetite A, Cornuet JM, Paetkau D, Baudouin L (2004) GENECLASS2: A software for genetic assignment and first-generation migrant detection. J Hered 95: 536–539.

33. Paetkau D, Slade R, Burdens M, Estoup A (2004) Genetic assignment methods for the direct, real-time estimation of migration rate: a simulation-based exploration of accuracy and power. Mol Ecol 13: 55–65.

34. Rannala B, Mountain JL (1997) Detecting immigration by using multilocus genotypes. Proc Natl Acad Sci USA 94: 9197–9201.

35. Cornuet JM, Luikart G (1997) Description and power analysis of two tests for detecting recent population bottlenecks from allele frequency data. G3 144: 2001–2014.

36. Luikart G, Allendorf FW, Cornuet JM, Sherwin WB (1998) Distortion of allele frequency distributions provides a test for recent population bottlenecks. J Hered 89: 238–247.

37. Luikart G, Cornuet JM (1998) Empirical evaluation of a test for identifying recently bottlenecked populations from allele frequency data. Conserv Biol 12: 228–237.

38. Waples RS (2006) A bias correction for estimates of effective population size based on linkage disequilibrium at unlinked gene loci. Conserv Genet 7: 167–184.

39. Waples RS, Do C (2008) LDNE: a program for estimating effective population size from data on linkage disequilibrium. Mol Ecol Resour 8: 753–756.

40. Wright S (1969) Evolution and the Genetics of Population. Vol 2. The Theory of Gene Frequencies. University of Chicago Press.

41. Lynch M, Ritland K (1999) Estimation of pairwise relatedness with molecular markers. G3 152: 1753–1766.

42. Kalinowski ST, Taper ML, Marshall TC (2007) Revising how the computer program CERVUS accommodates genotyping error increases success in paternity assignment. Mol Ecol 16: 1099–1006.

43. Jones OR, Wang J (2010) COLONY: a program for parentage and sibship inference from multilocus genotype data. Mol Ecol Resour 10: 551–555.

44. Wang J (2012) Computationally efficient sibship and parentage assignment from multilocus marker data. G3 191: 183–194.

45. Alho JS, Välimäki K, Merilä J (2010) Rhh: an R extension for estimating Multilocus heterozygosity and heterozygosity–heterozygosity correlation. Mol Ecol Resour 10: 720–72.

46. R Development Core Team (2011) R: A language and environment for statistical computing. R Foundation for Statistical Computing, Vienna, Austria.

47. Amos W, Worthington Wilmer J, Fullard K, Burg TM, Croxall JP (2001) The influence of parental relatedness on reproductive success. Proc R Soc Lond B - Biol Sci 268: 2021–2027.

48. Beier P (1993) Determining Minimum Habitat Areas and Habitat Corridors for Cougars. Conserv Biol 7: 94–108.

49. Crooks KR (2002) Relative Sensitivities of Mammalian Carnivores to Habitat Fragmentation. Conserv Biol 16: 488–502.

50. Logan K and Sweanor L (2001) Desert Puma: Evolutionary Ecology And Conservation Of An Enduring Carnivore. Island Press. 464.

51. Barbour E (2002) Metropolitan growth planning in California, 1900–2000. Public Policy Institute of California. San Francisco, California. 246.

52. Beier P (1996) Metapopulation modeling, tenacious tracking, and cougar conservation. In McCullough DR, editor, Metapopulations and wildlife management. Washington DC: Island Press. 293–323.

53. Riley SPD, Serieys LEK, Pollinger J, Sikich J, Dalbeck L, et al. Individual behaviors dominate the dynamics of an urban mountain lion population isolated by roads. Curr Biol 24: 1989–1994.

54. Roelke ME, Martenson JS, O'Brien SJ. (1993). The consequences of demographic reduction in the endangered Florida panther. Curr Biol 3: 340–350.

55. Culver M, Hedrick PW, Murphy K, O'Brien S Hornocker MG. (2008) Estimation of the bottleneck size in Florida panthers. Anim Conserv 11: 104–110.

56. Johnson WE, Onorato DP, Roelke ME, Land ED, Cunningham M, et al. (2010). Genetic restoration of the Florida panther. Science 329, 1641–1645.

9

Strong Endemism of Bloom-Forming Tubular *Ulva* in Indian West Coast, with Description of *Ulva paschima* Sp. Nov. (Ulvales, Chlorophyta)

Felix Bast*, Aijaz Ahmad John, Satej Bhushan

Centre for Biosciences, Central University of Punjab, Bathinda, Punjab, India

Abstract

Ulva intestinalis and *Ulva compressa* are two bloom-forming morphologically-cryptic species of green seaweeds widely accepted as cosmopolitan in distribution. Previous studies have shown that these are two distinct species that exhibit great morphological plasticity with changing seawater salinity. Here we present a phylogeographic assessment of tubular *Ulva* that we considered belonging to this complex collected from various marine and estuarine green-tide occurrences in a ca. 600 km stretch of the Indian west coast. Maximum Likelihood and Bayesian Inference phylogenetic reconstructions using ITS nrDNA revealed strong endemism of Indian tubular *Ulva*, with none of the Indian isolates forming part of the already described phylogenetic clades of either *U. compressa* or *U. intestinalis*. Due to the straightforward conclusion that Indian isolates form a robust and distinct phylogenetic clade, a description of a new bloom-forming species, *Ulva paschima* Bast, is formally proposed. Our phylogenetic reconstructions using Neighbor-Joining method revealed evolutionary affinity of this new species with *Ulva flexuosa*. This is the first molecular assessment of *Ulva* from the Indian Subcontinent.

Editor: Ross Frederick Waller, University of Cambridge, United Kingdom

Funding: Supported by INSPIRE Faculty award (IFA-LSPA-02) from Department of Science and Technology, Government of India. The funders had no role in study design, data collection and analysis, decision to publish, or preparation of the manuscript.

Competing Interests: The authors have declared that no competing interests exist.

* Email: felix.bast@cup.ac.in

Introduction

Genus *Ulva* (Linnaeus), commonly known as "Sea Lettuce", encompasses some of the most ubiquitous green seaweeds distributed throughout the world, with habitats ranging from marine to freshwater. This algal genus is both beneficial and disadvantageous; beneficial as some species of this genus, including *Ulva prolifera* [1] and *Ulva intestinalis* [2], are commercially cultivated worldwide for its culinary use and disadvantageous as this genus is notorious for its ability to cause massive green-tides [3] and marine fouling [4]. Species of this genus are well known for having highly plastic morphologies, and that the habits can change from tube-form to blade-form or *vice versa* in response to changing environmental conditions [5]. Therefore, morphology-based classifications, which have been routinely used since the inception of this genus, are now being replaced with molecular systematics [6,7]. For example, the genus *Enteromorpha*- which had been separated from *Ulva* based on tubular morphology by Link (Link in Nees 1820)- has recently been merged back to *Ulva* based on DNA sequence evidence [6–9]. Due to taxonomic confusions in morphology-based species delineation, concept of Operational Taxonomic Units (OTUs) have been used in recent phylogeographic assessments of *Ulva* from Hawaii [10] and USA [11].

Tubular *Ulva* in Indian Coast is believed to be comprised mainly of two species; *Ulva intestinalis* Linnaeus and *Ulva compressa* Linnaeus (personal observation). These two species are so closely related that they are regarded as cryptic species in a number of molecular phylogenetic studies [4,12]. These species are separated from each other based on microscopic and macroscopic morphological characters [4]. Microscopic characters include distinct cell arrangement in *U. compressa* consisting of rosettes of cells ("Cell islands") and longitudinal rows of cells in contrast to *U. intestinalis*, where there are no obvious arrangement of cells. Macroscopic characters include branching pattern and compression of thallus, in which *U. intestinalis* is mostly unbranched with hollow tubular monostromatic thalli, with very few branches for algae growing on low-saline environments, and *U. compressa* is highly branched with compressed thalli [13]. Taxonomic validity of these character states have been repeatedly questioned (see [4] for review). Morphological and phylogenetic variation in these two species has been investigated from the Baltic Sea Area [12] and the British Isles [4], and both of these reports concluded that *U. intestinalis* and *U. compressa* are distinct, monophyletic species. In recent phylogenetic assessments of *Ulva* from North Adriatic sea [14] and temperate Australia [8,9] these two species together formed strongly supported clade, confirming their evolutionary relatedness.

The genus *Ulva* from India has never been subjected to extensive taxonomic scrutiny to date. While phylogeographic assessments of *Ulva* have been conducted in various parts of the world, including Japan [15], Australia [8], China [16], North-East

Pacific [17] and Hawaii [16], sequence-based assessment of *Ulva* from Indian subcontinent have not yet been done. Objectives of the present study are to understand morphological and molecular variation of tubular *Ulva* occurring on the Indian west coast. Almost all of the previous phylogeographic assessments in genus *Ulva* were based on nucleoribosomal Internal Transcribed Spacer (ITS). ITS is one of the well-represented loci at Genbank and therefore we selected this locus for our molecular assessment.

Materials and Methods

Living Materials

During our 2012 expedition to the west coast of India, a particular tubular *Ulva* was detected causing massive blooms in a number of freely accessible locations (Table 1). Bloom specimens of tubular *Ulva*, either attached to intertidal substrates (including rocks, pebbles, wooden dinghies, mooring lines and breakwaters), or drifting while attached to a variety of floating objects were subsequently collected (Table 1). Collection coordinates were acquired with a handheld GPS device (eTrex 30, Garmin, USA). A map overlay of sampling locations with an accuracy of ± 10 meters is accessible at http://bit.ly/UlvaBloom. Photographs of the bloom were taken using a GPS-enabled digital camera (CyberShot DSC HX20V, Sony, Japan) and these photographs, with embedded GPS data, are available as online-only supplementary data (Figs. S1–10 in File S1). Seawater salinity was measured at the collection locations using a handheld salinometer (PCTTestr 35, Eutech Instruments, Singapore). Collected specimens were transported to the laboratory in zip-lock polythene bags under cold conditions (4–10°C). After washing the thalli in tap water to remove sediments and other contaminants, morphological characterization of the specimens was made using an upright microscope (BX53, Olympus, Japan) with an attached digital camera (E450, Olympus, Japan). Public domain software ImageJ (http://rsbweb.nih.gov/ij/) was used for scale calibration and size measurements. Pressed vouchers were prepared and deposited in the Central National Herbarium, Botanical Survey of India, Calcutta (*Index Herbariorum* code: CAL). Samples for molecular analyses were stored at -80°C awaiting further analysis.

DNA extraction, PCR amplification, purification and DNA sequencing

The frozen specimens were thawed in artificial sea water [18]. Total genomic DNA was extracted from the specimens using a HiPurA Algal Genomic Extraction Kit (HiMedia Laboratories Pvt. Ltd., Mumbai) following manufacturer's protocol. Tissues from the apical thalli were selected to increase DNA yield. The quality of DNA was checked on 0.8% agarose gel and the quantity of DNA was checked with spectrophotometer. Isolated DNA was stored at −20°C.

PCR amplification

A DNA working solution of 25 ng/μl was prepared for polymerase chain reaction (PCR) in a separate tube. The 20 μl PCR reaction mix contained 2 μl of 10× reaction buffer with 15 mM MgCl$_2$ (Applied Biosystems, India), 4 μl each of 10 μM primer, 2 μl of 1 μM dNTPs (Imperial Life sciences, India), 0.6 unit of rTaq DNA polymerase (Imperial Life sciences, India), 4 μl of template DNA and sterile water. The four universal primers used for amplifying the ITS regions and the 5.8S gene (fragment length = 639 bp) were: ITS1 (5'-TCCGTAGGT-GAACCTGCGG-3'), ITS2 (5'- GCTGCGTTCTTCATC-GATGC-3'), ITS3 (5'- GCATCGATGAAGAACGCAGC-3')

Table 1. Collected samples of tubular *Ulva* from algal bloom across West Coast of India.

Location (administrative state in parenthesis) and Isolate identifier	Morphospecies	Genbank accession #	CAL voucher accession #	Habitat	Salinity PSU	Cell size in μm^2 Mean±SD, n=20	Coordinates
Anjuna (Goa)-ANJ	*U. intestinalis*	KF385504	CAL-CUPVOUCHER-UI-2013-3	Attached, Exposed rocky shore	34	142.81±2.80	15.58419N, 73.73683E
Karwar (Karnataka)-KAR	*U. intestinalis*	KF385502	CAL-CUPVOUCHER-UI-2013-2	Drifted, Exposed rocky shore	35	77.59±1.75	14.8064N, 74.1174E
Kundapur (Karnataka)-KUN	*U. compressa*	KF385505	CAL-CUPVOUCHER-UC-2013-1	Drifted, Sheltered inlet	24	97.65±3.14	13.63804N, 74.68797E
Mangalore (Karnataka)-MAN	*U. intestinalis*	KF385506	CAL-CUPVOUCHER-UI-2013-3a	Attached, Sheltered river-mouth	30	133.18±3.90	12.84831N, 74.82924E
Kannur (Kerala)-KAN	*U. intestinalis*	KF385503	CAL-CUPVOUCHER-UI-2013-4	Attached, Exposed rocky shore	33	97.28±6.17	11.853439N, 75.376644E
Ponnani (Kerala)-PON	*U. intestinalis*	KF385501	CAL-CUPVOUCHER-UI-2013-5	Attached, Exposed rocky shore	31	52.19±3.23	10.78637N, 75.91265E

and ITS4 (5'- TCCTCCGCTTATTGATATGC-3') [19]. PCR amplifications were carried out in programmable thermal cycler (Veriti, ABI, USA) and reaction profile included an initial denaturation at 94°C for 5 minutes, followed by 35 cycles of 94°C for 1 minute, 52°C for 2 minutes and 72°C for 2 minutes, and a final extension of 72°C for 10 minutes.

Purification of PCR product

Amplicons were purified using ExoSAP-IT PCR clean-up kit following manufacturer's instructions (USB Corporation, Cleveland, OH, USA). A working solution of 1:10 (DNA: water) was prepared as sequencing template. PCR amplification reactions (as well as its sequencing) were carried out in duplicate for each target sequence of each isolate using the same set of primers as a quality control.

DNA sequencing

Purified PCR products were subjected to bidirectional Sanger sequencing using a dideoxy chain termination protocol with ABI BigDye Terminator Cycle Sequencing Ready Reaction Kit v3.1 (Applied Biosystems, Foster City, CA, USA) and a programmable thermal cycler (Veriti, ABI, USA), as per [20].DNA sequences were assembled using the computer program CodonCodeAligner (CodonCode Corporation, USA). Sequences were deposited in Genbank (Table 1).

Phylogenetic analysis

We followed the step-by-step protocol for phylogenetic analysis, including alignment construction, Maximum Likelihood test to find best-fitting substitution models [21], phylogeny reconstruction using Maximum Likelihood (ML), Bayesian Inference (BI) and distance analysis as outlined in Bast [22]. In summary, six sequences of tubular *Ulva* from India were aligned with other published accessions of *Ulva intestinalis* and *Ulva compressa* obtained from Genbank (Table S1) by MUSCLE algorithm inside computer program Geneious v6.1.6 (available at http://www.genious.com) and alignments were edited by eye. Phylogenetic analysis using ML algorithm was conducted in MEGA (www.megasoftware.net/) with starting tree generated by BioNJ. Substitution bias was modelled by Tamura-3-Parameter [23] (T3P) model with Gamma distribution (that was the best model in our test to find best fitting substitution models [21] with BIC (Bayesian Information Criterion) score of 5626.537).Heuristic searches were performed with tree bisection-reconnection, MULTREES and steepest descent options in effect. 1000 bootstrap replicates were performed under ML criterion to estimate interior branch support [24]. Phylogenetic analysis with BI was conducted using the MrBayes plug-in v3 [25] within Geneious. Analyses were run with four Markov chains using the T3P model with Gamma distribution for 10^6 generations with a tree saved every 100th generation. The first 1000 trees were discarded as burn-in, as determined by "burnin <number>" function of MrBayes plug-in. A consensus tree was constructed using the consensus tree builder within Geneious. In order to investigate relative phylogenetic position of our isolates in genus *Ulva*, a separate ITS dataset was constructed with 120 sequences obtained from Genbank, spanning all major species represented in the database. Due to the computational limitations, we used Neighbor-Joining (NJ) method for this dataset. All of our scientific datasets, including cell area measurements, DNA sequence alignment in FASTA format, results of ModelTest, T3P pairwise distances, tree in nexus format and original electropherograms of DNA sequences with contig assembly instructions are freely available at LabArchives (http://dx.doi.org/10.6070/H4639MP5).

Nomenclature

The electronic version of this article in Portable Document Format (PDF) in a work with an ISSN or ISBN will represent a published work according to the International Code of Nomenclature for algae, fungi, and plants, and hence the new names contained in the electronic publication of a PLOS ONE article are effectively published under that Code from the electronic edition alone, so there is no longer any need to provide printed copies. The online version of this work is archived and available from the following digital repositories: PubMed Central, LOCKSS and Research Gate.

Results

Seawater salinity ranged between 35PSU and 24PSU. As expected, exposed shores had higher salinity than inlets and river mouth areas. On external morphology, all six isolates had their own unique features (Fig 1, and Figs S1–S10 in File S1). All isolates were grass green in color, erect filamentous, and had a parietal chloroplast with more than three pyrenoids inside each cell (*Arrowheads* in Figs 1 D, H, L, P, T and X). Isolates ANJ, MAN and PON had some part of their thallus flattened (indicated by *Arrowheads* in Figs 1A, M and U, respectively) and had thicker thalli (Figs 1B, N and V), comparing with other isolates. In terms of thalli branching character state, isolate KUN was unique in that it was branched, up to two lateral branch orders (*Arrowheads* in Fig 1J), while the other isolates were unbranched. In terms of cell arrangement, isolates KAR and KAN were similar, with more or less linear arrangement of cells (cell layers accentuated with pairs of lines in Figs 1G and S, respectively). In terms of cell size, isolates ANJ and MAN were the largest and PON was the smallest (Table 1). A comparison of taxonomically relevant morphological characters for our isolates with *U. intestinalis*, *U. compressa* and *U. flexuosa* is presented (Table 2). As per thallus branching character and thallus compression character, isolates ANJ, KAR, MAN, KAN and PON were arbitrarily classified as *U. intestinalis* (unbranched, hollow) and isolate KUN as *Ulva compressa* (branched, compressed).

While we primarily employed ITS sequence data for barcoding, phylogenetic reconstruction using this locus revealed a number of evolutionary trends. Phylogenetic reconstruction using the ML (Fig 2) and BI (Fig 3) methods resulted in moderately-resolved phylograms, with three clades. All Indian isolates of tubular *Ulva* formed a single clade (highlighted "Paschima"). Our isolate KUN (*U. compressa* morphotype) seems to have been much diverged from Indian isolates of *U. intestinalis* morphotype as evidenced by long branch-length. This isolate clustered within *U. intestinalis* accessions from India in BI, but was basal to *U. intestinalis* accessions from India in ML. Other monophyletic clades included that of non-Indian isolates of *U. compressa* (highlighted "Compressa") and *U. intestinalis* (highlighted "Intestinalis"). Within-group mean T3P distance was 0.5352 for "Paschima" and 0.000 for both "Intestinalis" and "Compressa", which indicates a very high genetic heterogeneity for the "Paschima" clade. In one study [26], within-group JC (Jukes-Cantor) distance for *Enteromorpha* had been reported to range between 0.09 to 0.16. As within- group distance for "Paschima" clade observed in the present study being much higher, possibility that our isolate KUN belonging to another unique taxon from India cannot be ruled out.

Phylogenetic analysis with NJ conducted for 121 sequences of *Ulva* resulted in a moderately-resolved phylogram (Fig.4). Paschima clade showed evolutionary affinity to Flexuosa clade, albeit with weak bootstrap support (36, not shown in figure). *Ulva flexuosa*, *Ulva compressa*, and *Ulva intestinalis* formed respective

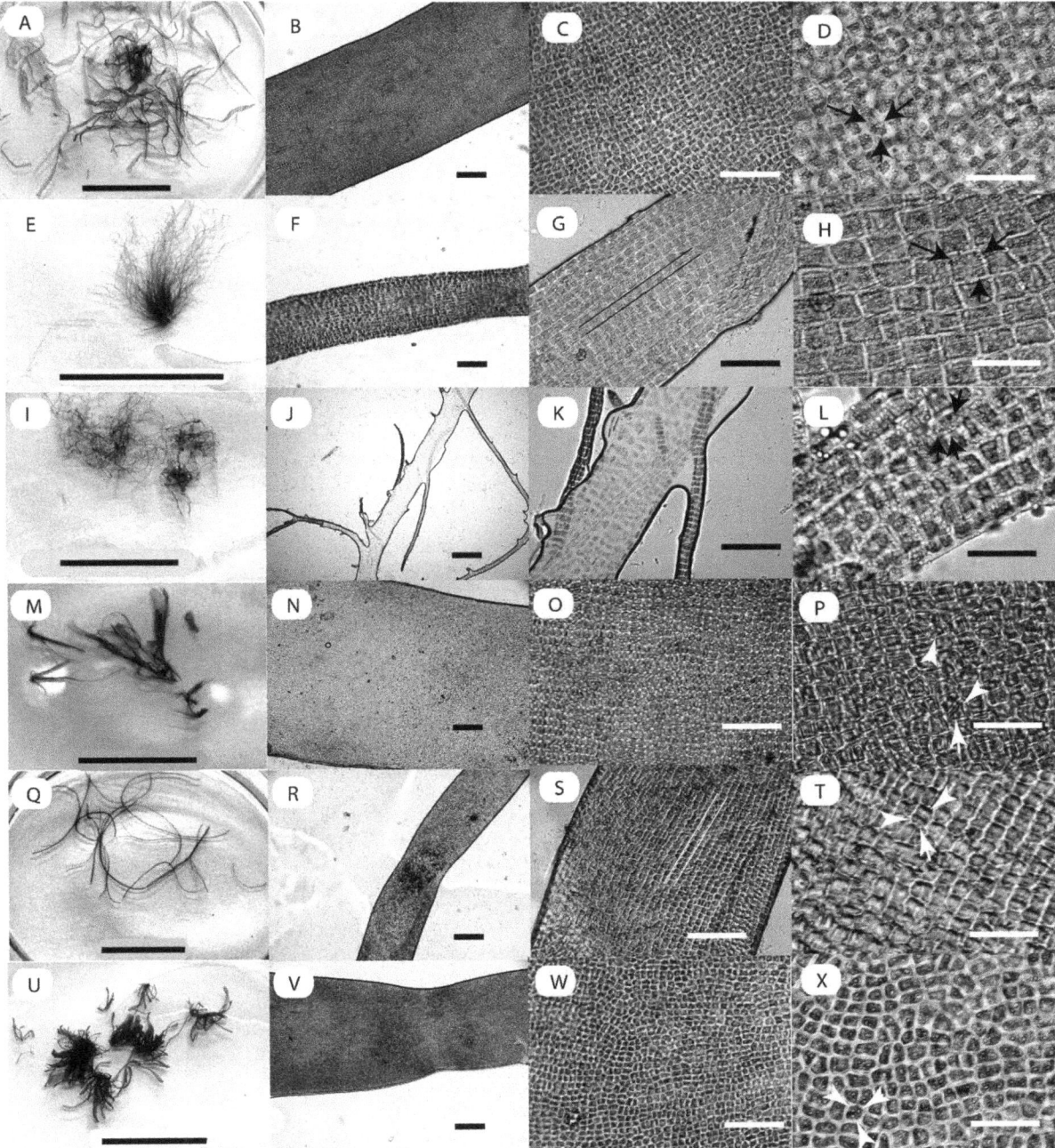

Figure 1. Morphology of tubular *Ulva* from India. A–D ANJ Isolate, E–H KAR isolate, I–L KUN isolate, M–P MAN isolate, Q–T KAN Isolate and U–X PON Isolate. *Arrowheads* in M and U indicate flat portions of thalli, J indicate branching pattern and D, H, L, P, T and X indicate pyrenoids. Scale bars are 2 mm for A, E, I, M, Q and U; 200 μm for B, F, J, N, R and V; 100 μm for C, G, K, O, S and W; and 50 μm for D, H, L, P, T and X.

monophyletic clades. A clade comprising of *Ulva rigida*, *Ulva laetevirens*, *Ulva scandinavica*, *Ulva fenestrata*, *Ulva armoricana* and *Ulva lactuca* had strong bootstrap support, indicating phylogenetic affinity of these species. *Ulva fasciata* formed a strongly supported clade with *Ulva ohnoi* (clade "fasciata") and *Ulva linza* clustered within a strongly supported clade comprising of *Ulva prolifera* (clade "prolifera"). For a definitive phylogenetic assessment of these taxa, additional genetic loci need to be employed.

Discussion

The present study made several interesting revelations, the most significant of which is the apparent endemism of a bloom-forming Indian tubular *Ulva* that is morphologically plastic and indistinguishable from *U. intestinalis* and *U. compressa*. Contrary to our expectations, the five isolates that had hollow, unbranched thalli did not group within the already described *U. intestinalis* clade in our phylogenetic analyses, nor did the single identified isolate that had compressed, branched thalli show affiliation to the *U.*

Table 2. Morphological characters of Indian isolates in comparison with *Ulva intestinalis*, *Ulva compressa* and *Ulva flexuosa* [29,30].

Character	Isolates		*Ulva intestinalis*	*Ulva compressa*	*Ulva flexuosa*
	ANJ, KAR, MAN, KAN, PON	**KUN**			
Tubular thallus branched or unbranched	Unbranched	Branched	Mostly unbranched	Branched	Mostly branched
Tubular thallus hollow or compressed	Hollow	Compressed	Hollow	Compressed	Hollow/Compressed
Cell arrangement: Linear or Nonlinear	Linear (Only for KAR and KAN)	Nonlinear	Nonlinear	Linear	Linear
Cell arrangement: Rosettes	Absent	Absent	Absent	Present	Absent

compressa clade. Instead, all of our isolates formed a strongly supported clade, which showed affinity to a previous sample identified as *Ulva intestinalis* from Gopnath, Gujarat, India (personal communication). This clearly indicates a high degree of endemism for the Indian tubular *Ulva*. Interestingly, Japanese isolates of either *Ulva compressa*, or *Ulva intestinalis*, were described to have very little pair-wise distance from European Isolates at ITS loci[27]. Given the vast geographical distance of ca.

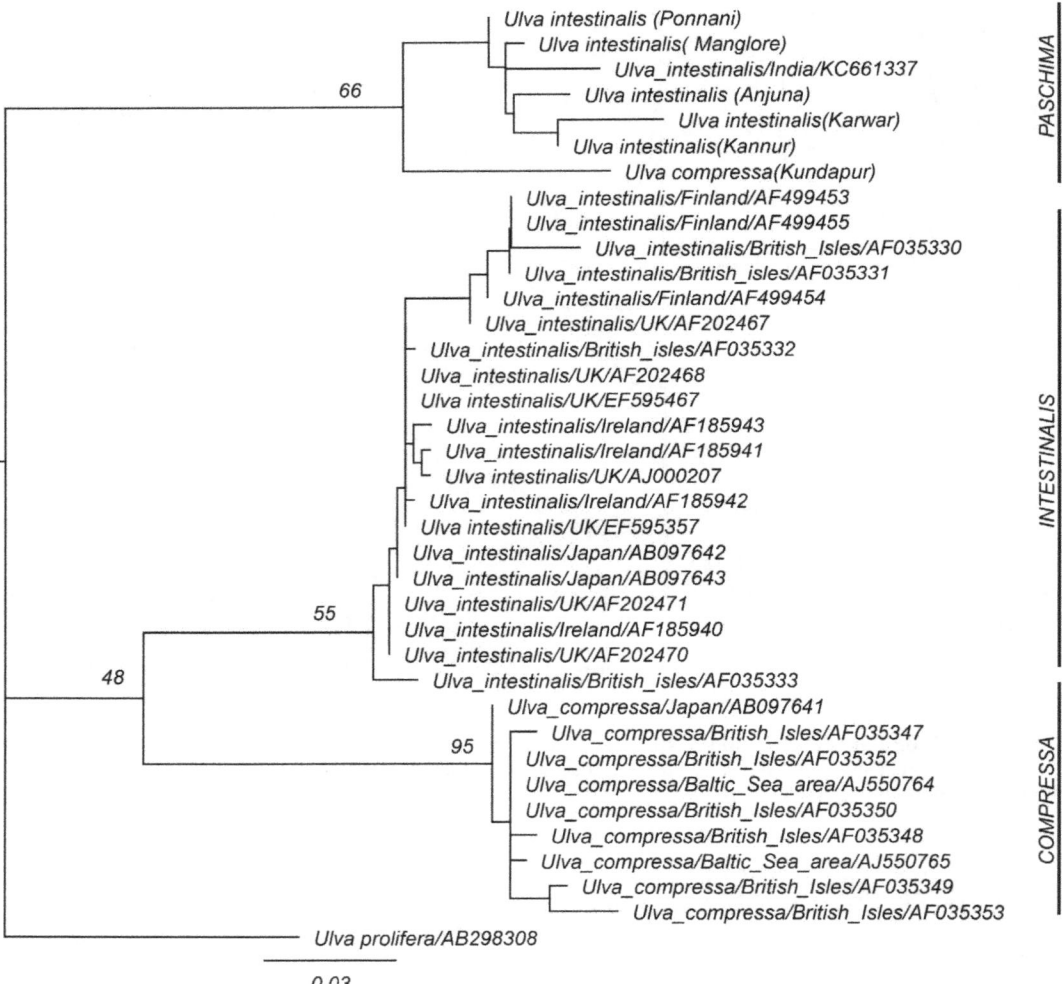

Figure 2. Phylogenetic position of tubular *Ulva* isolates from India among other tubular *Ulva* accessions in ITS dataset using Maximum Likelihood phylogenetic reconstruction (LnL = −2412.46) with T3P model of molecular evolution with gamma distribution (T3P+G). Numbers near nodes represent bootstrap support (1000 replicates), exceeding 50. This phylogram is rooted with *Ulva prolifera* as outgroup. Scale bar given on bottom is in the units of average nucleotide substitutions per site.

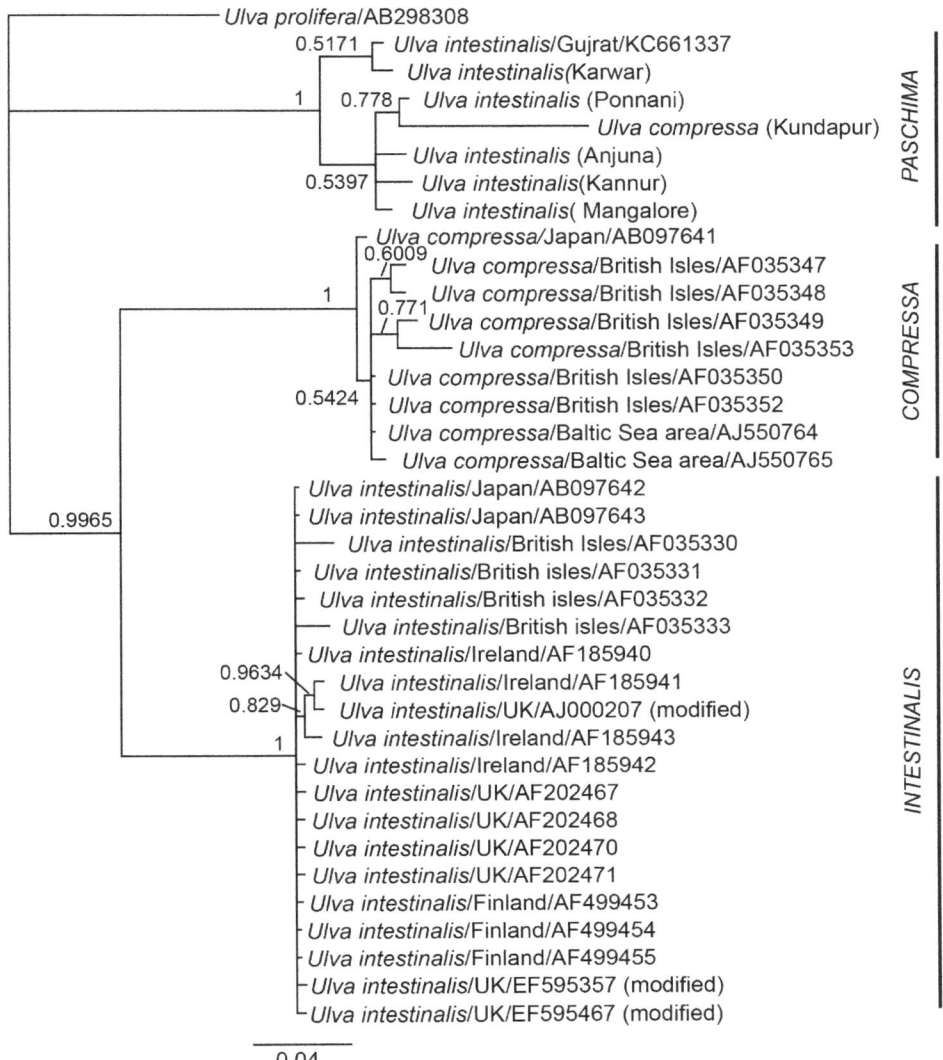

Figure 3. Phylogenetic position of tubular *Ulva* isolates from India among other tubular *Ulva* accessions in ITS dataset using Bayesian Inference phylogenetic reconstruction (LnL = −2628.193) with T3P model of molecular evolution with gamma distribution (T3P+G). Numbers near nodes represent Bayesian Posterior Probabilities, exceeding 0.5. This phylogram is rooted with *Ulva prolifera* as outgroup. Scale bar given on bottom is in the units of average nucleotide substitutions per site.

10,000 km, this earlier report could either be suggestive of a recent introduction of these species to either of these locations or existence of temperate haplotypes. A recent report on the molecular assessment of *Ulva* from Australia concluded that the genus encompasses a number of endemic potentially cryptic species in addition to cosmopolitan species [8]. In the light of these findings, assumptions of cosmopolitanism among certain species of *Ulva* can cause novel and endemic species to be overlooked.

Our identification of the KUN isolate as *U. compressa* was based on the previously described character states of branching pattern of thallus and compressed state of the filament [4]. However, phylogeny reconstruction clustered our *U. compressa* specimen within a clade comprised chiefly of hollow, unbranched tubular *Ulva* from India (similar to *U. intestinalis* morphospecies). These two tubular *Ulva* morphospecies from India might indeed be conspecific. Low salinity at the habitat of the KUN isolate might have influenced the species to acquire this morphotype as suggested by previous studies [4]. Alternately, non-Indian acces-

sions of *U. intestinalis* and *U. compressa* might indeed be unique species with yet-to-discover synapomorphic character state/s, as observed in our phylogenetic analyses. In summary, without molecular data, Indian species of bloom-forming *Ulva* most closely resemble with either/both *U. intestinalis* and *U. compressa*, two species that are shown in the literature to be difficult to distinguish between.

Results from our phylogenetic reconstructions strongly argue in favor of species-level taxonomic treatment for the OTUs from India, which is evolutionarily unrelated to either *Ulva intestinalis* or *Ulva compressa*. We therefore formally propose a new species of bloom-forming tubular *Ulva* as per the following description, congruent with Phylogenetic Species Concept [28]:

Ulva paschima Bast sp. nov. (Fig 1)

Description. Primary diagnosis is the phylogenetic affiliation of OTUs with ITS clade "Paschima" as per this report. Fronds erect filamentous and grass green in color; 5 cm–40 cm in length;

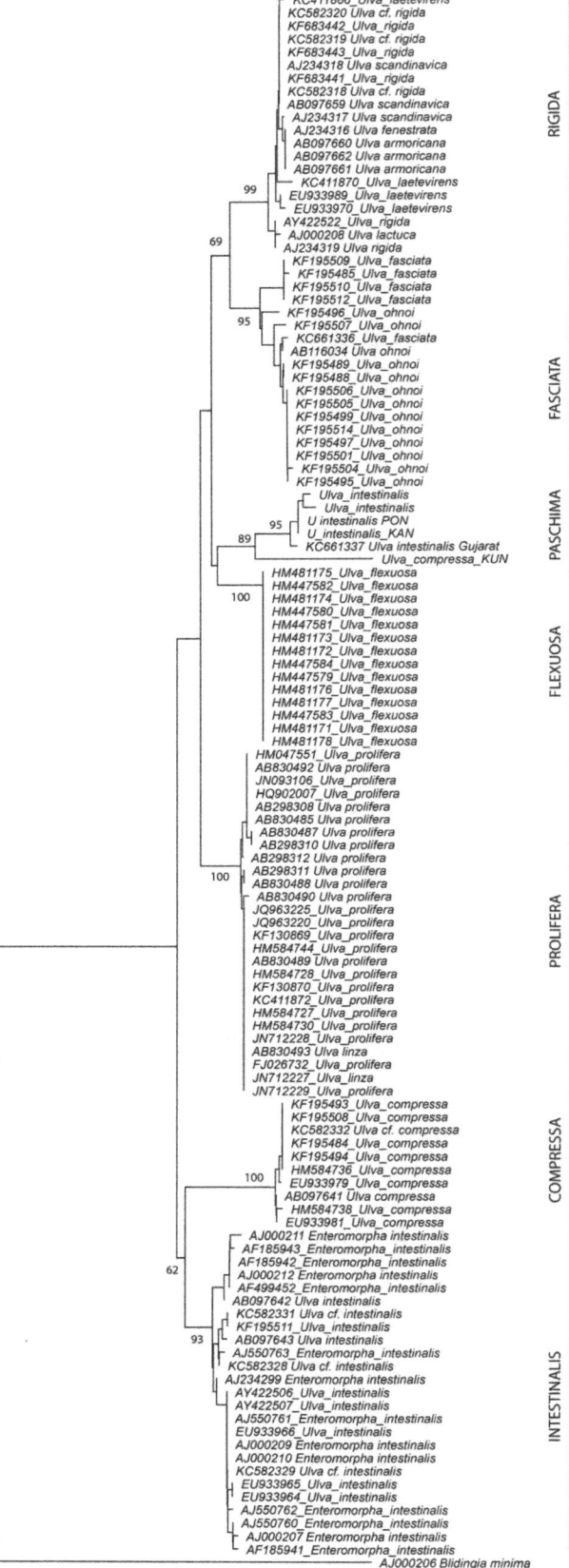

0.04

Figure 4. Phylogenetic position of tubular *Ulva* isolates from India among other accessions in ITS dataset using Neighbor-Joining Inference phylogenetic reconstruction (total tree length= 0.79153595) with T3P model of molecular evolution with gamma distribution (T3P+G). Numbers near nodes represent bootstrap support (100 replicates), exceeding 50. This phylogram is rooted with *Blidingia minima* as outgroup. Scale bar given on bottom is in the units of average nucleotide substitutions per site.

mostly unbranched tubular with some parts of the thalli compressed or flat, ribbon-like; tufts of filamentous thalli attached via rhizoid. Morphotype in low-saline inlets and estuaries might have branched, compressed thalli. Cells are more or less quadrilateral; some have linear cell arrangement. Parietal chloroplast with>2 pyrenoids per cell.

Holotype. Collected from intertidal rocks at a splash zone near Paraiso de Goa, Anjuna Beach, Goa, India (15.58419N, 73.73683E). Deposited at Central National Herbarium, Botanical Survey of India, Calcutta (*Index Herbariorum* code: CAL) under voucher # CAL-CUPVOUCHER-UP-2013-3. DNA sequences of nrDNA ITS1-5.8S-ITS2 complete region of the holotype deposited at Genbank under accession # KF385504.

Isotype. Deposited at Herbarium, the Central University of Punjab under voucher No.: CUPVOUCHER-UP-2013-3. Frozen voucher maintained at Centre for Biosciences, the Central University of Punjab under voucher No.: CUPFVOUCHER-UP-2013-1.

Etymology. Specific epithet in Sanskrit means "west" where the algae is first described in Indian Subcontinent.

Supporting Information

File S1   Compressed file containing Figure S1 to Figure S10. Figure S1, Photograph of algal bloom specimen of *Ulva intestinalis* isolate ANJ. **Figure S2,** Photograph of drifting algal bloom specimen of *Ulva intestinalis* isolate KAR, attached on mooring line. **Figure S3,** Photograph of algal bloom specimen of *Ulva compressa* isolate KUN. **Figure S4,** Photograph of algal bloom specimen of *Ulva compressa* isolate KUN. **Figure S5,** Photograph of algal bloom specimen of *Ulva compressa* isolate KUN. **Figure S6,** Photograph of algal bloom specimen of *Ulva intestinalis* isolate MAN. **Figure S7,** Photograph of algal bloom specimen of *Ulva intestinalis* isolate MAN, attached on wooden dinghy. **Figure S8,** Photograph of algal bloom specimen of *Ulva intestinalis* isolate MAN, attached on wooden dinghy. **Figure S9,** Photograph of algal bloom specimen of *Ulva intestinalis* isolate KAN. **Figure S10,** Photograph of algal bloom specimen of *Ulva intestinalis* isolate PON.

Acknowledgments

We thank Dr. Chris Yesson, Natural History Museum for his help with phylogenetics, especially refining the alignment. We are also thankful for the Vice chancellor, the Central University of Punjab for his support with respect to the execution of this research.

Author Contributions

Conceived and designed the experiments: FB. Performed the experiments: FB AAJ SB. Analyzed the data: FB. Contributed reagents/materials/analysis tools: FB. Wrote the paper: FB.

References

1. Hiraoka M, Oka N (2008) Tank cultivation of Ulva prolifera in deep seawater using a new "germling cluster" method. Journal of applied phycology 20: 97–102.
2. MacArtain P, Gill CI, Brooks M, Campbell R, Rowland IR (2007) Nutritional value of edible seaweeds. Nutrition reviews 65: 535–543.
3. Pang SJ, Liu F, Shan TF, Xu N, Zhang ZH, et al. (2010) Tracking the algal origin of the Ulva bloom in the Yellow Sea by a combination of molecular, morphological and physiological analyses. Marine environmental research 69: 207–215.
4. Blomster J, Maggs CA, Stanhope MJ (1998) Molecular and morphological analysis of Enteromorpha intestinalis and E. compressa (Chlorophyta) in the British Isles. Journal of Phycology 34: 319–340.
5. Tan IH, Blomster J, Hansen G, Leskinen E, Maggs CA, et al. (1999) Molecular phylogenetic evidence for a reversible morphogenetic switch controlling the gross morphology of two common genera of green seaweeds, Ulva and Enteromorpha. Molecular Biology and Evolution 16: 1011–1018.
6. Hayden HS, Blomster J, Maggs CA, Silva PC, Stanhope MJ, et al. (2003) Linnaeus was right all along: Ulva and Enteromorpha are not distinct genera. European Journal of Phycology 38: 277–294.
7. O'Kelly CJ, Kurihara A, Shipley TC, Sherwood AR (2010) Molecular assessment of Ulva spp. (Ulvophyceae, Chlorophyta) in the Hawaiian islands. Journal of Phycology 46: 728–735.
8. Kraft LG, Kraft GT, Waller RF (2010) Investigations into Southern Australian Ulva (ulvophyceae, chlorophyta) taxonomy and molecular phylogeny indicate both cosmopolitanism and endemic cryptic species. Journal of Phycology 46: 1257–1277.
9. Kirkendale L, Saunders GW, Winberg P (2013) A molecular survey of Ulva (Chlorophyta) in temperate Australia reveals enhanced levels of cosmopolitanism. Journal of Phycology 49: 69–81.
10. O'Kelly CJ, Kurihara A, Shipley TC, Sherwood AR (2010) Molecular assessment of Ulva spp.(ulvophyceae, chlorophyta) in the hawaiian islands Journal of Phycology 46: 728–735.
11. Guidone M, Thornber C, Wysor B, O'Kelly CJ (2013) Molecular and morphological diversity of Narragansett Bay (RI, USA) Ulva (Ulvales, Chlorophyta) populations. Journal of Phycology 49: 979–995.
12. Leskinen E, Rapaport Alstrom C, Pamilo P (2004) Phylogeographical structure, distribution and genetic variation of the green algae Ulva intestinalis and U. compressa (Chlorophyta) in the Baltic Sea area. Molecular ecology 13: 2257–2265.
13. De Silva M, Burrows EM (1973) An experimental assessment of the status of the species Enteromorpha intestinalis (L.) Link and Enteromorpha compressa (L.) Grev. J Marine Biol Assoc UK 53: 895–904.
14. Wolf MA, Sciuto K, Andreoli C, Moro I (2012) Ulva (Chlorophyta, Ulvales) biodiversity in the North Adriatic Sea (Mediterranean, Italy): cryptic species and new introductions. Journal of Phycology 48: 1510–1521.
15. Shimada S, Yokoyama N, Arai S, Hiraoka M (2009) Phylogeography of the genus Ulva (Ulvophyceae, Chlorophyta), with special reference to the Japanese freshwater and brackish taxa. Proceedings of Nineteenth International Seaweed Symposium: Springer. pp529–539.
16. Boo SM, Lee WJ (2010) Ulva and Enteromorpha (Ulvaceae, Chlorophyta) from two sides of the Yellow Sea: analysis of nuclear rDNA ITS and plastid rbcL sequence data. Chinese Journal of Oceanology and Limnology 28: 762–768.
17. Hayden HS, Waaland JR (2004) A molecular systematic study of Ulva (Ulvaceae, Ulvales) from the northeast Pacific. Phycologia 43: 364–382.
18. Kester DR, Duedall IW, Connors DN, Pytkowicz RM (1967) Preparation of artificial seawater. Limnology and Oceanography 12.
19. White TJ, Bruns T, Lee S, Taylor J (1990) Amplification and direct sequencing of fungal ribosomal RNA genes for phylogenetics. PCR protocols: a guide to methods and applications 18: 315–322.
20. Bast F, Rani P, Meena D (2014) Chloroplast DNA Phylogeography of Holy Basil (Ocimum tenuiflorum) in Indian Subcontinent. The Scientific World Journal 2014.
21. Tamura K, Peterson D, Peterson N, Stecher G, Nei M, et al. (2011) MEGA5: molecular evolutionary genetics analysis using maximum likelihood, evolutionary distance, and maximum parsimony methods. Molecular Biology and Evolution 28: 2731–2739.
22. Bast F (2013) Sequence Similarity Search, Multiple Sequence Alignment, Model Selection, Distance Matrix and Phylogeny Reconstruction. Nature Protocol Exchange.
23. Tamura K, Nei M (1993) Estimation of the number of nucleotide substitutions in the control region of mitochondrial DNA in humans and chimpanzees. Molecular Biology and Evolution 10: 512–526.
24. Felsenstein J (1985) Confidence limits on phylogenies: an approach using the bootstrap. Evolution: 783–791.
25. Ronquist F, Huelsenbeck JP (2003) MrBayes 3: Bayesian phylogenetic inference under mixed models. Bioinformatics 19: 1572–1574.
26. Blomster J, Bäck S, Fewer DP, Kiirikki M, Lehvo A, et al. (2002) Novel morphology in Enteromorpha (Ulvophyceae) forming green tides. Am J Bot 89: 1756–1763.
27. Shimada S, Yokoyama N, Arai S, Hiraoka M (2009) Phylogeography of the genus Ulva (Ulvophyceae, Chlorophyta), with special reference to the Japanese freshwater and brackish taxa. Springer. pp. 529–539.
28. Hennig W (1965) Phylogenetic systematics. Annual review of entomology 10: 97–116.
29. Gabrielson PW, Widdowson TB, Lindstrom SC (2006) Keys to the seaweeds and seagrasses of Southeast Alaska, British Columbia, Washington, and Oregon: University of British Columbia.
30. Brodie J, Maggs CA, John DM, Blomster J (2007) Green seaweeds of Britain and Ireland. London: British Phycological Society.

10

Genetics of the Pig Tapeworm in Madagascar Reveal a History of Human Dispersal and Colonization

Tetsuya Yanagida[1][*][¤], **Jean-François Carod**[2], **Yasuhito Sako**[1], **Minoru Nakao**[1], **Eric P. Hoberg**[3], **Akira Ito**[1]

1 Department of Parasitology, Asahikawa Medical University, Asahikawa, Hokkaido, Japan, 2 Institut Pasteur de Madagascar, Antananarivo, Madagascar, 3 US Department of Agriculture, Agricultural Research Service, US National Parasite Collection, Animal Parasitic Diseases Laboratory, Beltsville, Maryland, United States of America

Abstract

An intricate history of human dispersal and geographic colonization has strongly affected the distribution of human pathogens. The pig tapeworm *Taenia solium* occurs throughout the world as the causative agent of cysticercosis, one of the most serious neglected tropical diseases. Discrete genetic lineages of *T. solium* in Asia and Africa/Latin America are geographically disjunct; only in Madagascar are they sympatric. Linguistic, archaeological and genetic evidence has indicated that the people in Madagascar have mixed ancestry from Island Southeast Asia and East Africa. Hence, anthropogenic introduction of the tapeworm from Southeast Asia and Africa had been postulated. This study shows that the major mitochondrial haplotype of *T. solium* in Madagascar is closely related to those from the Indian Subcontinent. Parasitological evidence presented here, and human genetics previously reported, support the hypothesis of an Indian influence on Malagasy culture coinciding with periods of early human migration onto the island. We also found evidence of nuclear-mitochondrial discordance in single tapeworms, indicating unexpected cross-fertilization between the two lineages of *T. solium*. Analyses of genetic and geographic populations of *T. solium* in Madagascar will shed light on apparently rapid evolution of this organism driven by recent (<2,000 yr) human migrations, following tens of thousands of years of geographic isolation.

Editor: Yong-Gang Yao, Kunming Institute of Zoology, Chinese Academy of Sciences, China

Funding: This study was supported by the Institut (http://www.pasteur.mg/) Pasteur de Madagascar and by the Japan Society for Promotion of Science (JSPS: http://www.jsps.go.jp/) Asia/Africa Scientific platform (2006-2011), the Grant-in-Aid for Scientific Research from JSPS (21256003, 24256002) and the Special Coordination Fund for Promoting Science and Technology from the Ministry of Education, Japan (2010-2012) to A. Ito. The funders had no role in study design, data collection and analysis, decision to publish, or preparation of the manuscript.

Competing Interests: The authors have declared that no competing interests exist.

* Email: yanagi-t@yamaguchi-u.ac.jp

¤ Current address: Laboratory of Veterinary Parasitology, Joint Faculty of Veterinary Medicine, Yamaguchi University, Yoshida, Yamaguchi, Japan

Introduction

The pig tapeworm *Taenia solium* (Cestoda: Taeniidae) is an etiologic agent of cysticercosis, an important zoonosis and neglected tropical disease, and recently ranked as the most important food-borne parasites on a global scale [1]. The lifecycle of *T. solium* includes humans as the only definitive hosts and domestic pigs as principal intermediate hosts. Cysticercosis refers to infection of various tissues of swine or humans with cysticerci larvae due to ingestion of eggs released from people harboring adult worms in the intestine. Cysticercosis of the central nervous system (neurocysticercosis or NCC), warrants special attention because it is a major cause of seizures and epilepsy in endemic areas [2] and can be lethal especially in remote areas of developing countries [3]. *T. solium* is distributed worldwide where local people consume pork without meat inspection. We previously reported that *T. solium* can be divided into two mitochondrial (mtDNA) genetic lineages, Asian and Afro-American which differ in the clinical manifestations of human cysticercosis [4]. Their distributions are geographically disjunct in Asia or Africa and Latin America [4]. It has been postulated that *T. solium* emerged from Africa with early modern humans and through geographic

expansion became distributed initially across Eurasia prior to the advent of agriculture and domestication of swine [5–7]. Phylogenetic studies have suggested that divergence of the two lineages occurred in the Pleistocene [4,6,8]. Recently, sympatry of both mitochondrial lineages was confirmed in Madagascar [6,8].

Madagascar is a country known to be hyper-endemic for cysticercosis [9,10]. Cysticercosis in pigs results in condemnation of carcasses, particularly in heavy infections, and thus constitutes a considerable economic challenge. Understanding the current distribution for these parasites and the historical factors involved in geographic colonization of Madagascar can contribute insights of importance in developing a capacity for control and mitigation of infections in swine and human hosts.

Malagasy people are divided into 18 ethnic groups and have diverse cultures. Surprisingly, the first human settlement occurred approximately 2000 years ago as one endpoint of Austronesian migration. Linguistic and archeological evidence suggests that the Malagasy people have mixed ancestry from Island Southeast Asia (ISEA), especially Borneo, and from East Africa [11]; dual origins confirmed by analyses of mtDNA and nuclear DNA [12]. In addition, a contribution to the gene pool of Malagasy people from India has recently been suggested by mtDNA genetic analysis [13].

Prehistoric human migrations can also be traced by parasitological evidence. For example, archaeoparasitology of some intestinal parasites have indicated the existence of human migration routes into the New World other than those involving Bering Land Bridge [14]. Phylogenetic analysis suggested that *T. solium* has been introduced into Madagascar multiple times from a number of different areas [8], but the dynamics of these introductions and establishment were not fully elucidated. In the present study, reciprocal insights for the distributional history of hosts and parasites emerge from an exploration of *T. solium* and human occupation of Madagascar.

Historically disjunct populations of *T. solium* are now in sympatry in Madagascar, affording a unique opportunity to explore the possibility of cross-fertilization and hybridization as a fundamental process among cestodes, and concurrently reflect on the degree of isolation and distinct nature of these genotypes. Cestodes are hermaphrodites with two potential modes of reproduction, self- and cross-fertilization. *T. solium* has often been referred to as a self-fertilizer because it is nearly always found alone in the human intestine. However, random amplified polymorphic DNA showed heterozygosity in cysticerci of *T. solium*, suggesting cross-fertilization between different individual worms [15]. Consequently, it may be assumed that the two genotypes of *T. solium* can cross-fertilize in infections involving multiple adults, which may occur early in the infection process. Analysis of maternal inherited mtDNA alone, however, is not sufficient to examine putative hybridization events. Thus, we initially established nuclear DNA markers to differentiate geographic variation in *T. solium*. Secondarily, genetic polymorphism of *T. solium* in Madagascar was investigated to clarify whether hybridization occurs on the island.

Materials and Methods

Parasite isolates and DNA sequencing

During 2005 to 2008, 57 pigs slaughtered from 16 different localities in 5 provinces on Madagascar were found positive for *T. solium* cysticerci. No specific permissions were required for the field survey, and it did not involve endangered or protected species. Meat inspectors in each province were requested to collect infected pig meats at slaughterhouses from the various locations. Pigs were regularly slaughtered at the official slaughterhouses of each city (Table S1), and the slaughtering was controlled by meat inspectors according to the regulations of the Republic of Madagascar. Pigs were sacrificed for routine slaughterhouse purposes and not for research purposes. When positive for *Taenia* cysticerci, infected meats were cut and inserted into sterile containers, and sent to the Pasteur Institute of Madagascar within 24 hours. Then the cysticerci were extracted and washed at the laboratory, and frozen at −20°C until use. All samples were then fixed with 70% ethanol and shipped to Japan according to the research agreement between Pasteur Institute of Madagascar and Asahikawa Medical University. One or two cysts from each pig were subjected to molecular analysis. The genomic DNA of each cyst was extracted by DNeasy blood and tissue kit (Qiagen), and subsequently used as a template for polymerase chain reaction (PCR). For the mtDNA gene markers, the entire cytochrome *c* oxidase subunit I (*cox1*) and cytochrome *b* (*cob*) were amplified by PCR using previously reported primer pairs [4]. PCR products were treated with illustra ExoStar (GE Healthcare) to remove excess primers and dNTPs, and directly sequenced with a BigDye Terminator v3.1 and a 3500 DNA sequencer (Life Technologies).

Nuclear gene markers including RNA polymerase II second largest subunit (*rpb2*), phosphoenolpyruvate carboxykinase (*pepck*),

DNA polymerase delta (*pold*) and a low-molecular-weight glycoprotein antigen (*Ag2*) were amplified using primer pairs published previously [16,17]. These nuclear genes were chosen because they have been shown to be useful for the molecular phylogeny of taeniid tapeworms including species of *Taenia* (*rpb2*, *pepck* and *pold*) or for differentiating geographic genotypes of *T. solium* (*Ag2*). Initially, 41 geographic isolates of *T. solium* from 14 countries were used to investigate the geographical variability of nuclear gene markers. PCR products were sequenced with the same protocols as mtDNA gene markers. When geographical variations were found, new primers were designed to amplify the short fragments including mutation sites in order to reduce the cost and labor. PCR was performed in 20 μL volumes containing 0.5 units of Ex Taq Hot Start Version (TaKaRa, Japan), 0.2 mM of dNTP, 1×Ex Taq Buffer with a final MgCl$_2$ concentration of 2.0 mM, 15 pmol of each primer and 1.0 μL of genomic DNA. PCR amplification consisted of initial denaturation of 94°C for 2 min, 35 cycles of 94°C for 15 sec, 55°C for 15 sec and 72°C for 30 sec, and a terminal extension at 72°C for 1 min. In cases of double peaks in the sequencing of nuclear genes, PCR products were ligated into pGEM-T plasmid vector (Promega) and then introduced into *Escherichia coli* DH5α. At least 10 clonal colonies were picked from an agar plate and their insert DNAs were sequenced to confirm allelic polymorphism.

Data analysis

Nucleotide sequences of the mitochondrial *cob* (1068 sites) and *cox1* (1620 sites) were concatenated into a total sequence (2688 sites). They were aligned by Clustal W 2.0 [18] with those sequences available in public databases. Amino acid sequences were inferred with reference to the echinoderm mitochondrial genetic code [19]. Pairwise divergence values among the obtained nucleotide sequences were calculated using the MEGA5 package [20] using Kimura's two parameter model with a γ-shaped parameter ($\alpha = 0.5$). The identification of mtDNA haplotypes and the drawing of their network was computed by TCS 1.2 software [21] using statistical parsimony [22]. Evaluation of the rate of outcrossing was based on an estimate of the inbreeding coefficient for each nuclear locus and deviation from Hardy-Weinberg proportions as $F = 1\text{-}Hobs/Hexp$, where H is the actual population heterozygosity and *Hexp* is the expected heterozygosity under H − W equilibrium.

Results

Mitochondrial DNA phylogeography

In the present study, we collected 109 cysticerci larvae from 57 pigs across 5 provinces on Madagascar. In total, 8 haplotypes (MDG1 to MDG8) of concatenated *cox1* and *cob* genes were detected. When compared with individual genes, the numbers of haplotypes were reduced to 3 (*cob*) and 7 (*cox1*). All the nucleotide sequences of each haplotype are deposited in GenBank with accession numbers AB781355-AB781364. The frequency of the nucleotide substitution was 1.6% (17 sites/1068 sites) in *cob* and 1.4% (22/1620) in *cox1* (Tables S2 and S3). Among 39 point mutation sites identified, 24 (61.5%) were synonymous and 15 (38.5%) were non-synonymous substitutions. The maximum value of divergence among the 8 haplotypes was 1.4%. Among the mtDNA gene sequences of *T. solium* deposited in the public databases, 14 sets of the complete *cob* and *cox1* gene sequences were concatenated and used for the haplotype network analysis together with those from Madagascar (Table 1). These sequences were chosen because they had unequivocal published references

Table 1. Mitochondrial haplotypes of *T. solium* used for the phylogeographic analysis.

Haplotypes[a]	Localities	Accession numbers		References
		Cox1	*Cob*	
MDG1	Madagascar	AB781355	AB781362	This study
MDG2	Madagascar	AB781356	Same as MDG1	This study
MDG3	Madagascar	Same as MDG1	AB781363	This study
MDG4	Madagascar	AB781357	Same as MDG1	This study
MDG5	Madagascar	AB781358	Same as MDG1	This study
MDG6	Madagascar	AB781359	Same as MDG1	This study
MDG7	Madagascar	AB781360	AB781364	This study
MDG8	Madagascar	AB781361	Same as MDG7	This study
CHN1	China	AB066485	AB066570	Nakao et al. 2002[4]
CHN2	China	AB066486	AB066571	Nakao et al. 2002[4]
ID-BA	Bali, Indonesia	AB631045	Not determined	Swastika et al. 2012[24]
ID-PA	Papua, Indonesia	AB066488	AB066573	Nakao et al. 2002[4]
IND	India	AB066489	AB066574	Nakao et al. 2002[4]
NPL1	Nepal	AB491985	AB781746	Yanagida et al. 2010[23]
NPL2	Nepal	AB491986	Same as MDG1	Yanagida et al. 2010[23]
THA	Thailand	AB066487	AB066572	Nakao et al. 2002[4]
BRA	Brazil	AB066492	AB066577	Nakao et al. 2002[4]
CMR	Cameroon	Same as MEX1	AB066579	Nakao et al. 2002[4]
ECU	Ecuador	AB066491	AB066576	Nakao et al. 2002[4]
MEX1	Mexico	AB066490	AB066575	Nakao et al. 2002[4]
MEX2	Mexico	FN995657	FN995661	Michelet & Dauga 2012[6]
MEX3	Mexico	FH995658	FN995662	Michelet & Dauga 2012[6]
TZA	Tanzania	AB066493	AB066578	Nakao et al. 2002[4]

[a] The mitochondrial haplotypes were determined based on the concatenated nucleotide sequences of complete *cox1* (1620 bp) and *cob* (1068 bp), except for ID-BA.

allowing confirmation that the sequences of the two genes were obtained from one individual parasite.

Network analysis clearly showed these 8 haplotypes are divided into two genotypes (Fig. 1). Six haplotypes (MDG1-6) were the Asian genotype and the remaining two (MDG7-8) were the Afro-American genotype. Overall, 77% (84/109) of the Madagascan haplotypes were the Asian genotype (Table 2). The Asian genotype was found in all examined provinces and was generally dominant except in Toliara. The Afro-American genotype was identified in 4 of 5 examined localities. Among the haplotypes obtained, MDG1 was the major (62%), followed by MDG7 (23%). Among 52 pigs in which two cysts were examined, the different haplotypes were simultaneously obtained in 3 hosts; Asian and Afro-American haplotypes (MDG1 and MDG7) were identified from two hosts, and the different Asian haplotypes (MDG1 and MDG4) were obtained from one host. MDG1 was 100% identical to the haplotype obtained from a pig in Nepal [23], and one base different from the Indian haplotype. All Asian haplotypes from Madagascar are grouped with those from the Indian Subcontinent. On the other hand, these Asian haplotypes were distantly related to that from Papua, Indonesia. Further, the *cox1* haplotype of the isolate from Bali Island, Indonesia [23] was also distantly related. In contrast, MDG7 was one base different from MDG8 and the haplotypes from Mexico, Ecuador, Bolivia.

Nuclear DNA

Among the *Ag2*, *rpb2* and *pold* locus, two (*Ag2* and *rpb2*) or three (*pold*) alleles were confirmed from the 14 geographical isolates from 14 countries; no geographical variation was found in the *pepck* locus. Subsequently, *Ag2*, *rpb2* and *pold* were chosen as appropriate nuclear DNA markers to discriminate the Asian and Afro-American genotypes of *T. solium*. To amplify the target regions including the variable sites, new primers were designed for *rpb2* and *pold* (Table S4). An additional 27 geographical isolates were analyzed using these new primer sets, to confirm geographical variation. *Ag2A*, *rpb2A* and *poldA* (Asian alleles) were only found in Asia and *Ag2*B, *rpb2*B, *poldB* and *poldC* (Afro-American Alleles) were obtained from Latin American and African countries (Table 3). The sequence difference among the alleles was 1–3 bp. All the nucleotide sequences of each allele of *rpb2* and *pold* are deposited in GenBank with accession numbers AB781365-AB781369.

Establishment of nuclear DNA markers allowed us to investigate possible hybridization events in Madagascar. All three nuclear genes were amplified and sequenced for the same 109 cysts as mtDNA genes. Overall, the Asian alleles were the majority in Madagascar with frequencies of 0.81–0.84 (Table 2). Asian alleles were the majority in all the examined regions except for Toliara, and the frequencies of Afro-American alleles in the region were 0.60–0.72. No new alleles were identified among these three loci. Among 12 cysts, the nucleotide sequences of one or more loci could not be determined by direct sequencing because of double

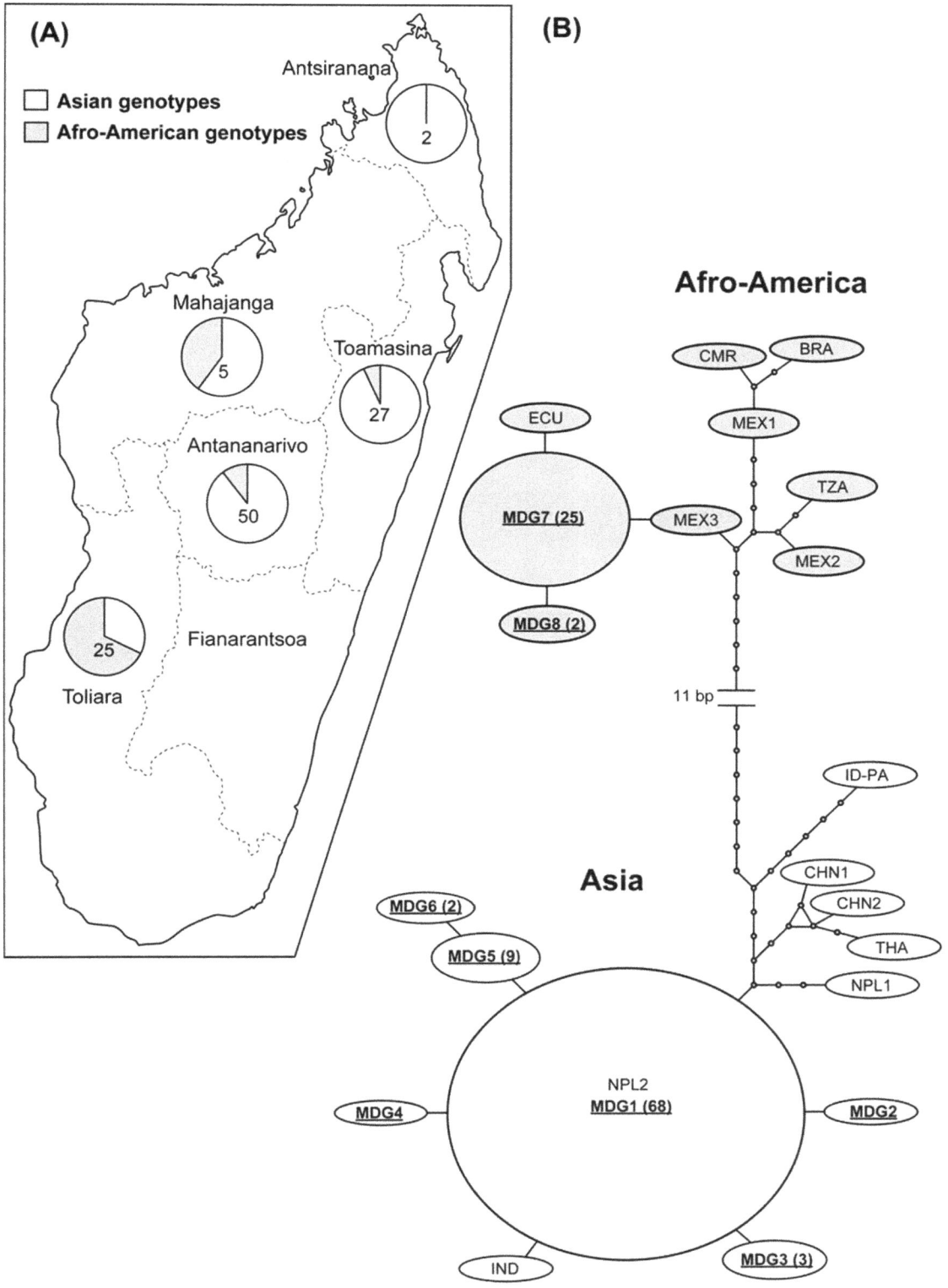

Figure 1. Mitochondrial genotypes of *T. solium* in Madagascar. (A) Pie charts illustrating the frequencies of the Asian and Afro-American mitochondrial genotypes of *T. solium* in each collection site. The numbers in the charts show the sample size for parasite isolates examined. Madagascar is divided into the 7 former provinces. (B) The haplotype network of concatenated mtDNA gene sequences. The size of the ellipses is roughly proportional to the haplotype frequency, and the actual numbers of haplotypes (>1) are enclosed in parentheses.

Table 2. Genotypes of *T. solium* in Madagascar at mitochondrial DNA and each nuclear locus.

| Provinces | No. of pigs | No. of cysts | No. of mtDNA haplotypes | | | | | | | | No. of genotypes at each nuclear locus | | | | | | | | |
| | | | Asian | | | | | Afro-American | | | Ag2 | | | rpb2 | | | pold | | |
			MDG1	MDG2	MDG3	MDG4	MDG5	MDG6	MDG7	MDG8	A/A	A/B	B/B	A/A	A/B	B/B	A/A	A/C	C/C
Antananarivo	26	50	44	0	0	1	0	0	5	0	46	1	3	49	1	0	47	2	1
Toamasina	14	27	13	0	1	0	9	2	2	0	27	0	0	27	0	0	27	0	0
Mahajanga	3	5	3	0	0	0	0	0	2	0	5	0	0	4	1	0	4	1	0
Toliara	13	25	8	1	0	0	0	0	14	2	6	3	16	6	2	17	7	5	13
Antsiranana	1	2	0	0	2	0	0	0	0	0	2	0	0	2	0	0	2	0	0
Total	57	109	68	1	3	1	9	2	23	2	86	4	19	88	4	17	87	8	14

peaks in the sequence electropherograms. As the result of cloning of the polymorphic PCR amplicons, two alleles were detected at an approximate ratio of 1:1. These cases were considered to be heterozygous in each locus. Two cysts obtained from one pig were heterozygous at the all three loci examined. Twenty-two cysts possessed discordant mitochondrial and nuclear genotypes, Asian and Afro-American, in at least one nuclear locus (Table 4). The inbreeding coefficient (F) was estimated only for the sub-population in Toliara because of the considerably biased allele frequency in the other sub-populations; at this locality, F was equal to 0.86 (*Ag2*), 0.90 (*rpb2*) and 0.79 (*pold*).

Discussion

The sympatric distribution of Asian and Afro-American mitochondrial genotypes was confirmed on Madagascar, corroborating a prior report [6,8]. Although the Afro-American mitochondrial genotype previously was identified only in Toliara [8], we confirmed the co-occurrence of Asian and Afro-American genotypes in 4 out of 7 provinces, indicating a widespread distribution for the two mitochondrial genotypes across the island. Major genotypes differed geographically and across provinces. The Asian genotype was generally dominant at all localities except in Toliara, where 64% of the parasite isolates were the Afro-American genotype.

Differences in the distribution of the dominant genotypes of *T. solium* among provinces can be attributed to disparate history and ethnic origins in each region and patterns of human dispersal and migration over the past several thousand years. Phylogenetic analyses of *Taenia* have suggested a relatively deep origin in Africa for *T. solium*, which may have initially parasitized hominin ancestors of modern humans in the early Pleistocene following a host-switching event from large carnivores [5,7,25]. It has been postulated that *T. solium* emerged from Africa with early modern humans and through geographic expansion became distributed initially across Eurasia prior to the domestication of swine which now represent a primary intermediate host [5,6]. Although there is no direct evidence, phylogenetic studies using mtDNA markers have suggested the divergence of the two genotypes, now associated respectively with Africa/America and with southern Asia/Indian Subcontinent occurred in the Pleistocene [4,6,8].

The dominant haplotype in Madagascar (MDG1) demonstrates Asian affinities and is genetically most similar to those from Nepal and India, but distantly related to that from Papua, Indonesia. Further, a *cox1* gene sequence of the isolate from Bali Island [24] was distantly related to MDG1 and other haplotypes from Madagascar. Consequently, it appears that the origin of the Asian genotype on Madagascar is not from ISEA, coincidental with the first human immigrants, but from the Indian Subcontinent. Although Asian origins of the Malagasy people have generally been linked to immigrants and populations from ISEA, our result and recent report on human mitochondrial genetics [13] indicate the importance of Indian influence on the diversity of people and culture in Madagascar consistent with and reflecting a history of human dispersal within the past 2,000 years.

On the other hand, the dominant Afro-American haplotype in Madagascar (MDG7) is closely related to those from Mexico and Ecuador. It does not imply a direct link for Madagascan and Latin American populations, because it is apparent that Afro-American haplotypes have been widely disseminated and the same haplotype can be obtained from both African and Latin American countries [4,8]. It was suggested that *T. solium* was introduced into Latin America from Europe or Africa coincidental with European expansion and development of maritime trade routes after the

Table 3. Distribution of alleles at each nuclear locus around the world.

Localities	No. isolates examined	Alleles		
		Ag2	rpb2	pold
China	4	Ag2A	rpb2A	poldA
Thailand	2	Ag2A	rpb2A	poldA
Papua, Indonesia	2	Ag2A	rpb2A	poldA
Nepal	3	Ag2A	rpb2A	poldA
India	4	Ag2A	rpb2A	poldA
Vietnam	1	Ag2A	rpb2A	poldA
Asian total	16			
Tanzania	7	Ag2B	rpb2B	poldB
Mozambique	7	Ag2B	rpb2B	poldB
South Africa	2	Ag2B	rpb2B	poldC
Cameroon	4	Ag2B	rpb2B	poldC
Mexico	1	Ag2B	rpb2B	poldB
Ecuador	2	Ag2B	rpb2B	poldC
Peru	1	Ag2B	rpb2B	poldC
Brazil	1	Ag2B	rpb2B	poldC
Afro-American total	25			

Table 4. Genotypes of *T. solium* showing nuclear-mitochondrial discordance.

ID of samples	MtDNA haplotype[a]	Genotype at each locus[a,b]			Localities
		Ag2	rpb2	pold	
TsolMDG21b	MDG1	B/B	B/B	A/A	Toliara
TsolMDG29a	MDG1	B/B	A/A	A/A	Toamasina
TsolMDG62a	MDG1	A/A	A/A	C/C	Antananarivo
TsolMDG62b	MDG1	A/A	A/A	A/C	Antananarivo
TsolMDG67a	MDG1	B/B	B/B	A/C	Toliara
TsolMDG68b	MDG1	B/B	B/B	C/C	Toliara
TsolMDG04a	**MDG7**	B/B	A/B	A/A	Antananarivo
TsolMDG04b	**MDG7**	B/B	A/A	A/A	Antananarivo
TsolMDG12b	**MDG7**	A/A	A/A	A/C	Antananarivo
TsolMDG13a	**MDG7**	A/B	A/A	A/A	Antananarivo
TsolMDG13b	**MDG7**	B/B	A/A	A/A	Antananarivo
TsolMDG25a	**MDG7**	A/B	A/B	A/C	Toliara
TsolMDG25b	**MDG7**	A/B	A/B	A/C	Toliara
TsolMDG28a	**MDG7**	A/A	B/B	C/C	Toliara
TsolMDG37a	**MDG7**	A/A	A/A	A/A	Toamasina
TsolMDG37b	**MDG7**	A/A	A/A	A/A	Toamasina
TsolMDG50a	**MDG7**	A/A	A/B	A/A	Mahajanga
TsolMDG50b	**MDG7**	A/A	A/A	A/C	Mahajanga
TsolMDG21a	**MDG7**	B/B	B/B	A/C	Toliara
TsolMDG68a	**MDG7**	B/B	B/B	A/C	Toliara
TsolMDG69a	**MDG7**	A/B	B/B	C/C	Toliara
TsolMDG69b	**MDG7**	B/B	B/B	A/C	Toliara

[a] Haplotypes and alleles in bold are Afro-American ones.
[b] Genotypes with underline indicate those at heterozygous loci.

15th century [4,26]. The dominance of the Afro-American genotype at Toliara, where the current populace is primarily of African descent, suggests that parasites were introduced to Madagascar, probably recurrently, with people and swine from coastal East Africa in a time frame within the past hundreds of years, although clarification requires further study of isolated populations in areas bordering the Mozambique Channel.

Both Asian and Afro-American genotypes on Madagascar showed a simple network with the major (MDG1 and MDG7) and satellite haplotypes. This result indicates a minimum of two independent events of anthropogenic introduction for *T. solium* from historically disjunct geographic regions in relatively shallow ecological time. It is not clear whether *T. solium* was introduced with infected pigs or humans, but it is reasonable to consider that establishment occurred after the first human settlement 2000 years ago because humans are the only definitive hosts. Phylogeography of swine has revealed the distribution of different haplogroups among South Asia, mainland Southeast Asia and ISEA, resulting from Neolithic, human-mediated translocation [27,28]. Thus, genetic analysis of the pigs in Madagascar may shed light on how the tapeworm dispersed across the Indian Ocean.

In the present study, nuclear-mitochondrial discordance was confirmed in all three loci examined, suggesting hybridization between individual worms possessing different genotypes in the recent past. Two cysts from a pig in Toliara were heterozygous at all three loci, suggesting these were F1 hybrids between Asian and Afro-American populations; this genotype could appear at the F2 or later generation by self-fertilization of a hybrid-derived individual worm. Nuclear-mitochondrial discordance in *T. solium* has been confirmed only in Madagascar to date, indicating the hybridization event occurred on the island. The inbreeding coefficient F of the sub-population in Toliara was about 0.8–0.9. If F is interpreted as the rate of selfing [29], it means that 10–20% of the parasite individuals in the subpopulation are outcrossing. The frequency of outcrossing is much less than that demonstrated in another taeniid tapeworm *Echinococcus granulosus*, which were estimated as 74% [30]. Such a contrast is consistent with extraordinarily large infrapopulations typical of *E. granulosus* in canid definitive hosts and thus the chance of mating is simply higher than that of *T. solium*. Nevertheless, the estimated rate of outcrossing for *T. solium* was unexpectedly high when considering that these tapeworms are nearly always found in single-worm infections in humans. However, we experienced a case of taeniasis involving 20 *T. solium* adults in China [31], and we assume that the multiple infection of *T. solium* tapeworms is not so rare in endemic areas. Our result suggests that the chance of outcrossing has been underestimated and establishes hybridization as a common outcome for the Asian and Afro-American genotypes in zones of contact or sympatry. Further epidemiological study on

taeniasis in Madagascar may contribute to a better understanding of the breeding systems of *T. solium*.

Conclusions

In the present study, we show that *T. solium* was introduced and established on Madagascar at least twice in the past 2000 years. An Asian origin, from the Indian Subcontinent, for some genotypes of *T. solium* contrasts with the established history and ancestry of the Malagasy culture primarily from ISEA. Our results demonstrate that tapeworms from geographically disjunct regions in Africa or Latin America and the Indian Subcontinent are now in secondary contact on Madagascar following a history of isolation for populations that may extend to the Pleistocene. Parasites with origins in Africa/Latin America or Asia reflect the complex history of development of the Malagasy culture, and in this case provide compelling evidence for the history of human occupation of the island. Our study highlights the importance of elucidating the determinants for distributions of human pathogens and is especially relevant given manifestation of distinct disease syndromes and socioeconomic impact associated with the two recognized genotypes of *T. solium* [4,32].

Supporting Information

Table S1 Location of the slaughterhouses and the numbers of pigs and cysts examined in each location.

Table S2 Nucleotide substitutions of mitochondrial *cob* gene in 22 haplotypes of *T. solium*.

Table S3 Nucleotide substitutions of mitochondrial *cox1* gene in 23 haplotypes of *T. solium*.

Table S4 PCR primer pairs used for the amplification of nuclear gene markers.

Acknowledgments

The authors are grateful to Ms. Toshiko Miura and Tomoe Nakayama for their kind support in molecular analyses.

Author Contributions

Conceived and designed the experiments: AI JC MN. Performed the experiments: TY. Analyzed the data: TY. Contributed reagents/materials/analysis tools: TY. Wrote the paper: TY JC YS MN EH AI.

References

1. Robertson LJ, van der Giessen WB, Batz MB, Kojima M, Cahill S (2013) Have foodborne parasites finally become a global concern? Trends Parasitol 29: 101–103.

2. Ndimubanzi PC, Carabin H, Budke CM, Nguyen H, Qian Y-J, et al. (2010) A systematic review of the frequency of neurocyticercosis with a focus on people with epilepsy. PLoS Negl Trop Dis 4: e870.

3. Ito A, Nakao M, Wandra T (2003) Human taeniasis and cysticercosis in Asia. Lancet 362: 1918–1920.

4. Nakao M, Okamoto M, Sako Y, Yamasaki H, Nakaya K, et al. (2002) A phylogenetic hypothesis for the distribution of two genotypes of the pig tapeworm *Taenia solium* worldwide. Parasitology 124: 657–662.

5. Hoberg E (2006) Phylogeny of *Taenia*: species definitions and origins of human parasites. Parasitol Int 55: S23–S30.

6. Michelet L, Dauga C (2012) Molecular evidence of host influences on the evolution and spread of human tapeworms. Biol Rev 87: 731–741.

7. Terefe Y, Hailemariam Z, Menkir S, Nakao M, Lavikainen A, et al. (2014) Phylogenetic characterisation of *Taenia* tapeworms in spotted hyenas and reconsideration of the "Out of Africa" hypothesis of *Taenia* in humans. Int J Parasitol 44: 533–541.

8. Michelet L, Carod JF, Rakontondrazaka M, Ma L, Gay F, et al. (2010) The pig tapeworm *Taenia solium*, the cause of cysticercosis: Biogeographic (temporal and spacial) origins in Madagascar. Mol Phylogenet Evol 55: 744–750.

9. Mafojane NA, Appleton CC, Krecek RC, Michael LM, Willingham AL III (2003) The current status of neurocysticercosis in Eastern and Southern Africa. Acta Trop 87: 25–33.

10. Rasamoelina-Andriamanivo H, Porphyre V, Jambou R (2013) Control of cysticercosis in Madagascar: Beware of the pitfalls. Trends Parasitol 29: 538–547.

11. Dewar RE, Wright HT (1993) The culture history of Madagascar. J World Prehist 7: 417–466.

12. Hurles M, Sykes B, Jobling M, Foster P (2005) The dual origin of the Malagasy in Island Southeast Asia and East Africa: evidence from maternal and paternal lineages. Am J Hum Genet 76: 894–901.
13. Dubut V, Cartault F, Payet C, Thionville M, Murail P (2010) Complete mitochondrial sequences for haplogroups M23 and M46: insights into the Asian ancestry of the Malagasy population. Hum Biol 81: 495–500.
14. Araujo A, Reinhard KJ, Ferreira LF, Gardner SL (2008) Parasites as probes for prehistoric human migrations? Trends Parasitol 24: 112–115.
15. Maravilla P, Gonzalez-Guzman R, Zuniga G, Peniche A, Dominguez-Alpizar JL, et al. (2008) Genetic polymorphism in *Taenia solium* cysticerci recovered from experimental infections in pigs. Infect Genet Evol 8: 213–216.
16. Knapp J, Nakao M, Yanagida T, Okamoto M, Saarma U, et al. (2011) Phylogenetic relationships within *Echinococcus* and *Taenia* tapeworms (Cestoda: Taeniidae): An inference from nuclear protein-coding genes. Mol Phylogenet Evol 61: 628–638.
17. Sato MO, Sako Y, Nakao M, Wandra T, Nakaya K, et al. (2011) A possible nuclear DNA marker to differentiate the two geographic genotypes of *Taenia solium* tapeworms. Parasitol Int 60: 108–110.
18. Larkin MA, Blackshields G, Brown NP, Chenna R, McGettigan PA, et al. (2007) Clustal W and Clustal X version 2.0. Bioinformatics 23: 2947–2948.
19. Nakao M, Sako Y, Yokoyama N, Fukunaga M, Ito A (2000) Mitochondrial genetic code in cestodes. Mol Biochem Parasit 111: 415–424.
20. Tamura K, Peterson D, Peterson N, Stecher G, Nei M, et al. (2011) MEGA5: molecular evolutionary genetics analysis using maximum likelihood, evolutionary distance, and maximum parsimony methods. Mol Biol Evol 28: 2731–2739.
21. Clement M, Posada D, Crandall K (2000) TCS: a computer program to estimate gene genealogies. Mol Ecol 9: 1657–1659.
22. Templeton AR, Crandall KA, Sing CF (1992) A cladistic analysis of phenotypic associations with haplotypes inferred from restriction endonuclease mapping and DNA sequence data. III. Cladogram estimation. Genetics 132: 619–633.
23. Yanagida T, Yuzawa I, Joshi DD, Sako Y, Nakao M, et al. (2010) Neurocysticercosis: assessing where the infection was acquired from. J Trav Med 17: 206–208.
24. Swastika K, Dewiyani CI, Yanagida T, Sako Y, Sudarmaja M, et al. (2012) An ocular cysticercosis in Bali, Indonesia caused by *Taenia solium* Asian genotype. Parasitol Int 61: 378–380.
25. Hoberg E, Alkire N, Queiroz A, Jones A (2001) Out of Africa: origins of the *Taenia* tapeworms in humans. P Roy Soc Lond B Bio 268: 781–787.
26. Martinez-Hernandez F, Jimenez-Gonzalez D, Chenillo P, Alonso-Fernandez C, Maravilla, et al. (2009) Geographical widespread of two lineages of *Taenia solium* due to human migrations: Can population genetic analysis strengthen this hypothesis? Infect Genet Evol 9: 1108–1114.
27. Larson G, Dobney K, Albarella U, Fang M, Matisoo-Smith E, et al. (2005) Worldwide phylogeography of wild boar reveals multiple centers of pig domestication. Science 307: 1618–1621.
28. Larson G, Cucchi T, Fujita M, Matisoo-Smith E, Robins J, et al. (2007) Phylogeny and ancient DNA of *Sus* provides insights into neolithic expansion in Island Southeast Asia and Oceania. Proc Natl Acad Sci USA 104: 4834–4839.
29. McCauley DE, Whittier DP, Reilly LM (1985) Inbreeding and the rate of self-fertilization in a grape fern, *Botrychium dissectum*. Am J Bot 72: 1978–1981.
30. Haag KL, Marin PB, Graichen DAS, De La Rue ML (2011) Reappraising the theme of breeding systems in *Echinococcus*: is outcrossing a rare phenomenon? Parasitology 138: 298–302.
31. Ito A, Li T, Chen X, Long C, Yanagida T, et al. (2013) Mini review on chemotherapy of taeniasis and cysticercosis due to *Taenia solium* in Asia, and a case report with 20 tapeworms in China. Trop Biomed 30: 164–173.
32. Campbell G, Garcia H, Nakao M, Ito A (2006) Genetic variation in *Taenia solium*. Parasitol Int 55: S121–S126.

11

Genetic Variability and Population Structure of *Disanthus cercidifolius* subsp. *longipes* (Hamamelidaceae) Based on AFLP Analysis

Yi Yu[1], Qiang Fan[1], Rujiang Shen[1], Wei Guo[3], Jianhua Jin[1], Dafang Cui[2]*, Wenbo Liao[1]*

1 Guangdong Key Laboratory of Plant Resources and Key Laboratory of Biodiversity Dynamics and Conservation of Guangdong Higher Education Institutes, School of Life Sciences, Sun Yat-Sen University, Guangzhou, China, 2 College of Forestry, South China Agriculture University, Guangzhou, China, 3 Department of Horticulture and Landscape Architecture, Zhongkai University of Agriculture and Engineering, Guangzhou, China

Abstract

Disanthus cercidifolius subsp. *longipes* is an endangered species in China. Genetic diversity and structure analysis of this species was investigated using amplified fragments length polymorphism (AFLP) fingerprinting. Nei's gene diversity ranged from 0.1290 to 0.1394. The AMOVA indicated that 75.06% of variation was distributed within populations, while the between-group component 5.04% was smaller than the between populations-within-group component 19.90%. Significant genetic differentiation was detected between populations. Genetic and geographical distances were not correlated. PCA and genetic structure analysis showed that populations from East China were together with those of the Nanling Range. These patterns of genetic diversity and levels of genetic variation may be the result of *D. c.* subsp. *longipes* restricted to several isolated habitats and "excess flowers production, but little fruit set". It is necessary to protect all existing populations of *D. c.* subsp. *longipes* in order to preserve as much genetic variation as possible.

Editor: Ting Wang, Wuhan Botanical Garden, Chinese Academy of Sciences, Wuhan, China

Funding: Support for this study was provided by a grant by the National Natural Science Foundation of China (Project 31170202) to WL, a grant by the National Natural Science Foundation of China (Project 30670141) to WL, a grant by the National Natural Science Foundation of China (Project 31100159) to QF, a grant by the National Natural Science Foundation of China (Project 40672017) to JJ. The funders had no role in study design, data collection and analysis, decision to publish, or preparation of the manuscript.

Competing Interests: The authors have declared that no competing interests exist.

* Email: lsslwb@mail.sysu.edu.cn (WL); cuidf@scau.edu.cn (DC)

Introduction

Disanthus Maxim. (Hamamelidaceae) is a monotypic genus endemic to China and Japan [1]. *Disanthus cercidifolius* subsp. *longipes* is the only subspecies in *Disanthus* which is distributed in China (southern Zhejiang, central and northwestern Jiangxi, and southern Hunan), while its sister, *D. c.* subsp. *cercidifolius*, is endemic to Japan [2]. *D. c.* subsp. *longipes* was first reported by Cheng [3] in 1938, and then revised by Chang [4] in 1948. In a systematic study of *Disanthus* [2], the author believed that the Chinese species of *Disanthus* was a sister population of *D. c.* subsp. *cercidifolius*, which is distributed in warm and humid forests of the Cathayan Land, and also was a Tertiary relic. *D. c.* subsp. *longipes* was characterized by its morphology and preference for humid, acid soils and shady habitats. The inflorescences are paired, axillary; Capitula are 2-flowers, purple; Leaves are heart shaped, green then turning to purple, orange, and red. It is usually a small tree, 2–3 m high, occasionally reaching heights of 6–8 m in forests when growing along streams. Because of severe habitat fragmentation that caused population decline, *Disanthus* was listed in the 1992 Red List of Endangered Plant Species of China [5], the 1994 IUCN Red List of Threatened Species (www.iucnredlist.org), the Key Wild Plants under State Protection [6] and the 2004 China Species Red List as bring a species at high risk of extinction in the wild [7].

Although study of *Disanthus* has attracted many investigators [8–22], little attention has been paid to its genetic analysis and population structure, except Xiao [23]. Xiao investigated the genetic diversity of *D. c.* subsp. *longipes* based on nine allozyme loci and found a higher genetic variation within populations as well as significantly lower variation among populations. However, sampling of Xiao's study was limited to part of the distribution area of *D. c.* subsp. *longipes* only. So a comprehensive study of the populations' genetic structure at different geographic scales is still needed.

Preserving the genetic diversity of endangered species is one of the primary goals in conservation planning. Because survival and evolution of species depended on the maintenance of sufficient genetic variability within and among populations to accommodate new selection pressures caused by environmental changes [24,25]. For endemic endangered species, intraspecific variation is a prerequisite for any adaptive changes or evolution in the future, and have profound implications for species conservation [26,27]. The knowledge of the levels and patterns of genetic diversity is important for designing conservation strategies for threatened and endangered species [28,29]. So identifying variations with molecular markers has provided the abundant information concerning genetic diversity in plant species [30–36]. Amplified fragment length polymorphism (AFLP) [37] is a PCR-based technique which has been successfully applied to the identification and estimation of molecular genetic diversity and population structure [32–34,38–41]. This technique can generate information on multiple loci in a single assay without prior sequence knowledge, and [42]. Using AFLP markers, we would know (1) the degree of genetic diversity within and among populations; (2) which factors might explain genetic variation; and (3) how to apply this information to develop recommendations for management of this endangered species.

Materials and Methods

Ethics Statement

Field studies were approved by Hunan Provincial Bureau of Forestry for collection in Xinning County (1XN), the Dupanglin National Nature Reserve (2DP), the Mangshan National Nature Reserve (3MS), and approved by Guangdong Provincial Bureau of Forestry for collection in Nanling National Nature Reserve (4NL), and approved by Zhejiang Provincial Bureau of Forestry for collection in QianJiang Source National Forestry Park (5QJ) and Zhulong town, Longquan City (6ZL), and approved by Jiangxi Provincial Bureau of Forestry for collection in Guanshan National Nature Reserve (7GS), and Mount Sanqingshan National Park (8SQ).

Specimen collection

The specimens were collected from eight populations from April to June of 2008. All populations but one grew in evergreen and deciduous broad-leaved mixed forests (often lived with species such as in genus *Cyclobalanopsis* and *Sorbus*, *etc.*) at altitudes of 450–1200 m. The exception (3MS) was the one growing in bamboo forests in the Mangshan Mountains of Yizhang. For molecular analysis, 10–11 individuals per population were sampled. The locations and information of populations are provide in Table 1 and Fig. 1. Before DNA extraction, all the dried leaves were preserved in silica gel [43]. All the voucher specimens are deposited at the Herbarium of Sun Yat-sen University (SYS).

DNA extraction and AFLP reactions

Genomic DNA was extracted from silica gel-dried leaves using the cetyl trimethylammonium bromide (CTAB) method [44]. The extracted DNA was dissolved in 100 μL of Tris-hydrochloride (TE) buffer [10 mmol/l Tris-HCl (pH 8.0), 1 mmol/EDTA (pH 8.0)] and used as a template for the polymerase chain reaction (PCR).

AFLP reactions were performed following the method reported by [37] with the following modifications. The restriction digest and ligation steps were done as separate reactions. For the digestion, approximately 500 ng of genomic DNA was incubated at 37°C (EcoRI) or 65°C (MseI) for 2 h in a 20 μL volume reaction containing 10× H Buffer (TOYOBO, Shanghai) and 10 U restriction enzymes EcoRI or MseI. For the ligation, 20 μL of a ligation mix consisting of 10× T4 DNA Ligase (TOYOBO, Shanghai), 1 μL EcoRI-adapter, 1 μL MseI-adapter, and 2 U T4 DNA Ligase was added to the sample and kept at 22°C for 3 h. After ligation, the samples were diluted 10-fold with sterile deionized water (dH₂O). A pre-selective polymerase chain reaction (PCR), using PTC-200 thermocycler (MJ research, Waltham, MA) was done using primer pairs with a single selective nucleotide extension. The reaction mix (total volume 20 μL) consisted of 4 μl template DNA from the restriction/ligation step, 1 μL primer (EcoRI/MseI), and 15 μL AFLP Core Mix (13.8 μL dH₂O, 1.6 μL MgCl₂, 1.6 μL dNTPs (2.5 mM), 1 U Taq DNA polymerase, and 10× H buffer). After an initial incubation at 94°C for 2 min, 20 cycles at 94°C for 20 s, 56°C for 30 s, and 72°C for 2 min, with a final extension at 60°C for 30 min, were performed. The PCR products of the amplification reaction were diluted 10-fold with dH₂O and used as a template for the selective amplification using two AFLP primers, each containing three selected nucleotides.

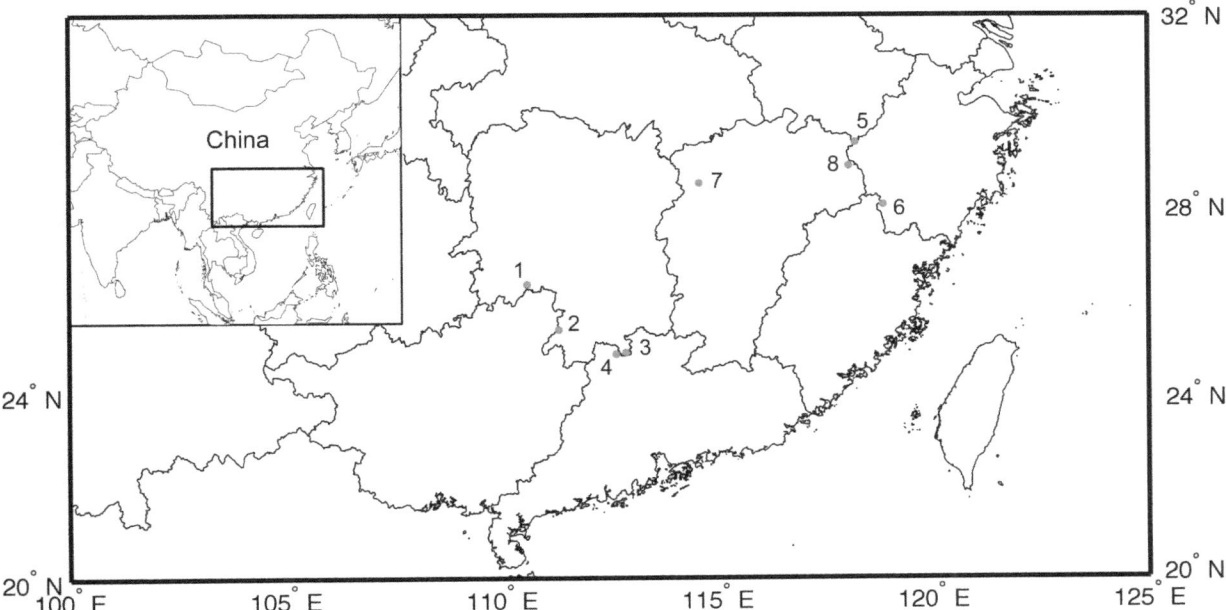

Figure 1. Location of eight populations in two groups sampled in a study of genetic diversity of *Disanthus cercidifolius* **subsp.** *longipes.* Populations are represented by black dots and located as Table 1. Note: 1:1XN, 2:2DP, 3:3MS, 4:4NL, 5:5QJ, 6:6ZL, 7:7GS, 8:8SQ.

Table 1. Information of location of populations.

ID	GPS co-ordinates	Location of populations	Location type	Altitude (m)	Collection voucher
1XN	26°25'08"N, 110°36'31"E	Xinning county, Hunan Province	mountain slope on the edge of forests	690–740	SHEN Rujiang and GUO Wei
2DP	25°27'43"N, 111°20'20"E	Dupanglin National Nature Reserve, Hunan Province	in bushes by stream	700–850	SHEN Rujiang and GUO Wei
3MS	24°58'12"N, 112°53'16"E	Mangshan National Nature Reserve, Hunan Province	in bamboo forests by stream	680–923	SHEN Rujiang and GUO Wei
4NL	24°56'18"N, 112°39'40"E	Nanling Naitional Nature Reserve, Guangdong Province	mountain slope	680–700	SHEN Rujiang
5QJ	29°23'59"N, 118°12'58"E	Qianjiang Source National Forestry Park, Zhejiang Province	mountain slope	650–700	SHEN Rujiang and GUO Wei
6ZL	28°06'28"N, 118°51'27"E	Zhulong town, Longquan City, Zhejiang Province	mountain slope by streams	1040	SHEN Rujiang and GUO Wei
7GS	28°33'22"N, 114°35'42"E	Guanshan National Nature Reserve, Jiangxi Province	mountain slope	560–600	CHEN Lin
8SQ	28°54'57"N, 118°03'52"E	Mount Sanqingshan National Park, Jiangxi Province	mountain slope in forests	620–1260	Observation team of Sun Yat-sen University

Nine primer combinations labeled with fluorescent 6-carboxy fluorescein (6-FAM) were probed for selective amplification, and only primer combinations with the greatest number of bands (EcoRI/MseI: AAC/CTG; ACA/CTG and ACT/CAT) were selected. Selective amplification was done using the following touchdown PCR conditions: 94°C for 2 min first, then 10 cycles at 94°C for 20 s, and 66°C for 30 s, with a 1°C decrease per cycle then extension at 72°C for 2 min; followed by 20 cycles at 94°C for 20 s, 56°C for 30 s, and 72°C for 2 min. After amplification, 3 μL of the samples were diluted 3-fold with sterile deionized water (dH$_2$O), and mixed with 10 μL formamide and 0.2 μL Size standard-600 (Beckman Coulter, Fullerton, CA). The mix was used for sequence analysis.

Raw data were collected on a CEQ8000 Sequencer (Beckman Coulter). AFLP products were resolved using a Beckman Coulter CEQ8000 genetic analyzer. Semi-automated fragment analysis was performed using the fragment analysis software of the CEQ8000. The chromatograms of fragment peaks were scored as present (1) or absent (0), and a binary qualitative data matrix was constructed. A total of 82 individuals were run twice with all primer combinations.

Data analysis

As a measure of population diversity, the binary data matrix was input to POPGENE version 1.32 [45], assuming Hardy–Weinberg equilibrium. The following indices were used to quantify the amount of genetic diversity within each population examined: the number of AFLP fragments (Frag$_{tot}$), the percentage of polymorphic fragments (Frag$_{poly}$), Nei's (1973) gene diversity (H), and Shannon's information index (I). In addition, the number of unique fragments (Frag$_{uni}$) and DWs, frequency-down-weighted marker values [46], were calculated as measures of divergence. For each population, the number of occurrences of each AFLP marker in that population was divided by the number of occurrences of that particular marker in the total dataset, then these values were summed up [46,47]. The value of DW was expected to be high in long-term isolated populations where rare markers should accumulate due to mutations, whereas newly established populations were expected to exhibit low values [46]. To even out the unequal sample sizes, DWs were calculated using five randomly

chosen individuals. The value of DW was calculated by AFLPdat [48], which is based on R version 2.9.0 (www.r-project.org).

Nei's genetic distance [49] of *D. c.* subsp. *longipes* populations were calculated using the software POPGENE version 1.32 [45], then cluster analysis was performed based on the unweighted pair group method with arithmetic averaging (UPGMA) [50] using NTSYS pc version 2.1e [51]. In addition, based on the genetic distance index devised by [52], the UPGMA dendrogram of individuals was drawn using TREECON version 1.3b [53]. The robustness of the branches was estimated by 1000 bootstrap replicates.

ARLEQUIN version 3.0 [54] was used to perform an analysis of molecular variance (AMOVA) [55] to assess the hierarchical genetic structure among populations and within populations. The AFLPdat program [48], based on R version 2.9.0 (www.r-project. org), was used to convert the AFLP data matrix to the ARLEQUIN input format. The AMOVA was performed by partitioning genetic variation among and within populations regardless of their geographic distribution. In this study, the traditional F-statistics [56] cannot be used in dominant marker AFLP, therefore Φ-statistics [55] replace F-statistics. The significance of Φ values was tested by 1000 permutations. Φ-statistics are computed from a matrix of Euclidean squared distances between every pair of individuals [40]. Two models of the eight populations were tested to investigate regional relationships. Firstly, we treat all populations as a single group to obtain a value for Φ_{st} as an overall measure of population divergence (a two-level analysis), and then we divided the populations into two groups, the East China region and the Nanling Range (three-level hierarchical analyses).

A distance matrix of Φ_{st} between every pair of populations was calculated in ARLEQUIN as a measure of interpopulation genetic differentiation, from which 1000 bootstrapped replicate matrices were then computed, so gene flow (Nm) based on Φ_{st} [56] could be calculated. Isolation-by-distance was investigated by computing the correlation between geographic distance and genetic distance (Φ_{st}) between every pair of populations and applying the Mantel test, using NTSYSpc version 2.1e [51]. The Mantel z-statistic value was tested non-parametrically by creating a null distribution of z using 1000 random permutations and comparing the observed z value [40].

Table 2. Number of loci evaluated for each of three AFLP primer combinations utilized in assays of 82 individuals of *Disanthus cercidifolius* subsp. *longipes* from eight populations.

Primer combination	Number of loci
EcoRI-AAC/MseI-CTG	144
EcoRI-ACA/MseI-CTG	152
EcoRI-ACT/MseI-CAT	157
Total	453

The AFLP data were also subjected to a principal components analysis (PCA), which may help reveal unexpected relationships among a large number of variables, reducing them to two or three new uncorrelated variables so they retain most of the original information [57]. We chose Jaccard's similarity coefficient [50] to calculate the eigenvalue and eigenvector. The standardized data were projected onto the eigenvectors of the correlation matrix and represented in a two-dimensional scatter plot [57]. Plots of samples in relation to the first three principal components were constructed with populations designated as either populations of the East China region or populations of the Nanling Range. The data from the PCA was analyzed using the computer program NTSYSpc version 2.1e [51]. A two-dimensional representation of genetic relationships among *D. c.* subsp. *longipes* genotypes was carried out using SPSS version 16.0 [58].

STRUCTURE version 2.3.1 [59] was used to investigate structure at the individual level. In this study, we inferred that each individual of *D. c.* subsp. *longipes* comes purely from one of the populations. It was applied using a "no admixture" model, 100 000 burn-in period, and 50 000 MCMC replicates after burn-in. The approach requires that the number of clusters K be predefined, and the analysis then assigns the individuals to clusters probabilistically [40]. We performed ten runs for each value of K (1 to 10). To determine the K value, we used both the LnP(D) value and Evanno's ΔK [60]. LnP(D) is the log likelihood of the observed genotype distribution in K clusters and can be output by STRUCTURE simulation [59]. Evanno's ΔK took consideration of the variance of LnP(D) among repeated runs and usually can indicate the ideal K. The suggested Δk = M(|L(k+1)-2L(k) +L(k-1)|)/S[L(k)], where L(k) represents the kth LnP(D), M is the mean of 10 runs, and S their standard deviation[60,61]. The output uses color coding to show the assignments of individuals in each population to the clusters.

Population structure was also investigated by HICKORY version 1.1 [62], in order to assess the importance of inbreeding in the data and the assumption of Hardy-Weinberg equilibrium. HICKORY makes it possible to evaluate departures from Hardy-Weinberg equilibrium in dominant as well as co-dominant markers [40]. The program AFLPdat [48] was used to convert the AFLP data matrix to the HICKORY input format. The deviance information criterion (*DIC*) criteria, *Dbar*, *Dhat*, and *pD*, for assessing the importance of inbreeding were computed using the default values: Burn-in 5000; sample 100 000; thin 20 [62].

Results

Population genetic diversity

In this study, three of the nine AFLP primer combinations were used (Table 2). The number of fragments for each primer combination (with percentage of polymorphisms within parenthesis), were: EcoRI-AAC/MseI-CTG: 144 (98.61%), EcoRI-ACA/

MseI-CTG: 152 (99.34%), and EcoRI-ACT/MseI-CAT: 157 (96.18%). The length of the fragments varied from 64 bp to 560 bp. The three AFLP primer combinations produced a total of 453 fragments in 82 individuals, of which 444 (98.01%) were polymorphic. The total number of fragments per population (Frag_tot) varied between 198 (7GS) and 227 (5QJ). The percentage of polymorphism (Frag_poly) across the eight populations ranged from 43.71% (7GS) to 50.11% (5QJ). The number of fragments that only occur in one population (Frag_uni) varied between 6 (7GS) and 16 (8SQ). The DW ranged from 56.85 (5QJ) to 111.19 (1XN), mean 76.88, and SD = 21.11. The diversity within a population (H) ranged from 0.1290 (2DP) to 0.1394 (5QJ) (Table 3).

AMOVA

The two-level AMOVA in ARLEQUIN gave a Φ_{st} value of 0.2328 (P<0.001), with 23.28% variation among populations and 76.72% within populations (Table 4). The three-level hierarchical AMOVA analyses of the two-group models shows that three-quarters of the variation (75.06%) was concentrated within populations, while the between group component (5.04%) was less than the between populations-within-group component (19.90%). All three Φ values were significant based on a 1000 permutation test (Table 4).

Pairwise genetic distance and gene flow

The pairwise genetic distance (Φst) and gene flow (Nm) matrix were used to establish the level of genetic divergence among the populations (Table 5). Estimates of pairwise genetic distance using AFLP date ranged from 0.1364 (P<0.001) for the most closely related populations (4NL and 6ZL), to 0.3428 (P<0.001) in the most divergent populations (3MS and 8SQ). The gene flow (Nm) among populations ranged from 0.4792 to 1.5830. Except for one population in Nanling National Nature Reserve (4NL), most of populations' Φst were above 0.2, and the corresponding Nm were below 1. These results indicate that the genetic distance of *D. c.* subsp. *longipes* is not close, yet nor dependent on geographic distance.

Relationship between geographic distance and genetic distance (The Mantel tests)

A Mantel test showed no correlation between the geographic distance and the Φst genetic distance (r = 0.38917, P = 0.9920, Fig. 2). This result implies that *D. c.* subsp. *longipes* does not demonstrate a historical pattern of isolation-by-distance.

Principal components analysis

A PCA of the AFLP-based distance data was performed to examine relationships among the populations of *D. c.* subsp. *longipes*. The PCA showed that populations from East China and the Nanling Range clearly do not cluster separately (Fig. 3). The

Table 3. Region, population (ID), sample size (N), total number of fragments per population (Frag$_{tot}$), percentage of polymorphic fragments (Frag$_{poly}$), number of fragments that only occur in one population (Frag$_{uni}$), frequency-down-weighted marker values (DW), Nei's (1973) gene diversity (H), Sannon's index (I) sampled in a study of *Disanthus cercidifolius* subsp. *longipes* (H. T. Chang) K. Y. Pan populations.

Region	ID*	N	Frag$_{tot}$	Frag$_{poly}$	Frag$_{uni}$	DW	H	I
Nanling range	1XN	10	216	47.68	12	111.19	0.1357±0.1779	0.2106±0.2570
	2DP	11	203	44.81	11	56.96	0.1290±0.1790	0.1991±0.2580
	3MS	10	204	45.03	10	93.96	0.1373±0.1839	0.2100±0.2652
	4NL	10	215	47.46	9	58.09	0.1383±0.1710	0.2136±0.2608
East China region	5QJ	11	227	50.11	10	56.85	0.1394±0.1778	0.2172±0.2564
	6ZL	10	209	46.14	12	77.45	0.1393±0.1822	0.2141±0.2631
	7GS	10	198	43.71	6	64.85	0.1353±0.1852	0.2062±0.2662
	8SQ	10	223	49.23	16	95.68	0.1373±0.1775	0.2138±0.2560
All	-	82	453	98.01	-	-	0.1781±0.1633	0.2918±0.2223

Table 4. Analysis of molecular variance (AMOVA) in *Disanthus cercidifolius* subsp. *longipes* for 82 individuals from eight populations.

Model	Source of variation	df	Sum of squares	Variance component	Percentage variance %	Φ-statistics	P
Two levels	among populations	7	1055.95	11.14	23.28	$\Phi_{st}=0.23283$	P<0.001
	within populations	74	2716.05	36.7	76.72		
	Total	81	3772	47.84			
three levels	between groups	1	237.66	2.46	5.04	$\Phi_{ct}=0.05040$	P<0.05
	between populations-within-groups	6	818.29	9.73	19.90	$\Phi_{sc}=0.20955$	P<0.001
	within populations	74	2716.05	36.70	75.06	$\Phi_{st}=0.24939$	P<0.001
	Total	81	3772	48.89			

Computed with ARLEQUIN version 3.0. The *P* values represent the probability of obtaining an equal or more extreme value by chance, estimated from 1000 permutations.

Table 5. Pairwise genetic distance (Φst, lower diagonal, $P<0.001$) and gene flow (Nm, upper diagonal) between eight populations of *Disanthus cercidifolius* subsp. *longipes* based on AFLP data.

	1XN	2DP	3MS	4NL	5QJ	6ZL	7GS	8SQ
1XN	-	1.4673	0.7122	1.1283	0.7637	0.7610	0.8506	0.6140
2DP	0.1456	-	0.8055	1.1586	0.6833	0.7813	0.7398	0.5624
3MS	0.2598	0.2369	-	1.4639	0.6874	0.7389	0.8037	0.4792
4NL	0.1814	0.1775	0.1459	-	1.0727	1.5830	1.2885	0.6644
5QJ	0.2466	0.2679	0.2667	0.1890	-	1.1441	0.7897	1.3148
6ZL	0.2473	0.2424	0.2528	0.1364	0.1793	-	1.1783	0.6917
7GS	0.2272	0.2526	0.2373	0.1625	0.2405	0.1750	-	0.5466
8SQ	0.2893	0.3077	0.3428	0.2734	0.1598	0.2655	0.3138	-

Significance levels are based on 1000 iterations and indicate that the probability that random Φst values are higher than the observed values. $Nm = (1 - Fst)/4*Fst$ [56].

natural divide is separated into three clusters: cluster 1 (1XN, 2DP), cluster 2 (5QJ, 8SQ), and cluster 3, a mixture that includes some individuals from cluster 1 or cluster 2. The first and second principal component axes, PC1 and PC2, accounted for 25.46% and 25.17% of the total variation, respectively.

Hierarchical and cluster analysis

Nei's genetic distances [49] between populations showed two main clusters of populations (Fig. 4). The cluster that comprises 5QJ and 8SQ is clearly differentiated from the others. Populations 1XN and 2DP are another cluster, and population 4NL and 6ZL are grouped with 7GS, although the UPGMA dendrogram of the 82 individuals based on Nei and Li's genetic distance is not clearly grouped (Fig. 5). Although most individuals in the same population cluster together, some individuals in different populations (except 1XN and 3MS) also are clustered together.

In the genetic structure analysis (STRUCTURE), the highest estimate of the likelihood of the data, given the number of clusters chosen, was $k = 10$ (ten clusters). We can't get a clear knee in Lnp(D) (Fig. 6). Evanno's ΔK took consideration of the variance of LnP(D) among repeated runs and usually can indicate the best k. See the Fig. 6, we can find that when the $k = 6$ and we get the highestΔK value. So the best $k = 6$. The diagram (Fig. 7) showing assignment of individuals to the clusters revealed a clinal structure to the data. In the six-cluster model ($k = 6$), gene pool 1 (including 5QJ, 8SQ) and gene pool 2 (including 1XN, 2DP) are restricted with a few from another pool, only being distributed in the southwest (gene pool 1) and northeast (gene pool 2). Gene pool 3 (including 3MS, 4NL), while Gene pool 4, pool 5 and pool 6, consist of 6ZL and 7GS. These 4 pools have the most widespread pattern.

The HICKORY results showed that there is inbreeding in populations of *D. c.* subsp. *longipes* (Table 6), because the *DIC* and *Dbar* parameter were lower in "Full model" than in other models. This pattern of results allows the "Full model" to be considered best, according to the HICKORY manual [62].

Discussion

Genetic diversity in endemic and endangered species

Accurate estimates of genetic diversity are useful for optimizing sampling strategies aiming at the conservation and management of genetic resources [26,63,64]. According to Hamrick and Godt (1989), there are strong associations between geographic range and genetic diversity [65]. Allozyme analyses concluded that endemic and geographically limited plant species generally possess less genetic variation, due to genetic drift and restricted gene flow [66–68]. However, in our study, the percentage of polymorphic fragments (Frag$_{poly}$) and Shannon's information index (I) are higher than those of endemic species based on allozyme [65]. This may account for the new technique can generate more genetic diversity information. Historical events have also been shown to be responsible for variations in genetic diversity [68]. Numerous allozyme studies and an increasing number of cpDNA and mtDNA studies now provide substantial evidence that putative refugee plant populations harbor higher levels of genetic diversity relative to their likely descendant populations [69]. *Disanthus cercidifolius* subsp. *longipes* is a Tertiary relic, which is distributed in warm and humid forests of the Cathayan Land in the Tertiary [2,70]. Lots of present endemics, several of which inhabit Pleistocene refugia during the Quaternary glacial period, were able to maintain higher levels of diversity because of population stability during the glacial cycle. *Disanthus cercidifolius* subsp. *longipes* may be one of them and preserved mutation during the

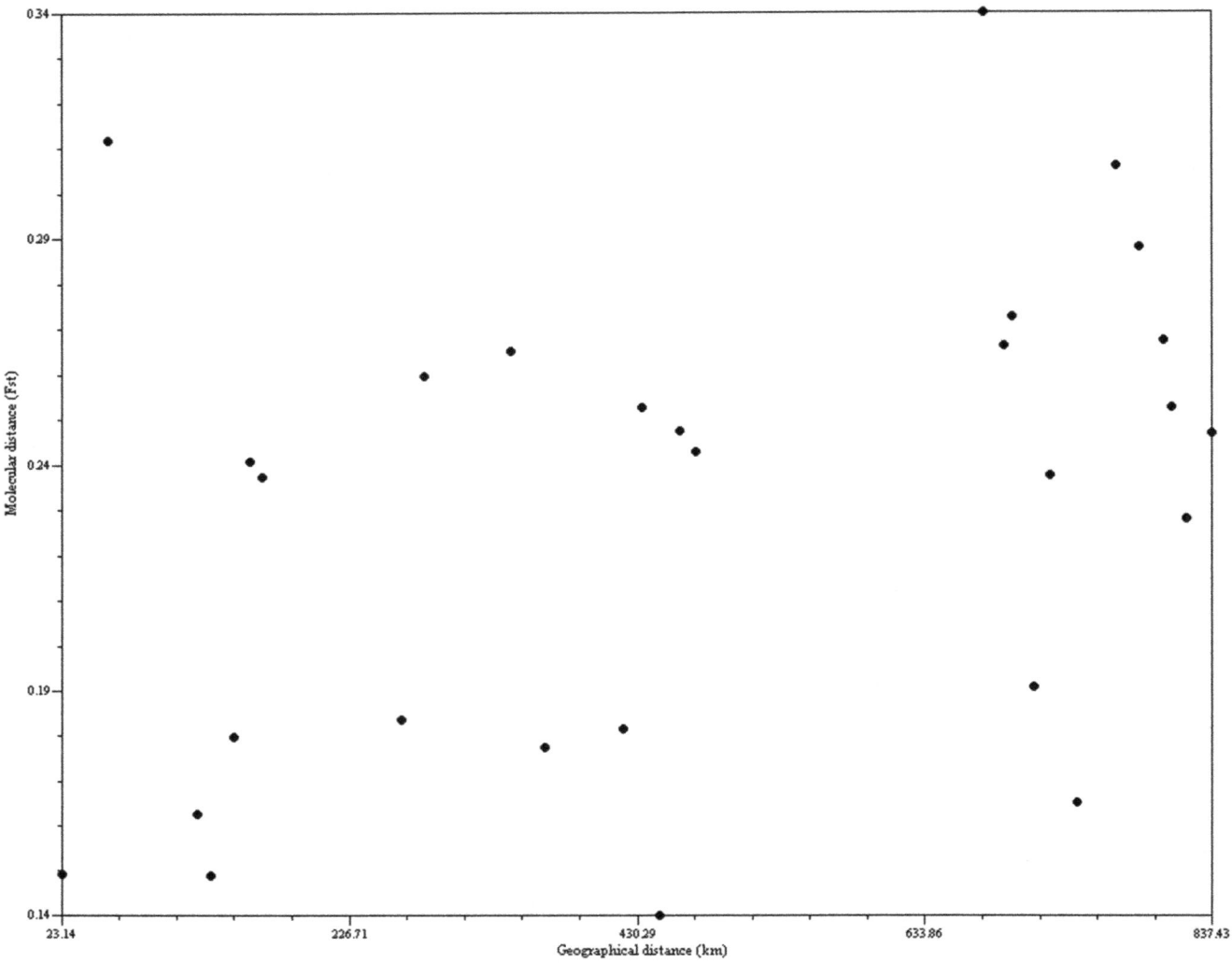

Figure 2. *Disanthus cercidifolius* **subsp.** *longipes,* **bivariate plot showing no correlation between matrices of pair-wise geographic (km) and genetic distance (Φst) in eight populations, comprising 82 individuals.** Computed with NTSYSpc ver. 2.1e.

relatively long glacial period. So it shows high genetic diversity [69,71]. Compared by AFLP, RAPD, and ISSR markers, the genetic diversity indices in this study revealed an intermediate level of genetic diversity of *D. c.* subsp. *longipes* compared to other endangered species [34,41,66,72,73].

Results of studies of genetic diversity within populations show that all populations were on the similar level, except for the Dupanglin National Nature Reserve (2DP) which is shown to be lower than the others. The lack of genetic diversity resulting from the total existing individual in population 2DP is limited (fewer than 100 individuals). In practice, a larger population often has higher genetic variation [72,74].

Genetic variation and gene flow

In this study AMOVA showed that the largest portion (76.72%) of genetic variance is contributed by genetic variation within populations (Φ_{st} value 0.23283) and only a small portion (23.28%) is due to differences among groups. Some other studies of endemic and endangered species show a similar pattern [34,41,66,72,73]. Moderate or high diversity and low population partitioning in rare plants have previously been attributed to a number of factors

[66,75–77], including insufficient length of time for genetic diversity to be reduced following a natural reduction in population size and isolation; adaptation of the genetic system to small population conditions; recent fragmentation (human disturbance) of a once-continuous range, i.e., genetic system; and extensive gene flow due to the combination of bird pollination and high outcrossing rates [66,78]. The direct estimate of gene flow (*Nm*) based on Φ_{st} 0.23283 was 0.82374, which means that the number of migrants per generation is lower than one. The present distribution of *D. c.* subsp. *longipes* is restricted to several isolated habitats [5,7]. However, fossil and palynological evidences suggested that, *Disanthus* was wide spread in warm and humid forests in the Tertiary [2,70]. Accordingly, a reasonable hypothesis is that the modern range of *Disanthus* species was the result of population fragmentation and contraction after the Quaternary glacial cycles. Considering that *D. c.* subsp. *longipes* has poor success in pollination, "excess flowers production, but little fruit set" [79], and fragmented habitat, inbreeding, which generally leads to decreased fitness, or inbreeding depression, of a population, some of the above causes result in the endangered status of *D. c.* subsp. *longipes.*

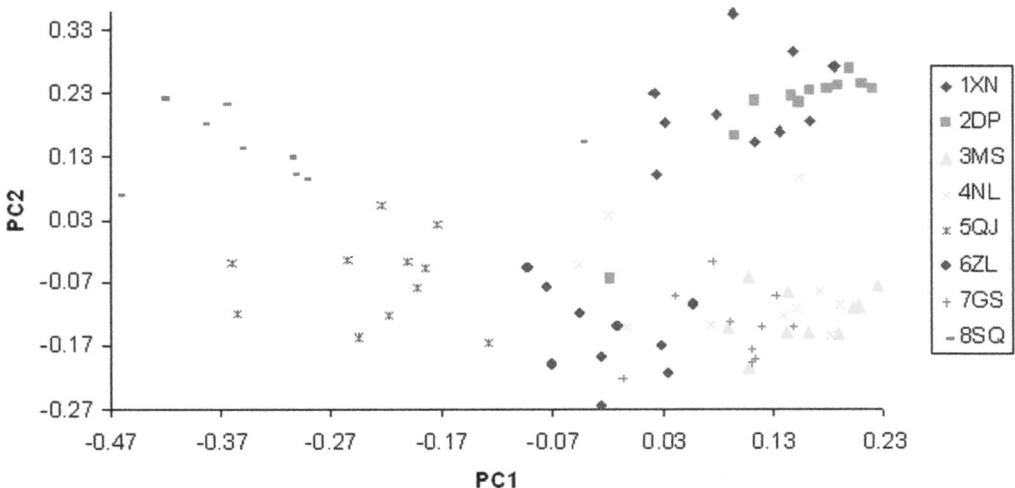

Figure 3. Principal coordinates analysis of 82 individuals from eight populations, based on dissimilarity matrix (Jaccard's coefficient). Accessions are plotted according to the values of first (x axis) and the second (y axis) components and with different symbols according to population. Principal coordinate axes shown (pc1 and pc2) represent 25.46% and 25.17% of respective variance in the dissimilarity matrix.

Genetic structure at different hierarchical levels

Although AFLP markers are dominant, they provide no information on heterozygote frequencies, and our investigation provides no direct information on the reproductive strategy of *D. c.* subsp. *longipes*. In general, outcrossing and long-lived seed plants maintain the most genetic variation within populations, while predominantly selfing, short-lived species harbor compara-

tively higher variation among populations [65,80]. According to data based primarily on allozyme analysis, $\Phi_{st} = 0.2$ for outcrossing species and 0.5 for inbreeders [65]. The overall Φ_{st} value 0.23283 and the gene flow value for *D. c.* subsp. *longipes* are similar to other species that outcross [81–84]. Permutation tests of the fixation index ($\Phi_{st} = 0.23$) indicated significant genetic structuring. The populations of *D. c.* subsp. *longipes* spread from east to west (Fig. 1). And in the field investigation of our study, we

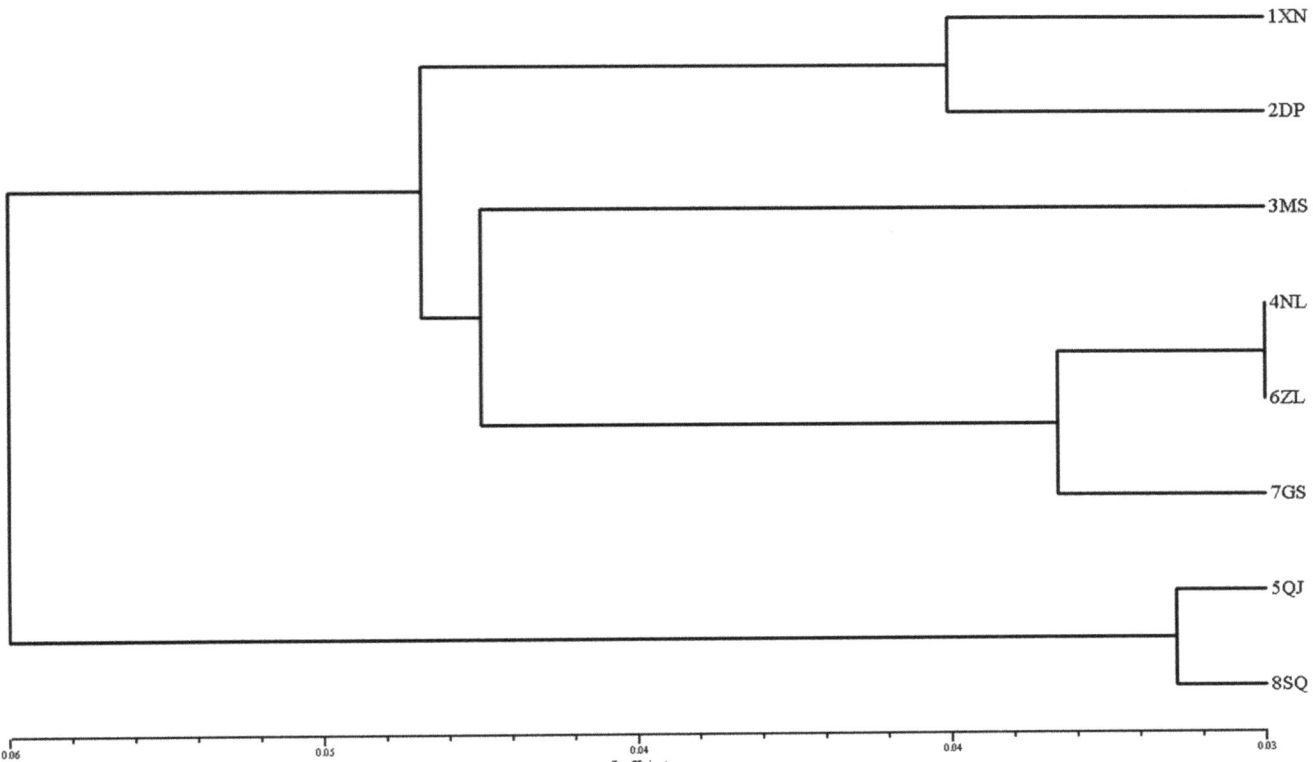

Figure 4. UPGMA dendrogram based on genetic distance. Note: A = 1XN, B = 2DP, E = 3MS, G = 4NL, H = 5QJ, I = 6ZL, J = 7GS, K = 8SQ.

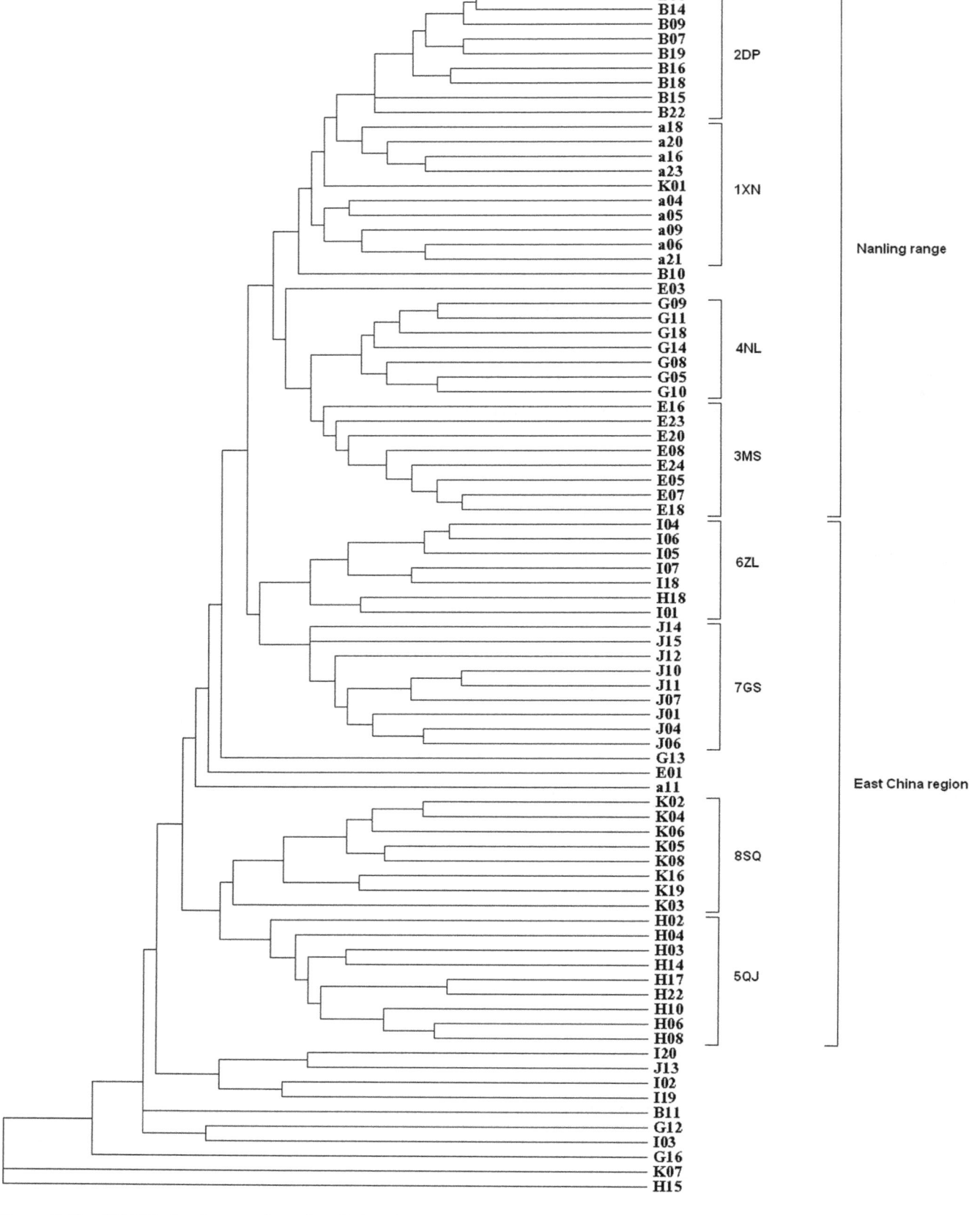

Figure 5. The UPGMA dendrogram based on Nei & Li's genetic distance.

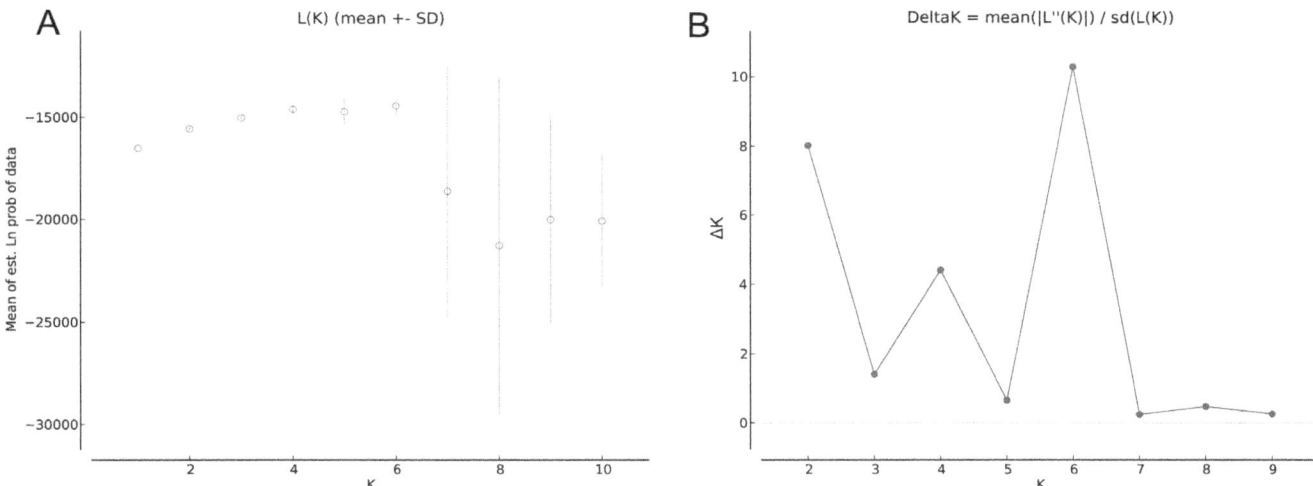

Figure 6. The mean LnP(D) and ΔK over 10 repeats of STRUCTRUE simulations. A. Mean lnP(D) value with k = 1–10. B. ΔK value with k = 2–9.

found that *D. c.* subsp. *longipes* is one of the dominant species in its habitat.

HICKORY software analysis suggests that *D. c.* subsp. *longipes* populations are not genetically differentiated. There is inbreeding in populations of *D. c.* subsp. *longipes* (Table 6), because the *DIC* and *Dbar* parameter were lower in "*Full model*" than that in other models [62]. While inbreeding generally leads to decreased fitness (inbreeding depression) of a population, it also can be advantageous, allowing plants to adapt to disadvantageous conditions [85–88]. *D. c.* subsp. *longipes* originally had an outcrossing system, but under some conditions (e.g., poor efficiency in wind or insect pollination), it would adopt inbreeding or mixed systems to reduce the risk of reproductive failure. This result is consistent with Xiao's conclusions [89] from his study of the reproductive ecology of *D. c.* subsp. *longipes*. If we use co-dominant markers, it most likely would provide an affirmative result and better understanding of the processes.

Hierarchical cluster analysis (Fig. 5) revealed that most individuals of a specific population were grouped together, despite the mixing-in of some individuals from different populations. Most geographically close populations tended to cluster together. The DW values show that the two close populations are clearly divergent, that one is long-term isolated (DW value higher) and the other is newly established (DW value lower). The DW value of 7GS is close to average, but its site is isolated. It might be that

there once had been other populations between 6ZL and 7GS. In fact, there is a population of *D. c.* subsp. *longipes* in the Junfeng Mountains, Jiangxi Province, according to Pan's data [2], but we have not obtained any specimens for analysis since 2008. Moreover, although *D. c.* subsp. *longipes* in the Jinggang Mountains, Jiangxi Province, might have been transplanted from Guanshan National Nature Reserve based on the data [90], we found a large population area of *D. c.* subsp. *longipes* in the Jinggang Mountains in October 2009, evidence that the present distribution of *D. c.* subsp. *longipes* appears to be relic from a once extensive range and population.

Genetic structure analysis of the individual samples using STRUCTURE shows that six gene pools are represented in the data, and each gene pool is relatively independent except a few mixtures from other clusters. 1XN and 2DP are the first cluster separated from the other populations; 5QJ and 8SQ form the distinct cluster 2 from the others; 3MS and 4NL also cluster with each other and become distinct from rest populations. These three clusters show high correlation with geographic distribution. And the rest three clusters mixed in 6ZL and 7GS. There is a close relationship between 6ZL and 7GS. It shows that 6ZL is related to 7GS rather than to 5QJ or 8SQ, although 6ZL is close to 5QJ and 8SQ in geographic location. It seems likely that there were some populations between current populations 6ZL and 7GS, or gene flow could have been accomplished by pollen or seed dispersal.

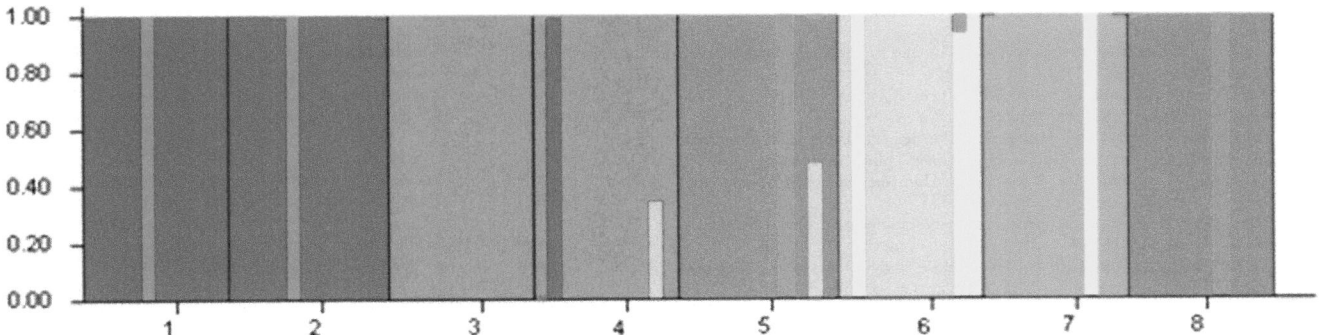

Figure 7. Diagram of results from STRUCTURE(k = 6).

Table 6. Genetic structure analysis using HICKORY ver. 1.1 [62].

Model	Parameter				
	Dbar	*Dhat*	*pD*	*DIC*	*f*
Full	6750.93	5395.03	1355.9	8106.82	0.980572
f=0	6780.4	5350.93	1429.47	8209.87	
Theta =0	12092.5	11716.4	376.128	12468.6	0.994715
f free	6844.98	5363.43	1481.55	8326.53	0.500214

Default values for computations were used as follows: Burn-in 5000; sample 100 000; thin 20.

The results of PCA based on AFLP data indicate a different result, which the eight populations form a triangle that has 1XN and 2DP at one angle, 5QJ and 8SQ at a second angle, but 3MS and 4NL mixed with 6ZL and 7GS in the third angle. The reasons for divergence of the East China region and the Nanling Range are not clear.

Guide for conservation measures

In the face of the species extinction generated by human beings [91,92], the urgent need for the conservation of biodiversity is global. In fact, many endemic species became endangered by the loss or fragmentation of habitats and change in natural conditions [26,29,64,93,94]. *D. c.* subsp. *longipes*, with its narrow archipelago-like distribution, habitat fragmentation, and reproductive-physiologic barriers, is typical species that can be easily affected. Although *D. c.* subsp. *longipes* is endangered in China, it has not yet been given high priority for protection. The original aim of this work is to provide some insight into the population biology of *D. c.* subsp. *longipes* in order to guide conservation measures. The guides which can be suggested by our study listed below. Firstly, Most of the genetic variation is located within populations, so we need to protect all the existing populations of *D. c.* subsp. *longipes* in order to preserve as much genetic variation as possible. Secondly, we should promote the further research and technology development to enable better reproduction of *Disanthus*. Finally, it is necessary to enforce the governmental law to forbid people stealing or purchasing wild *Disanthus*.

Conclusions

The results of the AFLP on populations of *Disanthus cercidifolius* subsp. *longipes* show a pattern of high within-population diversity and low among-population divergence. Although the distribution of *D. c.* subsp. *longipes* could be grouped as the East China region and the Nanling Range, the population divergence is clear. The estimates of genetic differentiation and gene flow suggest that the species primarily outcrosses, but can resort to mixed reproductive strategies. These patterns of genetic diversity and levels of genetic variation may be the result of *D. c.* subsp. *longipes* restricted to several isolated habitats and "excess flowers production, but little fruit set". The status of relic and their low reproductive success results in *D. c.* subsp. *longipes* being endangered.

Acknowledgments

We are grateful to the anonymous reviewers and the editor for their critical review; their comments substantially contributed to the revision and improvement of this work. We thank Dr. Sufang Chen for her valuable advice on writing.

Author Contributions

Conceived and designed the experiments: YY RS WL. Performed the experiments: YY RS. Analyzed the data: YY WG QF. Contributed reagents/materials/analysis tools: JJ DC. Wrote the paper: YY WL DC.

References

1. Mabberley DJ (1997) The Plant-book: A Portable Dictionary of the Vascular Plants. Cambridge University Press.
2. Pan K, Lu A, Wen J (1991) A systematic study on the genus *Disanthus* Maxim. (Hamamelidaceae). Cathaya 3: 1–28.
3. Cheng WC (1938) Observations on Mong-shan (Hunan). Science(Sci Soc China) 22: 400.
4. Chang HT (1948) Additions to the Hamamelidaceus flora of China. Sunyatsenia 7: 70.
5. Fu L (1992) China Plant Red Data Book: Rare and Endangered Plants. Beijing: Science Press.
6. State Forestry Administration and Ministry of Agriculture (1999) Key Wild Plants under State Protection. Beijing.
7. Wang S, Xie Y (2004) China Species Red List. Beijing: Higher Education Press.
8. Zavada M, Dilcher D (1986) Comparative pollen morphology and its relationship to phylogeny of pollen in the Hamamelidae. Ann Missouri bot Gard 73: 348–381.
9. Endress PK (1989) A suprageneric taxonomic classification of the hamamelidaceae. Taxon 38: 371–376.
10. Pan KY, Lu AM, Wen J (1991) A systematic study on the genus *Disanthus* Maxim. (Hamamelidaceae). Cathaya 3: 1–28.
11. Li J (1997) Systematics of the Hamamelidaceae based on morphological and molecular evidence. Ph.D. Thesis, University of New Hampshire.
12. Li J, Bogle AL, Donoghue MJ (1999) Phylogenetic relationships in the Hamamelidoideae inferred from sequences of trn non-coding regions of chloroplast DNA. Harvard Papers in Botany 4: 343–356.

13. Shi XH, Xu BM, Li NL, Sun YT (2002) Preliminary study on dormancy and germination of *Disanthus cercidifolius* Maxim var. *longipes* HT Chang seeds. Seed 6: 5–7.
14. Xiao YA, He P, Deng HP, Li XH (2002) Numerical analysis of population morphological differentiation of *Disanthus cercidiifolius* Maxim. var. *longipes* in Jinggangshan. Journal of Wuhan Botanical Research 20: 365–370.
15. Li K, Tang X (2003) The community characteristics and species diversity of the *Disanthus cercidifolius* var. *longipes* shrubland in the Guanshan nature reserve of Jiangxi Province. Journal of Nanjing Forestry University 27: 73–75.
16. Li X, Xiao Y, Hu W, Zeng J (2005) Isozymes analysis of genetic differentiation of the endangered plant *Disanthus cercidifolius* var. *longipes* HT Chang (Hamamelidaceae). Journal of Jianggangshan University (Natural Sciences) 26: 34–38.
17. Huang S, Fang Y, Tan X, Yan J, Fang S (2007) Effects of different concentrations of NAA on cutting regeneration of *Disanthus cercidifolius* var. *longipes*. Journal of Plant Resources and Environment 16: 74.
18. Gao P, Yang A, Yao X, Huang H (2009) Isolation and characterization of nine polymorphic microsatellite loci in the endangered shrub *Disanthus cercidifolius* var. *longipes* (Hamamelidaceae). Molecular Ecology Resources 9: 1047–1049.
19. Bogle A, Philbrick C (1980) A generic atlas of hamamelidaceous pollens. Contributions to the Gray Herbarium, Harvard University 210: 29–103.
20. Chang HT (1979) Hamaelidaceae. Flora Reipublicae Popularis Sinicae. Beijing: Science Press. pp.36–116.
21. Zhang ZY, Chang HT, Endress PK (2003) *Disanthus* Maximowicz. In: Z. Wu and P. H. Raven, editors. Flora of China. Beijing: Science Press and St. Louis: Missouri Botanical Garden Press.

22. Endress PK (1993) Hamamelidae. In: K. Kubitzki, J. G. Rohwer and V. Bittrich, editors. The Families and Fenera of Vascular Plants. Berlin: Springer-Verlag.

23. Xiao YA (2001) The study on population adaptability and genetic diversity of the endangered species *Disanthus cercidifolius* Maxim. var. *longipes* HT Chang. M.Sc. Thesis, Southwest China Normal University.

24. Soule M, Simberloff D (1986) What do genetics and ecology tell us about the design of nature reserves? Biological Conservation 35: 19–40.

25. Barrett S, Kohn J (1991) Genetic and evolutionary consequences of small population size in plants: implications for conservation. Genetics and conservation of rare plants: 3–30.

26. Ge S, Hong D, Wang H, Liu Z, Zhang C (1998) Population genetic structure and conservation of an endangered conifer, *Cathaya argyrophylla* (Pinaceae). International Journal of Plant Sciences 159: 351–357.

27. Millar C, LiBBY W (1991) Strategies for conserving clinal, ecotypic, and disjunct population diversity in widespread species. Genetics and conservation of rare plants: 149–170.

28. Francisco-Ortega J, Santos-Guerra A, Kim S, Crawford D (2000) Plant genetic diversity in the Canary Islands: a conservation perspective. American Journal of Botany 87: 909.

29. Qiu Y, Li J, Liu H, Chen Y, Fu C (2006) Population structure and genetic diversity of *Dysosma versipellis* (Berberidaceae), a rare endemic from China. Biochemical Systematics and Ecology 34: 745–752.

30. Song J, Murdoch J, Gardiner SE, Young A, Jameson PE, et al. (2008) Molecular markers and a sequence deletion in intron 2 of the putative partial homologue of LEAFY reveal geographical structure to genetic diversity in the acutely threatened legume genus *Clianthus*. Biological Conservation 141: 2041–2053.

31. Perez-Collazos E, Segarra-Moragues JG, Catalan P (2008) Two approaches for the selection of Relevant Genetic Units for Conservation in the narrow European endemic steppe plant *Boleum asperum* (Brassicaceae). Biological Journal of the Linnean Society 94: 341–354.

32. Yang J, Qian ZQ, Liu ZL, Li S, Sun GL, et al. (2007) Genetic diversity and geographical differentiation of *Dipteronia* Oliv. (Aceraceae) endemic to China as revealed by AFLP analysis. Biochemical Systematics and Ecology 35: 593–599.

33. Prohens J, Anderson GJ, Herraiz FJ, Bernardello G, Santos-Guerra A, et al. (2007) Genetic diversity and conservation of two endangered eggplant relatives (*Solanum vespertilio* Aiton and *Solanum lidii* Sunding) endemic to the Canary Islands. Genetic Resources and Crop Evolution 54: 451–464.

34. Jian SG, Zhong Y, Liu N, Gao ZZ, Wei Q, et al. (2006) Genetic variation in the endangered endemic species *Cycas fairylakea* (Cycadaceae) in China and implications for conservation. Biodiversity and Conservation 15: 1681–1694.

35. Pfosser M, Jakubowsky G, Schluter PM, Fer T, Kato H, et al. (2005) Evolution of *Dystaenia takesimana* (Apiaceae), endemic to Ullung Island, Korea. Plant Systematics and Evolution 256: 159–170.

36. Strand AE, Leebens-Mack J, Milligan BG (1997) Nuclear DNA-based markers for plant evolutionary biology. Molecular Ecology 6: 113–118.

37. Vos P, Hogers R, Bleeker M, Reijans M, van de Lee T, et al. (1995) AFLP: a new concept for DNA fingerprinting. Nucleic Acids Res 23: 4407–4414.

38. Armstrong TTJ, De Lange PJ (2005) Conservation genetics of *Hebe speciosa* (Plantaginaceae) an endangered New Zealand shrub. Botanical Journal of the Linnean Society 149: 229–239.

39. Kim SC, Lee C, Santos-Guerra A (2005) Genetic analysis and conservation of the endangered Canary Island woody sow-thistle, *Sonchus gandogeri* (Asteraceae). Journal of Plant Research 118: 147–153.

40. Andrade IM, Mayo SJ, Van den Berg C, Fay MF, Chester M, et al. (2007) A preliminary study of genetic variation in populations of *Monstera adansonii* var. *klotzschiana* (Araceae) from north-east brazil, estimated with AFLP molecular markers. Annals of Botany 100: 1143–1154.

41. Tang SQ, Dai WJ, Li MS, Zhang Y, Geng YP, et al. (2008) Genetic diversity of relictual and endangered plant *Abies ziyuanensis* (Pinaceae) revealed by AFLP and SSR markers. Genetica 133: 21–30.

42. Wolfe A, Liston A (1998) Contributions of PCR-based methods to plant systematics and evolutionary biology. Molecular systematics of plants II: 43–86.

43. Chase MW, Hills HH (1991) Silica gel: an ideal material for field preservation of leaf samples for DNA studies. Taxon 40: 215–220.

44. Doyle JJ, Doyle JL (1987) A rapid DNA isolation procedure for small quantities of fresh leaf tissue. Phytochemical bulletin 19: 11–15.

45. Yeh FC, Yang RC, Boyle TBJ, Ye ZH, Mao JX (1997) POPGENE, the user-friendly shareware for population genetic analysis. Molecular Biology and Biotechnology Centre, University of Alberta, Canada.

46. Schönswetter P, Tribsch A (2005) Vicariance and dispersal in the alpine perennial *Bupleurum stellatum* L.(Apiaceae). Taxon: 725–732.

47. Ortiz M, Tremetsberger K, Talavera S, Stuessy T, García-Castaño JL (2007) Population structure of *Hypochaeris salzmanniana* DC.(Asteraceae), an endemic species to the Atlantic coast on both sides of the Strait of Gibraltar, in relation to Quaternary sea level changes. Mol Ecol 16: 541–552.

48. Ehrich D (2006) AFLPdat: a collection of R functions for convenient handling of AFLP data. Mol Ecol Notes 6: 603–604.

49. Nei M (1978) Estimation of average heterozygosity and genetic distance from a small number of individuals. Genetics 89: 583–590.

50. Sneath PHA, Sokal RR (1973). Numerical Taxonomy, 2nd edn. San Francisco: Freeman.

51. Rohlf FJ (2000) NTSYS-pc: numerical taxonomy and multivariate analysis system, version 2.1. New York: Exeter Software.

52. Nei M, Li WH (1979) Mathematical model for studying genetic variation in terms of restriction endonucleases. Proc Natl Acad Sci USA 76: 5269–5273.

53. Van de Peer Y, De Wachter Y (1994) TREECON for Windows: a software package for the construction and drawing of phylogenetic trees for the Microsoft Windows environment. Comp Applic Biosci 10: 569–570.

54. Excoffier L, Laval G, Schneider S (2005) Arlequin (version 3.0): An integrated software package for population genetics data analysis. Evolutionary Bioinformatics 1: 47–50.

55. Excoffier L, Smouse P, Quattro J (1992) Analysis of molecular mariance inferred from metric distances among DNA haplotypes: Application to human mitochondrial DNA restriction data. Genetics 131: 479–491.

56. Wright S (1950) Genetical structure of populations. Nature 166: 247–249.

57. Nie Z-L, Wen J, SUN H (2009) AFLP Analysis of *Phryma* (Phrymaceae) Disjunct between Eastern Asia and Eastern North America. Acta Botanica Yunnanica 31: 289–295.

58. Inc S (2007) SPSS 16.0 for Windows. SPSS Inc, Chicago, IL.

59. Pritchard JK, Stephens M, Donnelly P (2000) Inference of population structure using multilocus genotype data. Genetics 155: 945–959.

60. Evanno G, Regnaut S, Goudet J (2005) Detecting the number of clusters of individuals using the software STRUCTURE: a simulation study. Mol Ecol 14: 2611–2620.

61. Zhang DL, Zhang HL, Wang MX, Sun JL, Qi YW, et al. (2009) Genetic structure and differentiation of *Oryza sativa* L. in China revealed by microsatellites. Theor Appl Genet 119: 1105–1117.

62. Holsinger K, Lewis P (2003) HICKORY: a package for analysis of population genetic data V. 1.0. University of Connecticut, Storrs, USA.

63. Cardoso M, Provan J, Powell W, Ferreira P, De Oliveira D (1998) High genetic differentiation among remnant populations of the endangered *Caesalpinia echinata* Lam.(Leguminosae-Caesalpinioideae). Molecular Ecology 7: 601–608.

64. Bouza N, Caujapé-Castells J, González-Pérez M, Batista F, Sosa P (2002) Population structure and genetic diversity of two endangered endemic species of the Canarian laurel forest: *Dorycnium spectabile* (Fabaceae) and *Isoplexis chalcantha* (Scrophulariaceae). International Journal of Plant Sciences 163: 619–630.

65. Hamrick J, Godt M (1989) Allozyme diversity in plant species. In: A. H. D. Brown, M. T. A. Clegg, L. Kahler and B. S. Weir, editors. Plant Population Genetics, Breeding and Genetic Resources. Sunderland, Mass: Sinauer. pp.43–63.

66. Ge XJ, Yu Y, Zhao NX, Chen HS, Qi WQ (2003) Genetic variation in the endangered Inner Mongolia endemic shrub *Tetraena mongolica* Maxim.(Zygophyllaceae). Biological Conservation 111: 427–434.

67. Hamrick JL, Godt MJW (1996): Conservation genetics of endemic plant species. In Conservation genetics: case histories from nature. In: Avise JC, Hamrick JL, Editors. New York: Chapman and Hall. pp.281–304.

68. Karron J (1991) Patterns of genetic variation and breeding systems in rare plant species. Genetics and Conservation of Rare Plants: 87–98.

69. Ge XJ, Yu Y, Zhao NX, Chen HS, Qi WQ (2003) Genetic variation in the endangered Inner Mongolia endemic shrub *Tetraena mongolica* Maxim. (Zygophyllaceae). Biol Conserv 111: 427–434.

70. Wu Z, Lu A, Tang Y, Chen Z, Li DZ (2003) The families and gerera of angiosperms in China. Beijing: Science Press.

71. Lewis PO, Crawford DJ (1995) Pleistocene Refugium Endemics Exhibit Greater Allozymic Diversity than Widespread Congeners in the Genus *Polygonella* (Polygonaceae). American Journal of Botany 82: 141–149.

72. Kwon J, Morden C (2002) Population genetic structure of two rare tree species (*Colubrina oppositifolia* and *Alphitonia ponderosa*, Rhamnaceae) from Hawaiian dry and mesic forests using random amplified polymorphic DNA markers. Mol Ecol 11: 991–1001.

73. Lacerda DR, Acedo MDP, Filho JPL, Lovato MB (2001) Genetic diversity and structure of natural populations of *Plathymenia reticulata* (Mimosoideae), a tropical tree from the Brazilian Cerrado. Molecular Ecology 10: 1143–1152.

74. Jian S, Zhong Y, Liu N, Gao Z, Wei Q, et al. (2006) Genetic variation in the endangered endemic species *Cycas fairylakea* (Cycadaceae) in China and implications for conservation. Biodiversity & Conservation 15: 1681–1694.

75. Schaal B, Hayworth D, Olsen K, Rauscher J, Smith W (1998) Phylogeographic studies in plants: problems and prospects. Mol Ecol 7: 465–474.

76. Slatkin M (1987) Gene flow and the geographic structure of natural populations. Science 236: 787.

77. Zawko G, Krauss S, Dixon K, Sivasithamparam K (2001) Conservation genetics of the rare and endangered *Leucopogon obtectus* (Ericaceae). Mol Ecol 10: 2389–2396.

78. Maguire T, Sedgley M (1997) Genetic diversity in *Banksia* and *Dryandra* (Proteaceae) with emphasis on Banksia cuneata, a rare and endangered species. Heredity 79: 394–401.

79. Xiao YA, Zeng JJ, Li XH, Hu WH, He P (2006) Pollen and resource limitations to lifetime seed production in a wild population of the endangered plant *Disanthus cercidifolius* Maxim. var. *longipes* H. T. Chang (Hamamelidaceae). Acta Ecologica Sinica 26: 496–502.

80. Huang Y, Zhang C, Li D (2009) Low genetic diversity and high genetic differentiation in the critically endangered *Omphalogramma souliei* (Primulaceae): implications for its conservation. Journal of Systematics and Evolution 47: 103–109.

81. Kothera L, Richards C, Carney S (2007) Genetic diversity and structure in the rare Colorado endemic plant *Physaria bellii* Mulligan (Brassicaceae). Conservation Genetics 8: 1043–1050.

82. Jacquemyn H, Honnay O, Galbusera P, Roldan-Ruiz I (2004) Genetic structure of the forest herb Primula elatior in a changing landscape. Molecular Ecology 13: 211–219.

83. Juan A, Crespo M, Cowan R, Lexer C, Fay M (2004) Patterns of variability and gene flow in *Medicago citrina*, an endangered endemic of islands in the western Mediterranean, as revealed by amplified fragment length polymorphism (AFLP). Mol Ecol 13: 2679–2690.

84. Morjan CL, Rieseberg LH (2004) How species evolve collectively: implications of gene flow and selection for the spread of advantageous alleles. Mol Ecol 13: 1341–1356.

85. Allard R, Jain S, Workman P (1968) The genetics of inbreeding populations. Adv Genet 14: 55–131.

86. Antonovics J (1968) Evolution in closely adjacent plant populations. V. Evolution of self-fertility. Heredity 23: 219–238.

87. Jain S (1976) The evolution of inbreeding in plants. Annual Review of Ecology and Systematics 7: 469–495.

88. Lloyd D (1980) Demographic factors and mating patterns in angiosperms.

89. Xiao YA (2005) Studies on Reproductive Ecology and photosynthetic adaptability of the Endangered Plant *Disanthus cercidifolius* var.*longipes* HT Chang. Ph.D. Thesis, Southwest China Normal University.

90. Liu R (1999) *Disanthus cercidifolius* var. *longipes*. Plants 4: 7.

91. Vitousek P, Mooney H, Lubchenco J, Melillo J (1997) Human domination of Earth's ecosystems. Science 277: 494.

92. Chapin F, Zavaleta E, Eviner V, Naylor R, Vitousek P, et al. (2000) Consequences of changing biodiversity. Nature 405: 234–242.

93. Palacios C, Gonzalez-Candelas F (1997) Lack of genetic variability in the rare and endangered *Limonium cavanillesii* (Plumbaginaceae) using RAPD markers. Molecular Ecology 6: 671–675.

94. Qiu Y, Luo Y, Comes H, Ouyang Z, Fu C (2007) Population genetic diversity and structure of *Dipteronia dyerana* (Sapindaceae), a rare endemic from Yunnan Province, China, with implications for conservation. Taxon 56: 427–437.

12

Demographics and Genetic Variability of the New World Bollworm (*Helicoverpa zea*) and the Old World Bollworm (*Helicoverpa armigera*) in Brazil

Natália A. Leite[1], Alessandro Alves-Pereira[2], Alberto S. Corrêa[1], Maria I. Zucchi[3], Celso Omoto[1]*

1 Departamento de Entomologia e Acarologia, Escola Superior de Agricultura "Luiz de Queiroz", Universidade de São Paulo, Piracicaba, São Paulo, Brazil, 2 Departamento de Genética, Escola Superior de Agricultura "Luiz de Queiroz", Universidade de São Paulo, Piracicaba, São Paulo, Brazil, 3 Agência Paulista de Tecnologia dos Agronegócios, Piracicaba, São Paulo, Brazil

Abstract

Helicoverpa armigera is one of the primary agricultural pests in the Old World, whereas *H. zea* is predominant in the New World. However, *H. armigera* was first documented in Brazil in 2013. Therefore, the geographical distribution, range of hosts, invasion source, and dispersal routes for *H. armigera* are poorly understood or unknown in Brazil. In this study, we used a phylogeographic analysis of natural *H. armigera* and *H. zea* populations to (1) assess the occurrence of both species on different hosts; (2) infer the demographic parameters and genetic structure; (3) determine the potential invasion and dispersal routes for *H. armigera* within the Brazilian territory; and (4) infer the geographical origin of *H. armigera*. We analyzed partial sequence data from the cytochrome c oxidase subunit I (COI) gene. We determined that *H. armigera* individuals were most prevalent on dicotyledonous hosts and that *H. zea* were most prevalent on maize crops, based on the samples collected between May 2012 and April 2013. The populations of both species showed signs of demographic expansion, and no genetic structure. The high genetic diversity and wide distribution of *H. armigera* in mid-2012 are consistent with an invasion period prior to the first reports of this species in the literature and/or multiple invasion events within the Brazilian territory. It was not possible to infer the invasion and dispersal routes of *H. armigera* with this dataset. However, joint analyses using sequences from the Old World indicated the presence of Chinese, Indian, and European lineages within the Brazilian populations of *H. armigera*. These results suggest that sustainable management plans for the control of *H. armigera* will be challenging considering the high genetic diversity, polyphagous feeding habits, and great potential mobility of this pest on numerous hosts, which favor the adaptation of this insect to diverse environments and control strategies.

Editor: João Pinto, Instituto de Higiene e Medicina Tropical, Portugal

Funding: This work was partially supported by Conselho Nacional de Desenvolvimento Científico e Tecnológico (CNPq) (Grant 308150/2009-0) and Comitê Brasileiro de Ação a Resistência a Inseticidas (IRAC-BR). The funders had no role in study design, data collection and analysis, decision to publish, or preparation of the manuscript.

Competing Interests: The authors have declared that no competing interests exist.

* Email: celso.omoto@usp.br

Introduction

The Heliothinae (Lepidoptera: Noctuidae) subfamily has 381 described species, many of which are important agricultural pests from the *Helicoverpa* Hardwick and *Heliothis* Ochsenheimer genera [1]. The *Helicoverpa* genus contains two of the primary Heliothinae pest species: *Helicoverpa armigera* (Hübner) (Old World bollworm) and *Helicoverpa zea* (Boddie) (New World bollworm). Although the exact evolutionary relationship between *H. armigera* and *H. zea* remains uncertain, these insects are considered to be 'twin' or 'sibling' species, and they are able to copulate and produce fertile offspring under laboratory conditions [2–5]. Some hypotheses propose that *H. zea* evolved from a small portion of the larger *H. armigera* population (i.e., a "founder effect") that reached the American continent approximately 1.5 million years ago, which is consistent with previous phylogeographic analyses of *H. armigera* and *H. zea* individuals [6,7].

H. armigera is considered to be one of the most important agricultural pests in the world. This insect is widely distributed throughout Asia, Africa, Europe, and Australia, and it has been shown to attack more than 100 host species from 45 different plant families [8–10]. In contrast, *H. zea* is restricted to the American continent and is of lesser economic importance; it is a secondary pest of cotton, tomato, and, most significantly, maize crops [11]. However, the scenario in Brazil changed in 2013 when *H. armigera* individuals, which are considered to be A1 quarantine pests, were officially reported within the Brazilian territory [12–14]. This situation increased in severity due to the great dispersal ability of this insect as well as the steady reports from several regions of the world that described new *H. armigera* lineages showing tolerance/resistance to insecticides and genetically modified plants [15,16]. It is estimated that *H. armigera* will cause a loss of more than US$2 billion to the 2013/14 Brazilian agriculture crop because of direct productivity losses and resources spent on phytosanitary products for soybean, cotton, and maize, which are the main crops of Brazilian agribusinesses. Therefore, *H. armigera* is now one of the most important pest species with respect to agriculture in Brazil [17].

High population densities of *Helicoverpa* spp. and the resulting economic damages to cultivated plants have been reported in different regions of Brazil, in particular in the Western state of Bahia [18]. Therefore, these reports suggest the existence of an invasion period prior to the first official report of *H. armigera* in Brazil. This atypical and confusing scenario was likely caused by the significant morphological similarities between *H. zea* and *H. armigera* [9,19] and by major changes in pest management programs over recent years. In addition, these population changes may have been related to the release and increased cultivation of crops that express *Bacillus thuringiensis* (Bt) genes in Brazil.

Aside from the identification of *H. armigera* individuals within the Brazilian territory, many basic pieces of information concerning this species, including its geographical distribution, the types of hosts it attacks, its invasion source, and its dispersal routes, remain poorly understood or completely unknown. Therefore, we attempted to address some of these outstanding questions using a phylogeographic approach by analyzing genetic sequence data from a portion of the cytochrome c oxidase subunit I (COI) gene of *Helicoverpa* spp. specimens isolated from different hosts and regions of Brazil. This study was performed with the following goals in mind: (1) to confirm and evaluate the occurrence of *H. armigera* and *H. zea* individuals from different hosts and regions of Brazil; (2) to assess the demographic parameters and genetic structure of *H. armigera* and *H. zea* populations within the Brazilian territory, with a focus on the region, season, and host; (3) to assess the potential invasion (single or multiple) and dispersal routes for *H. armigera* within the Brazilian territory; and (4) to determine the geographical origin of the *H. armigera* populations present in Brazil. This information will be essential for understanding the genetic diversity and population dynamics of these pests as well as for guiding both immediate control strategies (legal and/or phytosanitary) and subsequent long-term integrated management programs for the *Helicoverpa* spp. complex in Brazil.

Results

Identification of *Helicoverpa* spp., hosts, and geographic locations

One hundred thirty-nine individuals from the 274 *Helicoverpa* spp. specimens initially sampled were identified as *H. armigera* (98–100% homology) and 134 individuals were identified as *H. zea* (98–100% homology) (GenBank Accession numbers KM274936–KM275209 are listed in Table 1). *H. armigera* was primarily found on soybean, bean, and cotton crops, and these insects were widely distributed throughout the Midwest and Northeast of Brazil during both crop periods (winter and summer) (Figure 1). *H. armigera* was also found on sorghum, millet, and maize crops. However, for maize, *H. armigera* individuals were only found at one site during the summer growing season in Northeastern Brazil (state of Bahia). *H. armigera* was not found on maize crops in the Midwest, Southeast, or South of Brazil. *H. zea* was primarily found on maize crops and was present in all sampled regions during both the winter and summer growing seasons. Of the winter crops, millet and cotton were exceptional in that they could simultaneously support *H. zea* and *H. armigera* (Figure 1). We found no correlations between specific *H. armigera* mitochondrial lineages (haplotypes) and specific hosts (Figure 1).

Dataset assembly, haplotypes, and demographic analysis

Following alignment and editing, we were unable to identify indels or stop codons in the sequences from either species. However, using the most common haplotype for each species as a reference, eight non-synonymous substitutions were observed in 17 *H. armigera* individuals, and four non-synonymous substitutions were observed in eight *H. zea* individuals. However, considering the relatively high mutation rate reported for the COI gene in the *Helicoverpa* genus [20], as well the absence of indels and stop codons, it is unlikely that these sequences represent numts (nuclear mitochondrial DNA).

Twenty-six polymorphic sites were found among the 139 *H. armigera* individuals sampled, which yielded 31 haplotypes with a haplotype diversity (Hd) of 0.821 and a nucleotide diversity (Pi) of 0.0028. Sequence analysis of the 134 sampled *H. zea* individuals identified 19 polymorphic sites, which yielded 20 haplotypes with an Hd of 0.420 and a Pi of 0.0011 (Table 2). No significant differences in Hd or Pi were found for either species when the individuals were separated by growing season according to the sampled crops (Table 2). The results from Tajima's D test were only not significant for *H. armigera* individuals ($p = 0.07$) sampled on summer crops; however, Fu's Fs test was significant ($p < 0.01$). The Tajima's D and Fu's Fs test results for both *H. armigera* and *H. zea* were negative and significant when the individuals were tested as a single group and when the individuals were split into groups based on the crop on which they were sampled (summer or winter; temporally). These results indicate an excess of low frequency polymorphisms and are consistent with either population expansion or purifying selection (Table 2). In addition, the model of sudden expansion [21] did not reject the hypothesis of expansion demographics for *H. armigera* (SSD = 0.0012, $p = 0.48$; Raggedness = 0.0433, $p = 0.61$) or *H. zea* (SSD = 0.0002, $p = 0.90$; Raggedness = 0.1492, $p = 0.72$).

Statistical analysis of population structure

The results of the analysis of molecular variance (AMOVA) with two hierarchical levels showed that the greatest amount of total variation was accounted for by differences among individuals within populations: 92.89% for *H. armigera* ($\Phi_{ST} = 0.071$) and 94.22% for *H. zea* ($\Phi_{ST} = 0.058$) (Table S1). For the AMOVA with three hierarchical levels for *H. armigera*, the largest percentage of variation occurred within populations, separating individuals into groups by time (winter and summer crops; 93.17%, $\Phi_{CT} = 0.006$; $\Phi_{SC} = 0.074$; $\Phi_{ST} = 0.068$), host group (mono- and dicotyledonous; 99.24%, $\Phi_{CT} = -0.01$; $\Phi_{SC} = 0.018$; $\Phi_{ST} = 0.007$), and each host type (crop; 93.19%, $\Phi_{CT} = -0.042$; $\Phi_{SC} = 0.105$; $\Phi_{ST} = 0.068$) (Table S1). The group separation for *H. armigera* was not significant for any of the three tested groups ($p > 0.10$). The AMOVA with three hierarchical levels divided the *H. zea* individuals into groups by time (winter and summer crops), which showed a larger variation within populations (93.76%, $\Phi_{CT} = 0.010$; $\Phi_{SC} = 0.052$; $\Phi_{ST} = 0.062$); the group division was not significant ($p > 0.10$) (Table S1).

Network analysis and Bayesian phylogeny

Analysis of the genetic connections between the *Helicoverpa* spp. represented in the haplotype network revealed a close genetic relation between *H. armigera* and *H. zea*, which were separated by only 13 mutational steps (Figure 2). By separately analyzing the connections between the genetic haplotypes of each species, we inferred the existence of two predominant maternal lineages for *H. armigera*: H1 (31.65%) and H3 (23.02%), which were located at the center of the haplotype network. The other haplotypes of *H. armigera*, with the exception of haplotype H2 (15.83%), all had

Table 1. Sampling sites for *Helicoverpa armigera* and *Helicoverpa zea* in Brazil, including the sites where these insects were sampled for this study, abbreviations, sample sizes for the mitochondrial genes (COI), crops sampled, geographic coordinates, dates sampled, and GenBank Accession.

Sites (City, State)	Abbreviation (Site, Crop)	Crop	Sample size H. armigera	Sample size H. zea	Lat. (S)	Lon. (W)	Date	GenBank Accession
Winter cropping								
Barreiras, Bahia	BA1Co	Cotton	3	-	12°08'54"	44°59'33"	05.22.12	KM274936–KM274938
Luís E. Magalhães, Bahia	BA2Co	Cotton	11	1	12°05'58"	45°47'54"	05.24.12	KM274939–KM274950
Balsas, Maranhão	MA1Co	Cotton	10	-	07°31'59"	46°02'06"	06.23.12	KM274987–KM274996
Luís E. Magalhães, Bahia	BA3Be	Bean	23	-	12°05'58"	45°47'54"	06.12.12	KM274979–KM274986, KM275038–KM275052
Luís E. Magalhães, Bahia	BA4Mi	Millet	6	3	12°05'58"	45°47'54"	05.10.12	KM274951–KM274959
Luís E. Magalhães, Bahia	BA5Sr	Sorghum	16	-	12°05'58"	45°47'54"	05.10.12	KM274960–KM274975
Capitólio, Minas Gerais	MG1Ma	Maize	-	14	20°36'17"	46°04'19"	06.08.12	KM274997–KM275010
Luís E. Magalhães, Bahia	BA6Ma	Maize	-	13	12°05'58"	45°47'54"	06.12.12	KM274976–KM274978, KM275053–KM275062
Itapira, São Paulo	SP1Ma	Maize	-	7	22°26'11"	46°49'20"	06.12.12	KM275011–KM275017
Assis, São Paulo	SP2Ma	Maize	-	7	22°39'40"	50°23'58"	06.15.12	KM275018–KM275024
São Gabriel do Oeste, Mato Grosso do Sul	MS1Ma	Maize	-	13	19°23'37"	54°33'49"	06.27.12	KM275025–KM275038
Rondonópolis, Mato Grosso	MT1Ma	Maize	-	7	16°28'17"	54°38'14"	08.01.12	KM275063–KM275069
Summer cropping								
Riachão das Neves, Bahia	BA7Sy	Soybean	8	-	12°08'54"	44°59'33"	10.21.12	KM275070–KM275077
Luís E. Magalhães, Bahia	BA8Sy	Soybean	5	-	12°05'58"	45°47'54"	10.31.12	KM275078–KM275082
Rondonópolis, Mato Grosso	MT2Sy	Soybean	13	-	16°28'17"	54°38'14"	11.08.12	KM275083–KM275092, KM275156–KM275158
Chapadão do Sul, Mato Grosso do Sul	MS2Sy	Soybean	6	-	18°46'44"	52°36'59"	11.29.12	KM275097–KM275102
Balsas, Maranhão	MA2Sy	Soybean	10	-	07°31'59"	46°02'06"	01.06.13	KM275103–KM275112
São Desidério, Bahia	BA9Sy	Soybean	10	-	12°21'08"	44°59'03"	01.15.13	KM275127–KM275136
Limoeiro do Norte, Ceará	CE1Co	Cotton	-	4	05°08'56"	38°05'52"	10.08.12	KM275093–KM275096
São Desidério, Bahia	BA10Co	Cotton	14	-	12°21'08"	44°59'03"	01.15.13	KM275147–KM275155, KM275202–KM275206
Cândido Mota, São Paulo	SP3Ma	Maize	-	7	22°44'46"	50°23'15"	01.14.13	KM275113–KM275119
Jardinópolis, São Paulo	SP4Ma	Maize	-	7	21°03'47"	47°45'05"	03.04.13	KM275120–KM275126
Barreiras, Bahia	BA11Ma	Maize	4	-	11°33'33"	46°19'47"	02.21.13	KM275137–KM275140
Luís E. Magalhães, Bahia	BA12Ma	Maize	-	9	12°05'58"	45°47'54"	03.28.13	KM275141–KM275146, KM275207–KM275209
Rolândia, Paraná	PR1Ma	Maize	-	12	23°19'13''	51°29'01''	01.24.13	KM275159–KM275170
Passo Fundo, Rio Grande do Sul	RS1Ma	Maize	-	10	28°16'08''	52°37'15''	01.30.13	KM275171–KM275180
Montividiu, Goiás	GO1Ma	Maize	-	10	17°19'19''	51°14'51''	02.05.13	KM275181–KM275190
Capitólio, Minas Gerais	MG2Ma	Maize	-	11	20°36'17"	46°04'19"	03.10.13	KM275191–KM275201
Total			**139**	**135**				

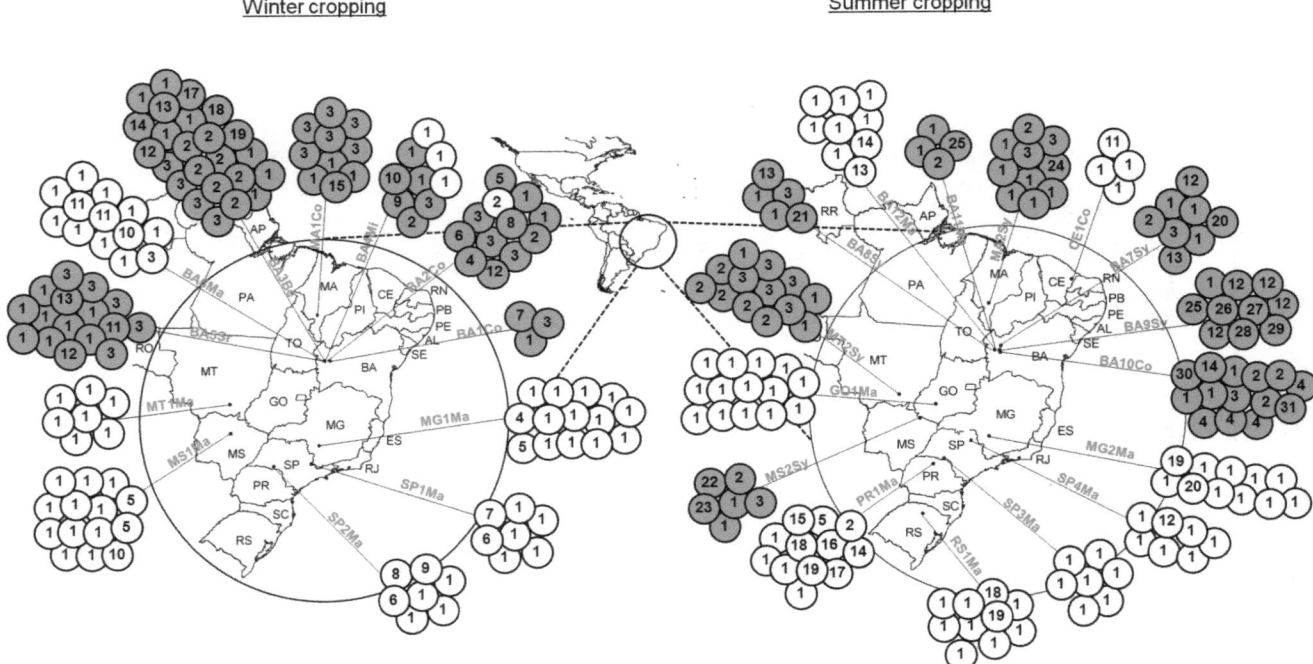

Figure 1. Geographic distributions of COI haplotypes of *H. armigera* and *H. zea*. One hundred and thirty nine and 135 COI haplotypes were analyzed for these species, respectively. The samples were separated into two temporal groups (winter crops and summer crops). Each circle represents the haplotypes identified in a given population; a number within a circle denotes the COI haplotypes identified in that population. Colored circles refer to *H. armigera* specimens, and white circles refer to *H. zea* specimens. The abbreviations refer to the sampled locations and crops (Table 1).

frequencies below 5%. Haplotypes H19, H18, H16, H12, H21, and H25 formed an outer cluster within the haplotype network of *H. armigera* (Figure 2). The haplotype network for *H. zea* revealed a genetic haplotype relationship with a single central high-frequency lineage (H1 = 76.30%) surrounded by low-frequency haplotypes (<5%) (Figure 2).

The optimal nucleotide substitution model identified by the MODELTEST 2.3 software program was the GTR+I+G model (Generalized time reversible + Proportion of invariable sites + Gamma distribution model). The estimated model parameters were based on empirical base frequencies (A = 0.3092, C = 0.1463, G = 0.1312, and T = 0.4133), with the proportion of invariable sites (I) set to 0.7393 and the gamma distribution shape parameter set to 0.5778. The consensus tree generated by the Bayesian analysis divided the *Helicoverpa* spp. specimens sampled in Brazil into two monophyletic clades (*H. armigera* and *H. zea*) with an associated probability of 99% (Figure 3; Figure S1). The probabilities separating the *H. zea* individuals into groups within this species were not significant. A single *H. armigera* individual (MS2Sy6) was separated from the other individuals with an associated probability of 98%. Finally, *Helicoverpa gelotopoeon* showed a closer phylogenetic relationship to *H. armigera* and *H. zea* compared with *H. assulta* (Figure 3; Figure S1).

Network analysis: Brazilian vs. Old World *Helicoverpa armigera*

The haplotype network constructed using the edited sequences collected in Brazil, along with numerous Old World sequences, identified 38 distinct haplotypes (Figure 4). H1 (28%) and H2 (24%), which are widely distributed throughout Brazil, Europe, and China, were the most frequent haplotypes and occupied the

central region of the haplotype network. All other haplotypes, with the exception of H3 and H10, showed frequencies below 5%. Finally, the majority of haplotypes with low frequencies represented by singletons were located at the network extremities (Figure 4).

Discussion

Our results indicate a widespread distribution for *H. armigera* throughout the Midwest and Northeast of Brazil on a variety of crops, particularly dicotyledons, beans, soybeans, and cotton as well as, to a lesser extent, millet, sorghum, and maize. This pest was not found on maize crops in the Midwest, Southeast, or South of Brazil, despite the fact that these crops were initially identified as sources of *H. armigera* in this system. *H. armigera* individuals associated with maize crops were only found at a single sampling site in the Northeast (state of Bahia) during February 2013. In contrast, *H. zea* individuals were essentially found only on maize crops, with the exception of a few individuals collected from millet and cotton crops, where *H. zea* individuals were found alongside *H. armigera* individuals. Before the documentation of *H. armigera* in Brazil in 2013, we had hypothesized that major source of *Helicoverpa* spp. attacking different host plant was maize crops. However, our findings showed that targeting the control of *H. armigera* on maize crops may not be effective because *H. zea* was the predominant species in this host plant. The possibility of the formation of hybrid individuals between these two species, which has been reported under laboratory conditions [3,4], needs to be investigated under field conditions to improve our pest management programs.

Table 2. Number of individuals, haplotype designation, and genetic diversity for the sampled populations grouped according to geographical origin.

Group	N. Individuals (samples)	N. haplotypes	Distribution of Haplotypes (n)	Haplotype Diversity (Hd)	Nucleotide diversity (Pi)	Tajima's D test (p value)	Fu's Fs test (p value)
H. armigera							
Pooled	139 (14)	31	-	0.821	0.0028	-1.729 (<0.01)	-26.361 (<0.01)
Winter cropping	69 (6)	19	H1(22); H2(9); H3(21); H4(1); H5(1); H6(1); H7(1); H8(1); H9(1); H10(1); H11(1); H12(2); H13(2); H14(1); H15(1); H16(1); H17(1); H18(1); H19(1).	0.805	0.0028	-1.608 (=0.03)	-11.891 (<0.01)
Summer cropping	70 (8)	19	H1(22); H2(13); H3(11); H4(4); H12(4); H13(2); H14(1); H20(1); H21(1); H22(1); H23(1); H24(1); H25(2); H26(1); H27(1); H28(1); H29(1); H30(1); H31(1).	0.835	0.0028	-1.353 (=0.07)	-11.254 (<0.01)
H. zea							
Pooled	135 (16)	20	-	0.420	0.0011	-2.190 (<0.01)	-22.912 (<0.01)
Winter cropping	65 (8)	11	H1(50); H2(1); H3(1); H4(1); H5(3); H6(2); H7(1); H8(1); H9(1); H10(2); H11(2).	0.408	0.0009	-2.156 (<0.01)	-9.735 (<0.01)
Summer cropping	70 (8)	13	H1(53); H2(1); H5(1); H11(1); H12(1); H13(1); H14(2); H15(1); H16(1); H17(1); H18(2); H19(3); H20(2).	0.427	0.0012	-1.967 (<0.01)	-10.411 (<0.01)

Demographic analyses using neutrality tests and a Mismatch Distribution Analysis indicated an expansion of the *H. armigera* and *H. zea* populations within the Brazilian territory. Population expansions were also consistent with the Haplotype network structure, which was characteristic of species undergoing processes of demographic expansion [22]. Brazilian *H. armigera* individuals showed two primary maternal lineages, whereas *H. zea* showed a single primary lineage, all of which were surrounded by numerous lower-frequency haplotypes. Therefore, these central high-frequency haplotypes represent the ancestral haplotypes, with the low-frequency haplotypes more recently derived [23]. Furthermore, signs of the *H. armigera* population expansion are likely because of the recent introduction of this pest into Brazil. Following the founder event, during which a portion of the overall genetic diversity of the species was introduced to Brazil, the *H. armigera* population further propagated. According to Nibouche et al. [24], *H. armigera* can migrate as far as 2,000 km, which likely facilitated the colonization of a variety of crops. The migration and colonization of crop areas by a small group of individuals can cause bottleneck effects, which, combined with plague population-suppression strategies (e.g., insecticide use that kills all but a small portion of the population), can lead to the types of demographic expansions observed for *H. zea* and *H. armigera* in Brazil [25–27]. In addition, the expansion of maize, soybean, and cotton crops into the North and Northeast of Brazil over the previous decade may also be responsible, in part, for the demographic expansion of these species, specifically *H. zea*. Additionally, assuming that not all COI variation is neutral, *Helicoverpa* spp. populations could be suffering selection, especially considering that populations have colonized new environments recently. However, further studies using a larger number of molecular markers from nuclear and mitochondrial genome regions would answer these questions. The *H. armigera* and *H. zea* population genetics were not structured according to space, time (winter and summer crops), or host (crops). Unstructured genetic networks have been reported for other populations of these two pest species in other parts of the world, which were based on several molecular markers, including mtDNA, allozymes, and microsatellites [7,24,25,28,29,30]. Both species showed wide spatial haplotype distributions, and no genetic relationships were identified using a haplotype network analysis or an AMOVA. This scenario may be because these populations have a polyphagous feeding habit and migratory characteristics.

The unstructured population of *H. armigera* and the wide distribution of the two ancestral maternal lineages within the Brazilian territory did not allow us to infer any hypothetical invasion or dispersal routes for this species within the region. However, we noted that the haplotype and nucleotide diversities found for *H. armigera* in Brazil are similar to or greater than those reported for natural *H. armigera* populations in the Old World [7,20]. For example, one outer branch of the *H. armigera* haplotype network, formed by haplotypes H19, H18, H16, H12, H21, and H25, is noteworthy for having the greatest genetic distance from the central haplotypes (H1 and H3), and these haplotypes have yet to be identified in Old World populations [7,20]. In addition, joint analysis of the haplotypes from Brazil and the Old World yielded an overall structure that was similar to the haplotype network obtained only from the Brazilian individuals. In particular, the two most frequent haplotypes were identified throughout Brazil, Europe, China, and India, whereas the majority of the singletons were from Brazil and China. The cited literature, along with our results that showed a wide geographic distribution for *H. armigera* during the first half of 2012, support the hypothesis of an invasion period prior to the first reports of this

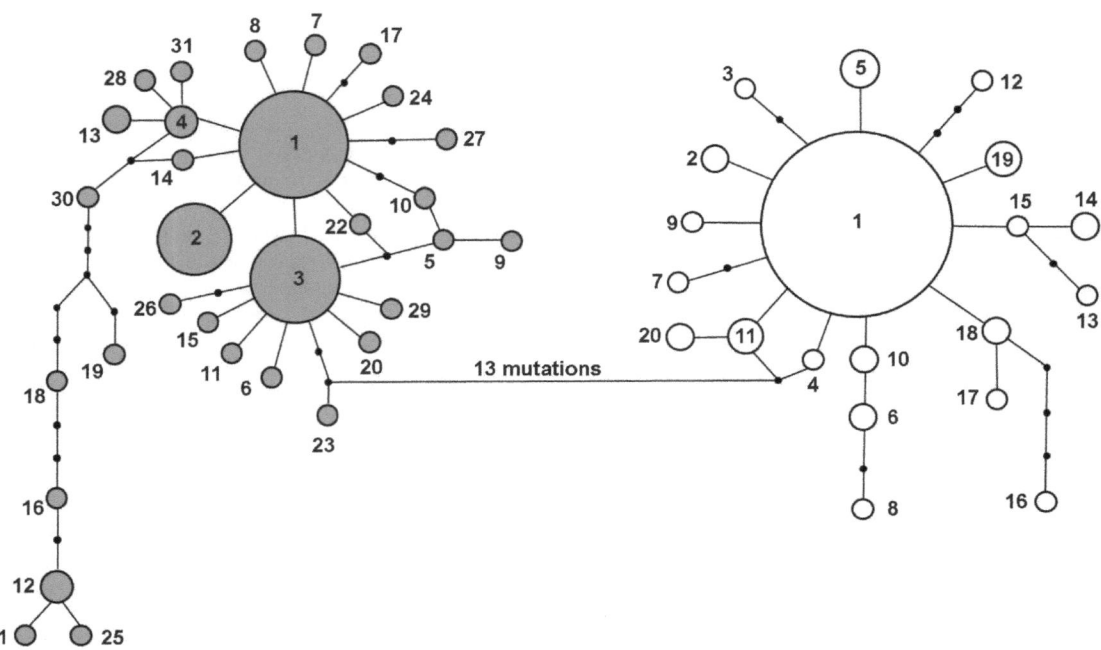

Figure 2. Haplotype network based COI sequences from *H. armigera* and *H. zea* samples collected in Brazil. Partial mtDNA COI (658 bp) sequences from *H. armigera* (colored circles) and *H. zea* (white circles) were analyzed from samples collected in Brazil. Each haplotype is represented by a circle and is identified by a number from 1–31. The *H. armigera* and *H. zea* COI haplotypes are shown as described in Table 2. The numbers of nucleotide substitutions between the haplotypes are indicated by black circles. The total number of nucleotide substitutions separating the *H. armigera* specimens from the *H. zea* specimens is shown.

species in Brazil. Alternatively, these findings are also consistent with a more recent invasion that involved a large gene pool, multiple invasion events, or some combination of these events.

The low genetic divergence observed between *H. armigera* and *H. zea* in the haplotype network analysis and the Bayesian phylogeny confirms the close genetic relatedness of these two species. Therefore, the reported co-occurrence of these species in time and space, as well as on the same hosts (as described here),

could allow for the formation of hybrid individuals, which has been reported under laboratory conditions [3,4]. Although the existence of hybrids in the wild remains unconfirmed, this scenario is of significant concern. In particular, recombination or introgression phenomena between *H. armigera*, which is reportedly resistant to control methods, and *H. zea*, which has adapted to the environmental conditions of the American continent, may enable gene transfer and fixation in some individuals. Therefore, hybridization may enable the selection of breeds with enhanced hybrid vigor and the ability to rapidly adapt to current management and suppression methods.

The population studies described in this study indicate a recent demographic expansion and a high mitochondrial genetic diversity for *H. armigera* and *H. zea* in Brazil. Therefore, the sustainable management of *H. armigera* will likely become a significant challenge for Brazilian entomology in the coming years, especially considering the polyphagous feeding habit, the great dispersal ability, and the numerous reports of resistance to insecticides and Bt crops for this insect [8,24,31–35]. This scenario requires immediate attention, as there is an imminent risk of *H. armigera* expanding throughout the American territory and perhaps reaching agricultural areas in Central and North America. However, it was not possible to trace the invasion and dispersal routes of *H. armigera* in the Brazilian territory. Nevertheless, the hypotheses of an invasion period prior to the first reports in the literature and/or an invasion that involved a diverse gene pool are both consistent with the observed high incidence and rapid adaptation of *H. armigera* in the Brazilian territory. Our confirmation that the predominant maternal lineages in the Brazilian territory are the same compared with those in Europe and Asia may represent a starting point to guide *H. armigera* management programs. Indeed, control strategies have a greater

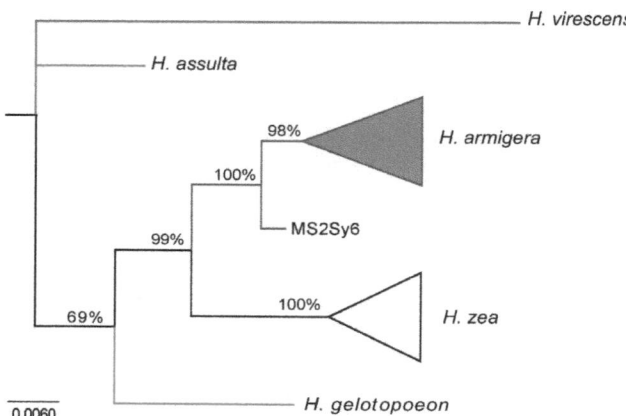

Figure 3. Bayesian phylogenetic tree of *H. armigera* and *H. zea* individuals sampled in Brazil. This phylogenetic tree is based on partial COI haplotype sequences and includes *H. assulta* and *H. gelotopoeon* sequences. Numbers near the interior branches indicate posterior probability (×100) values. The outgroup used was *Heliothis virescens*. *H. armigera* COI haplotypes and Genbank Accession numbers can be found in Table S2.

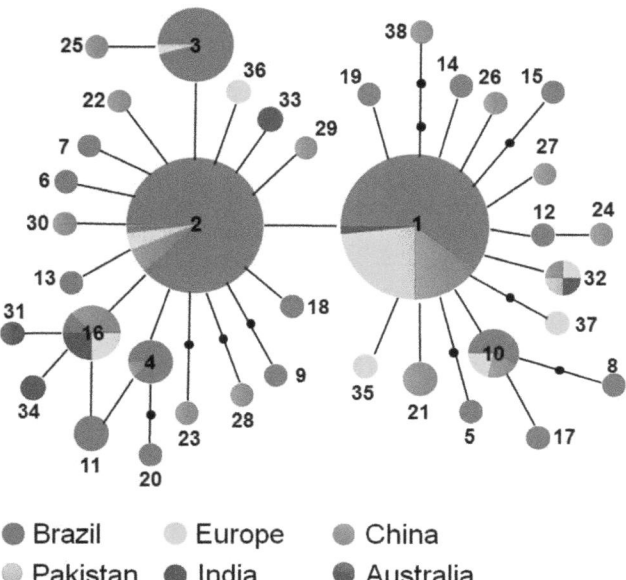

Figure 4. Haplotype network based COI sequences from *H. armigera* samples from Brazil and Old World specimens. Partial mtDNA COI (590 bp) sequences from this species were analyzed. Thirty-eight haplotypes were identified from 212 individuals sampled from China (n = 35), Thailand (n = 1), Australia (n = 1), Pakistan (n = 2), Europe (n = 28), India (n = 6), and Brazil (n = 139). *H. armigera* COI haplotypes are shown as described in Table S2. Each circle represents a haplotype and its number. The colors represent the frequency of each haplotype in the country/continent, with dark green (Brazil), light green (Pakistan), yellow (Europe), brown (India), light blue (China), and dark blue (Australia).

chance of success when reliable information is gathered in the regions where the pests, their hosts, and their natural enemies have co-evolved over a significant period of time.

Materials and Methods

Sampling procedures

Permit access to collect material used in our research at various crop sites was granted by respective growers. GPS coordinates of each location are listed in Table 1.

Brazilian agriculture has shown successive and overlapping crops in space and time, and these crops can be largely separated into two harvest groups that are primarily characterized by their rainfall needs. In particular, winter crops are grown between May and September and require low rainfall, whereas summer crops are grown between October and April and require high rainfall. Our initial sampling design was directed at understanding the *H. zea* population dynamics and primarily involved maize fields. However, attacks on soybean, cotton, bean, sorghum, and millet crops were also reported between May 2012 and April 2013 (Brazilian agricultural year). Therefore, we directed our sampling efforts towards a variety of crops and regions throughout Brazil. We also focused on the Western region of Bahia State, Brazil, which was the site of numerous *Helicoverpa* spp. attacks, to determine whether maize crops were the main source of *H. zea* in the Brazilian agricultural system. A total of 274 *Helicoverpa* caterpillars were collected at 19 sampling sites from six different crops (Table 1). In the absence of morphological characters or nuclear markers to reliably distinguish between *H. zea* and *H. armigera*, species identification was carried out using the sequence

fragment of COI mitochondrial gene by comparing with *H. zea* and *H. armigera* species barcodes [7,18,19,39] and determining homology with BlastN tool.

DNA extraction, PCR amplification, and gene sequencing

Genomic DNA was isolated from the thorax of each adult using an Invisorb Spin Tissue Kit (STRATEC Molecular, Berlin, Germany), according to the manufacturer's protocol. A fragment of the COI mitochondrial gene was amplified by polymerase chain reaction (PCR) with the primers LCO(F) (5' - GGT CAA CAA ATC ATA AAG ATA TTG G - 3') and HCO(R) (5' - TAA ACT TCA GGG TGA CCA AAA AAT CA - 3') [36]. Amplification reactions were performed using 10 ng genomic DNA, 50 mM $MgCl_2$, 0.003 $mg.mL^{-1}$ BSA, 6.25 mM dNTPs, 10 pmol each primer, 1 U Taq DNA Polymerase (Life Technologies, Carlsbad, CA, USA), and 10% 10× Taq Buffer in a final volume of 25 μL. The PCR program consisted of an initial denaturation step at 94°C for 3 min, followed by 35 cycles of denaturation at 94°C for 30 s, annealing at 45°C for 30 s, and polymerization at 72°C for 1.5 min, with a final extension step at 72°C for 10 min. Following amplification, the aliquots were visually inspected using agarose gel (1.5% w/v) electrophoresis. The amplicons were purified by ethanol precipitation, and a second round of amplification was performed using the Big Dye Terminator v3.1 Cycle Sequencing system (Applied Biosystems, Foster City, CA, USA), which was followed by further purification. DNA sequencing was performed using the ABI3500xl automated genetic analyzer (Applied Biosystems, Foster City, CA, USA) at the State University of Campinas (Universidade Estadual de Campinas, Campinas, São Paulo, Brazil).

Dataset assembly, haplotypes, and demographic analysis

All sequences were manually edited using the Chromas Lite version 2.01 [37] software program and were aligned using the ClustalW tool from the BioEdit version 7.0 [38] software program. After editing and aligning the COI sequences, we determined the 658 bp consensus sequence, which was then posteriorly compared with the *H. zea* and *H. armigera* species barcodes [41] to determine homology using the BlastN tool, which is available online at NCBI [40].

The MEGA version 4 [41] software program was used to inspect the COI sequences from each species individually for the presence of numts [42]. In particular, we searched for the following numt signatures: (i) insertions/deletions (*indels*); (ii) stop codons leading to premature protein termination; and (iii) increased rates of non-synonymous mutations. The presence of signatures (i) and (ii) was considered sufficient to regard a sequence as a COI numt. In the presence of signatures (i) or (ii), signature (iii) was used to confirm the sequence as a numt. The presence of signature (iii) alone was not considered sufficient to define a sequence as a numt.

Haplotype and nucleotide diversity parameters for each species were estimated using the DnaSP version 5 [43] software program. Neutrality tests using Tajima's D [44] and Fu's Fs [45] were performed using the Arlequin version 3.1 [46] software program, and significance was determined using 1,000 random samples in coalescent simulations. Based on the recommendations in the Arlequin manual, we activated the "Infer from distance matrix" option under "Haplotype definition", and the Fu's Fs statistical values were considered to be significant at a level of 5% only when the *P*-value was below 0.02. The diversity estimates and neutrality tests were performed using all sampled individuals from each species, which were divided into winter-crop and summer-crop groups. A Mismatch Distribution Analysis using a spatial expansion model [21] was also performed using the Arlequin version 3.1 software program, and significance was determined using 1,000 bootstrap replicates. We used the goodness-of-fit of the observed mismatch distribution to the expected distribution from the spatial expansion model and the sum of square deviations (SSD) as a test statistic (P-value support).

Population structure analysis

Using Arlequin 3.1, we also performed an AMOVA at the two- and three-hierarchy levels [47]. For the three-hierarchy AMOVA, we first separated the samples depending on whether they were collected on winter or summer crops and then further divided them by host plant (monocotyledonae or dicotyledonae).

Network analysis and Bayesian phylogenies

Genetic differences and connections among *Helicoverpa* spp. haplotypes were determined by constructing a maximum parsimony network [48] using the TCS 1.21 software program [49]. To resolve ambiguities present in the haplotype network, we used the criteria of coalescence theory and population geography proposed by Crandall and Templeton [23].

We used the distance matrix option in the PAUP *4.0 software program to calculate the inter- and intra-species genetic distances, which were inferred using the nucleotide substitution model and the Akaike Information Criteria [50] selected by MODELTEST 2 [51]. The MrBayes v3.2 software program [52] was used to estimate Bayesian phylogenies. In particular, the Bayesian analysis was performed with 10 million generations using one cold and three heated chains. *Helicoverpa assulta* (Guenée) (GenBank Accession number: EU768937), *H. gelotopoeon* Dyar (EU768938), and *H. virescens* (IN799050) sequences were included as outgroups for the Bayesian analysis. We obtained a 50%-majority-rule consensus tree with posterior probabilities that were equal to the bipartition frequencies.

Network analysis: Brazil vs. Old World

Seventy-three sequences from a variety of Old World sites that were present in GenBank were included with the 139 *H. armigera* sequences we collected in Brazil. In particular, 73 sequences were obtained from specimens collected in China (N = 35) [GenBank Accession numbers GQ892840 - GQ892855, GQ995232 - GQ995244 [20], HQ132369 (Yang, 2010), JX392415, and JX392497 (not published)], Thailand (1) [(EU768935)], Australia (1) [(EU768936) [5]], Pakistan (2) [(JN988529 and JN988530) (not published)], Europe (28) [(FN907979, FN907980, FN907988, FN907989, FN907996 - FN907999, FN908000 - FN908003, FN908005, FN908006, FN908011, FN908013 - FN908018, FN908023, FN908026, GU654969, GU686757, GU686955, and JF415782) (not published)] and India (6) [(HM854928-HM854932 and JX32104) (not published)] (Table S2). This new data set was edited and aligned as follows. The sequences were different

lengths; thus, the editing and alignment processes generated a total of 212 sequences 590 bp in length, excluding indels. The sequences from individuals collected in Brazil, which were previously analyzed using a fragment length of 658 bp, as entered into GenBank (see Table 1), were edited by removing the first 36 bp and the last 32 bp. Using the TCS 1.21 software program [49], we subjected this data set to haplotype network analysis using a maximum parsimony network [48] to investigate the genetic connections between haplotypes from Brazil and the Old World as well as to infer the origins of maternal lineages within *H. armigera* populations in Brazil.

Supporting Information

Figure S1 Bayesian phylogenetic tree of *H. armigera* and *H. zea* individuals sampled in Brazil. This phylogenetic tree is based on partial COI haplotype sequences and includes *H. assulta* and *H. gelotopoeon* sequences. Numbers near the interior branches indicate the posterior probability ($\times 1,000$) values. The outgroup used was *Heliothis virescens*. *H. armigera* COI haplotypes and Genbank Accession numbers can be found in Table S2.

Table S1 Hierarchical analysis of molecular variance (AMOVA), for population genetics structure of *Helicoverpa armigera* and *H. zea* with a mithocondrial (COI) region marker.

Table S2 Global *Helicoverpa armigera* including the Brazilian *H. armigera* haplotypes, and relevant GenBank Accession numbers. Numbers of individuals sequenced from each locality are indicated in parentheses.

Acknowledgments

We thank Celito Breda, Diego Miranda, Fábio Wazne, Germison Tomquelski, José Wilson de Souza, Marcos Michelotto, Milton Ide, Paulo Saran, Pedro Brugnera, Pedro Matana Junior, Rodrigo Franciscatti, Rodrigo Sorgatto, Rubem Staudt, Sérgio de Azevedo, and SGS Gravena (SISBIO License #18018-1) for helping to collect insect samples in different Brazilian regions. We also thank Jaqueline Campos for technical assistance and Prof. José Baldin Pinheiro for providing laboratory space and equipment.

Author Contributions

Conceived and designed the experiments: NAL AAP MIZ CO. Performed the experiments: NAL AAP. Analyzed the data: NAA ASC AAP MIZ CO. Contributed reagents/materials/analysis tools: CO MIZ ASC NAL AAP. Wrote the paper: NAL ASC AAP MIZ CO.

References

1. Pogue MG (2013) Revised status of *Chloridea* Duncan and (Westwood), 1841, for the *Heliothis virescens* species group (Lepidoptera: Noctuidae: Heliothinae) based on morphology and three genes. Syst Entomol 38: 523–542.
2. Mitter C, Poole RW, Matthews M (1993) Biosystematics of the Heliothinae (Lepidoptera: Noctuidae). Annu Rev Entomol 38: 207–225.
3. Laster ML, Hardee DD (1995) Intermating compatibility between north american *Helicoverpa zea* and *Heliothis armigera* (Lepidoptera: Noctuidae) from Russia. J Econ Entomol 88: 77–80.
4. Laster ML, Sheng CF (1995) Search for hybrid sterility for *Helicoverpa zea* in crosses between the north american *H. zea* and *H. armigera* (Lepidoptera: Noctuidae) from China. J Econ Entomol 88: 1288–1291.
5. Cho S, Mitchell A, Mitter C, Regier J, Matthews M, et al. (2008) Molecular phylogenetics of heliothine moths (Lepidoptera: Noctuidae: Heliothinae), with

comments on the evolution of host range and pest status. Syst Entomol 33: 581–594.
6. Mallet J, Korman A, Heckel D, King P (1993) Biochemical genetics of *Heliothis* and *Helicoverpa* (Lepidoptera: Noctuidae) and evidence for a founder event in *Helicoverpa zea*. Ann Entomol Soc Am 86: 189–197.
7. Behere GT, Tay WT, Russell DA, Heckel DG, Appleton BR, et al. (2007) Mitochondrial DNA analysis of field populations of *Helicoverpa armigera* (Lepidoptera: Noctuidae) and of its relationship to *H. zea*. BMC Evol Biol 7: 117.
8. Fitt GP (1989) The ecology of *Heliothis* species in relation to agroecosystems. Annu Rev Entomol 34: 17–52.

9. Pogue M (2004) A new synonym of *Helicoverpa zea* (Boddie) and differentiation of adult males of *H. zea* and *H. armigera* (Hübner) (Lepidoptera: Noctuidae: Heliothinae). Ann Entomol Soc Am 97: 1222–1226.

10. Wu KM, Lu YH, Feng HQ, Jiang YY, Zhao JZ (2008) Suppression of cotton bollworm in multiple crops in China in areas with Bt toxin-containing cotton. Science 321: 1676–1678.

11. Degrande PE, Omoto C (2013) Estancar prejuízos. Cultivar Grandes Culturas Abril: 32–35.

12. Czepack C, Albernaz KC, Vivan LM, Guimarães HO, Carvalhais T (2013) Primeiro registro de ocorrência de *Helicoverpa armigera* (Hübner) (Lepidoptera: Noctuidae) no Brasil. Pesq Agropec Trop 43: 110–113.

13. Specht A, Sosa-Gómez DR, Paula-Moraes SV de, Yano SAC (2013) Identificação morfológica e molecular de *Helicoverpa armigera* (Lepidoptera: Noctuidae) e ampliação de seu registro de ocorrência no Brasil. Pesq Agropec Bras 48: 689–692.

14. Agropec Consultoria (2013) Pragas quarentenárias: consulta a dados sobre pragas quarentenárias presentes e ausentes no Brasil. Available: http://spp.defesaagropecuaria.com/. Accessed 2014 Jan 29.

15. Yang Y, Li Y, Wu Y (2013) Current status of insecticide resistance in *Helicoverpa armigera* after 15 years of Bt cotton planting in China. J Econ Entomol 106: 375–381.

16. Martin T, Ochou GO, Djihinto A, Traore D, Togola M, et al. (2005) Controlling an insecticide-resistant bollworm in West Africa. Agric Ecosyst Environ 107: 409–411.

17. MAPA (2014) Combate à praga *Helicoverpa armigera*. Brasilia: MAPA.

18. Tay WT, Soria MF, Walsh T, Thomazoni D, Silvie P, et al. (2013) A brave new world for an old world pest: *Helicoverpa armigera* (Lepidoptera: Noctuidae) in Brazil. Plos One 8: e80134.

19. Behere GT, Tay WT, Russell DA, Batterham P (2008) Molecular markers to discriminate among four pest species of *Helicoverpa* (Lepidoptera: Noctuidae). Bull Entomol Res 98: 599–603.

20. Li QQ, Li DY, Ye H, Liu XF, Shi W, et al. (2011) Using COI gene sequence to barcode two morphologically alike species: the cotton bollworm and the oriental tobacco budworm (Lepidoptera: Noctuidae). Mol Biol Rep 38: 5107–5113.

21. Rogers AR, Harpending H (1992) Population growth makes waves in the distribution of pairwise genetic differences. Mol Biol Evol 9: 552–569.

22. Excoffier L, Hofer T, Foll M (2009) Detecting loci under selection in a hierarchically structured population. Heredity 103: 285–298.

23. Crandall KA, Templeton AR (1993) Empirical tests of some predictions from coalescent theory with applications to intraspecific phylogeny reconstruction. Genetics 134: 959–969.

24. Nibouche S, Bues R, Toubon JF, Poitout S (1998) Allozyme polymorphism in the cotton bollworm *Helicoverpa armigera* (Lepidoptera: Noctuidae): comparison of African and European populations. Heredity 80: 438–445.

25. Endersby NM, Hoffmann AA, McKechnie SW, Weeks AR (2007) Is there genetic structure in populations of *Helicoverpa armigera* from Australia? Entomol Exp Appl 122: 253–263.

26. Albernaz KC, Silva-Brandao KL, Fresia P, Consoli FL, Omoto C (2012) Genetic variability and demographic history of *Heliothis virescens* (Lepidoptera: Noctuidae) populations from Brazil inferred by mtDNA sequences. Bull Entomol Res 102: 333–343.

27. Domingues FA, Silva-Brandão KL, Abreu AG, Perera OP, Blanco CA, et al. (2012) Genetic structure and gene flow among Brazilian populations of *Heliothis virescens* (Lepidoptera: Noctuidae). J Econ Entomol 105: 2136–2146.

28. Zhou XF, Faktor O, Applebaum SW, Coll M (2000) Population structure of the pestiferous moth *Helicoverpa armigera* in the Eastern Mediterranean using RAPD analysis. Heredity 85: 251–256.

29. Han Q, Caprio MA (2002) Temporal and spatial patterns of allelic frequencies in cotton bollworm (Lepidoptera: noctuidae). Environ Entomol 31: 462–468.

30. Asokan R, Nagesha S, Manamohan M, Krishnakumar N, Mahadevaswamy H, et al. (2012) Molecular diversity of *Helicoverpa armigera* Hübner (Noctuidae: Lepidoptera) in India. Orient Insects 46: 130–143.

31. Gunning RV, Dang HT, Kemp FC, Nicholson IC, Moores GD (2005) New resistance mechanism in *Helicoverpa armigera* threatens transgenic crops expressing *Bacillus thuringiensis* Cry1Ac toxin. Appl Environ Microbiol 71: 2558–2563.

32. Zhang X, Liang Z, Siddiqui ZA, Gong Y, Yu Z, et al. (2009) Efficient screening and breeding of *Bacillus thuringiensis* subsp. kurstaki for high toxicity against *Spodoptera exigua* and *Heliothis armigera*. J Ind Microbiol Biotechnol 36: 815–820.

33. Martin T, Ochou GO, Hala-N'Klo F, Vassal J-M, Vaissayre M (2000) Pyrethroid resistance in the cotton bollworm, *Helicoverpa armigera* (Hübner), in West Africa. Pest Manag Sci 56: 549–554.

34. Achaleke J, Brevault T (2010) Inheritance and stability of pyrethroid resistance in the cotton bollworm *Helicoverpa armigera* (Lepidoptera: Noctuidae) in Central Africa. Pest Manag Sci 66: 137–141.

35. Nair R, Kalia V, Aggarwal KK, Gujar GT (2013) Variation in the cadherin gene sequence of Cry1Ac susceptible and resistant *Helicoverpa armigera* (Lepidoptera: Noctuidae) and the identification of mutant alleles in resistant strains. Curr Sci 104: 215.

36. Folmer O, Black M, Hoeh W, Lutz R, Vrijenhoek R (1994) DNA primers for amplification of mitochondrial cytochrome c oxidase subunit I from diverse metazoan invertebrates. Mol Mar Biol Biotechnol 3: 294–299.

37. Technelysium Pty Ltd (1998–2005) Chromas lite version 2.01. Available: http://www.technelysium.com.au/chromas_lite.html.

38. Hall TA (1999) Bioedit: a user-friendly biological sequence alignment editor and analysis program for Windows 95/98/NT. Nucleic Acids Symposium Series.

39. Encyclopedia of Life. Available: http://www.eol.org. Accessed 2013 July 4.

40. Matten T (2002) The BLAST Sequence Analysis Tool. In: McEntyre J, Ostell J, editors. The NCBI Handbook [Internet]. Bethesda (MD): National Center for Biotechnology Information (US).

41. Tamura K, Dudley J, Nei M, Kumar S (2007) MEGA4: Molecular Evolutionary Genetics Analysis (MEGA) software version 4.0. Mol Biol Evol 24: 1596–1599.

42. Lopez JV, Yuhki N, Masuda R, Modi W, O'Brien SJ (1994) Numt, a recent transfer and tandem amplification of mitochondrial DNA to the nuclear genome of the domestic cat. J Mol Evol 39: 174–190.

43. Librado P, Rozas J (2009) DnaSP v5: A software for comprehensive analysis of DNA polymorphism data. Bioinformatics 25: 1451–1452.

44. Tajima F (1989) Statistical method for testing the neutral mutation hypothesis by DNA polymorphism. Genetics 123: 585–595.

45. Fu YX (1997) Statistical tests of neutrality of mutations against population growth, hitchhiking and background selection. Genetics 147: 915–925.

46. Excoffier L, Laval G, Schneider S (2005) Arlequin v. 3.0: an integrated software package for population genetics data analysis. Evol Bioinform Online 1: 47–50.

47. Excoffier L, Smouse PE, Quattro JM (1992) Analysis of molecular variance inferred from metric distances among DNA haplotypes - application to humam mitochondrial - DNA restriction data. Genetics 131: 479–491.

48. Templeton AR, Crandall KA, Sing CF (1992) A cladistic analysis of phenotypic associations with haplotypes inferred from restriction endonuclease mapping and DNA sequence data. III. Cladogram estimation. Genetics 132: 619–633.

49. Clement M, Posada D, Crandall KA (2000) TCS: a computer program to estimate gene genealogies. Mol Ecol 9: 1657–1659.

50. Akaike H (1974) A new look at the statistical model identification. IEEE T Automat Contr 19: 716–723.

51. Nylander JAA (2004) MrModeltest v2. Program distributed by the author. Uppsala University: Evolutionary Biology Centre.

52. Ronquist F, Teslenko M, van der Mark P, Ayres DL, Darling A, et al. (2012) MrBayes 3.2: efficient Bayesian phylogenetic inference and model choice across a large model space. Syst Biol 61: 539–542.

13

First Record of the Myrmicine Ant Genus *Meranoplus* Smith, 1853 (Hymenoptera: Formicidae) from the Arabian Peninsula with Description of a New Species and Notes on the Zoogeography of Southwestern Kingdom Saudi Arabia

Mostafa R. Sharaf*, Hathal M. Al Dhafer, Abdulrahman S. Aldawood

Plant Protection Department, College of Food and Agriculture Science, King Saud University, Riyadh, Saudi Arabia

Abstract

The ant genus *Meranoplus* is reported for the first time from the Arabian Peninsula (Kingdom of Saudi Arabia) by the new species *M. pulcher* sp. n., based on the worker caste. Specimens were collected from Al Sarawat and Asir Mountains of southwestern Kingdom of Saudi Arabia using pitfall traps. *Meranoplus pulcher* sp. n. is included in the Afrotropical *M. magretii*-group, with greatest similarity to *M. magrettii* André from Sudan. A key to the Afrotropical species of the *M. magretii*-group is presented. A brief review of the ant taxa with Afrotropical affinities in southwestern region of the Kingdom of Saudi Arabia is given.

Editor: Ulrike Gertrud Munderloh, University of Minnesota, United States of America

Funding: This project was supported by the NSTIP Strategic technologies program, grant number (12-ENV2484-02), in the Kingdom of Saudi Arabia. The funders had no role in study design, data collection and analysis, decision to publish, or preparation of the manuscript.

Competing Interests: The authors have declared that no competing interests exist.

* Email: antsharaf@gmail.com

Introduction

Since Smith [1] established the genus *Meranoplus*, 87 species and subspecies are currently included [2]. This genus is broadly distributed in the Old World tropics, ranging through the Afrotropical, Malagasy, Oriental, and Indo-Australian regions [3–5]. The genus is absent from the Palaearctic and Oceania regions, except for a single species, *M. levellei* Emery, 1883 from New Caledonia [6,7].

Meranoplus has been historically placed in the tribe Meranoplini, which also includes the genera *Dicroaspis* Emery, 1908 and *Calyptomyrmex* Emery, 1887 [8]. Recently, the tribe Meranoplini was restricted to two genera, *Meranoplus* [9] and a fossil genus *Parameranoplus* Wheeler, 1915 [10]. The *Meranoplus* fauna has been revised for most zoogeographical realms including the Afrotropical [3], Oriental [11,12], Australian [13–17], and Malagasy [5]. Among the previous regions, the Australian region is the most taxa rich in regard to the number of species [3,13].

Available information on the habitats and biology of *Meranoplus* species is limited. Most species nest directly in the soil, under stones, in rotten wood or in leaf litter [3,4,13]. African species have been reported to nest either in the ground with a crater type of entrance, or at the base of plants. Nests of one or two species are constructed among roots with workers ascending trees or low shrubs [3]. The majority of species are generalized omnivores [13], whereas others are considered seed harvesters, *e. g. M. diversus*-group [4,13,18].

The specialized morphology of the genus *Meranoplus* is related to a specialized behaviors, thanatosis, "playing dead" and and becoming cryptic. Several species when disturbed, fold their legs beneath the promesonotal shield and quickly accumulate dirt on body hairs and remain motionless in a fetal-like position [19,20].

A thorough diagnosis of the genus *Meranoplus* has been provided by Bolton [3,21]. The genus can be recognized by the combination of the following traits, masticatory margin of mandibles armed with four to five teeth; palp formula 5, 3; clypeus large, with median portion usually carinate at each side; antennal scrobes well-developed, usually long; antennae nine-segmented with a three segmented club; eyes large, located behind midlength of head, sometimes close to the posterior corners of head; pronotum and mesonotum fused into a plate or shield that extends posteriorly and laterally so the mesosomal sides, and usually also the propodeum, are concealed in dorsal view; lateral and/or posterior margins of promesonotal shield usually armed with spines, lobes, foliacious processes; petiole sessile usually cuneate in lateral view.

The *M. magrettii*-group can be distinguished by the following characters [3]: masticatory margin of mandibles armed with four or five teeth; mesosoma when seen from above with the propodeum concealed by the broad promesonotal shield;

propodeal spines present; petiole cuneate in lateral view with unarmed dorsal surface; postpetiole broad and nodiform.

The *M. magrettii*-group has only two species known from the Afrotropical region, *M. magrettii* a species that is broadly distributed in the region and inhabiting savannah, grassland and dry woodland, and *M. peringueyi* apparently confined to South Africa [3].

The southwestern mountainous region of the Kingdom of Saudi Arabia (KSA) is one of the most diverse in terms of species diversity and relative abundances of insects. Taxonomic and biogeographical studies have indicated that the insects of this region have strong affinities with the Afrotropical Region [21–33]. A recent and comprehensive study of the Afrotropical relationships of the region is by El-Hawagryi *et al.* [32]. These authors recorded 17 orders of insects representing 129 families and at least 582 species and subspecies from Al-Baha Province. Their biogeographic analysis of the species composition clearly revealed that this region has a close affinity with the insect fauna of the Afrotropical Region.

In the present study, the myrmicine ant genus *Meranoplus* is recorded for the first time from KSA by the new species *M. pulcher* sp. n. and represents a new generic record for the Arabian Peninsula. In addition, a synopsis of the similarity of the Afrotropical fauna with the southwestern region of KSA is presented based on available records of ants.

Materials and Methods

Study area

Shada Al Ala Mountain is a parallel extension of Hijaz Mountains to the west. This region is managed as a natural protectorate in southwestern of Al Baha Province, 20 km northwest of Al Mukhwah Governorate. The region has a substantial high diversity of wild plants including *Albizia lebbeck* (L.) Benth. (Fabaceae), *Solenostemon* Thonn.(Lamiaceae), *Juniperus procera* Hochst. ex Endlicher (Cupressaceae), *Santalum* L. (Santalaceae), *Pimpinella anisum* L. (Apiaceae), *Rhamnus frangula* L. (Rhamnaceae), *Opuntia ficus-indica* (L.) Mill. (Cactaceae), *Ricinus communis* L. (Euphorbiaceae), *Olea europaea* ssp. *africana* (Mill.) P. Green. (Oleaceae), *Prunus dulcisn* (Mill.) D.A.Webb (Rosaceae), *Maerua crassifolia* Forssk. (Capparceae), *Pandanus tectorius* Parkinson (Pandanaceae), *Panicum Turgidum* Forssk.

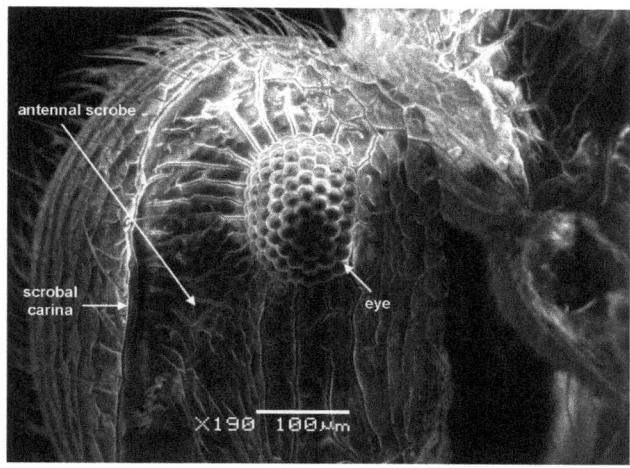

Figures 2. Head in profile.

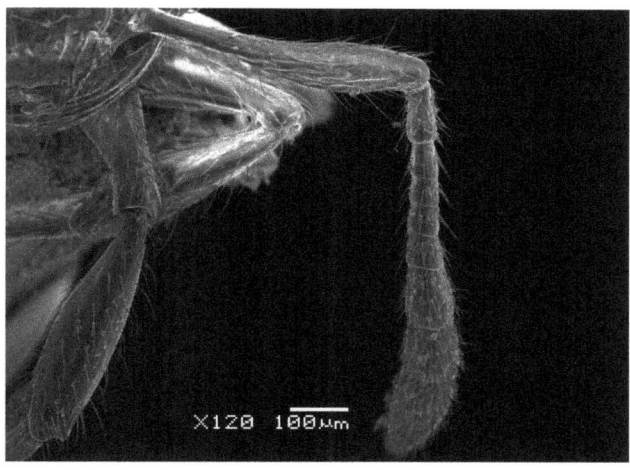

Figures 3. Antenna.

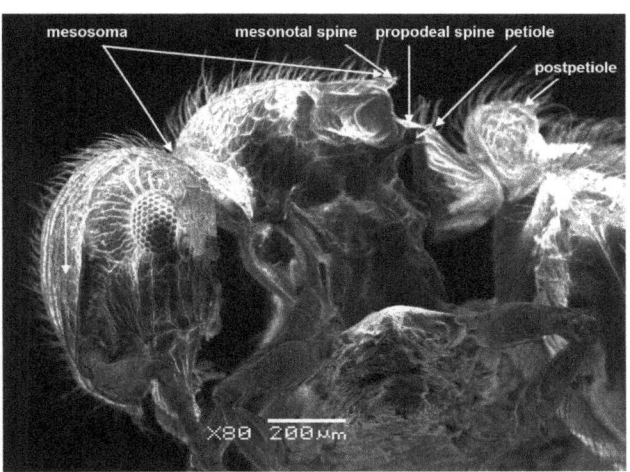

Figures 1. *Meranoplus pulcher* **sp. n., paratype worker, body in profile.**

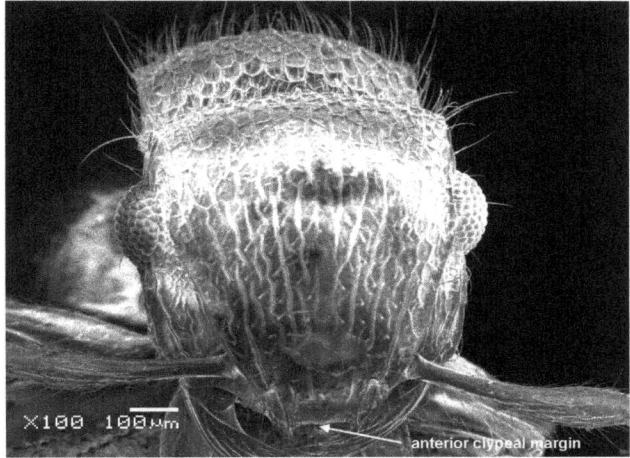

Figures 4. Head in full-face view.

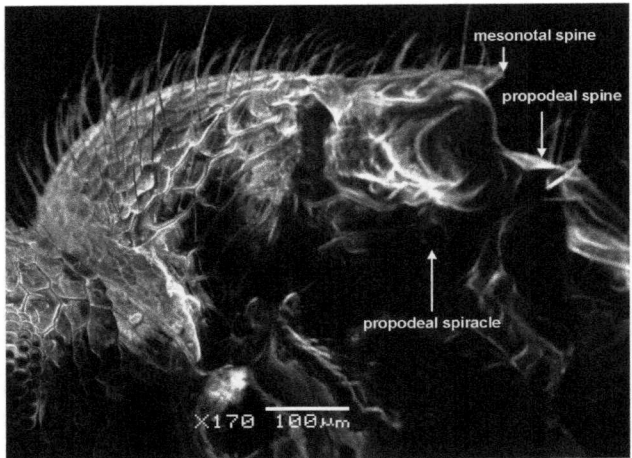

Figures 5. *Meranoplus pulcher* **sp. n., paratype worker, meso-soma in profile.**

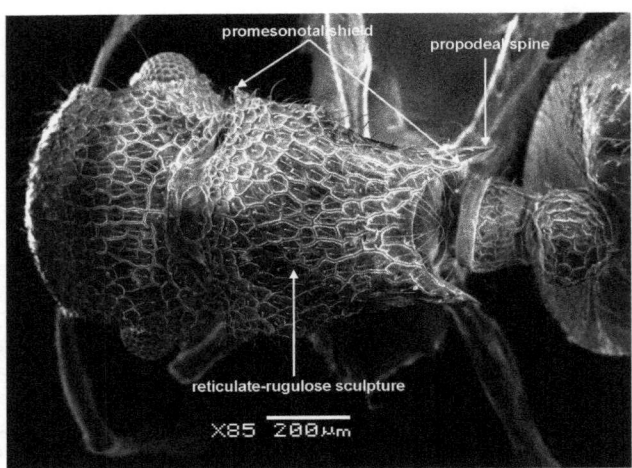

Figures 6. Mesosoma in dorsal view.

(Poaceae), *Coffea arabica* L. (Rubiaceae), *Opuntia ficus-indica* (L.) Mill. (Cactaceae), *Breonadia salicina* (Vahl) Hepper & J.R.I.Wood (Rubiaceae), *Haloxylon salicornicum* (Moq.) Bunge ex Boiss. (Chenopdiaceae), *Lycium shawii* Roem. & Schult (Solanaceae), *Cactus* (Cactaceae), and *Acacia* spp. (Memosaceae).

Sampling procedures

Materials listed in this work, the holotype and 29 paratype specimens were collected during an insect inventory of the southwestern region of KSA by using more than 900 pitfall traps. A single specimen of the new species was collected by an aspirator from a *Cactus* sp. All the materials is deposited in King Saud University Museum of Arthropods, College of Food and Agriculture Sciences (KSMA), King Saud University, Riyadh, Kingdom of Saudi Arabia, except a single paratype specimen in California Academy of Sciences (CASC), San Francisco, USA.

All measurements and indices are expressed in millimeters and follow the standards of Boudinot & Fisher [5].

Measurements

ATL: *Abdominal Tergum IV Length*. Maximum length of fourth abdominal tergum measured with anterior and posterior margins in same plane of focus.

ATW: *Abdominal Tergum IV Width*. Maximum width of fourth abdominal tergum with anterior, posterior, and lateral borders in same plane of focus.

CDD: *Clypeal Denticle Distance*. Distance between clypeal denticle apices, measured in full-face view.

CW: *Clypeus Width*. Distance between the apices of the frontal lobes across the clypeus.

EL: *Eye Length*. Maximum eye length in profile view.

EW: *Eye Width*. Maximum eye width in profile view.

HL: *Head Length*. Maximum length of head capsule, excluding mandibles, measured from anterior margin of clypeus to nuchal carina, with both in same plane of focus.

HLA: *Head Length, Anterior*. Distance between the anterior edges of the eyes to the mandible bases in full-face view.

HW: *Head Width*. Maximum width of head capsule behind the eyes, in full-face view.

PML: *Promesonotum Length*. Maximum length of promesonotum from posterior spine/denticle apices to anterolateral denticle apices; all four apices in same plane of focus. (= PMD, [15])

PPH: *Postpetiole Height*. Measured from sternal process base to postpetiole apex in lateral view.

PPL: *Postpetiole Length*. Measured from anterior to posterior inflections of postpetiole node in lateral view.

PWA: *Promesonotal Width, Anterior*. Maximum width of promesonotal shield between anterolateral denticle apices in dorsal view. (= PW, [15])

PWP: *Promesonotal Width, Posterior*. Distance between posterior-most promesonotal spine or denticle apices.

PTH: *Petiole Height*. Measured from petiole sternum to apex in lateral view.

PTL: *Petiole Length*. Measured from anterior to posterior inflections of petiole node.

SL: *Scape Length*. Maximum length of the scape excluding basal constriction.

SPL: *Propodeal Spine Length*. Workers: distance from inner posterior margin of propodeal spiracle to propodeal spine apex. Gynes: maximum propodeal spine length from basal inflection of spine, to spine apex.

WL: *Weber's Length*. Maximum diagonal length of mesosoma from anterior inflection of pronotum to posterolateral corner of the metapleuron or the metapleural lobes, whichever is most distant.

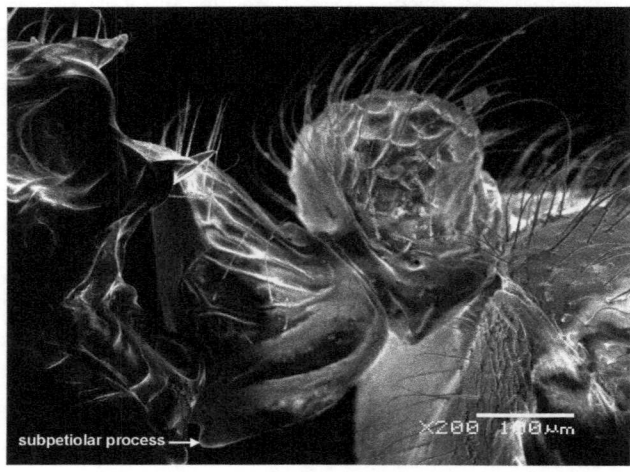

Figures 7. Waist in profile.

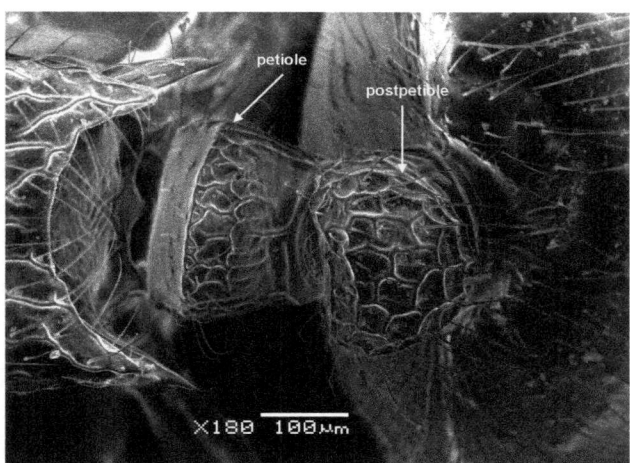

Figures 8. Waist in dorsal view.

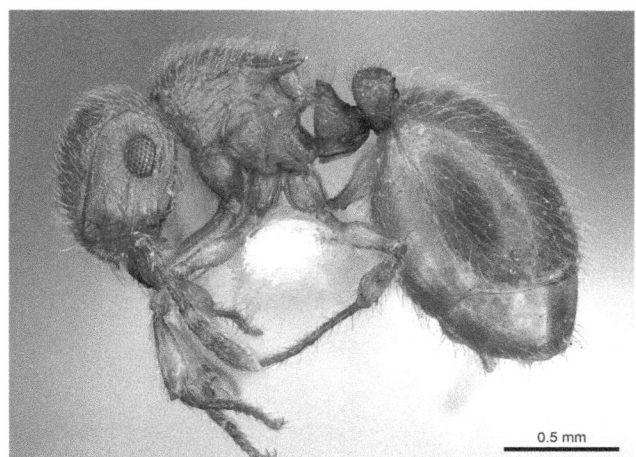

Figures 9. *Meranoplus pulcher* **sp. n., paratype worker, body in profile.**

Indices

CDI: *Clypeal Denticle Index*. CDD×100/CML
CI: *Cephalic Index*. HW×100/HL
CS: *Cephalic Size*. (HW+HL)/2
EYE: *Eye Index*. 100× (EL+EW)/CS
OMI: *Ocular-Mandibular Index*. EL×100/HLA
PMI: *Promesonotum Index 1*. PWA×100/PML (= PMI2, [15])
PPI: *Postpetiole Index*. PPL×100/PPH
PTI: *Petiole Index*. PTL×100/PTH
PWI: *Promesonotum Index 2*. PWP×100/PML
SEI: *Scape-Eye Index*. EL×100/SL
SI: *Scape Index*. SL×100/HW

Illustrations

Specimens were photographed by Michele Esposito (CASC) using a JVC KYF70B3CCD digital camera attached to a Leica M420 stereomicroscope. All digital images were processed using Auto-Montage (Syncroscopy, Division of Synoptics Ltd, USA) software. Images of the specimens are available in full color on www.antweb.org. The map was created by the ArcGIS 9.2 program, with the help of Prof. Mahmoud S. Abdel-Dayem (King Saud University).Specimens were examined and imaged using scanning electron microscope (SEM) (JSM-6380 LA) visualization to record morphological details of the new species. The JSM-6380 LA is a high-performance scanning electron microscope with a high resolution of 3.0 nm.

No specific permits were required for the described field studies or for the surveyed locations which are not privately-owned or protected in any way or do not have endangered or protected species.

Nomenclatural acts

The electronic edition of this article conforms to the require-ments of the amended International Code of Zoological Nomen-clature, and hence the new names contained herein are available under that Code from the electronic edition of this article. This published work and the nomenclatural acts it contains have been registered in ZooBank, the online registration system for the ICZN. The ZooBank LSIDs (Life Science Identifiers) can be resolved and the associated information viewed through any standard web browser by appending the LSID to the prefix "http://zoobank.org/". The LSID for this publication is: urn:

lsid:zoobank.org:pub:80A78374-53CF-4A19-90CB-4A392E62699 9. The electronic edition of this work was published in a journal with an ISSN, and has been archived and is available from the following digital repositories: PubMed Central, LOCKSS.

Results

Meranoplus pulcher SHARAF sp. n. (Figs. 1–11)

urn:lsid:zoobank.org:act:A5035370-971E-451B-B15E-23F8BBC E5847.

Holotype worker

SAUDI ARABIA, Al-Baha Province, Shada Al Ala, 19°51.066'N, 41°18.037'E, 1325 m, 23.IV.2014, P. T. (*Al Dhafer et al. Leg.*), deposited in KSMA, King Saud Museum of Arthropods, College of Food and Agriculture Sciences, King Saud University, Riyadh, Kingdom of Saudi Arabia.

Paratypes workers

All the following paratype specimens are deposited in KSMA, 3 workers, same locality and data as the holotype; 1 worker with

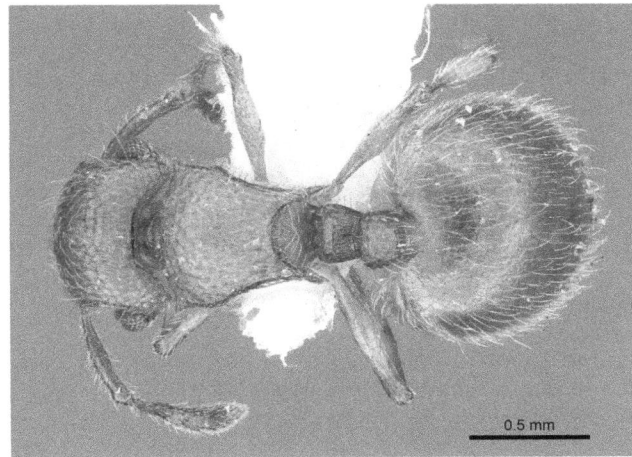

Figures 10. Body in dorsal view.

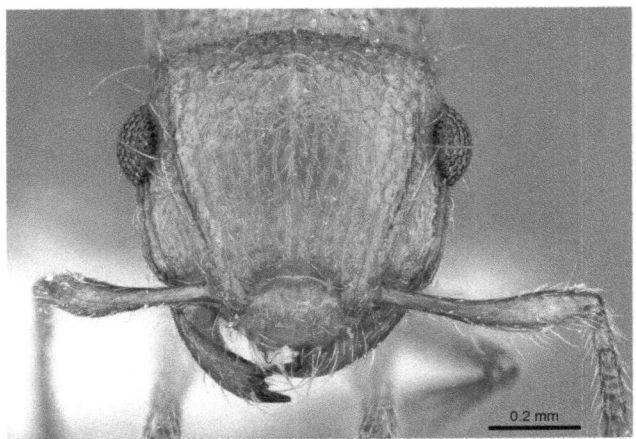

Figures 11. Head in full-face view. CASENT 0914336, Photographer: Michele Esposito, copyright www.antweb.org.

Figures 12. *Meranoplus margettii,* **body in profile.**

same data as the holotype except the collecting data 15.II.2014; 1 worker, SAUDI ARABIA, Asir Province, Raydah, 18°11.749'N, 42°23.345'E, 1614 m, 28.IV.2014, P.T. *(Al Dhafer et al. Leg.),* 3 workers, Shada Al Ala, 19°50.575'N, 41°18.691'E, 1666 m, 23.VIII.2014, P. T. *(Al Dhafer et al. Leg.);* 9 workers, Shada Al Ala, 19°50.411'N, 41°18.686'E, 1611 m, 23.VIII.2014, P. T. *(Al Dhafer et al. Leg.);* 3 workers, Shada Al Ala, 19°50.329'N, 41°18.604'E, 1563 m, 23.VIII.2014, P. T. *(Al Dhafer et al. Leg.);* 5 workers, Shada Al Ala, 19°50.710'N, 41°18.267'E, 1474 m, 23.VIII.2014, P. T. *(Al Dhafer et al. Leg.);* 5 workers, Shada Al Ala, 19°51.066'N, 41°18.037'E, 1325 m, 23.VIII.2014, P. T. *(Al Dhafer et al. Leg.);* 4 workers, Asir Province, Raydah, 18°11.618'N, 42°23.420'E, 1772 m, 26.VIII.2014, P. T. *(Al Dhafer et al. Leg.);* 1 workers, Asir Province, Raydah, 18°11.749'N, 42°23.345'E, 1614 m, 26.VIII.2014, P. T. *(Al Dhafer et al. Leg.);* unique specimen identifier CASENT 0914336, in (CASC) California Academy of Science Collection, San Francisco, California, USA.

Worker measurements

Maximum and minimum based on all specimens, n = 5, (holotype): TL 3.20-3.70 (3.27), HL 0.77–0.87 (0.80), HW 0.67–0.82 (0.72), HLA 0.25–0.30 (0.25), CW 0.22–0.27 (0.30), CDD 0.12–0.15 (0.12), SL 0.47–0.62 (0.60), EL 0.17–0.22 (0.17), EW 0.12–0.15 (0.15), PML 0.40–0.52 (0.47), PWA 0.62–0.75 (0.70), PWP 0.37–0.47 (0.45), SPL 0.17–0.22 (0.22), WL 0.75–0.87 (0.77), PTL 0.12–0.17 (0.20), PTH 0.30–0.42 (0.37), PPL 0.15–0.22 (0.22), PPH 0.25–0.35 (0.32), ATW 1.02–1.22, (1.12) ATL 0.97–1.15 (1.05), CI 87–94 (90), SI 67–82 (83), OMI 63–80 (68), CDI 0.44--0.68 (40), SEI 31–43 (28), PMI 144–155 (149), PPI 60–80 (69), PTI 40–46 (54), PWI 82–93 (96), CS 0.72–0.84 (0.76), EYE 38–47 (42) (n = 5).

Diagnosis

Although *M. pulcher* sp. n. is superficially similar to *M. magrettii* (Figs. 12, 13), it can be readily distinguished by the following contrasting characters: Colour: *M. pulcher* is yellow, *M. magrettii* is light to dark brown; anterior clypeal margin: distinctly concave in *M. pulcher,* more or less flat to shallowly concave in *M. magrettii;* subpetiolar process: in *M. pulcher* short and triangular, in *M. magrettii* the process is more developed forming a short finger or a less developed process; petiolar sculpture: posterior face

of petiolar node areolate-rugose in *M. pulcher* and smooth in *M. magrettii;* sculpture of first gastral tergite: superficially and finely shagreenate in *M. pulcher,* in *M. magrettii* the sculpture of first gastral tergite varies from dense shagreenate or reticulate punctate.

Description

Head. Head slightly longer than broad with convex sides and straight posterior margin; anterior clypeal margin distinctly concave with well-developed clypeal carinae; mandibles armed with four teeth; eyes relatively large (EL 0.25–0.26 x HW; EYE 38–47) with 12 ommatidia in the longest row; scapes when laid back from their insertions just reach posterior margin of eyes; scrobal carinae well-developed.

Mesosoma. Anterior pronotal corners armed with a pair of short triangular teeth; promesonotal shield distinctly broader than long (PMI 144–155) widening behind pronotum; promesonotal suture absent; posterior corners of mesonotum armed with a pair of sharp spines; posterior mesonotal margin between spines strongly concave and without secondary armament; propodeal spines long and sharp originating at level of propodael spiracles and curved upwards; propodeal lobes well-developed.

Waist. Petiole cuneate in profile, sessile, with a broad anterior margin and a narrow acute dorsum; petiolar and postpetiolar anteroventral processes present; postpetiole nodiform, subrectangular in profile, taller than broad (PPI 60–80).

Sculpture. Mandibles longitudinally striated; cephalic dorsum densely and finely longitudinally rugulose, posterior margin

Figures 13. *Meranoplus margettii,* **body in dorsal view.**

Figures 14. Type locality of *Meranoplus pulcher* **sp. n., Shada Al Ala Protectorate, Saudi Arabia.**

areolate-rugose; promesonotal shield, posterior face of petiolar node and postpetiole dorsum reticulate rugulose, anterior petiolar face smooth and sides transversally rugulose; first gastral tergite finely and densely shagreenate.

Pilosity. All body surface covered with fine, pale, profuse hairs.

Colour. Colour unicolorous yellow, in some specimens, postpetiole and posterior margin of first gastral tergite brownish. The six examined specimens showed a clear size variation.

Etymology

The species name is derived from the Greek word "pulcher" that means "beautiful" referring to the attractive appearance of this ant species.

Habitat

Twenty five workers of the new species were collected from Al-Baha Province, Shada Al Ala Protectorate (Fig. 14) and six workers from Raydah Protectorate. Both collections were from pitfall traps placed next to *Acacia* trees. The soil was extremely dry with abundant dry seeds of shrubs. Despite several hours of

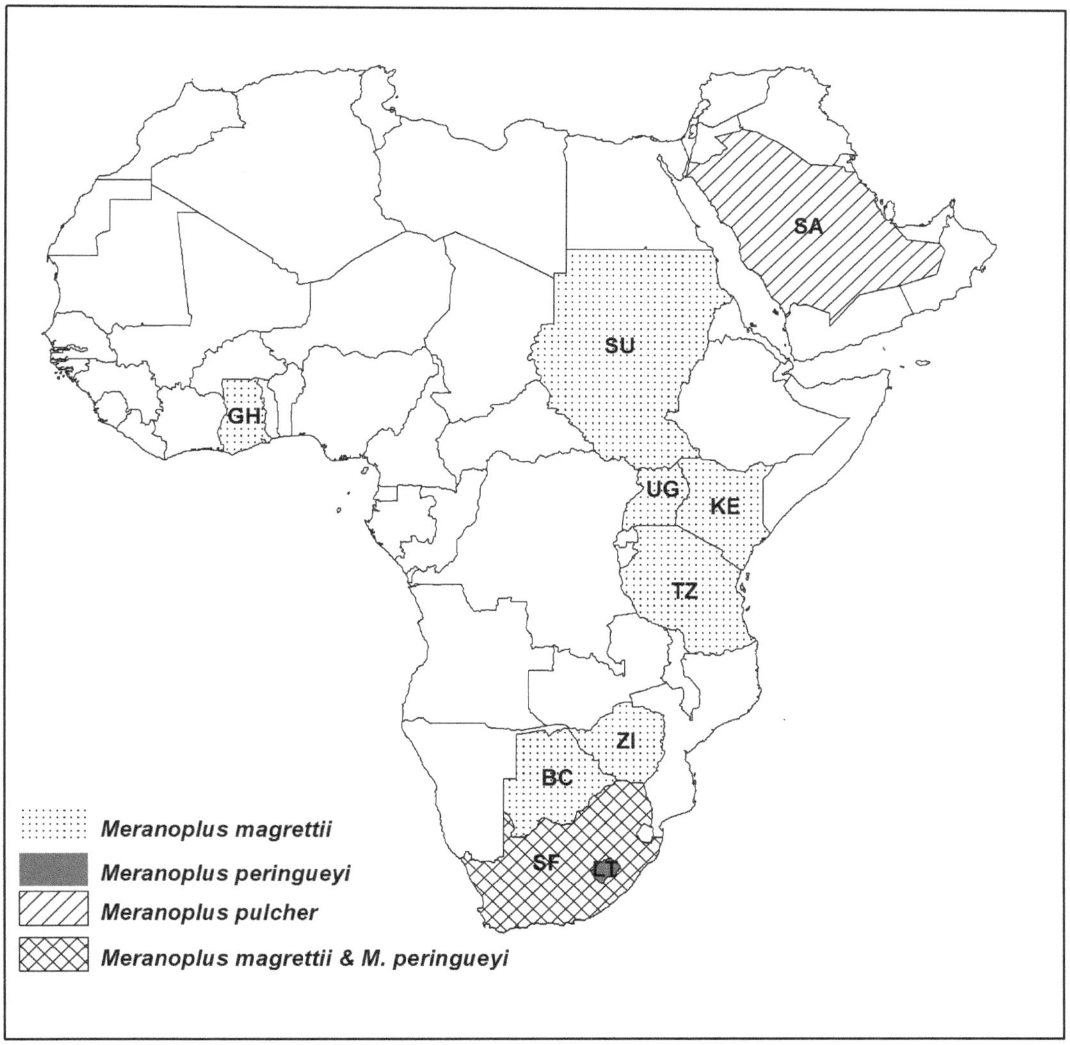

Meranoplus magrettii
Meranoplus peringueyi
Meranoplus pulcher
Meranoplus magrettii & M. peringueyi

Figures 15. Geographic distribution of species of the *Meranoplus* **margettii-group.**

Table 1. Ants (Hymenoptera: Formicidae) species recorded from southwestern Kingdom of Saudi Arabia with a known Afrotropical distribution.

Species	Locality in KSA	Date	Type locality
Anochetus traegaordhi Mayr, 1904	Fayfa	30.iii.1983	Sudan
Camponotus empedocles Emery, 1920	Anamas	18.ix.1983	Zimbabwe
	Wadi Azizah	18.ix.1983	Zimbabwe
	Wadi al Amar	18.ix.1983	Zimbabwe
	Wadi Majarish	22.iii.1982	Zimbabwe
	Tanuma	8.iv.1983	Zimbabwe
Camponotus ilgii Forel	Fayfa	28-30.iii.1983	Ethiopia
Camponotus flavomarginatus Mayr, 1862	Anamas	8.iv.1983	Ghana
	Sawdah Mountain	9.iv.1983	Ghana
Camponotus sericeus (Fabricius, 1798)	Wadi Majarish	7.i.83	Senegal
Camponotus thales Forel, 1910	Anamas	8. iv.83	Lesotho
	Sawdah Mountain	9. iv.83	Lesotho
Cataglyphis abyssinica (Forel, 1904)	Abu Arish	3.iv.1983	Ethiopia
Crematogaster senegalensis Roger, 1863	Sug al Ahad (Asir)	26.iii.83	Senegal
Crematogaster luctans Forel, 1907	Fayfa	29.iii.1983	Kenya
	Fayfa	6.iv.2013	Kenya
Lepisiota incisa (Forel, 1913)	Anamas	8. iv.83	Democratic Republic of Congo
Lepisiota obtusa (Emery, 1901)	Abu Arish	25.iii.83	Ethiopia
Leptogenys maxillosa (Smith, 1858)	Fayfa	30.iii.1983	Mauritius
Melissotarsus emeryi Forel, 1907	Fayfa	29.iii.1983	Ethiopia
Monomorium afrum André, 1884	Abha-Najran RD	10.iv.1983	Sudan
Monomorium schultzei Forel, 1910	Wadi Majarish	3.i.1983	South Africa
Platythyrea modesta Emery, 1899	Fayfa	30.iii.1983	Cameroun
	Raydah (Asir)	26.viii.2014	Cameroun
Polyrhachis viscosa Smith, 1858	Fayfa	31.iii.1983	South Africa
Strumigenys arnoldi Forel, 1913	Dhi Ayn (Al Baha)	20.ix.2011	Zimbabwe
Syllophopsis cryptobium Santschi, 1921	Wadi Bagara	10.xi.2012	Democratic Republic of Congo
Tetramorium doriae Emery, 1881	Wadi Elzaraeb (Al Baha)	15.v.2010	Ethiopia
Tetramorium khyarum Bolton, 1980	Dhi Ayn (Al Baha)	23.ix.2011	Nigeria
Tetramorium sericeiventre Emery, 1877	Al Tawlah	8.iv.1983	Ethiopia
	Anamas	8.iv.1983	Ethiopia

Table 2. Formicidae species described from Kingdom of Saudi Arabia with Afrotropical congeners.

Species	Type Locality	Coordinates	Altitude	Date
Tetramorium amalae Sharaf & Aldawood, 2012	Al Bahah, Amadan Forest	20.20000°N 41.21667°E	1881 m	19.V.2010
Tapinoma wilsoni Sharaf & Aldawood, 2012	Al Baha, Al Sarawat Mountains, Dhi Ayn	19.92972°N 41.44278°E	741 m	15.v.2011
Aenictus arabicus Sharaf & Aldawood, 2012	Al Baha-Mukhwah Aqaba RD	20.00000°N '41.43758°E	1300 m	19.IV.2012
Monomorium dryhimi Sharaf & Aldawood, 2011	Al Bahah province, Amadan forest	20.20000°N 41.21667°E	1881 m	19.V.2010
Monomorium kondratieffi Sharaf & Aldawood, 2013	Al Bahah province, AlUrdiya Governorate, Wadi Qonouna	19.42936°N 41.60503°E	353 m	12.v.2011
M. sarawatensis Sharaf & Aldawood, 2013	Al Baha-Mukhwah Aqaba RD	20.00000°N '41.43758°E	1300 m	19.IV.2012

observing the nest no additional specimens were found. Pitfall trapping is apparently an efficient method for collecting this group of ants.

Key to the Afrotropical species of *Meranoplus magrettii*-group based on workers

In the key to Afrotropical species (Bolton [3]:48), *Meranoplus pulcher* sp. n. will key to couplet 7 along with *peringueyi* and *magrettii*. Couplet 7 is modified here to separate the three species of the *M. magrettii*-group.

7. Mandibles armed with 5 teeth   *peringueyi* Emery
- Mandible armed with 4 teeth   8
8 Colour moderately dark brown, Subpetiolar process more developed, in the form of a short finger (Fig. 12), Posterior face of petiolar node smooth (Fig. 13). (Ghana, Sudan, Uganda, Kenya, Tanzania, Zimbabwe, Botswana, South Africa, Fig. 15)   *magrettii* André
- Colour yellowish, Subpetiolar process present but short (Fig. 7), Posterior face of petiolar node areolate-rugose. (Fig. 8) (Saudi Arabia, Fig. 15)   *pulcher* sp. n.

Zoogeography of the southwestern region of Kingdom of Saudi Arabia

The KSA is located at the interchange of three major biogeographical realms, the Palaearctic, Afrotropical and Oriental. Much mixing of these faunal elements has occurred. Geologically, the Ethiopian and Arabian Peninsula highlands and mountains separated approximately 13 mybp producing the Great Rift Valley through a rifting process as the African continental crust separated [34,35]. Biogeographically, the southwestern region of the KSA belongs to the Afrotropical region [21–23,25–33].

Studies of other taxonomic groups of insects have revealed similar results, for example, the Scythrididae (flower moths) (Lepidoptera: Gelechioidea) were treated for the Palaearctic region [36] and the Afrotropical species *Scythris albocanella* Bengtsson 2002 was recorded from various localities in southwestern region of KSA. In addition, Marnert *et al.* [37] studied the pseudoscorpion arachnids (Pseudoscorpiones) of the region and concluded that the southwestern region of KSA has a clear faunal similarity with the Afrotropical region.

At least ten pantropical ant genera have been recorded from this region of KSA including *Anochetus* Mayr, 1861, *Cryptopone* Emery, 1893, *Cerapachys* Smith, 1857 *Dorylus* Fabricius, 1793, *Hypoponera* Santschi, 1938, *Leptogenys* Roger, 1861, *Melissotarsus* Emery, 1877, *Platythyrea* Roger, 1863, *Polyrhachis* F. Smith, 1857 and *Tetraponera* F. Smith, 1852. Strong biogeographical affinities are with the Afrotropical Region [38]. The sole faunistic work carried out for the knowledge of the ants of KSA [38] recorded 89 species from southwestern region, 18 of which having an Afrotropical distribution (table 1). Despite limited amount of specimens collected from this region mentioned in the above work, preliminary conclusions supported an Afrotropical faunal relationship of the region. Recently, six new species were described from the region [28–33], taxa more closely related to Afrotropical congeners (Table 2). An approximate estimate of the relative percentage of the Afrotropical faunal similarity of the region is

31%. Taking into account the unidentified materials accumulated over the last ten years from the region by the authors and also the vast area of the region, some of which has not been surveyed, it is expected this number will increase.

Discussion

Meranoplus pulcher sp. n. is the first member of the genus recorded from KSA and from the vast Arabian Peninsula. Following Bolton [3], it belongs to *M. magrettii*-group and cannot be identified using the available keys to species of Afrotropical [3], Malagasy [5], Oriental [11] or Australian [13,17] regions. *Meranoplus pulcher* sp. n. is similar to *M. magrettii* André from Sudan to which it will key to in Bolton [3], sharing the following characters: mandibles striate and armed with four teeth, anterior pronotal corners armed with a pair of short triangular teeth, promesonotal shield narrowing behind pronotum, posterior corners of mesonotum armed with a pair of short spines, posterior mesonotal margin concave and unarmed, petiole cuneate in profile and postpetiole nodiform.

The habitats of *M. pulcher* sp. n. and *M. magrettii* apparently are not similar. The latter species is restricted to sub-Saharan Africa and has been collected from savannah, open-woodland and grassland habitats [3]. *Meranoplus pulcher* sp. n.is apparently restricted to juniper woodlands of southwestern mountains of KSA. The author (MRS) has made extensive collections of ants from the southwestern KSA. Typical Afrotropical ant genera mentioned above (e.g. *Strumigenys*, *Anochetus*, *Pachycondyla*, *Cerapachys*, *Dorylus* etc.) were commonly encountered. For example, the Afrotropical species, *S. arnoldi* Forel was reported from Al-Baha Province [39] providing evidence of faunal similarities with the Afrotropical Region. Additional future collections from this area of KSA will no doubt provide further evidence of this biographical connection.

The record of *M. pulcher* sp. n. of the Afrotropical *margettii*-group is an additional evidence of the Afrotropical faunal similarities of the southwestern mountains of KSA which is consistent with other faunal influences for the Region [21–23,25–33].

Acknowledgments

We are grateful to Prince Bandar Bin Saud Al Saud, Head of the Saudi National Commission for Wildlife Conservation and Development for the appreciated support during the study. The authors are grateful to Barry Bolton and Brendon Boudinot for valuable suggestions that improved the manuscript. Special thanks to Brian Fisher and Michele Esposito (California Academy of Sciences, San Francisco, USA) for assistance in photographing the new species, to Loutfy El-Juhany for identifying plants of the studies area, and to Mahmoud Abdel-Dayem for making the map. The authors are indebted to Boris Kondratieff (Colorado State University) for his valuable comments and critical editing, actually, without his help this work could not have been completed.

Author Contributions

Conceived and designed the experiments: MRS HMAD ASA. Performed the experiments: MRS HMAD ASA. Analyzed the data: MRS HMAD ASA. Contributed reagents/materials/analysis tools: MRS HMAD ASA. Wrote the paper: MRS HMAD ASA.

References

1. Smith F (1853) Monograph of the genus *Cryptocerus*, belonging to the group Cryptoceridae - family Myrmicidae - division Hymenoptera Heterogyna. Transactions of the Entomological Society of London (2) 2 (1854): 213–228.

2. Bolton B (2013) An online catalog of the ants of the World. Version 1 January 2013 Available from: http://www.antcat.org/catalog/(retrieved on 5 May 2014).

3. Bolton B (1981) A revision of the ant genera *Meranoplus* F. Smith, *Dicroaspis* Emery and *Calyptomyrmex* Emery (Hymenoptera: Formicidae) in the Ethiopian

zoogeographical region. Bulletin of the British Museum of Natural History, 42: 43–81.

4. Brown WL Jr (2000) Diversity of ants. In: Agosti et al. (Eds) Ants. Standard methods for measuring and monitoring biodiversity. Biological diversity hand book series. Smithsonian Institution Press, Washington and London, 280 pp.

5. Boudinot BE, Fisher BL (2013) A taxonomic revision of the *Meranoplus* F. Smith of Madagascar (Hymenoptera: Formicidae: Myrmicinae) with keys to species and diagnosis of the males. Zootaxa, 3635 (4): 301–339.

6. Wheeler WM (1935) Check list of the ants of Oceania. Bernice P. Bishop Museum Occasional Papers, 11: 2–56.

7. Fisher BL (2010) Biogeography. *In*: Lach, L., Parr, C.L. & Abbot, K.L. (Eds), *Ant Ecology*. Oxford University Press, Oxford, pp. 402.

8. Kugler C (1978) A comparative study of the myrmicine sting apparatus. Studia Entomologica 20: 413–548.

9. Bolton B (2003) Synopsis and classification of Formicidae. Memoirs of the American Entomological Institute 71: 370 pp.

10. Engel MS (2001) A monograph of the Baltic amber bees and evolution of the Apoidea (Hymenoptera). Bulletin of the American Museum of Natural History, 259: 1–192.

11. Schödl S (1998) Taxonomic revision of Oriental *Meranoplus* F. Smith, 1853 (Insecta: Hymenoptera: Formicidae: Myrmicinae). Annalen des Naturhistorischen Museums in Wien, 100: 361–394.

12. Schödl S (1999) Description of *Meranoplus birmanus* sp. nov. from Myanmar, and the first record of *M. bicolor* from Laos (Hymenoptera: Formicidae). Entomological Problems, 30: 61–65.

13. Anderson AN (2006) A systematic overview of Australian species of the myrmicine ant genus *Meranoplus* F. Smith, 1853 (Hymenoptera: Formicidae). Myrmecologische Nachrichten, 8: 157–170.

14. Schödl S (2004) On the taxonomy of *Meranoplus puryi* Forel, 1902 and *Meranoplus puryi curvispina* Forel, 1910 (Insecta: Hymenoptera: Formicidae). Annalen des Naturhistorischen Museums in Wien, 105: 349–360.

15. Schödl S (2007) Revision of Australian *Meranoplus*: the *Meranoplus diversus* group. *In*: Snelling R, Fisher B, Ward P (Eds), *Advances in Ant Systematics: Homage to E.O. Wilson*. The American Entomological Institute, Gainesville, Florida, pp. 370–424.

16. Taylor RW (1990) The nomenclature and distribution of some Australian and New Caledonian ants of the genus *Meranoplus* Fr. Smith (Hymenoptera: Formicidae: Myrmicinae). General and Applied Entomology, 22: 31–40.

17. Taylor RW (2006) Ants of the genus *Meranoplus* F. Smith, 1853 (Hymenoptera: Formicidae): three new species and others from northeastern Australian rainforests. Myrmecologische Nachrichten, 8: 21–29.

18. Anderson AN, Azcárate FM, Cowie ID (2000) Seed selection by an exceptionally rich community of harvester ants in the Australian seasonal tropics. Journal of Animal Ecology, 69: 975–984. http://dx.doi.org/10.1046/j.1365-2656.2000.00452.x

19. Dornhaus A, Powell S (2010) Foraging and defense strategies. *In*: Lach L, Parr CL, Abbot KL (Eds.), *Ant Ecology*. Oxford University Press, Oxford, 402 pp.

20. Hölldobler B (1988) Chemical communication in *Meranoplus* (Hymenoptera: Formicidae). Psyche, 95: 139–151. http://dx.doi.org/10.1155/1988/74829

21. Bolton B (1994) Identification Guide to the Ant Genera of the World: Cambridge, Mass, 222 pp.

22. Eig A (1938) Taxonomic studies on the Oriental species of the genus *Anthemis*. Palestine Journal of Botany, Jerusalem, 1: 161–224.

23. Zohary M (1973) Geobotanical foundations of the Middle East. Vols. 1–2. G. Fischer, Stuttgart, Swets & Zeitlinger, Amsterdam, 738 pp.

24. Robertson HG (2000) Afrotropical ants (Hymenoptera: Formicidae): taxonomic progress and estimation of species richness. Journal of Hymenoptera Research. 9: 71–84.

25. Lehrer AZ, Abou-Zied EM (2008) Une espèce nouvelle du genre Engelisca Rohdendorf de la faune d'Arabie Saoudite (Diptera, Sarcophagidae). Fragmenta Dipterologica, 14: 1–4.

26. Doha SA (2009) Phlebotomine sand flies (Diptera, Psychodidae) in different localities of Al-Baha province, Saudi Arabia. Egyptian Academic Journal of Biological Sciences, 1: 31–37.

27. ALdawood AS, Sharaf MR, Taylor B (2011) First record of the myrmicine ant genus *Carebara* Westwood, 1840 (Hymenoptera, Formicidae) from Saudi Arabia with description of a new species *C. abuhurayri* sp. n. ZooKeys, 92: 61–69. http://dx.doi.org/10.3897/zookeys.92.770

28. Sharaf MR, Aldawood AS (2011) *Monomorium dryhimi* sp. n., a new ant species (Hymenoptera, Formicidae) of the *M. monomorium* group from Saudi Arabia, with a revised key to the Arabian species of the group. ZooKeys, 106: 47–54. http://dx.doi.org/10.3897/zookeys.106.1390

29. Sharaf MR, Aldawood AS (2012) A new ant species of the genus *Tetramorium* Mayr, 1855 (Hymenoptera, Formicidae) from Saudi Arabia, including a revised key to the Arabian species. PLoS ONE, 7 (2), e30811. http://dx.doi.org/10.1371/journal.pone.0030811

30. Sharaf MR, Aldawood AS, El-Hawagry MS (2012a) A new ant species of the genus *Tapinoma* (Hymenoptera, Formicidae) from Saudi Arabia with a key to the Arabian species. ZooKeys, 212: 35–43. http://dx.doi.org/10.3897/zookeys.212.3325

31. Sharaf MR, Aldawood AS, El-Hawagry MS (2012b) First record of the ant subfamily Aenictinae (Hymenoptera, Formicidae) from Saudi Arabia, with the description of a new species. ZooKeys, 228: 39–49. http://dx.doi.org/10.3897/zookeys.228.3559

32. El-Hawagryi MS, Khalil MW, Sharaf MR, Fadl HH, Aldawood AS (2013) A preliminary study on the insect fauna of Al-Baha Province, Saudi Arabia, with descriptions of two new species. Zookeys, 274: 1–88. http://dx.doi.org/10.3897/zookeys.274.4529

33. Sharaf MR, Aldawood AS (2013) First occurrence of the *Monomorium hildebrandti*-group (Hymenoptera: Formicidae), in the Arabian Peninsula, with description of a new species *M. kondratieffi* n. sp. Proceedings of the Entomological Society of Washington, 115 (1): 75–84.

34. Davison I, Al-Kadasi M, Al-Khirbash S, Al-Subbary AK, Baker J, et al. (1994) Geological evolution of the southeastern Red Sea Rift margin, Republic of Yemen. Geological Society of America Bulletin, 106: 1474–1493.

35. Bosworth W, Huchon P, Mcclay K (2005) The red sea and Gulf of Aden basins. Journal of African Earth Sciences, 43: 334–378.

36. D'entrèves PP, Roggero A (2004) Four new species, a new synonymy and some new records of Scythris Hübner, [1825] (Gelechioidea: Scythrididae). Nota lepidopterologica, 26 (3/4): 153–164.

37. Mahnert V, Sharaf MR, Aldawood AS (2014) Further records of Pseudoscorpions (Arachnida, Pseudoscorpions) from Saudi Arabia. Zootaxa, 3764 (3): 387–393.

38. Collingwood CA (1985) Hymenoptera: Fam. Formicidae of Saudi Arabia. Fauna of Saudi Arabia 7: 230–301.

39. Sharaf MR, Fisher BL, Aldawood AS (2014) Notes on Ants of the genus *Strumigenys* F. Smith, 1860 (Hymenoptera, Formicidae) in the Arabian Peninsula, with a key to species. Sociobiology 61 (3): x–xx.

14

Distance-Decay and Taxa-Area Relationships for Bacteria, Archaea and Methanogenic Archaea in a Tropical Lake Sediment

Davi Pedroni Barreto[1], Ralf Conrad[3], Melanie Klose[3], Peter Claus[3], Alex Enrich-Prast[2,4]*

1 Instituto de Microbiologia Prof. Paulo de Góes, Universidade Federal do Rio de Janeiro, Rio de Janeiro, Brazil, **2** Instituto de Biologia, Universidade Federal do Rio de Janeiro, Rio de Janeiro, Brazil, **3** Max-Planck Institute for Terrestrial Microbiology, Marburg, Hessen, Germany, **4** Department of Water and Environmental Studies, Linköping University, Linköping, Sweden

Abstract

The study of of the distribution of microorganisms through space (and time) allows evaluation of biogeographic patterns, like the species-area index (z). Due to their high dispersal ability, high reproduction rates and low rates of extinction microorganisms tend to be widely distributed, and they are thought to be virtually cosmopolitan and selected primarily by environmental factors. Recent studies have shown that, despite these characteristics, microorganisms may behave like larger organisms and exhibit geographical distribution. In this study, we searched patterns of spatial diversity distribution of bacteria and archaea in a contiguous environment. We collected 26 samples of a lake sediment, distributed in a nested grid, with distances between samples ranging from 0.01 m to 1000 m. The samples were analyzed using T-RFLP (Terminal restriction fragment length polymorphism) targeting *mcrA* (coding for a subunit of methyl-coenzyme M reductase) and the genes of Archaeal and Bacterial 16S rRNA. From the qualitative and quantitative results (relative abundance of operational taxonomic units) we calculated the similarity index for each pair to evaluate the taxa-area and distance decay relationship slopes by linear regression. All results were significant, with *mcrA* genes showing the highest slope, followed by Archaeal and Bacterial 16S rRNA genes. We showed that the microorganisms of a methanogenic community, that is active in a contiguous environment, display spatial distribution and a taxa-area relationship.

Editor: Jonas Waldenström, Linneaus University, Sweden

Funding: This study was funded by the National Council of Scientific and Technological Development (CNPq), trough the project 477260/2011-0. (Brazil - www.cnpq.br) and by the Max Planck Society (Germany - www.mpg.de). The funders had no role in study design, data collection and analysis, decision to publish, or preparation of the manuscript.

Competing Interests: The authors have declared that no competing interests exist.

* Email: aeprast@biologia.ufrj.br

Introduction

The biogeography concept is defined as the study of the distribution and the range of living organisms across space and time. Most studies in this field were traditionally performed targeting macro-organisms such as plants and animals [1]. Since the development of molecular tools, the concepts of biogeography started to be also studied in microorganisms [2–5].

A long-held concept in microbial ecology is that microorganisms are ubiquitously distributed and can be found in any habitat with favorable environmental conditions. This concept was introduced by Martinus Willem Beijerinck and concisely summarized by Lourens Gerhard Marinus Baas Becking in the quote, "Everything is everywhere, the environment selects" [6]. This statement is based on some traits of the microorganisms, such as the small size of individuals and the consequent ease of their dispersal across long distances, high rates of reproduction, short generation times, and large population sizes, leading to a small chance of local extinction.

Free-living eukaryotic microorganisms are often described as occurring ubiquitously. When they are not dominant in some specific environment, it is possible to reanimate the cryptic diversity by changing the environmental conditions *in vitro* [3,7].

A study showed that it is possible to find nearly 80% of all known species of the flagellate genus *Paraphysomonas* in just a small sample of sediment [8], meaning that the global diversity of this genus is well represented by its local diversity. This observation is mostly explained by the high dispersal rate of the flagellates (due to their low size), their extremely short generation times (leading to a low rate of extinction) and also their capacity to generate resistant forms when the environmental conditions are unfavorable [8]. The authors suggested that if eukaryotic microorganisms were ubiquitous, then prokaryotic microorganisms should be ubiquitous as well, since they have an even smaller size and larger populations. Indeed, some studies on prokaryotes suggested global distribution, for example, psychrophilic polar bacteria were found at both the South and the North poles [9].

Recent studies, however, showed that the distribution of microorganisms is not random, and that biogeography patterns of distribution are established [10,11]. For example, the genetic distances between populations of microorganisms were shown to increase with geographic distance, which might represent a speciation process driven by the geographic isolation of the microorganisms [12]. Other studies were able to identify endemic microorganisms, and true geographic isolation in extreme environments like hot springs, pristine soils, salt lakes, and hot

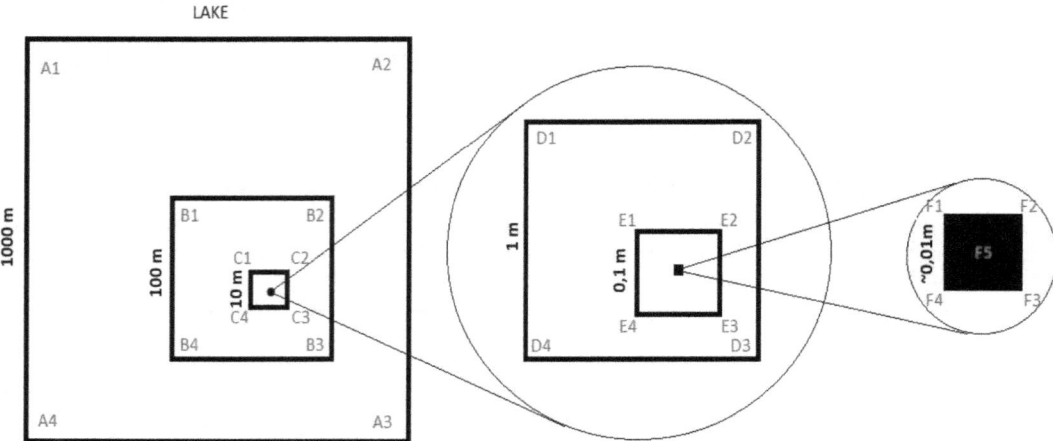

Figure 1. Sampling grid. Designed inside the Lago Negro area, A1-F5 represents the sampling points. (Adapted from (Horner-Devine 2004)).

and cold deserts around the world, all being strong evidence for non-cosmopolitan distribution [10,13–17]. The main difficulty is to evaluate the factors determining the geographical distribution, whether historic evolutionary events (geographic barriers for example) or contemporary ecological environmental factors. [5].

The taxa-area relationship is one of the most consistent general patterns in ecology and well described for macro-organisms [1]. It is represented by the equation $S = c\,A^z$, where S is the number of species, A is the area sampled, c is a constant that is empirically derived from the taxon and the specific location studied, and the exponent z, the power law index (i.e. z-value), represents the rate of the increasing number of species along the increasing sampling area (graphical slope). When significant, the z-value may be strong evidence for biographical distribution of the species.

Values for the z exponent have already been described for microorganisms. Interestingly, z-values for microorganisms were often smaller than for macro-organisms. This result may be attributed: to the larger capacity of dispersion; to the lack of a clear "species" resolution; and to the use of molecular fingerprint or sequencing techniques [18–20]. Molecular fingerprint methods, such as T-RFLP (Terminal restriction fragment length polymorphism) and DGGE (Denaturing gradient gel electrophoresis), proved to be an important tool for accessing the diversity of microorganisms in different environments with relatively low costs and little time consumed [21]. However, fingerprinting methods usually are limited as they detect the most common species and thus underestimate the total diversity in a sample [22]. That is mostly because fingerprinting techniques lump different closely related "species" into a single taxonomic unit (often called Operational Taxonomical Unity – OTU), and usually ignore rare species. Nevertheless, fingerprinting techniques are still extremely valuable for rapidly comparing the microbial community composition in different environments [2].

Another parameter of species distribution through space is the distance-decay relationship, which consists of the decay of similarity between different communities as a function of the distance separating them [23], and can also be seen as evidence of a biogeographical pattern. The main difference between the distance-decay approach and the species-area relationship is based on the consideration of the relative abundance of the species in addition of their presence or absence. Bell et al. showed that bacteria in water-filled tree holes, found at the same place, displayed a significant distance-decay relationship [24].

Little is known about the geographical distribution of methanogenic archaea. So far, hot desert soil methanogenic Archaea were shown to be widely spread between different parts of the globe, and this could be found mostly by reactivation of the cryptic methanogenic process *in vitro* [25]. To our best knowledge, there is no description of distance-decay or species-area relationship for non-extremophilic archaea in a contiguous environment. Given the high ecological stability of methanogenic sediments and soils, with a continuous anaerobic environment and a regular input of organic matter, the microbial communities related to the methanogenesis processes tend to be stable through time [26].

Among the processes commonly underlying the biogeographical patterns of distribution of organisms and communities - selection, drift, dispersion and diversification - selection and dispersion are closely related with geographical distances, given the increase of habitat heterogeneity and dispersion limitations with increasing area [27]. Thus, we hypothesized that if there is a change of diversity patterns among different spots of a contiguous lake sediment, it should have a strong correlation with the geographical distance between them, and this change should also present itself differently depending on which part of the microbial community is considered. For this purpose we were targeting three different genes. The 16S rRNA genes of Bacteria and Archaea that are transcribed to generate the structural RNA of the small-subunit ribosomes are universal and strongly conserved genes and are therefore widely used as taxonomic markers [28]. The *mcrA* is a functional gene coding for the alpha subunit of the methyl-coenzyme M reductase, an enzyme being essential and characteristic for the methanogenesis biochemical pathway in Archaea [29]. Thus, we were targeting different groups of prokaryotes. By that we expected to see if there is a differential influence of geographic distance as a factor driving distribution of these three different groups of microorganisms.

We were able to show a geographical distribution pattern of the composition of Bacteria, Archaea and methanogenic Archaea in a contiguous tropical lake sediment and the presence of a significant **z-value** for all three groups. We hypothesized that intrinsic ecological differences between the communities, general diversity profiles, and technical particularities and limitations could be defining the differences in the distribution of these taxonomical groups or at least how we perceived it.

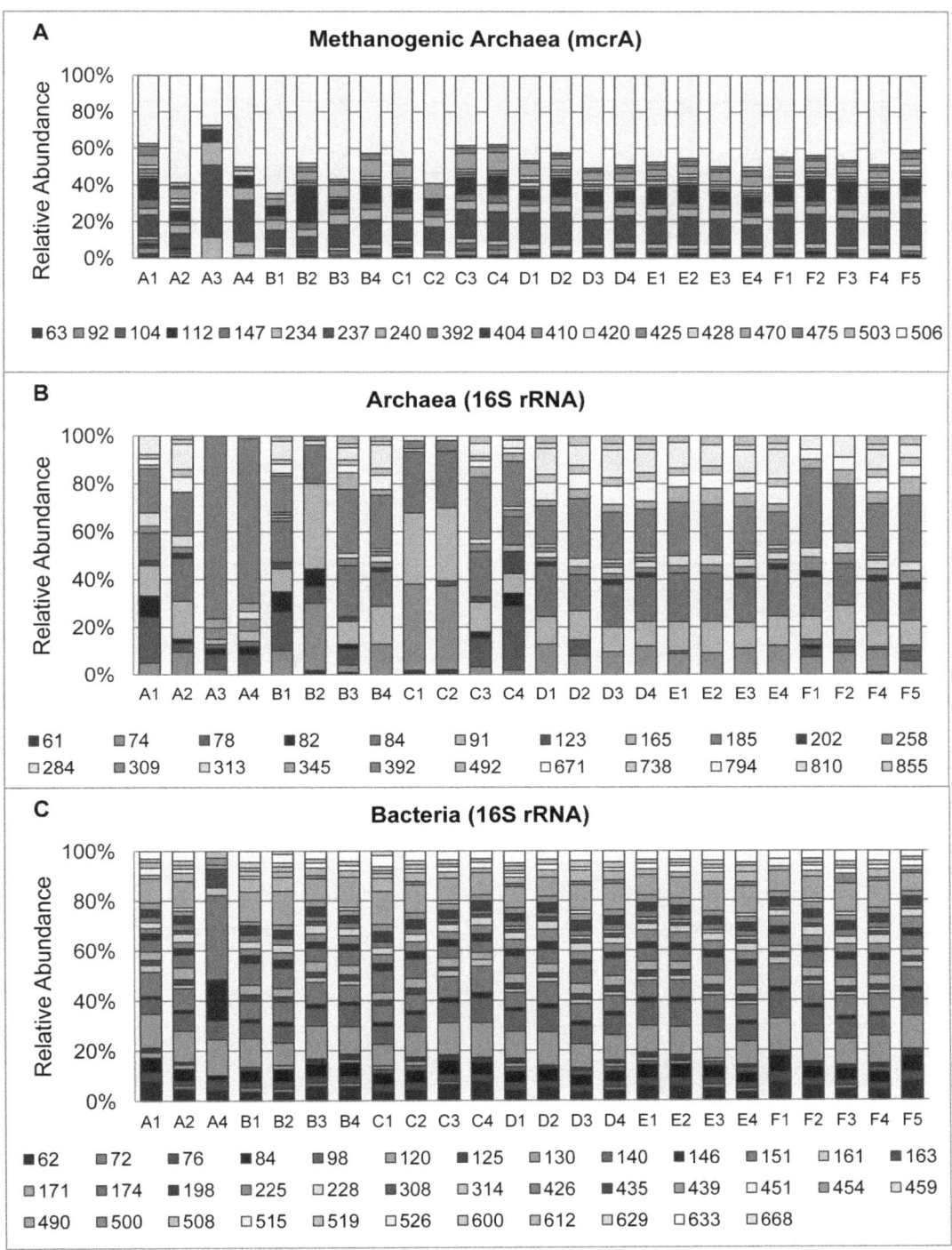

Figure 2. T-RFLP profiles of the three different genes studied in sediment inside the Lago Negro area. A1-F5 represent the sampling points (see Figure 1 for more details), the bar size represents the relative abundance of each OTU, which are defined by their size in base pair (bp). (A) *mcrA*, (B) Archaeal 16S rRNA gene, and (C) Bacterial 16S rRNA gene.

Materials and Methods

Sampling and Study Area

The Pantanal consists of the largest floodplain in South America and is periodically flooded by the Paraguay River and its tributaries. Altitude above sea level varies from 80 to 120 m and the total estimated area is around 138.123 km². The water flow

continuously carries organic material, and during the flood period the sediment spreads all over the plain constituting the most important source of carbon and nutrients for the methanogenic archaea [30]. The climate is hot and wet in the summer, and cold and dry in the winter. The maximum temperature often surpasses 40°C. Between the months of May and July, the average

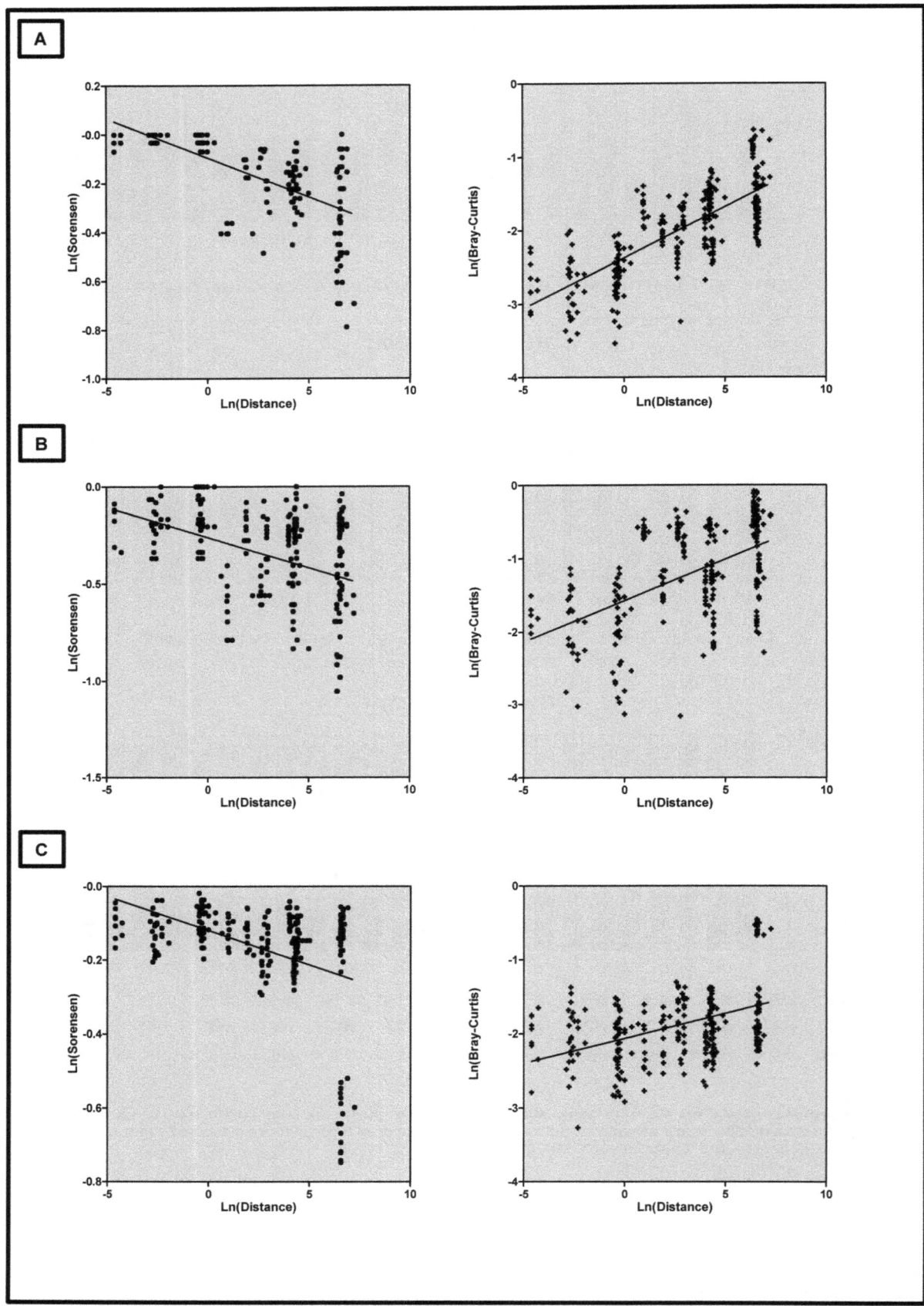

Figure 3. Distance decay and taxa-area relationships. Log transformed Bray-Curtis and Sørensen indices plotted against the distance separating the points, for the three different genes, lines represent a simple linear regression. (A) *mcrA*; (B) Archaeal 16S rRNA gene and (C) Bacterial 16S rRNA gene.

temperature drops below 20°C, and the minimum temperature may reach 0°C [31].

The study was conducted in a single lake (19°02.651′S 57°30.254′W) called *Lagoa Negra*. No special permission is needed for sediment sampling in non-protected areas in Brazil, so no specific permissions were required for these locations. This study did not involve endangered or protected species.

The lake is located at the west margin of the Paraguay River and a few kilometers east from the cities of Corumbá and Ladário, close to the Bolivia-Brazil border. *Lagoa Negra* is a perennial shallow freshwater lake, with water depth varying between 2 m and 3 m approximately, getting lower or higher following the flood regime. The total area of the lake is 10.8 km^2 (maximum length of 4.1 km and maximum width of 3.5 km, but it can also vary following the Paraguay River flood regime), water pH ranging from 6.9 to 8.3 and conductivity ranging from 140 to 240 µS cm^{-1} at 25°C. The top layer of sediment is mostly composed of a thin clayey substrate with approximately 2% of sand (>63 µm) and total carbon and organic matter concentrations of 1.8% and 2% respectively [32]. Twenty one sediment cores (7 cm diameter) were collected in the lake in July 2011. The cores were distributed on a nested square grid with distance between points varying from 0.01 m to approximately 1,400 m (Figure 1). The 10 cm top layer of the sediment was homogenized and then 1 ml was sampled and frozen for molecular biology analyses. The central core E was sub sampled within distances of 10 cm and 1 cm between points inside the core area (also 10-cm deep with polyethylene straws). The sediment samples were immediately frozen in liquid nitrogen and then air-shipped to Germany for later molecular biology analyses at the Max-Planck Institute for Terrestrial Microbiology.

T-RFLP

The frozen sediment samples were thawed, and the total DNA was extracted using the FastDNA SPIN Kit for Soil (MP Biomedicals) following the manufacturer's protocol and a three times treatment with 5.5 M guanidine thyocianate as an additional cleaning step during the matrix binding step. The total DNA was quantified using spectophotometry. Preparations were considered to be of good quality when resulting in more than 30 ng/µl of DNA.

Bacterial and Archaeal 16S rRNA gene fragments were PCR-amplified using the primer pairs Eub 9/27f (5′–GAG TTT GAT CMT GGC TCA G–3′) and Eub 907/926r (5′–CCG TCA ATT CMT TTR AGT TT–3′), as described by Lane and Weisburg [33] for Bacteria, and the primer pairs A109f (5′–ACK GCT

CAG TAA CAC GT–3′) and A934b (5′–GTG CTC CCC CGC CAA TTC CT–3′), as described by Großkopf et al. [34] for Archaea. For the T-RFLP analyses the forward Bacteria and the backward Archaea primers were labeled with FAM (5-carboxyfluorescein). Each 50-µl volume of PCR reaction contained 1× GoTaq Flexi Green Buffer (Promega); 1.5 mM MgCl$_2$ (Promega); 200 µM of dNTP's Mix (Fermentas); 0.33 µM of each primer described (Sigma); 1 U GoTaq Flexi DNA Polymerase (Promega) and 10 µg of Bovine Serum Albumin BSA (Roche). Diluted total DNA extract was added as template (1 µl). The reaction was started with an initial denaturation step (94°C for 3 min), followed by 24–28 cycles of denaturation (94°C for 45 sec), annealing (52°C for 45 sec) and extension (72°C for 80 sec), and a final extension step (72°C for 5 min).

For amplification of *mcrA* the primer pair MCRf (5′–TAY GAY CAR ATH TGG YT–3′) and MCRb (5′–ACR TTC ATN GCR TAR TT–3′) was used as described by Springer et al. [29], the forward primer being labeled with FAM for T-RFLP analyses. Each 50-µl of PCR reaction contained 1× MasterAmp PCR PreMix B (Biozym); 0.33 µM of each primer described; 1 U GoTaq Flexi DNA Polymerase (Promega); 10 µg BSA (Roche) and 1 µl of DNA template. The reaction started with an initial denaturation step (94°C for 3 min), followed by 32 cycles of denaturation (94°C for 45 sec), annealing (50°C for 45 sec) and extension (72°C for 90 sec), finished by a final extension step (72°C for 5 min).

The quality of the PCR products was controlled using agarose gel (1.5%) electrophoresis. The DNA was then purified using the GenElute PCR Clean-Up Kit (Sigma) following the manufacturer's instructions, and stored at −20°C.

For T-RFLP, the PCR product was digested using the following restriction enzymes: MspI incubated overnight at 37°C for Bacterial 16S rRNA genes, TaqI incubated for 3 h at 65°C for Archaeal 16S rRNA genes, and Sau96 incubated for 3 h at 37°C for *mcrA*. After the incubation, the digested products were once again cleaned with the SigmaSpin Sequencing Reaction Clean-Up, Post-Reaction Purification Columns (Sigma). The samples were denatured at 94°C for 2 min and loaded into an ABI 3100 automated gene sequencer (Applied Biosystems) for separation of the TRFs. T-RFLP data were retrieved by comparison with an internal standard using GeneScan 3.71 software (Applied Biosystems).

Taxa-area and distance-decay relationships

The T-RFLP profiles were analyzed and standardized as described in Dunbar et. al [35] resulting in T-RFs of 60 to 855

Table 1. Distance-decay regression coefficients based on the Bray-Curtis dissimilarity index **(Y)**; Regression coefficient based on the Sørensen Similarity index; and the exponent **z** calculated by the distance-decay approach values for each one of the target genes calculated by -(regression coefficient)/2.

Gene	Distance-decay (Y)	Sørensen Regr. Coefficient	(z)
mcrA	0.14	−0.032	0.016
Archaea 16S rRNA	0.11	−0.031	0.0155
Bacteria 16S rRNA	0.067	−0.018	0.009

Table 2. Z-values previously described for microorganisms, 99% or 95% represent the taxonomic resolution applied for definition of the OTUs (sequence similarity).

Microbial Community	z-value	Diversity Access	Habitat Org.	Reference
Desert Soil Fungi	0.074	fingerprint (ARISA)	Contiguous	[19]
Salt-Marsh Bacteria (99% OTU)	0.04	sequencing	Contiguous	[18]
Salt-Marsh ß-proteobacteria (99% OTU)	0.019	sequencing	Contiguous	[18]
Salt-marsh Bacteria (95% OTU)	0.019	sequencing	Contiguous	[18]
Salt-Marsh ß-proteobacteria (95% OTU)	0.008*	sequencing	Contiguous	[18]
Water-filled tree holes Bacteria	0.26	fingerprint (DGGE)	Island	[20]
Metal-cutting fluid sump Bacteria	0.26-0.29	fingerprint (DGGE)	Island	[38]
Soil Bacteria	0.03	Fingerprint (T-RFLP)	Noncontiguous	[39]
Tropical Forest Soil Bacteria	0.42 and 0.47	Fingerprint (T-RFLP)	Contiguous	[40]
Freshwater lake sediment Bacteria	0.009	fingerprint (T-RFLP)	Contiguous	this study
Freshwater lake sediment Archaea	0.0155	Fingerprint (T-RFLP)	Contiguous	this study
Freshwater lake sediment methanogenic Archaea	0.016	fingerprint (T-RFLP)	Contiguous	this study

*not significant different from zero.

base pairs (bp) size, each representing more than 1% of the total fluorescence of that sample.

We used the 25 resulting profiles for pair wise calculation of the Bray-Curtis dissimilarity indices [36] and the Sørensen similarity indices resulting in 300 different pairs for each targeted gene. A simple linear regression of the log-transformed data of the Bray-Curtis indices plotted against the distances between the sediment samples from which the pairs originated was used to estimate the slope of the distance-decay relationship [23], in this work we are using the resulting regression coefficients (**Y**) as a parameter for discussion.

The same log-transformed linear regression was used to calculate the slope of the Sørensen indices plotted against the distance between points. The resultant similarity slope was used to calculate the **z-value** of the taxa-area relationship with the formula $\log(S_S) = \text{constant} - 2z \log(D)$, where S_S is the pair-wise similarity between communities and D is the distance between two samples used in the distance-decay approach, where the **z-value** is determined by **−(regression coefficient)/2** as described by Harte et al. [37].

By using both, the taxa-area relationship based on the Sørensen index and the distance-decay based on the Bray-Curtis dissimilarity index we addressed the subject not only by "presence/absence" of a taxon (Sørensen index), but also the relative abundance of a taxon (Bray-Curtis index) provided by the T-RFLP.

In order to avoid randomization patterns that could influence the results, the same calculations were performed utilizing distances smaller than 200 m and 20 m and they showed similar slopes as the complete data set (data not shown).

Results

The analysis of the *mcrA* T-RFLP profile resulted in a total of 18 different OTUs of methanogens throughout the lake area (Figure 2A) of which 5 OTUs were found at all sampling points (237 bp, 240 bp, 404 bp, 470 bp and 506 bp) and one OTU (63 bp), which was the least frequently retrieved OTU, was found at only 6 sampling points. The OTU with 506 bp showed the highest relative abundance in all the samples. The Archaeal 16S rRNA T-RFLP profile showed 22 different OTUs (Figure 2B)

with none of them showing dominance over the others, only 2 OTUs were found at all sampling points (91 bp and 392 bp) and the OTUs with 165 bp and 345 bp were found at only 1 and 2 sampling points, respectively. The Bacterial 16S rRNA gene T-RFLP profiles showed the largest number of OTUs among the different genes targeted with a total of 37 different OTUs found at all the sampling points across the lake area, showing a high diversity of dominant groups (Figure 2C). Eight OTUs were recovered from all the sampling points (62 bp, 76 bp, 84 bp, 130 bp, 140 bp, 146 bp, 151 bp and 439 bp); in contrast, 3 other OTUs (459 bp, 526 bp and 612 bp) were recovered from only 3 sampling points. In a literature based affiliation it was possible to relate some of the observed OTUs with previously described ones, keeping in mind that such affiliation can only be tentative (Table S1).

Linear regressions were performed to evaluate the slope coefficient of the correlations between the geographical distance and the communities' similarities, and Mantel tests (with 9999 permutations) were performed to test the significance of these correlations.

All the three different T-RFLP profiles showed a significant distance-decay relationship based on the Bray-Curtis dissimilarity (**Y**) (Figure 3). The *mcrA* gene showed the largest slope of 0.14 (Mantel test $P < 0.001$, $R = 0.69$), followed by the archaeal 16S rRNA gene with a slope of 0.11 (Mantel test $P = 0.005$, $R = 0.30$), while the slope for the bacterial 16S rRNA gene was only 0.067 (Mantel test $P = 0.006$, $R = 0.34$).

The regression coefficient based on the Sørensen similarity indices calculated for all the three genes also showed significant values (Figure 3) of -0.032 ($P < 0.001$, $R = -0.60$) for *mcrA*, -0.031 (Mantel test $P = 0.03$, $R = -0.45$) for archaeal and -0.018 (Mantel test $P = 0.006$, $R = -0.34$) for bacterial 16S rRNA genes. However the calculated z-values were small, thus indicating relatively flat taxa-area relationships (Table 1).

Discussion

Our study indicates a small but significant biogeographical distribution of the microbial community composition in the lake sediment for all the three different genes targeted, which represent the Bacteria, the Archaea, and the methanogens. The conclusion is based on the distance decay relationship of operational taxonomic units of three different microbial genes. All dominant operational taxa (more than 10% relative abundance) from the three genes analyzed in this study were observed at all the sampling points within the lake. Nevertheless, a significant taxa-area relationship could be observed. These values were small if compared to others described for macro-organisms, which was expected given the high capacity of dispersion of the targeted microorganisms. However, the similarity distance-decay, which takes into consideration also the relative abundance of the groups, shows that they were not homogeneously distributed within the entire lake area. We therefore conclude that the most distant samples were the ones, which were most different from each other.

The **z-values** and the distance decay regression coefficients (**Y**) observed in this study were lower than others previously observed for microorganisms in the literature [17–19][36–38], but are the first ones described for Bacteria, Archaea and methanogenic archaea in a single contiguous environment. The existence of significant **z-values** shows that all these microbial groups have a biogeographical distribution. The low values that we found were expected given that we used a fingerprint technique for accessing the microbial diversity of our target environment, and not a high resolution ribotyping technique. At an OTU resolution of 95% sequence similarity of the 16S rRNA gene, Horner-Devine et al. described similar z-values for Bacteria ($z = 0.019$) and Beta-proteobacteria ($z = 0.008$) in salt-marshes [18]. At higher resolutions the z presented higher values (0.04 for Bacteria and 0.019 for Beta-proteobacteria at 99% sequences similarity). It is interesting to note that the lower values described by the mentioned study were presented by the lower taxonomical levels reached, while in this study we observed higher values for *mcrA* and the Archaea than for the Bacteria. Other **z-values** found for other groups of microorganisms are displayed in the Table 2.

The differences between **z-values** and distance decay values (**Y**) described for the three different genes, representing three different groups of microorganisms, may be explained by two hypotheses. The first one is that Bacteria (16S rRNA) have a larger capacity for dispersal than the Archaea (16S rRNA) in general and the methanogenic archaea (*mcrA*) in particular. By sampling the first 10 centimeters of the sediment we were able to access different communities living at different depths at the same time. We assume that the methanogens (all Archaea) were preferable located in the deeper sediment layers, since their activity is inhibited by the presence of other electron acceptors like oxygen, nitrate, iron and sulfate, potentially present in the surface layers of the sediment. On the other hand, Bacteria were not restricted to deeper sediment layers [41–43]. The Bacteria domain showed weaker species-area and distance-decay relationships than the methanogens (*mcrA*) and the Archaea (16S rRNA). The profile of Archaea and Bacteria OTUs was well distributed throughout the lake area, showing the large diversity of micro-habitats that can be exploited by these groups, and that the ability of dispersion seems not to be compromised at smaller scales.

The second hypothesis is based on the taxonomic sensitivity of the T-RFLP method. OTUs can represent a large variety of different taxonomic groups. The universal primers targeting 16S rRNA genes do not represent the entire set of conceivable species, which are many more than the OTUs derived from T-RFLP. Hence, the total diversity of the community is underestimated by T-RFLP [44]. A study targeting 16S rRNA genes of the bacterioplankton of temperate lakes showed that the sequence similarity within a single OTU varied from 73 to 100%, however, sequence homology of 97% is generally used to define bacterial species [5,45]. When utilizing a functional gene such as the *mcrA*,

the probability of reaching lower taxonomic levels is higher, because translated genes show diminished conservation and can present a higher variability of codons, and thus a higher chance for being differentiated by terminal fragment size methods [46].

This may explain why the **z-values** were higher for T-RFLP of *mcrA* than of 16S rRNA genes, but it does not explain why the Archaea distance decay and **z-values** were higher than those of the Bacterial since both of them targeted 16S rRNA genes. Several authors described that Bacteria diversity tends to be higher than Archaea diversity at different environments [47–49], thus the T-RFLP fingerprinting technique could be more efficient in accessing a larger part of the total diversity of Archaea 16S rRNA and consequentially being closer to the actual taxa distribution of the environment, thus showing higher values of **z** and **Y**.

Our work shows some important results that contribute to a better understanding of similar ones developed within different environments. Studies of bacterioplankton distribution in lakes showed similar patterns between different and not connected lakes in distant geographic regions, but with variations in their relative abundance [50]. On the other hand bacterioplankton communities distance dissimilarity inside one single lake were described as being weaker than between different lakes in North America, and mostly influenced by different water regimes and partial geographic isolation [51]. Some studies in saline lakes in China, Mongolia and Argentina showed that Bacterial biogeography in these environments was based on contemporary environmental factors (Na^+, CO_3^{2-}, and HCO_3^- ion concentrations, pH and temperature) and geographic distance, while Archaeal biogeography was influenced only by environmental factors [17].

Geographical distances have been previously described as an important factor related with microbial spatial distribution corroborating our findings [52–54]. Rosselló-Mora et al. used metabolic compounds as a comparison parameter between different populations of the extremophilic bacterium *Salinibacter rubium*. They discovered that the divergence among different phenotypes was related to different geographical locations, and geographical distance between the sites[55]. Environmental and geographical factors also influenced magnetotactic bacteria biogeographical distribution [56].

Our distance-decay and taxa-area relationships can possibly be a result of some degree of dispersal limitation coupled with ecological drift, thus maintaining the taxa-area relationship and distance-decay pattern significant. It is believed that drift plays a important role for the distance-decay patterns found among microorganisms living in some degree of dispersal limitation (i.e. in subsurface habitats)[5]. A high dispersal rate is expected inside this contiguous environment with no significant geographical barriers imposing any kind of mobility limitation to microorganisms, and that could be a factor flattening the curves. But as we already stated, a group preferentially found in deeper layers of sediment (as the methanogenic archaea) could be more easily restricted.

This study did not focus on scanning environmental factors that could be driving the observed biogeographical pattern, so selection as a possible driving factor cannot be excluded, and selection processes are also related with geographical distances, as the diversity of habitats tends to increase with an increasing area [5,27]. However, given the small scale of the study and the probable absence of a clear environmental gradient within the Lagoa Negra we don't think it as a major factor. We believe that the correlation that we described between geographical distance and communities' structures are representative of the biogeographical pattern of distribution of the lake microbial community.

In conclusion, we showed that there was a significant geographic distribution of methanogenic archaea, Archaea and even Bacteria, related with geographic distance in a contiguous environment, i.e. the sediment of a tropical lake.

Supporting Information

Table S1 Tentative genetic affiliation of the OTUs. For the three studied genes there are in the literature some genetic affiliation with different phylogenetic groups, we show some of these in this table. * The affiliation with the phylogenetic groups was based on literature data that used the same primers and restriction enzymes for TRFLP. All the affiliations are only tentative, and can only be interpreted as the probable main groups of the lake sediment community.

Acknowledgments

We thanks, the colleagues Roberta Peixoto, Juliana Valle, Ruan Andrade and the Federal University of Mato Grosso do Sul - Brazil (UFMS) for the support during the sampling.

Author Contributions

Conceived and designed the experiments: DPB RC AEP. Performed the experiments: DPB PC MK. Analyzed the data: DPB PC MK. Contributed reagents/materials/analysis tools: RC PC MK AEP. Wrote the paper: DPB RC AEP.

References

1. Begon M, Townsend CR, Harper JL (2006) Ecology: from individuals to ecosystems. 4th ed. Blackwell Publishing Ltd.
2. Martiny JBH, Bohannan BJM, Brown JH, Colwell RK, Fuhrman JA, et al. (2006) Microbial biogeography: putting microorganisms on the map. Nat Rev 4: 102–112.
3. Fenchel T, Esteban GF, Finlay BJ (1997) Local versus Global Diversity of Microorganisms: Cryptic Diversity of Ciliated Protozoa. Oikos 80: 220–225.
4. Fenchel T (2003) Biogeography for bacteria. Science 301: 925.
5. Hanson Ca, Fuhrman Ja, Horner-Devine MC, Martiny JBH (2012) Beyond biogeographic patterns: processes shaping the microbial landscape. Nat Rev Microbiol 10: 497–506.
6. Baas-Becking L, Becking LGMB (1934) Geobiologie of inleiding tot de milieukunde. 18.
7. Finlay BJ (2002) Global dispersal of free-living microbial eukaryote species. Science 296: 1061.
8. Finlay BJ, Clarke KJ (1999) Ubiquitous dispersal of microbial species. Nature 400: 1999.
9. Staley JT, Gosink JJ (1999) Poles apart: biodiversity and biogeography of sea ice bacteria. Annu Rev Microbiol 53: 189–215.
10. Papke RT, Ramsing NB, Bateson MM, Ward DM (2003) Geographical isolation in hot spring cyanobacteria. Environ Microbiol 5: 650–659.
11. Papke RT, Ward DM (2004) The importance of physical isolation to microbial diversification. FEMS Microbiol Ecol 48: 293–303.
12. Diniz-Filho JAF, Telles MPDC (2000) Spatial pattern and genetic diversity estimates are linked in stochastic models of population differentiation. Genet Mol Biol 23: 541–544.
13. Cho J-CC, Tiedje JM (2000) Biogeography and Degree of Endemicity of Fluorescent Pseudomonas Strains in Soil. Appl Environ Microbiol 66: 5448–5456.
14. Oda Y, Star B, Huisman LA (2003) Biogeography of the purple nonsulfur bacterium Rhodopseudomonas palustris. Appl Environ Microbiol 69: 5186–5191.
15. Takacs-Vesbach C, Mitchell K, Jackson-Weaver O, Reysenbach A-L (2008) Volcanic calderas delineate biogeographic provinces among Yellowstone thermophiles. Environ Microbiol 10: 1681–1689.
16. Bahl J, Lau MCY, Smith GJD, Vijaykrishna D, Cary SC, et al. (2011) Ancient origins determine global biogeography of hot and cold desert cyanobacteria. Nat Commun 2: 163.
17. Pagaling E, Wang H, Venables M, Wallace A, Grant WD, et al. (2009) Microbial biogeography of six salt lakes in Inner Mongolia, China, and a salt lake in Argentina. Appl Environ Microbiol 75: 5750–5760.
18. Horner-Devine MC, Lage M, Hughes JB, Bohannan BJM (2004) A taxa-area relationship for bacteria. Nature 432: 750–753.
19. Green JL, Holmes AJ, Westoby M, Oliver I, Briscoe D, et al. (2004) Spatial scaling of microbial eukaryote diversity. Nature 432: 747–750.
20. Bell T, Ager D, Song J-I, Newman Ja, Thompson IP, et al. (2005) Larger islands house more bacterial taxa. Science 308: 1884.
21. Head IM, Saunders JR, Pickup RW (1998) Microbial evolution, diversity, and ecology: a decade of ribosomal RNA analysis of uncultivated microorganisms. Microb Ecol 35: 1–21.
22. Woodcock S, Curtis TP, Head IM, Lunn M (2006) Taxa-area relationships for microbes: the unsampled and the unseen. Ecology 9: 805–812.
23. Nekola JC, White PS (2004) The distance decay of similarity in biogeography and ecology. J Biogeogr 26: 867–878.
24. Bell T (2010) Experimental tests of the bacterial distance-decay relationship. ISME J 4: 1357–1365.
25. Angel R, Claus P, Conrad R (2012) Methanogenic archaea are globally ubiquitous in aerated soils and become active under wet anoxic conditions. ISME J 6: 847–862.
26. Conrad R (2007) Microbial ecology of methanogens and methanotrophs. Adv Agron 96: 1–63.
27. Nemergut DR, Schmidt SK, Fukami T, O'Neill SP, Bilinski TM, et al. (2013) Patterns and processes of microbial community assembly. Microbiol Mol Biol Rev 77: 342–356.
28. Stackebrandt E, Goebel BM (1994) Taxonomic Note: A Place for DNA-DNA Reassociation and 16S rRNA Sequence Analysis in the Present Species Definition in Bacteriology. Int J Syst Bacteriol 44: 846–849.
29. Springer E, Sachs MS, Woese CR, Boone DR (1995) Partial gene sequences for the A subunit of methyl-coenzyme M reductase (mcrI) as a phylogenetic tool for the family Methanosarcinaceae. Int J Syst Bacteriol 45: 554–559.
30. Marani L, Alvalá PCC (2007) Methane emissions from lakes and floodplains in Pantanal, Brazil. Atmos Environ 41: 1627–1633.
31. Guerrini V (1978) Bacia do alto rio Paraguai: estudo climatológico. Brasília: EDIBAP/SAS.
32. Bezerra M de O, Mozeto A (2008) Deposição de carbono orgânico na planície de inundação do rio Paraguai durante o Holoceno Médio. Oecol Bras 12: 155–171.
33. Weisburg WG, Barns SM, Pelletier Da, Lane DJ (1991) 16S ribosomal DNA amplification for phylogenetic study. J Bacteriol 173: 697–703.
34. Großkopf R, Janssen PH, Liesack W (1998) Diversity and structure of the methanogenic community in anoxic rice paddy soil microcosms as examined by cultivation and direct 16S rRNA gene sequence retrieval.
35. Dunbar J, Ticknor LO, Kuske CR (2001) Phylogenetic specificity and reproducibility and new method for analysis of terminal restriction fragment profiles of 16S rRNA genes from bacterial communities. Appl Environ Microbiol 67: 190–197.
36. Roger Bray J, Curtis JTT, Bray JRR (1957) An ordination of the upland forest communities of southern Wisconsin. Ecol Monogr 27: 325–349.
37. Harte J, McCarthy S, Taylor K, Kinzig A, Fischer ML (1999) Estimating Species-Area Relationships from Plot to Landscape Scale Using Species Spatial-Turnover Data. Oikos 86: 45.
38. Van der Gast CJ, Lilley AK, Ager D, Thompson IP (2005) Island size and bacterial diversity in an archipelago of engineering machines. Environ Microbiol 7: 1220–1226.
39. Fierer N, Jackson RB (2006) The diversity and biogeography of soil bacterial communities. Proc Natl Acad Sci U S A 103: 626.
40. Noguez AMM, Arita HTT, Escalante AEE, Forney LJJ, Garcia-Oliva F, et al. (2005) Microbial macroecology: highly structured prokaryotic soil assemblages in a tropical deciduous forest. Glob Ecol Biogeogr 14: 241–248.
41. Falz KZ, Holliger C, Großkopf R, Liesack W, Nozhevnikova AN, et al. (1999) Vertical distribution of methanogens in the anoxic sediment of Rotsee (Switzerland). Appl Environ Microbiol 65: 2402–2408.
42. Lovley DR, Klug MJ (1986) Model for the distribution of sulfate reduction and methanogenesis in freshwater sediments. Geochim Cosmochim Acta 50: 11–18.
43. Chan OC, Claus P, Casper P, Ulrich A, Lueders T, et al. (2005) Vertical distribution of structure and function of the methanogenic archaeal community in Lake Dagow sediment. Environ Microbiol 7: 1139–1149.
44. Liu WT, Marsh TL, Cheng H, Forney LJ (1997) Characterization of microbial diversity by determining terminal restriction fragment length polymorphisms of genes encoding 16S rRNA. Appl Environ Microbiol 63: 4516–4522.
45. Eiler A, Bertilsson S (2004) Composition of freshwater bacterial communities associated with cyanobacterial blooms in four Swedish lakes. Environ Microbiol 6: 1228–1243.
46. Marsh TL (1999) Terminal restriction fragment length polymorphism (T-RFLP): an emerging method for characterizing diversity among homologous populations of amplification products. Curr Opin Microbiol 2: 323–327.
47. Inagaki F, Nunoura T, Nakagawa S, Teske A, Lever M, et al. (2006) Biogeographical distribution and diversity of microbes in methane hydrate-bearing deep marine sediments on the Pacific Ocean Margin. Proc Natl Acad Sci U S A 103: 2815–2820.
48. Ochsenreiter T, Selezi D, Quaiser A, Bonch-Osmolovskaya L, Schleper C (2003) Diversity and abundance of Crenarchaeota in terrestrial habitats studied by 16S RNA surveys and real time PCR. Environ Microbiol 5: 787–797.
49. Bowman JP, McCuaig RD (2003) Biodiversity, community structural shifts, and biogeography of prokaryotes within Antarctic continental shelf sediment. Appl Environ Microbiol 69: 2463.

50. Lindström ES, Leskinen E (2002) Do neighboring lakes share common taxa of bacterioplankton? Comparison of 16S rDNA fingerprints and sequences from three geographic regions. Microb Ecol 44: 1–9.

51. Yannarell AC, Triplett EW (2004) Within and between Lake Variability in the Composition of Bacterioplankton Communities: Investigations Using Multiple Spatial Scales. Appl Environ Microbiol 70: 214.

52. Schauer R, Bienhold C, Ramette A, Harder J (2010) Bacterial diversity and biogeography in deep-sea surface sediments of the South Atlantic Ocean. ISME J 4: 159–170.

53. Galand PE, Potvin M, Casamayor EO, Lovejoy C (2010) Hydrography shapes bacterial biogeography of the deep Arctic Ocean. ISME J 4: 564–576.

54. Xiong J, Liu Y, Lin X, Zhang H, Zeng J, et al. (2012) Geographic distance and pH drive bacterial distribution in alkaline lake sediments across Tibetan Plateau. Environ Microbiol 14: 2457–2466.

55. Rosselló-Mora R, Lucio M (2008) Metabolic evidence for biogeographic isolation of the extremophilic bacterium Salinibacter ruber. ISME J 2: 242–253.

56. Lin W, Wang Y, Gorby Y, Nealson K, Pan Y (2013) Integrating niche-based process and spatial process in biogeography of magnetotactic bacteria. Sci Rep 3: 1643.

15

Genetic Diversity and Population Structure of the Pelagic Thresher Shark (*Alopias pelagicus*) in the Pacific Ocean: Evidence for Two Evolutionarily Significant Units

Diego Cardeñosa[1], John Hyde[2], Susana Caballero[1]*

1 Laboratorio de Ecología Molecular de Vertebrados Acuáticos-LEMVA, Departamento de Ciencias Biológicas, Universidad de Los Andes, Bogotá, Colombia, **2** Southwest Fisheries Science Center, National Marine Fisheries Service, La Jolla, California, United States of America

Abstract

There has been an increasing concern about shark overexploitation in the last decade, especially for open ocean shark species, where there is a paucity of data about their life histories and population dynamics. Little is known regarding the population structure of the pelagic thresher shark, *Alopias pelagicus*. Though an earlier study using mtDNA control region data, showed evidence for differences between eastern and western Pacific populations, the study was hampered by low sample size and sparse geographic coverage, particularly a lack of samples from the central Pacific. Here, we present the population structure of *Alopias pelagicus* analyzing 351 samples from six different locations across the Pacific Ocean. Using data from mitochondrial DNA COI sequences and seven microsatellite loci we found evidence of strong population differentiation between western and eastern Pacific populations and evidence for reciprocally monophyly for organelle haplotypes and significant divergence of allele frequencies at nuclear loci, suggesting the existence of two Evolutionarily Significant Units (ESU) in the Pacific Ocean. Interestingly, the population in Hawaii appears to be composed of both ESUs in what seems to be clear sympatry with reproductive isolation. These results may indicate the existence of a new cryptic species in the Pacific Ocean. The presence of these distinct ESUs highlights the need for revised management plans for this highly exploited shark throughout its range.

Editor: Michael Hart, Simon Fraser University, Canada

Funding: Funding for this project was provided by Universidad d elos Andes (Proyecto semilla Profesor Asociado y Proyecto semilla estudiante) and NOAA. The funders had no role in study design, data collection and analysis, decision to publish, or preparation of the manuscript.

Competing Interests: The authors have declared that no competing interests exist.

* Email: sj.caballero26@uniandes.edu.co

Introduction

Sharks of the open ocean are wide-ranging, highly migratory species that routinely cross national borders. Such characteristics may lead to the misconception that these marine resources are unlimited and resistant to localized depletion. When population subdivision exists, these misconceptions lead to an unsustainable use of the resource, ending in the depletion of the population under pressure. If severe enough, these population depletions may result in reduced genetic diversity and a concomitant reduction in the ability of a population to adapt to environmental or disease stressors [1,2]. To identify management units a variety of tools are available for sharks (e.g. conventional and electronic tagging, life history studies, population genetic analyses, fishery catch data). Over the last 20 years genetic data have become increasingly important for delineating management units and understanding population connectivity in the marine realm.

Highly migratory and broadly distributed open ocean shark species are expected to show little to no population heterogeneity. As expected, previous population genetic studies on epipelagic sharks, such as the shortfin mako shark *Isurus oxyrinchus* [3], the basking shark *Cetorhinus maximus* [4], the whale shark *Rhincodon typus* [5,6], and the blue shark *Prionace glauca* [7] show low to no genetic structuring among ocean basins.

The pelagic thresher shark (*Alopias pelagicus*) is a large (up to 330 cm TL) aplacental viviparous epipelagic shark with a distribution restricted to the Indian and Pacific Oceans. Fecundity is very low with litters of only 1–2 pups [8] with an unknown gestation period, but assumed to be around a year or less [9]. The pelagic thresher shark is one of the most abundant open ocean sharks in the Eastern Tropical Pacific (ETP) and Western Pacific Ocean (WP), and one of the most exploited shark species in commercial, artisanal and illegal fisheries of these regions [10–12]. Despite a high level of exploitation little is known regarding their life history, population dynamics, and overall abundance.

Between 2009 and 2011, tissue samples, collected in Colombian fishing ports and during seizures of illegal shark finning vessels, were sent to the Universidad de Los Andes for molecular identification. We found that more than 95% of those samples were *A. pelagicus*, suggesting that perhaps exploitation rates in this region are higher than previously thought and that there is a need to gather more information on this species for its effective management [12]. Furthermore, Tsai et al. [11] suggested, using a stochastic stage-based model, that the northwest Pacific stock of *A.*

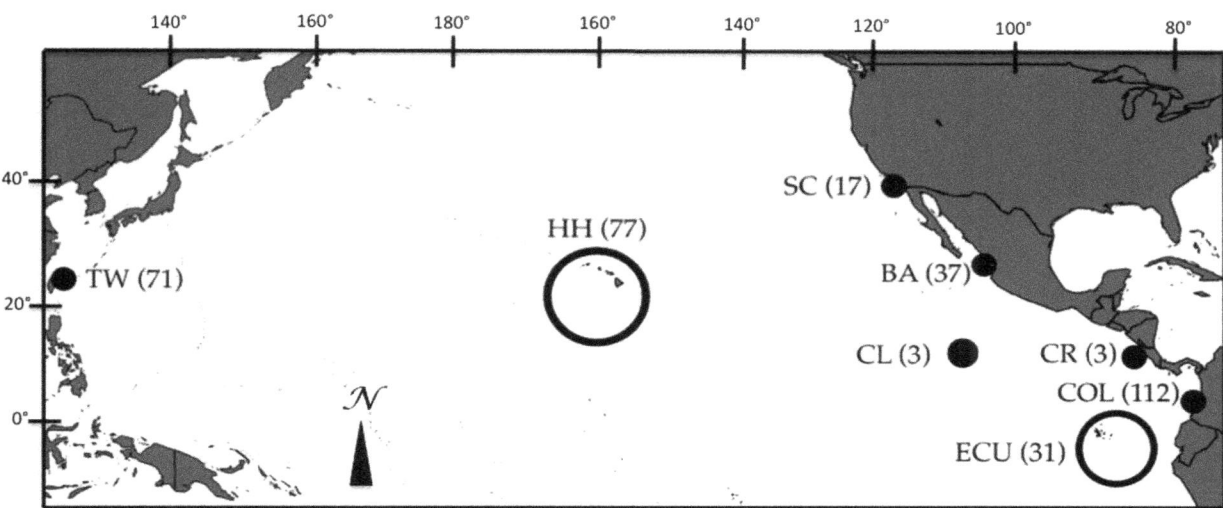

Figure 1. Map indicating sampling locations in the Pacific Ocean. TW = Taiwan, HH = Hawaii, BA = Baja, SC = Southern California, CL = Clipperton Island, CR = Costa Rica, COL = Colombia, ECU = Ecuador. Sample numbers for each location are shown in parentheses.

pelagicus is overexploited. The combined levels of exploitation, low fecundity and lack of population information led this species to be classified as Vulnerable on the IUCN Red List.

Despite their common occurrence and importance to fisheries, little is known about the genetic diversity of thresher sharks. Early work by Eitner [13], used allozyme data to examine phylogenetic relationships among the three species and suggested the existence of an unrecognized taxon in the Eastern Pacific. Trejo [14] studied the global population structure and genetic diversity of all three thresher shark species by analyses of DNA sequence data from the mitochondrial control region, finding different phylogeographic patterns for each species. Genetic heterogeneity was present among sampled populations of *A. vulpinus* but with no clear biogeographic signal. Slight population differentiation existed between samples of *A. superciliosus* from locations in the Atlantic and the Indo-Pacific, but not among samples from within the Indo-Pacific. In contrast to the other two species, *A. pelagicus* populations exhibited a very marked biogeographic pattern with strong population structure between Eastern and Western Pacific locations indicating limited gene flow across the Pacific Ocean. An interesting finding was the presence of two well-differentiated clades corresponding roughly to sampling region. One haplotype clade representing samples predominantly from the Western Pacific and the other grouping haplotypes found exclusively in the Eastern Pacific.

Though a strong biogeographic signal was detected for *A. pelagicus*, Trejo's study was limited by the number of samples both overall and within sampled areas, by a lack of geographically intermediate samples from the Central Pacific, and by the use of a single mitochondrial marker. Despite their limitations, the results of both Trejo [14] and Eitner [13] identify interesting patterns of genetic heterogeneity that deserve further evaluation. Here, we build upon previous examination of the molecular ecology of *A. pelagicus* through increased sample number, inclusion of additional sampling areas spanning the range of *A. pelagicus* in the Pacific Ocean, and analyses of both mitochondrial and nuclear molecular markers. The generated data will allow us to evaluate several key questions: 1.) Are the patterns observed by Trejo similar when bi-parental nuclear data are included? 2.) Given the strong vicariance between Eastern and Western Pacific locations observed by Trejo [14], are intermediate haplotypes present in the Central Pacific? 3.) Are the distinct clades identified by Trejo [14] more likely the result of population subdivision with restricted geneflow or do they support the existence of an unrecognized taxon as suggested by Eitner [13]?

Methodology

Tissue collection and DNA extraction

A total of 351 samples were collected between 1997 and 2011 (Fig. 1). The majority of these samples were collected by fishery observers from dead animals captured in regional fisheries using a variety of different capture techniques (i.e. longline, purse seine, gillnet). Additional samples were collected during NOAA research cruises and from sampling landings made by artisanal fishers. Collaborating institutions that aided in the collection of samples used in this study include; the National Marine Fisheries Service (USA), Incoder Subdirección de Pesca (Colombia), the Inter-American Tropical Tuna Commission, the Ministerio de Agricultura, Ganadería, Acuacultura, y Pesca (Ecuador), the Fisheries Research Institute (Taiwan). No Ethical Approval was required for this project since the samples were collected from dead animals already captured in commercial and recreational fishing activities.

Samples were identified by experts in the field and for the samples for which identification was uncertain, we implemented the molecular identification protocol described by Caballero *et al.* [12]. DNA was extracted by heating small pieces of tissue in 200 µl of 10% Chelex solution (BioRad) at 60°C for 20 minutes, then 103°C for 25 minutes followed by a brief centrifugation and storage at 4°C [15].

Cytochrome Oxidase I amplification and analyses

A portion of the mitochondrial COI gene (655 bp) was amplified using the universal primers FishCoxI F (5′ TCWAC-CAACCACAAAGAYATYGGCAC) and FishCoxI R (TAR-ACTTCWGGGTGRCCRAAGAATCA) modified from Ward *et al.* [16]. The PCR profile was as follows: 94°C for 2 min followed by 35 cycles of 94°C for 30 s, 55°C for 45 s and 72°C for

Table 1. Twenty-four variable sites over 655 bp of the mitochondrial COI gene determining 19 Pacific *Alopias pelagicus* haplotypes.

Haplotypes	Variable Sites																							
	40	43	82	91	107	145	167	169	175	247	265	280	370	371	376	425	454	497	529	541	592	613	628	637
H1	G	G	A	T	A	A	G	A	T	T	T	T	G	G	C	T	T	C	C	A	C	T	G	T
H2	.	.	.	.	G	.	.	.	.	.	.	.	.	.	.	.	.	.	.	.	.	.	.	.
H3	A	A	.	.	.	.	.	G	.	.	.	.	A	.	.	.	.	T	T	.	.	.	A	.
H4	.	A	.	.	.	.	.	.	.	.	.	.	.	.	.	.	.	.	.	.	.	.	.	.
H5	.	.	.	.	.	.	.	.	.	.	.	C	.	.	.	.	.	.	.	.	.	.	.	.
H6	A	A	.	.	.	.	.	G	.	.	.	.	A	.	.	.	.	T	T	G	.	.	A	C
H7	A	A	.	.	.	.	.	.	.	.	.	.	A	.	T	C	.	T	T	.	T	.	A	.
H8	.	.	.	.	.	.	.	.	C	.	.	.	.	.	.	.	.	.	.	.	.	.	.	.
H9	A	A	.	.	.	.	.	G	.	.	.	.	A	.	.	.	C	T	T	.	.	.	A	.
H10	?	?	?	?	?	.	.	.	.	.	.	.	.	.	.	.	.	.	.	.	.	.	?	?
H12	.	.	.	.	G	G	.	.	.	C	.	.	.	.	.	.	.	.	.	.	.	.	.	.
H13	.	.	.	.	.	.	.	.	.	.	.	.	.	.	.	.	.	.	.	.	.	C	?	?
H14	A	A	.	.	.	.	.	G	.	.	.	.	A	A	.	.	.	T	T	.	.	.	A	.
H15	.	.	.	.	.	.	A	.	.	.	.	.	.	.	.	.	.	.	.	.	.	.	.	.
H16	.	A	.	C	.	.	.	.	.	.	.	.	.	.	.	.	.	.	.	.	.	.	.	.
H17	.	.	.	.	.	.	.	.	.	.	C	.	.	.	.	.	.	.	.	.	.	C	.	.
H18	.	A	.	.	.	.	.	.	.	.	C	.	.	.	.	.	.	.	.	.	.	.	.	.
H19	.	.	G	.	.	.	.	.	.	.	.	.	.	.	.	.	.	.	.	.	.	.	.	.

(?) denotes missing data.

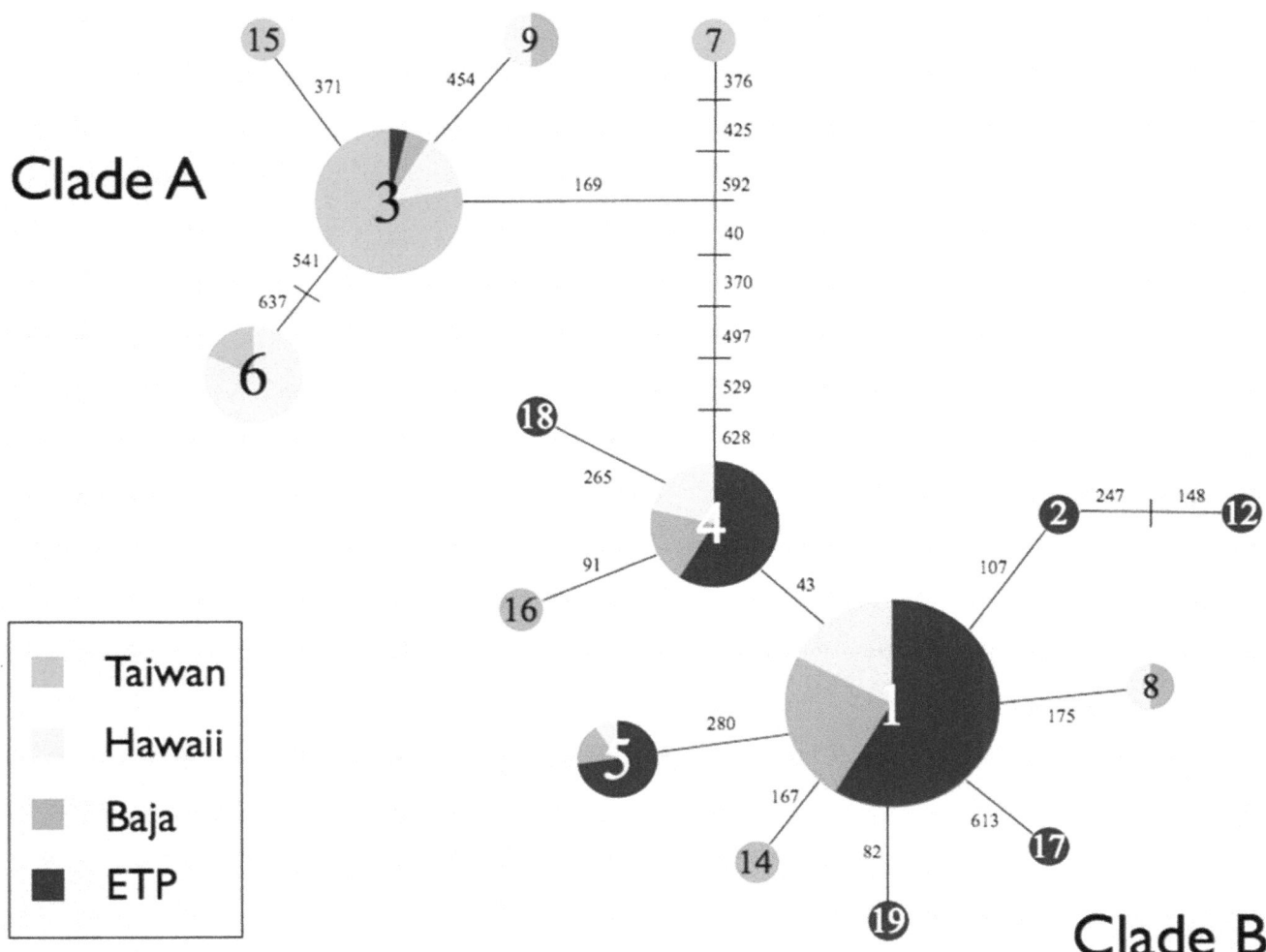

Figure 2. Haplotype network obtained from the TCS analysis. The size of the circle represents the frequency of each haplotype. Numbers represent substitutions between haplotypes and small lines represent hypothetical haplotypes not observed in this study.

40 s, with a final extension of 72°C for 10 min. A no-template negative control was included with each PCR batch to monitor for reagent contamination. PCR products were visualized on 2% agarose gels stained with ethidium bromide and single-band PCR products were enzymatically cleaned using ExoSapIT (Affymetrix). Cleaned products were cycle sequenced using BigDye Terminator v3.1 chemistry and run on an ABI3730 Genetic Analyzer.

All forward and reverse sequences were checked and edited manually using Geneious Pro v. 3.6.1 (http://www.geneious.com/) and aligned using Seaview 4.4.1 software [17]. Haplotypes were defined using MacClade [18]. A statistical parsimony network was constructed using the software TCS v. 1.21 [19], providing a 95% plausible set for all haplotype linkages. The optimal model of DNA substitution was chosen using JModelTest v.2.3.1 [20]. To understand phylogenetic patterns among haplotypes the best fit model (GTR+I) was used for phylogenetic reconstructions performed in Beast v.1.7.5 [21].

Haplotype and nucleotide diversity calculations as well as pairwise comparisons of both F_{ST} and Φ_{ST} were performed using Arlequin v.3.5.1 [22] using the pairwise nucleotide difference model to calculate genetic distance. As Trejo [14] showed

evidence for two distinct mtDNA clades and Eitner [13] suggested the existence of a possible unrecognized taxon, we chose to partition the samples in two distinct manners. In the first partition, samples were grouped solely by geographic region (i.e. Taiwan, Hawaii, southern California, Baja, Central America, Colombia, Ecuador). In the second partition, samples were first grouped by the mtDNA clade their COI haplotype belonged to (i.e. Clade A, Clade B) and subsequently grouped by geographic region. Analysis of molecular variance (AMOVA), using 10,000 random permutations, was performed to identify optimal groupings and to test biogeographic and phylogenetic hypotheses.

Microsatellite loci amplification and analyses

Nine microsatellite loci (Iox-01, Iox-30, Iox-12 [3]) and (AV-H8, AV-H110, AV-H138, AV-I11, Iox-M36 and Iox-M115). Primer sequence and repeat information for primers AV-H8, AV-H10, AV-H138, AV-111, Iox-M36 and Iox-M115 are available on GenBank (Table S1). Microsatellite loci were amplified separately in 30 μl PCR reactions. Two thermocycling profiles were implemented. The first profile consisted of an initial denaturation at 94°C for 2 min, followed by 35 cycles of 94°C for 30 s, 60 s at specified annealing temperature (Table S1), and

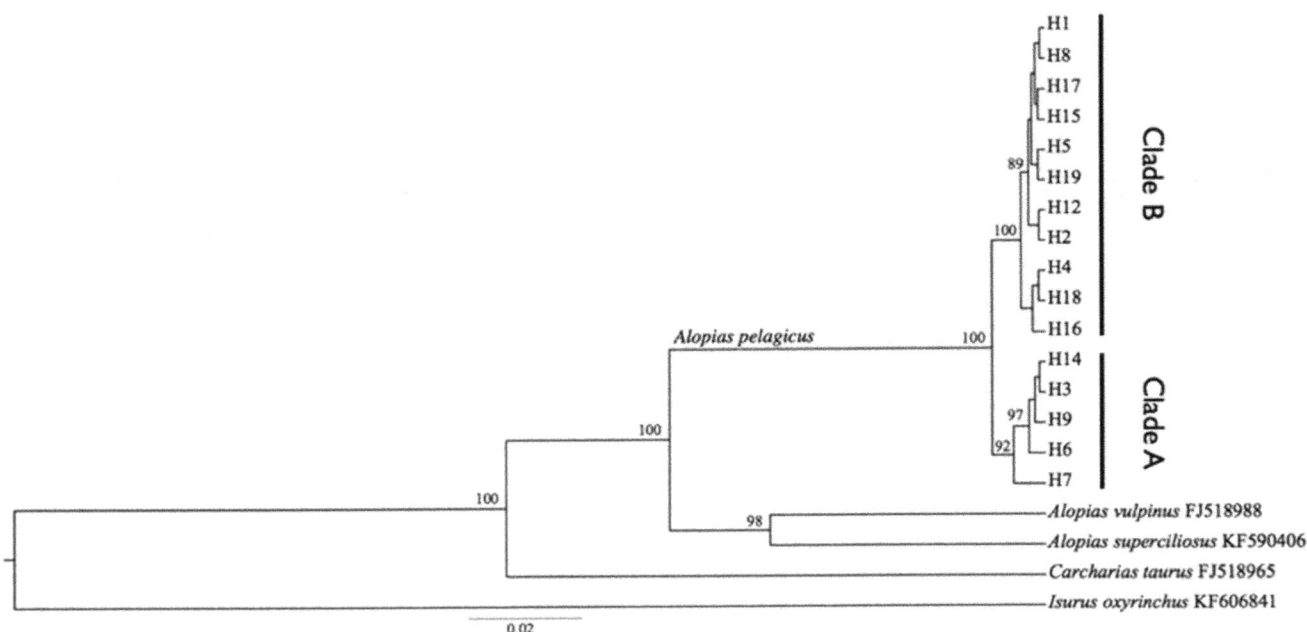

Figure 3. Coalescent tree showing reciprocal monophyly between haplotypes from the A and B clades. Posterior values are shown as branch labels.

72°C for 90 s, with a final extension of 72°C for 30 min. The second profile consisted of the same temperatures with a final extension of 60 min (Table S1).

PCR products were run on either an ABI 3500 automated sequencer at Universidad de Los Andes (Colombia) or an ABI 3730 Genetic Analyzer at the NOAA SWFSC (La Jolla, CA) using the internal ROX 500 size standard. Electropherograms were visualized and allele size calling was done using ABI PRISM GeneMapper Software v4.1 (Life Technologies). A reference set of samples were used to cross-calibrate scores between analysis platforms and labs. All authors scored all samples for all loci individually in order to minimize scoring errors across readers.

Patterns of genetic structure were evaluated using Structure v2.3.4 [23,24], which assigns individuals to groups using a Bayesian model-based method that minimizes linkage disequilibrium and deviations from Hardy-Weinberg expectations. The admixture model with correlated allele frequencies was selected and models were run that both did and did not employ the sampling location prior [25]. The sampling location prior was used in two distinct ways. To better elucidate localized genetic structure individuals were coded by region of capture, using it as a prior.

Additionally, to test for possible reproductive isolation between the two mtDNA clades identified by Trejo [14], individuals were coded by which mtDNA clade their COI haplotype belonged to using both geographic location and mtDNA clade as priors in this analysis. To infer the most likely number of groups (K) we compared the log probability LnP(D) of different values for K using an *ad hoc* statistic (ΔK, [26]) that calculates the second order rate of change of LnP(D). Runs were performed with a 10,000 step burn-in followed by 100,000 MCMC steps to test $K = 1–10$ with 20 repetitions each.

To test for population differentiation and genetic diversity, samples with three or more loci with missing data were excluded. Differentiation and genetic diversity among population units were assessed utilizing several methods. Pairwise F_{ST} and R_{ST} values were calculated using Arlequin v.3.5.1 [22], population differentiation was tested by exact test as implemented in Genepop [27], and a genetic differentiation index D [28] was calculated using the R-package DEMEtics [29]. Departures from expectations of both linkage disequilibrium and Hardy-Weinberg equilibrium (HWE) and were tested for using Arlequin v.3.5.1 [22].

Table 2. Pairwise Fst (below diagonal) and Φst (above diagonal) values for the COI gene from the Pacific Ocean populations of *Alopias pelagicus*.

$\frac{\Phi_{ST}}{F_{ST}}$	ETP (n = 121)	Baja (n = 56)	Hawaii (n = 75)	Taiwan (n = 64)
ETP	π = 0.104% h = 0.586±0.036	0.0624 (*0.0014±0.000*)	0.4766 (***0.000±0.000***)	0.8964 (***0.000±0.000***)
Baja	−0.0026 (*0.432±0.005*)	π = 0.299% h = 0.631±0.061	0.2608 (***0.000±0.000***)	0.7882 (***0.000±0.000***)
Hawaii	0.1458 (***0.000±0.000***)	0.1131 (***0.000±0.000***)	π = 0.688% h = 0.755±0.024	0.36255 (***0.000±0.000***)
Taiwan	0.5283 (***0.000±0.000***)	0.5092 (***0.000±0.000***)	0.3728 (***0.000±0.000***)	π = 0.130% h = 0.306±0.071

Probability values based on 10,000 permutations are shown in italic. Significant scores after Bonferroni correction are in bold. Haplotype (h) and nucleotide (π) % ± standard deviation (SD) diversity values are shown in the diagonal of each population unit. Numbers of samples per location are shown in parentheses.

Table 3. Pairwise Fst (below diagonal) and Φst (above diagonal) values for the COI gene when only Clade A haplotypes of *Alopias pelagicus* are analyzed.

F_{ST} \ Φ_{ST}	Baja (n = 6)	Hawaii (n = 38)	Taiwan (n = 71)
Baja	π = 0.081% h = 0.533±0.172	0.4971 (*0.000±0.000*)	0.4355 (*0.000±0.000*)
Hawaii	0.3916 (*0.003±0.001*)	π = 0.151% h = 0.472±0.072	0.0768 (*0.111±0.003*)
Taiwan	0.4728 (*0.000±0.000*)	0.1239 (*0.079±0.003*)	π = 0.130% h = 0.306±0.071

Probability values based on 10,000 permutations are shown in italic. Significant different values (p<0.05) in bold. Haplotype (h) and nucleotide (π) % ± standard deviation (SD) diversity values are shown in the diagonal of each population unit. Numbers of samples per location are shown in parentheses.

Results

Mitochondrial DNA COI analyses

A total of 323 sequences of the 655 bp COI fragment were successfully obtained from samples analyzed. Analysis of these data identified 24 variable sites that defined 19 unique haplotypes, although H11 was removed due to excessive missing data (Table 1), and H10 and H13 were removed from TCS and Beast analyses for the same reason. All haplotype sequences were submitted to GenBank under accession numbers KM218907–KM218923. The haplotype network derived from the TCS analysis and a phylogenetic tree showed two well differentiated and well supported clusters, hereafter referred to as clade A and clade B (Figures 2 and 3). Nine haplotypes were unique to individual regions (H2, H12, H17, H18, and H19 in the ETP region, H14 and H16 in Baja, and H7 and H15 in Taiwan) with the rest being observed across multiple regions. Haplotype H1 showed the highest frequency in the eastern Pacific, and the analysis suggested it to be the ancestral haplotype in this region. Haplotype H3 was the most common in the western Pacific and only seven samples from the eastern Pacific shared that haplotype. Haplotype H9 was shared between Hawaii and Baja. Of the sequences obtained from GenBank, the Indo-Pacific samples grouped within clade A and the single sample from Mexico grouped within clade B. Sampling locations at the edge of the sampling distribution (Taiwan and ETP) were composed entirely of haplotypes from a single clade, clade A haplotypes in Taiwan and clade B haplotypes in the ETP. Samples from Baja and Hawaii contained a mixture of haplotypes from both clades (Figure 2).

Comparisons of F_{ST} and Φ_{ST} between geographically sampled populations produced very high values of Φ_{ST} for most pairwise comparisons (Φ_{ST} 0.0624–0.8964) with most populations being significantly different from each other (Table 2), though comparison between Baja and ETP populations was not significant after Bonferroni correction. When comparisons were performed separating locations by mtDNA clade, comparison of clade A

populations produced a broad range of values of Φ_{ST} (Φ_{ST} 0.0768–0.4971) with significant differentiation for all comparisons involving Baja (Table 3). Baja had a very low clade A sample size so the power to differentiate this population is low and values of F_{ST} and Φ_{ST} are likely inaccurate. For the comparisons involving only Clade B populations, we found low values of Φ_{ST} (Φ_{ST} − 0.0066–0.0131) and non-significant differences in pairwise comparisons for both F_{ST} and Φ_{ST} (Table 4).

Haplotype diversity was similar in each sampled area with the highest nucleotide diversity found in Hawaii. Taiwan showed the lowest haplotype diversity compared to all other regions (Table 3).

Microsatellite analyses

Microsatellite fragments were analyzed for nine microsatellite loci for 331 individuals. Initial scoring identified two loci, Iox-01 and Iox-30 that were monomorphic in this species and were subsequently eliminated from downstream analyses. The number of alleles ranged between two (AV-I11) and 39 (AV-H8) across all sampled populations.

Genetic diversity values including expected (H_E) and observed heterozygosity (H_O), were obtained for seven loci in all populations units, along with deviations from H-W equilibrium (Table 5). H_E and H_O varied among population units at different loci. When samples were partitioned strictly by location, HWE analyses showed the Hawaii and Baja regions were significantly out of HWE at most loci while the Taiwan and ETP regions were mostly in HWE. The alternate sample partitioning that separated out clade A and B individuals by location showed groups to be in HWE for all loci except locus AVH8 in the sample set from Colombia and locus Iox-12 in samples from southern California (Table 5).

Evaluation of the K values produced by Structure using the ΔK method [26] identified K = 2 as the most likely number of groups present in the data both with and without consideration of the location prior used for either geographic location or mtDNA clade (Figure 4a, 4b and Figure 1S). In both Baja and Hawaii where the samples were composed of a mixture of individuals from both

Table 4. Pairwise Fst (below diagonal) and Φst (above diagonal) values for the COI gene when only Clade B haplotypes of *Alopias pelagicus* are analyzed.

F_{ST} \ Φ_{ST}	ETP (n = 121)	Baja (n = 47)	Hawaii (n = 37)
ETP	π = 0.104% h = 0.586±0.036	−0.0092 (*0.788±0.004*)	−0.0131 (*0.666±0.005*)
Baja	−0.0085 (*0.679±0.005*)	π = 0.095% h = 0.536±0.066	−0.0066 (*0.554±0.005*)
Hawaii	−0.0128 (*0.619±0.005*)	−0.0104 (*0.691±0.004*)	π = 0.088% h = 0.536±0.055

Probability values based on 10,000 permutations are shown in italic. Significant different values (p<0.05) in bold. Haplotype (h) and nucleotide (π) % ± standard deviation (SD) diversity values are shown in the diagonal of each population unit. Numbers of samples per location are shown in parentheses.

Table 5. Genetic diversity for seven microsatellite loci in all sampling locations analyzed.

loci		TW	HHWP	HHEP	BA	SC	CA	COL	ECU
		N = 71	N = 32	N = 36	N = 10	N = 18	N = 5	N = 92	N = 31
lox-12	n	n = 18	n = 15	n = 23	n = 14	**n = 14**	n = 8	n = 36	n = 25
	Ho	Ho = 0.929	Ho = 0.813	Ho = 0.778	Ho = 1.00	**Ho = 0.529**	Ho = 0.800	Ho = 0.900	Ho = 0.900
	He	He = 0.918	He = 0.897	He = 0.941	He = 0.963	**He = 0.941**	He = 0.956	He = 0.953	He = 0.953
	p	p = 0.0283	p = 0.3899	p = 0.0098	p = 1.000	**p = 0.0000**	P = 0.262	p = .0075	p = 0.0574
lox-M36	n	n = 3	n = 3	n = 3	n = 2	n = 2	n = 2	n = 2	n = 3
	Ho	Ho = 0.471	Ho = 0.219	Ho = 0.265	Ho = 0.111	Ho = 0.056	Ho = 0.000	Ho = 0.315	Ho = 0.138
	He	He = 0.405	He = 0.249	He = 0.415	He = 0.111	He = 0.056	He = 0.533	He = 0.360	He = 0.194
	p	p = 0.2479	p = 0.4805	p = 0.0419	p = 1.000	p = 1.000	p = 0.0474	p = 0.2489	p = 0.1061
AV-I11	n	n = 3	n = 2	n = 2	n = 2	n = 2	n = 2	n = 2	n = 3
	Ho	Ho = 0.296	Ho = 0.188	Ho = 0.472	Ho = 0.500	Ho = 0.222	Ho = 0.400	Ho = 0.609	Ho = 0.581
	He	He = 0.306	He = 0.222	He = 0.488	He = 0.395	He = 0.514	He = 0.533	He = 0.494	He = 0.539
	p	p = 0.4527	p = 0.3912	p = 1.000	p = 1.000	p = 0.0215	P = 1.000	p = 0.0328	p = 0.0161
AV-H8	n	n = 31	n = 25	n = 29	n = 12	n = 19	n = 6	**n = 39**	n = 21
	Ho	Ho = 0.939	Ho = 0.931	Ho = 0.879	Ho = 0.778	Ho = 0.875	Ho = 0.800	**Ho = 0.965**	Ho = 0.900
	He	He = 0.953	He = 0.960	He = 0.971	He = 0.922	He = 0.958	He = 0.844	**He = 0.967**	He = 0.953
	p	p = 0.2991	p = 0.4292	p = 0.0957	p = 0.0562	p = 0.0152	p = 0.7962	**p = 0.0009**	p = 0.0025
AV-H138	n	n = 16	n = 10	n = 19	n = 9	n = 11	n = 9	n = 32	n = 24
	Ho	Ho = 0.958	Ho = 0.750	Ho = 0.861	Ho = 0.900	Ho = 0.944	Ho = 0.800	Ho = 0.868	Ho = 0.935
	He	He = 0.872	He = 0.826	He = 0.889	He = 0.826	He = 0.887	He = 0.978	He = 0.902	He = 0.937
	p	p = 0.3662	p = 0.0952	p = 0.4408	p = 0.9976	p = 0.4609	p = 0.1204	p = 0.1349	p = 0.0481
AV-H110	n	n = 16	n = 15	n = 18	n = 10	n = 16	n = 5	n = 21	n = 18
	Ho	Ho = 0.9	Ho = 0.903	Ho = 0.944	Ho = 0.700	Ho = 0.938	Ho = 0.800	Ho = 0.921	Ho = 0.900
	He	He = 0.913	He = 0.888	He = 0.908	He = 0.916	He = 0.952	He = 0.822	He = 0.906	He = 0.905
	p	p = 0.5505	p = 0.4710	p = 0.9680	p = 0.1221	p = 0.7795	p = 0.8999	p = 0.1407	p = 0.1942
lox-M115	n	n = 9	n = 8	n = 8	n = 5	n = 5	n = 4	n = 8	n = 7
	Ho	Ho = 0.818	Ho = 0.667	Ho = 0.548	Ho = 0.500	Ho = 0.333	Ho = 1.00	Ho = 0.698	Ho = 0.643
	He	He = 0.808	He = 0.812	He = 0.644	He = 0.683	He = 0.601	He = 0.733	He = 0.718	He = 0.641
	p	p = 0.3055	p = 0.2376	p = 0.1459	p = 0.3804	p = 0.0388	p = 0.3956	p = 0.1785	p = 0.7269

N = sample size for each population; n = total number of alleles. Ho = observed heterozygosity. He = expected heterozygosity. Significant scores after Bonferroni correction for loci out of equilibrium are shown in bold.

A

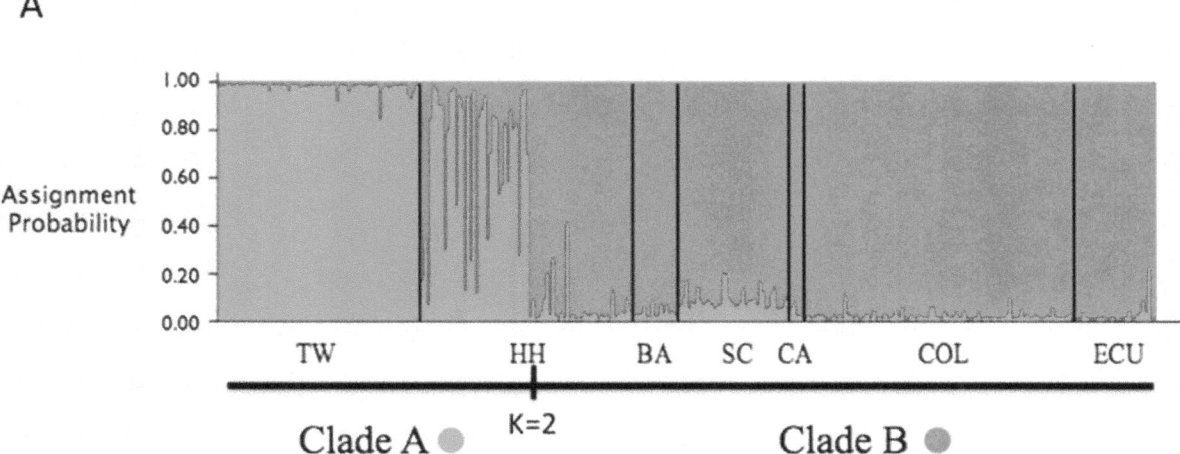

B

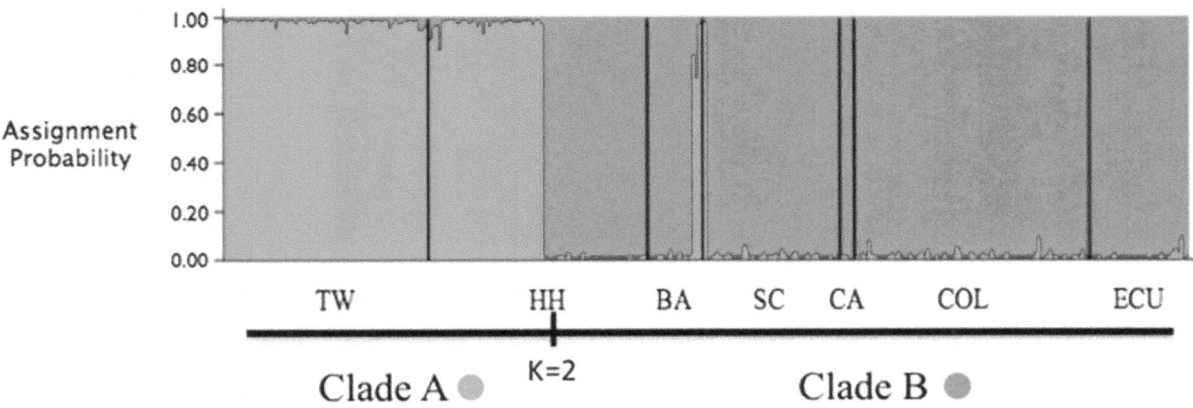

Figure 4. Structure bar plot showing the assignment probabilities (K = 2) of each genotyped individual of *A. pelagicus* from the different sampling locations in the Pacific Ocean. A) sampling region used as location prior. B) COI haplotype clade (A or B) used as the location prior.

clades, use of the clade ID as the location prior resulted in very high assignment of individuals to their respective group (Figure 4a and 4b).

When samples were grouped strictly by location, F_{ST} (0.000–0.075) and R_{ST} (−0.032–0.135) comparisons were significant for most comparisons with Taiwan and between Hawaii and Colombia and Ecuador (Table 6). After separating samples by region and haplotype clade, all F_{ST} (−0.003–0.109) and R_{ST} (−0.032–0.338) comparisons remained large and significant between clade A and B groups while comparisons within clade B group were low and not significant (Table 7). Comparisons within the clade A (e.g. Taiwan and Hawaii Clade A) were significant for F_{ST} and non-significant for R_{ST} values (Table 7). Non-significant values of R_{ST} between Taiwan, Baja, southern California and Central America were found, though this may be an artifact of low sample size for some of these comparisons (Table 6, 7).

As both F_{ST} and R_{ST} are inappropriate measures at deeper levels of divergence, both Jost's D and exact test analyses were performed (Table 8). Jost's D showed a high degree of differentiation between clade A and B groups (0.151–0.357) as well as within clade A locations (0.089). The exact test results identified additional differentiation within clade B locations (Table 8).

Comparisons between sample partitions

Structure likelihood values were compared over multiple K's both with no location prior as well as considering the location prior for both data partitions (geographic location and mtDNA clade). In all analyses K = 2 had the highest likelihood (Table S2). Among the three run variants for K = 2, utilization of the mtDNA clade as location prior had the highest likelihood (−8756.49+−7.88) followed by geographic location as location prior (−8774.73+−7.26) (Table S2).

Table 6. Population differentiation based on Fst and Rst values between pairwise populations with seven microsatellite loci.

	TW	HH	BA	SC	CA	COL	ECU
TW (n=71)	------	**0.037**	0.124	0.038	0.231	**0.115**	**0.135**
HH (n=68)	**0.020**	------	0.101	0.007	0.199	**0.071**	**0.081**
BA (n=13)	**0.066**	0.018	------	0.042	−0.011	−0.019	−0.009
SC (n=18)	0.043	0.013	0.024	------	0.130	0.036	0.027
CA (n=5)	**0.075**	0.039	0.045	0.059	------	0.003	−0.032
COL(n=91)	**0.047**	**0.008**	0.009	0.009	0.021	------	−0.014
ECU(n=31)	**0.041**	0.009	0.017	−0.000	0.037	0.003	------

Significant scores after Bonferroni correction are in bold and probability values are based on 10,000 permutations. Below diagonal Fst values and above diagonal Rst values. Numbers of samples per location are shown in parentheses.

Table 7. Population differentiation based on Fst and Rst values between pairwise populations with seven microsatellite loci with populations split by mtDNA clade.

	TW (A)	HH (A)	HH (B)	BA (B)	SC (B)	CA (B)	COL (B)	ECU (B)
TW (A) (n=71)	------	0.008	**0.087**	0.171	0.038	0.231	**0.115**	**0.135**
HH (A) (n=32)	**0.012**	------	**0.118**	**0.277**	0.092	**0.338**	**0.150**	**0.156**
HH (B) (n=36)	**0.056**	**0.063**	------	0.098	0.001	0.093	0.018	0.020
BA (B) (n=10)	**0.094**	**0.100**	0.013	------	0.103	0.014	0.021	0.023
SC (B) (n=18)	0.043	**0.053**	0.012	0.027	------	0.130	0.036	0.027
CA (B) (n=5)	0.075	**0.109**	0.009	0.070	0.059	------	0.003	−0.032
COL (B) (n=91)	**0.047**	**0.055**	−0.003	0.020	0.009	0.021	------	−0.014
ECU (B) (n=31)	**0.041**	**0.046**	0.007	0.025	0.000	0.037	0.003	------

Significant scores after Bonferroni correction are in bold and probability values are based on 10,000 permutations. Below diagonal Fst values and above diagonal Rst values. Mitochondrial DNA clade and numbers of samples per location are shown in parentheses.

Table 8. Population differentiation based on Jost's D and Exact test values between pairwise populations with seven microsatellite loci.

	TW (A)	HH (A)	HH (B)	BA (B)	SC (B)	CA (B)	COL (B)	ECU (B)
TW (A) (n=71)	-----	+	++	++	++	++	++	++
HH (A) (n=32)	**0.089** 0.079±0.148	-----	++	++	++	++	++	++
HH (B) (n=36)	**0.245** 0.202±0.274	**0.158** 0.118±0.216	-----	-	-	-	-	-
BA (B) (n=10)	**0.303** 0.209±0.367	**0.189** 0.082±0.277	0.065 -0.049±0.149	-----	-	+	+	-
SC (B) (n=18)	**0.213** 0.187±0.293	**0.190** 0.105±0.248	0.036 0.015±0.154	0.144 0.032±0.281	-----	+	-	-
CA (B) (n=5)	**0.338** 0.159±0.496	**0.357** 0.239±0.576	0.085 -0.087±0.251	0.334 0.109±0.544	0.237 0.095±0.515	-----	-	-
COL(B) (n=92)	**0.211** 0.226±0.269	**0.151** 0.122±0.195	0.002 0.022±0.088	0.136 0.059±0.266	0.031 -0.021±0.097	0.150 0.052±0.41	-----	+
ECU(B) (n=31)	**0.241** 0.205±0.275	**0.167** 0.149±0.251	0.027 -0.016±0.082	0.141 0.038±0.239	0.010 -0.005±0.1448	0.168 -0.040±0.345	0.004 0.027±0.099	-----

Significant scores after Bonferroni correction are in bold. Below diagonal Jost's D values with 95% confidence intervals and above diagonal exact test significance. Degrees of significance: $p<0.0018$ (+). $p<0.000001$ (++). Mitochondrial DNA clade and numbers of samples per location are shown in parentheses.

AMOVA comparisons of microsatellite data were performed to compare hypothesized groups (Table 9). Grouping samples by geographic region (Western Pacific, Central Pacific, Eastern Pacific) explained 1.8% of the variance ($\Phi_{CT} = 0.018$ p = NS), but significant within group variance remained ($\Phi_{SC} = 0.018$ p< 0.05). Samples grouped by mtDNA clade (i.e. clade A and B) explained 4.9% of the variance ($\Phi_{CT} = 0.049$ p<0.05), while within group variance was fairly low ($\Phi_{SC} = 0.008$ p<0.05) but still significant.

Discussion

This study presents the first extensive analyses of the molecular ecology of *A. pelagicus* in the Pacific Ocean using both mitochondrial and nuclear molecular markers. Though an earlier study using mtDNA control region data [14], showed evidence for differences between eastern and western Pacific populations, the study was hampered by low sample size and geographic coverage, particularly a lack of samples from the central Pacific. The COI data from this study is largely concordant with this previous study, which is not surprising as the entire mtDNA molecule is a single locus. Both datasets found well-defined phylogenetic clades that were strongly separated by geography with samples on either side of the Pacific almost entirely composed of a single regionally specific clade. Samples collected in the Central Pacific represented a ~50:50 mix of these two clades offering a unique opportunity to test for reproductive isolation between these clades in an area of sympatry using nuclear markers.

A core task of this study was to evaluate and extend earlier findings by Eitner [13] and Trejo [14] regarding the pelagic thresher shark. By addition of multiple nuclear microsatellite loci and a broader geographic sampling we were able to test whether the distinct biogeographic pattern observed by Trejo [14] was maintained and evaluate the level of gene flow between eastern and western Pacific populations. We were also able to test whether the data better supported geographically separated populations or whether these distinct mtDNA clades represent reproductively isolated units. Microsatellite loci were mostly out of HWE in populations (i.e. Hawaii, Baja) that contained a mixture of animals with clade A and clade B haplotypes. When individuals within these populations were separated by mtDNA clade, deviations from HWE were mostly eliminated (Table 5). Structure analyses using the location prior to compare likelihoods of geographic versus mtDNA partitions indicated that grouping samples by mtDNA clade had a higher likelihood than grouping samples by geographic location (Table S2). However, in both cases, the K value was the same, with two as the most likely number of groups. Similarly, when AMOVA was run to compare these two hypothesized groupings, grouping samples by mtDNA clade explained three times the genetic variance as grouping samples geographically, while also reducing the within group variance. Structure results using geographic location as a prior resulted in two distinct groups, largely concordant with patterns observed with mtDNA haplotypes (Figure 4a). When the Structure analyses considered the mtDNA clade as the location prior there was very strong assignment of individuals to two distinct groups with little evidence for introgression (Figure 4b). Together these analyses support that the mtDNA clades first observed by Trejo are reproductively isolated from each other, even in areas of sympatry.

These results support the existence of two groups on separate evolutionarily trajectories, conforming to the concept of Evolutionarily Significant Units (ESUs). ESUs have been described based on the presence of reciprocal monophyly for organelle haplotypes (e.g. mitochondrial DNA) and significant divergence of

Table 9. AMOVA results for both mtDNA and microsatellite data for alternative grouping of samples.

Groups	mtDNA			Microsatellites		
	Φct	Φsc	Φst	Φct	Φsc	Φst
Regional Grouping	0.643	**0.009**	**0.646**	0.018	**0.023**	**0.041**
mtDNA Clade Grouping	**0.879**	**0.190**	**0.902**	**0.049**	**0.008**	**0.057**

Significant values in bold (p<0.05).

allele frequencies at nuclear loci (e.g. microsatellites, [30]). In this study, we found two reciprocally monophyletic mtDNA clades with significant divergence of allele frequencies at microsatellite loci and a significant level of reproductive isolation between both clades in the Pacific Ocean based on the Structure analysis (Figure 4), both when this analysis was run using the geographic sampling location as prior or the COI clade as prior. The reason for this marked population division remains unclear.

Observed levels of H_O and H_E were similar across all locations and were generally higher than values observed in other oceanic species [6,31,32], but similar to those found in *I. oxyrinchus* [33]. The lower haplotype diversity in Taiwan could be a consequence of the sampling methodology as most samples came from a two sampling events, which could increase the chance of sampling related individuals. Though we sampled most of the geographic range of *A. pelagicus* in the Pacific, they are commonly found throughout the Indian Ocean so samples from the rest of the species range should be analyzed in order to assess the true population structure and genetic diversity within this ESU.

The overall findings in this study are in support of the assertion that a cryptic taxon exists within *Alopias* [13]. During the course of this study, several of the samples that Eitner [13] used in his study were evaluated for inclusion. Our initial analyses indicated that the *A. pelagicus* samples used in his study were most likely *A. superciliosus* and his unrecognized taxon was most likely *A. pelagicus*. Because of this discrepancy these samples were not included in this study.

Both our study and that of Trejo [14] identified regionally specific mtDNA clades on both sides of the Pacific. Samples from the central Pacific identified animals with haplotypes from both ESUs at high frequencies with little evidence for introgression in the microsatellite loci analyses. These results leave some new questions open; are both ESUs found year-round in this region or do they seasonally migrate here? Is there spatial or temporal separation between these groups that allow for reproductive isolation? Future movement studies should be conducted using electronic tags, especially in this zone of overlap.

Conclusions

The results of this study provide important information to scientists, resource managers and governmental agencies regarding management and conservation of pelagic thresher sharks. The existence of two ESUs of *A. pelagicus* in the Pacific Ocean and the genetic differentiation presented here is the highest found in the

literature for a large epipelagic shark. Moreover, considering the slow mutation rate of sharks compared to other vertebrates [34], the strong genetic differentiation found in *A. pelagicus* in the Pacific Ocean, and the almost nonexistent geneflow, this is likely an indicator of the existence of a cryptic species complex. Due to conflicts with the species concept found in the literature, we choose to leave our findings at the ESU level, although future research should include analyses of both morphological data and additional genetic markers. Regardless of the taxonomic label, the two ESUs described in this study warrant attention as they effectively form regional eastern and western Pacific populations which given their restricted geographic distribution makes them especially susceptible to overexploitation. These findings should be considered in management plans and initiatives such as the different National Actions Plans for the Conservation of Elasmobranch Species.

Supporting Information

Figure S1 Plot of the second order rate of change of the likelihood (ΔK) showing the true value of K after testing $K = 1-10$ with 20 repetitions each.

Table S1 Name, PCR profile number, annealing temperature, reference or Genbank accession numbers for primers used in this study.

Table S2 Structure Likelihood values for multiple Ks with No Location Prior, Geographic Location Prior and mtDNA Clade Prior.

Acknowledgments

We want to thank all the members of the LEMVA (Universidad de Los Andes) and the Fisheries Resources Division (NOAA SWFSC) for their support and help provided during this research. Thanks to Mike Musyl (NMFS-PIFSC), Wei-Chuan Chiang (Fisheries Research Institute Taiwan) for providing samples, and all the people from INCODER, NOAA and IATTC involved in the tissue collection.

Author Contributions

Conceived and designed the experiments: SC JH DC. Performed the experiments: DC JH SC. Analyzed the data: DC JH SC. Contributed reagents/materials/analysis tools: JH SC. Wrote the paper: DC SC JH.

References

1. Kuussaari M, Kankare M, Vikman P, Fortelius W (1998) Inbreeding and extinction in a butterfly metapopulation. Nature 392: 491–494. doi:10.1038/33136

2. Westemeier RL (1998) Tracking the Long-Term Decline and Recovery of an Isolated Population. Science 282: 1695–1698. doi:10.1126/science.282.5394.1695

3. Schrey AW, Heist EJ (2002) Microsatellite markers for the shortfin mako and cross-species amplification in lamniformes. Conserv Genet 3: 459–461.

4. Hoelzel RA, Shivji MS, Magnussen J, Francis MP (2006) Low worldwide genetic diversity in the basking shark (*Cetorhinus maximus*). Biology Letters 2: 639–642. doi:10.1098/rsbl.2006.0513

5. Castro ALF, Stewart BS, Wilson SG, Hueter RE, Meekan MG, et al. (2007) Population genetic structure of Earth's largest fish, the whale shark (*Rhincodon typus*). Molecular Ecology 16: 5183–5192. doi:10.1111/j.1365-294X.2007.03597.x

6. Schmidt JV, Schmidt CL, Ozer F, Ernst RE, Feldheim KA, et al. (2009) Low Genetic Differentiation across Three Major Ocean Populations of the Whale Shark, *Rhincodon typus*. PLoS ONE 4: e4988. doi:10.1371/journal.pone.0004988.t006

7. Ovenden JR, Kashiwagi T, Broderick D, Giles J, Salini J (2009) The extent of population genetic subdivision differs among four co-distributed shark species in the Indo-Australian archipelago. BMC Evol Biol 9: 40. doi:10.1186/1471-2148-9-40

8. Liu K, Chen C, Liao T, Joung S (1999) Age, growth, and reproduction of the pelagic thresher shark, *Alopias pelagicus* in the northwestern Pacific. Copeia 1: 68–74.

9. Camhi MD, Pikitch EK, Babcock E (2008) Sharks of the Open Ocean. Camhi MD, Pikitch EK, Babcock E, editors Blackwell Publishing Ltd. 1 pp.

10. Caldas JP, Castro-González E (2010) Plan de Acción Nacional para la Conservación y Manejo de Tiburones, Rayas y Quimeras de Colombia (PAN-Tiburones Colombia).

11. Tsai W-P, Liu K-M, Joung S-J (2010) Demographic analysis of the pelagic thresher shark, *Alopias pelagicus*, in the north-western Pacific using a stochastic stage-based model. Mar Freshwater Res 61: 1056. doi:10.1071/MF09303

12. Caballero S, Cardeñosa D, Soler G, Hyde J (2011) Application of multiplex PCR approaches for shark molecular identification: feasibility and applications for fisheries management and conservation in the Eastern Tropical Pacific. Molecular Ecology Resources 12: 233–237. doi:10.1111/j.1755-0998.2011.03089.x

13. Eitner BJ (1995) Systematics of the genus *Alopias* (Lamniformes: Alopiidae) with evidence for the existence of an unrecognized species. Copeia 3: 562–571.

14. Trejo T (2005) Global Phylogeography of Thresher Sharks (*Alopias* spp.) Inferred From Mitochondrial DNA Control Region Sequences. A thesis submitted to the faculty of Caifornia State University Monterey Bay in partial fulfillment of the requirements for the degree: 1–58.

15. Hyde JR, Lynn E, Humphreys R Jr, Musyl M, West AP, et al. (2005) Shipboard identification of fish eggs and larvae by multiplex PCR, and description of fertilized eggs of blue marlin, shortbill spearfish, and wahoo. Mar Ecol Prog Ser 286: 269–277.

16. Ward RD, Zemlak TS, Innes BH, Last PR, Hebert PDN (2005) DNA barcoding Australia's fish species. Philosophical Transactions of the Royal Society B: Biological Sciences 360: 1847–1857. doi:10.1093/molbev/msg133

17. Gouy M, Guindon S, Gascuel O (2010) SeaView Version 4: A Multiplatform Graphical User Interface for Sequence Alignment and Phylogenetic Tree Building. Mol Biol Evol 27: 221–224. doi:10.1093/molbev/msp259

18. Maddison D, Maddison W (2000) MacClade: Analysis of Phylogeny and Character Evolution. Sinauer, Sunderland, MA, USA.

19. Clement M, Posada D, Crandall KA (2000) TCS: a computer program to estimate gene genealogies. Molecular Ecology 9: 1657–1659.

20. Posada D (2008) jModelTest: phylogenetic model averaging. Mol Biol Evol 25: 1253–1256. doi:10.1093/molbev/msn083

21. Drummond AJ, Rambaut A (2007) BEAST: Bayesian evolutionary analysis by sampling trees. BMC Evol Biol 7: 214. doi:10.1186/1471-2148-7-214

22. Excoffier L, Laval G, Schneider S (2005) Arlequin (version 3.0): an integrated software package for population genetics data analysis. Evol Bioinform Online 1: 47–50.

23. Pritchard JK, Stephens M, Donnelly P (2000) Inference of population structure using multilocus genotype data. Genetics 155: 945–959.

24. Hubisz MJ, Falush D, Stephens M, Pritchard JK (2009) Inferring weak population structure with the assistance of sample group information. Molecular Ecology Resources 9: 1322–1332. doi:10.1111/j.1755-0998.2009.02591.x

25. Hubisz MJ, Falush D, Stephens M, Pritchard JK (2009) Inferring weak population structure with the assistance of sample group information. Molecular Ecology Resources 9: 1322–1332. doi:10.1111/j.1755-0998.2009.02591.x

26. Evanno G, Regnaut S, Goudet J (2005) Detecting the number of clusters of individuals using the software STRUCTURE: a simulation study. Molecular Ecology 14: 2611–2620. doi:10.1111/j.1365-294X.2005.02553.x

27. Raymond M, Rousset F (1995) GENEPOP (version 1.2): population genetics software for exact tests and ecumenicism. J Hered 86: 248–249.

28. Jost L (2008) Gst and its relatives do not measure differentiation. Molecular Ecology 17: 4015–4026. doi:10.1111/j.1365-294X.2008.03887.x

29. Gerlach G, Jueterbock A, Kraemer P, Deppermann J, Harmand P (2010) Calculations of population differentiation based on GST and D: forget GST but not all of statistics! Molecular Ecology 19: 3845–3852.

30. Moritz C (1994) Defining "Evolutionarily Significant Units" for conservation. Trends in Ecology & Evolution 9: 373–375. doi:10.1016/0169-5347(94)90057-4

31. Pardini AT, Jones CS, Scholl MC, Noble LR (2000) Isolation and characterization of dinucleotide microsatellite loci in the Great White Shark, *Carcharodon carcharias*. Molecular Ecology 9: 1176–1178.

32. Daly-Engel TS, Seraphin KD, Holland KN, Coffey JP, Nance HA, et al. (2012) Global Phylogeography with Mixed-Marker Analysis Reveals Male-Mediated Dispersal in the Endangered Scalloped Hammerhead Shark (*Sphyrna lewini*). PLoS ONE 7: e29986. doi:10.1371/journal.pone.0029986.t004

33. Schrey AW, Heist EJ (2003) Microsatellite analysis of population structure in the shortfin mako (*Isurus oxyrinchus*). Canadian Journal of Fisheries and Aquatic Sciences 60: 670–675. doi:10.1139/f03-064

34. Martin AP (1995) Mitochondrial DNA sequence evolution in sharks: rates, patterns, and phylogenetic inferences. Mol Biol Evol 12: 1114–1123.

16

Dealing with Discordant Genetic Signal Caused by Hybridisation, Incomplete Lineage Sorting and Paucity of Primary Nucleotide Homologies: A Case Study of Closely Related Members of the Genus *Picris* Subsection *Hieracioides* (Compositae)

Marek Slovák[1]*, Jaromír Kučera[1], Eliška Záveská[2], Peter Vďačný[3]

1 Institute of Botany, Slovak Academy of Sciences, Bratislava, Slovakia, **2** Department of Botany, Charles University, Praha, Czech Republic, **3** Department of Zoology, Comenius University, Bratislava, Slovakia

Abstract

We investigated genetic variation and evolutionary history of closely related taxa of *Picris* subsect. *Hieracioides* with major focus on the widely distributed *P. hieracioides* and its closely related congeners, *P. hispidissima*, *P. japonica*, *P. olympica*, and *P. nuristanica*. Accessions from 140 sample sites of the investigated *Picris* taxa were analyzed on the infra- and the inter-specific level using nuclear (ITS1-5.8S-ITS2 region) and chloroplast (*rpl32-trnL*[(UAG)] region) DNA sequences. Genetic patterns of *P. hieracioides*, *P. hispidissima*, and *P. olympica* were shown to be incongruent and, in several cases, both plastid and nuclear alleles transcended borders of the taxa and genetic lineages. The widespread *P. hieracioides* was genetically highly variable and non-monophyletic across both markers, with allele groups having particular geographic distributions. Generally, all gene trees and networks displayed only a limited and statistically rather unsupported resolution among ingroup taxa causing their phylogenetic relationships to remain rather unresolved. More light on these intricate evolutionary relationships was cast by the Bayesian coalescent-based analysis, although some relationships were still left unresolved. A combination of suite of phylogenetic analyses revealed the ingroup taxa to represent a complex of genetically closely related and morphologically similar entities that have undergone a highly dynamic and recent evolution. This has been especially affected by the extensive and recurrent gene flow among and within the studied taxa and/or by the maintenance of ancestral variation. Paucity of phylogenetically informative signal further hampers the reconstruction of relationships on the infra- as well as on the inter-specific level. In the present study, we have demonstrated that a combination of various phylogenetic analyses of datasets with extremely complex and incongruent phylogenetic signal may shed more light on the interrelationships and evolutionary history of analysed species groups.

Editor: Simon Joly, Montreal Botanical Garden, Canada

Funding: Funding for this study was provided by The Research and Development Support Agency, Bratislava, Slovakia (grants no. APVV-0239-09 and APVV-0320-10 to Karol Marhold). The funder had no role in study design, data collection and analysis, decision to publish, or preparation of the manuscript.

Competing Interests: The authors have declared that no competing interests exist.

* Email: marek.slovak@savba.sk

Introduction

Identifying patterns of genetic variation within and among closely allied congeners is crucial in understanding their evolutionary relationships and processes involved in their diversification. Exploration of genetic patterns across the entire geographic ranges of each of the constituent species not only enables clarification of modes of their diversification and historical changes in spatial distribution, but may also shed light on their potential reciprocal interactions (e.g. hybridisation and introgression).

In many instances, however, multi-gene phylogenies of closely related and recently diversified taxa are incongruent, often with alleles transcending species or genetic groups, and finally cause the genetic patterns to exhibit a diffusive or mosaic-like structure [1,2,3,4,5]. A suite of various factors and processes might lead to discordant topologies of genetic trees and complex genetic patterns. The most commonly detected evolutionary events affecting genetic variation are hybridization and introgression, gene duplication, incomplete lineage sorting and/or recombination [6,7,8,9]. Eventually, the low level of variation in the involved genetic markers and the different molecular nature of employed coding and non-coding regions in nuclear, mitochondrial and chloroplast DNA may be a source of gene trees incongruence. Subjective taxonomic delimitation of studied taxa not corresponding with real evolutionary patterns must be also taken into account [10,11]. It may prove difficult to correctly diagnose the relative influence of those factors on the studied species, because they all result in very similar complex genetic patterns and discordant phylogenetic inferences [9,12,13].

Studies that have carefully and explicitly analyzed multiple reasons of genetic incongruence are not common [14,15,16,17]. On the contrary, the overwhelming majority of studies were focused on only a subset of possible reasons of incongruence, however, using a variety of different approaches [1,2,3,5]. Reconstruction of species relationships, i.e. species trees based on multiple gene trees, is recently performed using coalescent models that can handle incongruence caused by incomplete lineage sorting [18]. However, these models are applicable only to datasets where no gene flow (hybridization) among species is to be expected [19,20]. Indeed, it is fairly difficult to distinguish between effects caused by hybridisation and incomplete lineage sorting [9]. One way how to overcome these difficulties has been proposed by Joly et al. [9] who developed an approach enabling to test the influence of hybridization in the presence of incomplete lineage sorting, as implemented in the software JML [20].

Members of *Picris* subsect. *Hieracioides* Vassiliev (Compositae) [21] represent an interesting group for investigation of genetic patterns, diversification and reciprocal interactions within and among closely related species that differ in ecological requirements and geographic distribution [22,23,24]. According to the current taxonomic concepts [21,23,24], *Picris* subsect. *Hieracioides* comprises seven morphologically very similar taxa from Europe, Asia and Africa: *P. abyssinica* Sch. Bip., *P. hieracioides* L. including subspecies *P. hieracioides* subsp. *umbellata* (Schrank) Ces. and *P. hieracioides* L. subsp. *hieracioides* (referred as *P. h.* subsp. *umbellata* and *P. h.* subsp. *hieracioides* in the text), *P. hispidissima* (Bartl.) W. D. J. Koch, *P. japonica* Thunb., *P. nuristanica* Bornm., and *P. olympica* Boiss [21,24,25]. Information on overall relationships within the genus *Picris* subsect. *Hieracioides* is still limited. Only three taxa from the subsection *Hieracioides*, namely *P. abyssinica*, *P. hieracioides* and *P. nuristanica*, were marginally analysed in a study focused on the genus *Leontodon*. The two latter *Picris* species were shown to be closely related, while *P. abyssinca* was genetically much more distant and clustered together with other North African *Picris* taxa [26]. All members of the subsection *Hieracioides* are reported to be perennials with homocarpic achenes (i.e. with same morphological type of achenes within capitulum) and two-hooked anchor hairs on the stem and peduncles [21]. Here, we focus on species from Europe and Asia Minor, with main attention paid besides of the widespread and highly variable *P. hieracioides* also to *P. hispidissima*, *P. olympica* and marginally also to *P. japonica* and *P. nuristanica*.

Although our previous investigations on *P. hieracioides*, *P. hispidissima*, *P. japonica*, and *P. nuristanica* revealed that these taxa are very similar morphologically and karyologically as well as they are closely related genetically (AFLP), their evolutionary relationships and mutual interactions were left unresolved [22–24,27,28]. Additionally, our previous studies [23,24] did detect morphologically intermediate plants in mixed populations of *P. hispidissima* and *P. hieracioides* and also found populations that represented genetic admixtures of different *P. hieracioides* subspecies and/or genetic groups, mostly accompanied by transitional morphology. In all cases, the observed intermediate genetic and/or morphological states were hypothetically attributed to hybridisation although this heterogeneity may be also a result of retention of ancestral variation. No studies collecting the relevant data on *P. olympica* have been done prior to the present study. Likewise, overall relationships within *Picris* subsect. *Hieracioides* have not been investigated yet. Thus the specific goals of the present study are:

(1) Clarifying the phylogenetic relationships among *P. hieracioides*, *P. hispidissima*, *P. japonica*, *P. olympica*, and *P. nuristanica*.

(2) Exploring the overall genetic pattern of *P. hieracioides* in Europe and Asia Minor, the distribution of its infraspecific genetic groups, and comparing its genetic variation with that of *P. hispidissima* and *P. olympica*.

(3) Shed more light on potential interactions within the studied group and on the processes acting during evolutionary history.

Material and Methods

Studied species

Picris hieracioides is a biennial to perennial, morphologically highly variable species that includes two infraspecific taxa, *P. h.* subsp. *umbellata* and *P. h.* subsp. *hieracioides*. The subspecies of *P. hieracioides* are morphologically distinguishable by a combination of several morphological traits on both vegetative and reproductive organs. This was definitely proved also by cultivation experiments under environmentally homogeneous conditions [24]. Beside this, they also differ in longevity, ecological requirements, and geographic distribution [22,24]. Additionally, each subspecies harbours two highly allopatric genetic (AFLP) and morphologically cryptic groups [24]. *Picris hispidissima* is a biennial species (M. Slovák, unpublished data) that differs morphologically from *P. hieracioides* in having a pectinate-ciliate indumentum of involucral bracts and inflation of the peduncle [23]. The last studied species, *P. olympica*, is a perennial characterized by a low caespitose growth with quite short ascending scapose-like stems bearing only a few capitula. It is considered to be morphologically closely related to *P. hieracioides* [21,29]. The investigated species differ conspicuously in their ecological requirements and geographic ranges, with *P. hieracioides* being the most widespread (distributed across the major part of Europe and Asia Minor) and having wide ecological amplitude [21,24]. The nominate *P. h.* subs. *hieracioides* is mainly biennial, growing on open, dry, xerothermous biotopes with no bedrock preference and thriving in lowlands or lower mountains throughout all of Europe. In contrast, *P. h.* subsp. *umbellata* is a short-lived perennial herb inhabiting humid tall herb communities from montane forests to sub-alpine communities across European calcareous mountain ranges. Both *P. hispidissima* and *P. olympica* have restricted distributions. The former occurs in dry, calcareous rocky slopes and crevices in the Adriatic coastal mountains from Croatia to Montenegro, while the latter is restricted to a few high mountain ranges in western Anatolia where it inhabits open, craggy grasslands with alkaline bedrocks at alpine levels [21,29]. All species, but especially *P. hieracioides*, have spread from their natural biotopes into man-made habitats and migrated along anthropogenic corridors ([22,24]; M. Slovák unpublished data).

Sampling strategy and plant material

In order to perform the reconstruction of phylogenetic relationships among the studied species and to reveal their genetic patterns, we aimed to include as many populations as possible from the entire geographic ranges of *P. hieracioides*, *P. hispidissima*, and *P. olympica* (Table S1). This, however, was at the expense of sample size at the population level [30,31,32]. We also co-analyzed a limited number of samples of the closely related Asian taxa *P. japonica* (2 populations) and *P. nuristanica* (1 population). We did not include samples of *P. abyssinica* since it was proved that it is genetically highly divergent from *P. hieracioides* and *P. nuristanica* but is more closely related to other African taxa [26]. Each population of the investigated species had more than several hundreds of individuals, and the taxa under study are neither endangered nor protected and no specific permits were required to collect plant samples at the study sites. We sequenced one to three accessions per population for the nuclear ITS and the plastid $rpl32$-$trnL^{(UAG)}$ regions (Table S1). To cover as high proportion of genetic variation of analysed taxa as possible, but simultaneously with respect to sampling limitation at population level, we decided to analyse more than one individual for populations where non-homogenized ITS sequences indicated existence of more than a single allele. The widespread *P. hieracioides* including populations morphologically corresponding to both subspecies, *P. h.* subsp. *hieracioides* and *P. h.* subsp. *umbellata*, was collected from 122 sample sites covering both subspecies' entire geographic distributions, and henceforth these sites are referred to as populations (Table S1). Twelve populations of *P. hispidissima* from its entire distribution area were sampled [23]. Three of the five currently recognized populations of *P. olympica* were also collected for the purposes of the present study.

Intermediate morphological variation was identified in a few populations of *P. hieracioides* which were mostly located in zones where the two subspecies or their lineages meet [24]. Since the extent of intermediate morphological variation across these populations was wide, it was impossible to unequivocally sort some of them into discrete categories. Therefore, in order to avoid formation of artificial entities, which might lead to subsequent biasness of results and interpretations, we assigned particular populations a priori not to accepted subspecies, but we considered all of them as *P. hieracioides* instead.

Selection of appropriate outgroup taxa was based on previously published studies [21,26] and our preliminary phylogenetic analyses. Thus a priori selected outgroup comprised members of all sections of the genus *Picris* (*P. galilaea* (Boiss.) Eig., *P. capuligera* (Durieu) Walp., *P. pauciflora* Willd., *P. rhagodioloides* (L.) Desf., *P. scaberrima* Ten., *P. sinuata* (Lam.) Lack, and *P. strigosa* M. Bieb.) and two members of the closely related genus *Helminthotheca* (*H. aculeata* (Vahl) Lack and *H. echioides* (L.) Holub). Voucher specimens were deposited in the herbarium of the Institute of Botany at the Slovak Academy of Science (SAV).

DNA sequence markers, DNA extraction and PCR amplification

We utilised sequences of the nuclear ribosomal internal transcribed spacer (ITS1-5.8S-ITS2 region) and the non-coding cpDNA intergenic spacer $rpl32$-$trnL^{(UAG)}$ [33]. Several single-copy nuclear genes (A25, A28 and B12) [34] and another non-coding chloroplast regions (3'$trnV^{(UAC)}$-$ndhC$, $ndhF$-$rpl32$ and $trnH^{(GUG)}$-$psbA$) [33,35] were tested during preliminary screening. However, these were proved to be unsuitable for the present study due to low levels of variation and/or problems encountered during their PCR amplification.

Total genomic DNA was extracted from silica gel-dried leaf tissue by the DNeasy plant mini kit (Qiagen, Hilden, Germany), following the manufacturer's protocol. The ITS region and the plastid $rpl32$-$trnL^{(UAG)}$ spacer (referred as cpDNA henceforth in the text) were amplified using the standard PCR reaction, employing the following universal primer pairs: P1A, P4 and internal ones, ITS2 and ITS3, if necessary [36,37] and $trnL$-$retF$ and $rpl32$-$retR$ [39].

Amplifications were carried out in a Mastercycler ep Gradient S thermal cycler (Eppendorf, Hamburg, Germany), using PCR reaction volumes of 25 µL consisting of 0.75 U of *Pfu* polymerase (Fermentas, St. Leon-Rot, Germany), 0.2 mM of each dNTP, 0.2 µM of each primer, 1 µL of DNA template, and reaction buffer containing 2 mM of $MgSO_4$ (Fermentas, St. Leon-Rot, Germany). For ITS amplification, we used the following thermo-cycler program: preheating at 94°C for 3 min, then running it for 35 cycles at 94°C for 30 sec, then at 50°C for 30 sec, and 72°C for 60 sec, with the final extension at 72°C for 10 min, and cooling at 4°C. The cpDNA region was amplified using the "$rpl16$" program, as described by [38]. PCR products were purified using the NucleoSpin Extract II kit (Macherey-Nagel, Düren, Germany). Cycle sequencing reactions were carried out using the same primers, and sequencing was performed on an ABI PRISM 3130*xl* sequencer at BITCET Consortium, Comenius University in Bratislava. The sequencing of a few accessions repeatedly failed for the ITS region and these were therefore excluded from further analyses (Table S1).

Sequence alignments and phylogenetic analysis

Sequences were edited and aligned manually using the BioEdit program version 7.0.4.1 [39]. In case of ITS sequences, the electroferograms were carefully inspected for intra-individual polymorphic sites (IPS) having more than one signal (cf [40]). These were labelled with NC-IUPAC ambiguity codes. Polymorphic positions within ITS sequences, in which both bases were detected also separately in different accessions elsewhere in the alignment, were considered additive polymorphic sites (APS) [40].

We believe that the observed intra-individual variation did not arise from PCR errors, because numerous accessions of *P. hieracioides* and also other taxa possessed homogenized ITS sequences. The geographic pattern of homogeneous and heterogeneous ITS sequences was obviously not random, indicating a genuine variation rather than erroneous signals caused by inaccurate amplification. Although the intra-individual ITS variation might be revealed by cloning procedure, this approach, on the other hand, may result in amplification of numerous ITS sequence types which do not represent relevant variability [41]. Incomplete concerted evolution and/or recombination, processes operating on the multiple-copy regions like ITS, increase the number of unwanted sequence types making the ITS phylogenetic analysis even more complex. Furthermore, with our current sampling involving hundreds of accessions exhibiting intra-individual ITS polymorphism, the cloning procedure would be out of the reasonable solution. As long as most of the sequences were possible to obtain by direct sequencing and polymorphic sites could be designated by IUPAC ambiguity codes, we decided to separately analyse two ITS datasets, one including and one excluding the individuals with APS (see below).

Altogether we generated three alignments: the first one was designated as ITS_1 and involved 117 sequences without APS (Alignment S1. ITS_1). The second one was designated as ITS_2 and comprised 216 sequences coming from all accessions analysed for the ITS region and including all sequences with APS (Alignment S2. ITS_2). The second alignment was constructed after considering that the intra-individual polymorphic sites, especially those with APS, may significantly influence the hierarchical structure of the phylogenetic trees.

Finally, the third cpDNA alignment comprised all plastid sequences, in particular 219 accessions (Alignment S3. cpDNA). In order to detect possible incongruence between particular markers, the incongruence length difference (ILD) test [42], as implemented in the "partition homogeneity" test of PAUP* (1000 replicates, Mulpars off, outgroups included), was employed. Although topologies of the ITS and cpDNA trees were shown to be incongruent, statistical support especially at their basal nodes was comparatively poor, causing their main branching patterns to be unsupported. Therefore, we decided to take the advantage of synergistic effect of combining datasets and generated also concatenated alignments. The concatenated alignment,

ITS_cpDNA, contained 117 ITS sequences without APS and corresponding cpDNA sequences (Alignment S4. ITS_cpDNA).

Indels were not coded separately but treated as missing data. The aligned datasets were analysed independently using the following phylogenetic approaches:

(1) Maximum parsimony (MP) phylogenetic analysis was performed with the heuristic search option in PAUP* version 4.0b10 [43]. The following settings were utilised: accelerated character transformation (ACCTRAN), gaps treated as missing data, single-site polymorphisms determined uncertainties, tree construction with stepwise addition, 1000 bootstrap replicates with random taxon addition, tree bisection-reconnection (TBR) branch swapping, and retention of multiple trees found during branch swapping (MULTREES option in effect). The identical sequences were merged in McClade version 4.0 PPC [44] to reduce computation time. Clade support was calculated via bootstrap analyses using 10000 re-samplings done with the fast heuristic search in PAUP*. Bootstrap support was categorized according to the following criteria: strong (>85%), moderate (70%–85%), weak (50%–69%), or poor (<50%).

(2) Bayesian inference (BI) was run in MrBayes version 3.1.2, using the Markov Chain Monte Carlo algorithm (MCMC) [45]. Bayesian analyses were performed on the CIPRES Portal version 1.15 [46]. Prior to Bayesian analyses, the most appropriate nucleotide substitution models were chosen, using the Akaike Information Criterion (AIC) in jMODELTEST version 0.0.1 [47,48]. Evolutionary models were calculated for each part of the datasets separately. Specifically, the ITS datasets contained partitions corresponding to the ITS1 and ITS2 spacers and the 5.8S rRNA gene, while concatenated datasets included besides the three ITS region partitions also a fourth partition represented by the cpDNA region. The following models or model combinations were found to be the most appropriate for the datasets studied: (1) ITS_1 and ITS_2 datasets – the SYM + G model for ITS1 and ITS2 sequences and the K80 model for 5.8S rRNA gene sequences; (2) cpDNA dataset – the TVM + I + G model; (3) concatenated ITS_cpDNA – the SYM + G model for ITS1 and ITS2, the K80 model for the 5.8S rRNA gene sequences, and the TVM + I + G model for the cpDNA partition.

All BI analyses were run with four independent Metropolis-coupled MCMC chains (three heated and one cold chain) for ten to twenty five million generations and sampled every 1000th generation. The first 25% of sampled trees were regarded as 'burn-in' trees and were discarded prior to reconstruction of a 50% majority-rule consensus tree. Stationarity was confirmed by checking convergence diagnostic parameters. Specifically, the average standard deviation of split frequencies was lower than 0.01 in all cases; the plots of generations versus log probability of the observed data showed no obvious trends; and the Potential Scale Reduction Factor (PSRF) approached 1. Finally, topologies and node posterior probability values were compared among the runs. The topologies were stabilized among all datasets with only minute differences in branching pattern of terminal clades. Nodes with posterior probability (PP) values of 0.90 and above were regarded as significant and those with PP values below 0.90 regarded as non-significant.

(3) Net-like approaches were used to identify and display potential contradictory signals in the datasets. All alignments including ITS_2 containing intra-individual polymorphic sites were analysed using the neighbour-net analysis of [49] in SplitsTree version 4.10 with uncorrected P-distance and default settings. To visualize the relationships among cpDNA haplotypes and to detect possible ancestral polymorphism, the cpDNA dataset was subjected to haplotype network analysis based on the parsimony method of [50] using TCS version 1.21 [51] limited to 30 steps of parsimonious connection in creating the network.

(4) In order to precisely specify the amount of phylogenetically informative signal in the datasets, we analysed spectrum of supporting nucleotide positions. There are three groups of supporting positions recognized by Wägelle and Rödding [52]: (1) symmetrical or binary positions have two different character states in functional outgroup and ingroup and thus support both group of a split equally; (2) asymmetrical positions support only one group which possesses the same nucleotide at particular position, while the other group harbours different and more than one character state at this position; (3) noisy positions include same character states present in all sequences of the functional ingroup but also at least in one sequence of the functional out-group and thus represent convergences or chance similarities between ingroup and outgroup, or alternatively ingroup autapomorphies.

(5) We also apply a Bayesian coalescent-based approach to estimate a species tree employing *BEAST as implemented in the program BEAST version 1.7.4 [18]. Two input files, one with 117 homogeneous ITS sequences (Alignment S1. ITS_1) and the second one containing corresponding cpDNA sequences, were used for the BEAST analysis. Populations assigned to particular taxa were used as OTU's. The input file for *BEAST was created in BEAUti version 1.7.4, with the following settings: two data partitions (corresponding to the two loci), the best-fit evolutionary model for each partition as determined by jMODELTEST, uncorrelated lognormal clock, a Yule process model for the species tree prior, and other parameters as default. Four independent MCMC analyses were run each for 120 million generations, sampling every 1000th generation. Another MCMC analysis was run with settings suitable for a subsequent JML analysis (see below), i.e. with piecewise constant population size model and 40 million generations sampling every 1000th generation. The computer program Tracer version 1.5 [53] was used to check convergence of all parameters to the stationary distribution in each run and TreeAnnotater version 1.7.4 was employed to set the burn-in (discarding the first 30000 trees) and to calculate the maximum clade credibility tree.

(6) We performed statistical tree topology tests on ML gene trees inferred from the ITS_1, cpDNA and ITS_cpDNA alignments to find out whether discrepancies between topologies shown in particular gene trees and coalescent species tree are statistically significant. To this end, looking at the topology of the coalescent species tree, we enforced the following constrains on gene trees: basal position of *P. olympica* within the subsection *Hieracioides*; sister relationship of *P. nuristanica* and *P. japonica*; and monophyly of *P. hispidissima* and *P. hieracioides* (Table 1). Constrained trees were built under the evolutionary substitution model as specified for each alignment above, using the maximum likelihood (ML) criterion and heuristic search with TBR swapping algorithm and 10 random sequence addition replicates. The site-wise likelihoods for the best unconstrained ML tree and all constrained trees were calculated in the computer program raxmlGUI version 1.3 [54] and consequently were compared

using the approximately unbiased, weighted Shimodaira-Hasegawa, and weighted Kishino-Hasegawa tests as implemented in the computer package CONSEL version 0.1j [55–57]. A *p*-value of <0.05 was chosen for rejection of the null hypothesis that the log likelihoods of the constrained and best unconstrained trees are not significantly different.

(7) To test whether hybridization influenced species relationships and could be the source of gene tree incongruence, we employed the program JML [20]. This software uses a posterior distribution of species trees, population sizes and branch lengths to simulate replicate sequence datasets under the coalescent with no migration. The minimum pairwise sequence distance between sequences of two species is evaluated on the simulated datasets and compared to the one estimated from the original data (i.e. from the ITS or cpDNA dataset). This procedure, the posterior predictive test, is a good predictor of hybridization events that disturb the bifurcating species tree model. Two separate JML analyses were run to simulate sequence replicates in the ITS and cpDNA datasets. For these analyses, 40,000 species trees resulting from the JML-specified *BEAST analysis were used. Settings for particular simulations involved: (1) relative mutation rate as inferred from the log file generated during the *BEAST analyses (set to 1.016 and 0.385 for the ITS and the cpDNA simulation, respectively); (2) heredity scalar (2 and 1 for the ITS and cpDNA simulation, respectively); and (3) appropriate model of sequence evolution for both markers. In each analysis, 9,000 trees were removed as burn-in and every 10th tree was used for simulations. Based on the original sequence data files, minimum pairwise sequence distances between all pairs of species and exact probabilities of observing these distances in simulations under assumption of no migration were calculated. All pairwise sequence distances with *p*-value <0.05 were recorded as potential cases of hybridization.

Results

A total of 202 sequences of the ITS region and 205 of the cpDNA spacer of taxa a priori considered as ingroup were obtained. Results of phylogenetic analyses, however, revealed that beside of the 202 ingroup ITS sequences, another three accessions that were a priori considered as outgroup taxa (*P. scaberrima* and *P. strigosa*), appeared to be a part of ingroup. In contrast, all accessions a priori considered as outgroup taxa were shown to be part of real outgroup in the plastid region analysis (see below). Thus, sequences of 11 accessions (7 species) are considered outgroup for the ITS region while 14 accessions (9 species) for the cpDNA spacer. Information on the DNA datasets and details on maximum parsimony analyses are summarized in Table 2.

The ITS region

In addition to homogeneous sequences, we identified numerous accessions containing intra-individual single nucleotide polymorphic sites in the ITS dataset (Table S2). Out of 653 nucleotide positions, 148 were detected to be polymorphic in at least one individual of the subsection *Hieracioides*, with up to 16 intra-individual polymorphic sites per sequence. Forty two polymorphic sites showed an additive pattern, ranging between 1 and 16 APS per individual sequence (Table S2). The heterogeneous sequences varied conspicuously in terms of number of APS with respect to potentially parental ITS variants, ranging from those with fully additive patterns to those comprising almost completely homogenized contigs.

Results of BI, MP and split decomposition analyses based on the ITS_1 dataset (117 sequences without APS) did not show the analysed *Picris* taxa from the subsection *Hieracioides* to form a monophyletic group (Figures 1 and 2A). Specifically, accessions of the presumed outgroup species, *P. scaberrima* and *P. strigosa*, were clustered within the ingroup taxa (Figures 1 and 2A).

The Bayesian 50% majority-rule consensus tree and the strict consensus MP tree based on the ITS_1 dataset (117 sequences without APS) displayed essentially identical topologies with only small differences in terminal positions and/or in clade supports

Table 1. Log likelihoods and *p*-values of AU, WSH, and WKH tests for tree comparisons considering different topological scenarios.

Topology	Alignment	Log likelihood (–ln L)	Δ (–ln L)[a]	AU[b]	WSH[c]	WKH[d]
Best maximum likelihood tree (unconstrained)	ITS_1	2408.7451	–	0.843	0.948	0.794
	cpDNA	1973.5581	–	0.777	0.903	0.650
	ITS_cpDNA	4977.4663	–	0.798	0.922	0.682
Basal position of *P. olympica* within the subsection *Hieracioides*	ITS_1	2414.1421	5.40	0.212	0.319	0.201
	cpDNA	1973.5582	0.00	0.514	0.726	0.350
	ITS_cpDNA	4979.1512	1.68	0.385	0.582	0.293
Sister relationship of *P. nuristanica* and *P. japonica*	ITS_1	2408.7453	0.00	0.250	0.709	0.206
	cpDNA	1973.5582	0.00	0.377	0.706	0.308
	ITS_cpDNA	4977.4664	0.00	0.511	0.840	0.318
Monophyly of *P. hispidissima* and *P. hieracioides*	ITS_1	2428.4956	19.75	**0.013**	0.077	0.057
	cpDNA	1981.1009	7.54	**0.033**	0.139	0.139
	ITS_cpDNA	4992.7916	15.33	0.141	0.205	0.114

Significant differences (*p*<0.05) between the best unconstrained and constrained topologies are indicated in bold.
[a]Δ (–ln L) - differences in log likelihoods between the best gene tree and alternative gene trees.
[b]AU - approximately unbiased test.
[c]WSH - weighted Shimodaira-Hasegawa test.
[d]WKH - weighted Kishino-Hasegawa test.

Table 2. Information for the DNA datasets and details on maximum parsimony analyses.

Alignment	ITS_1[a]	cpDNA[b]	concat1[c]
No. of sequences/No. of sequences in merged dataset	117	219/112	117
Characters in the alignment	653	960	1613
Length variation (including outgroup)	640–648	702–908	1352–1546
No. of variable sites	167	101	263
No. of parsimony informative characters	138	60	207
No. of most parsimonious trees	1515	1512	4130
Tree length in parsimony	252	127	430
Consistency index	0.774	0.898	0.702
Retention index	0.935	0.955	0.895

[a]ITS1 – analysis based on the ITS_1 Alignment S1. ITS_1.
[b]cpDNA - analysis based on the Alignment S3. cpDNA.
[c]concat - analysis based on the concatenated Alignment S4. ITS_cpDNA.

(Figure 1). All ingroup taxa including both aforementioned presumed outgroup species formed a large strongly supported clade (BS = 90%, PP = 1.00), with shallow hierarchical structure comprising numerous sub-clades with various levels of support (Figure 1). All non-*hieracioides* species, *P. hispidissima*, *P. japonica*, *P. olympica*, *P. nuristanica*, *P. scaberrima*, and *P. strigosa*, formed their own strongly supported sub-clades, with *P. olympica* and *P. scaberrima* being the most divergent (Figure 1). *Picris hieracioides* is not monophyletic and forms four ITS groups with allopatric or parapatric spatial distributions (Figures 1, 2A and 3): ITS_1 composed of two sub-clades: ITS_1A – mostly lowland populations morphologically corresponding to the *P. h.* subsp. *hieracioides* morphotype from the Apennine Peninsula, the Balkan Peninsula, western Turkey, rarely from Central and northwestern Europe and a single population morphologically assignable to the *P. h.* subsp. *umbellata* morphotype from northwestern Europe (blue in Figures 1, 2A and 3); ITS_1B – mostly lowland populations of the *P. h.* subsp. *hieracioides* morphotype from Central, northwestern Europe and Balkan Peninsula and a single population of the *P. h.* subsp. *umbellata* morphotype from the Western Carpathians (pale blue in Figures 1, 2A and 3); ITS_2 – mountain populations of the *P. h.* subsp. *umbellata* morphotype from Alps and the Western Carpathians (red in Figures 1, 2A and 3); ITS_3 – mostly mountain populations corresponding to the *P. h.* subsp. *umbellata* morphotype from the Sierra Nevada, Pyrenees, Jura Mts., Alps, Germany and Sweden (orange in Figures 1, 2A and 3); and ITS_4 – an unresolved group of populations assignable to both of *P. hieracioides* morphotypes from the Apennine and the Iberian Peninsulas, Belgium and Poland (brown in Figures 1, 2A and 3). We also endeavoured to analyse the dataset ITS_2 (all 216 ITS sequences including those with APS) using MP and BI. Both approaches, however, resulted in a completely unresolved clade with all accessions from the subsection *Hieracioides* placed in a basal polytomy (data not shown).

The split decomposition diagram based on the ITS_1 dataset (117 ITS sequences) showed four main splits having various levels of bootstrap support (Figure 2A): (1) *P. olympica* separated from the rest by a set of long parallel and strongly supported edges; (2) lowland populations of *P. hieracioides*, morphologically corresponding to *P. h.* subsp. *hieracioides*, are clearly separated from all ingroup and outgroup taxa (ITS 1A and 1B). The exception represented few populations of the *P. h.* subsp. *hieracioides* morphotype that clustered together with populations attributable

to the *P. h.* subsp. *umbellata* morphotype in an unresolved ITS4 group; (3) a large heterogeneous group associating several smaller well supported splits corresponding to *P. hispidissima*, *P. japonica*, *P. nuristanica*, and two lineages composed of mountainous populations of the *P. h.* subsp. *umbellata* morphotype (ITS_2 and ITS_3); and (4) a group of both morphotypes of *P. hieracioides* (ITS_4) and *P. strigosa*, a presumably outgroup species. *Picris scaberrima* formed a long edge linked with outgroup taxa by short parallel splits. Importantly, the second neighbour-net analysis based on the ITS_2 dataset (206 ITS sequences) revealed essentially the same pattern as the previous one (figure not shown). This apparently indicates that heterogeneous sequences did not significantly affect the overall split pattern. Individuals with un-homogenized ITS sequences and numerous APS appeared mostly in the central position of the network, while those possessing more homogenized sequences, with lower number of APS, were preferentially incorporated in terminal positions (figure not shown).

Analyses of spectrum of supporting positions revealed that majority of phylogenetically informative positions are shared by outgroup taxa (Figure 2B). The entire functional ingroup, involving all accessions of the subsection *Hieraciodes* plus *P. strigosa* and *P. scaberrima*, are supported only by 7 nucleotide positions. Within the functional ingroup, the highest support obtained *P. olympica* (9 positions), *P. japonica* (4 positions), the *P. hieracioides* clade with lowland populations corresponding to the *P. h. hieracioides* morphotype (3 positions) and *P. nuristanica* (2 positions). All other above mentioned splits/subclades were supported at maximum by a single position.

The cpDNA spacer

MP and BI analyses yielded trees with different topologies to a certain extent. However, there was no strongly supported incongruence between them. In order to maintain consistency in the presentation of our results we discuss only the topology based on the BI phylogenetic tree.

In contrast to the ITS trees, *P. scaberrima* and *P. strigosa* appeared clustered together with other outgroup taxa. All accessions of the *Hieracioides* group appeared in a largely unresolved clade with five essentially supported cpDNA subclades and two unsupported groups (cpDNA_A–G) placed in a basal polytomy (Figure 4). The topology and variation of non-*hieracioides* taxa in the cpDNA spacer tree were in some aspects

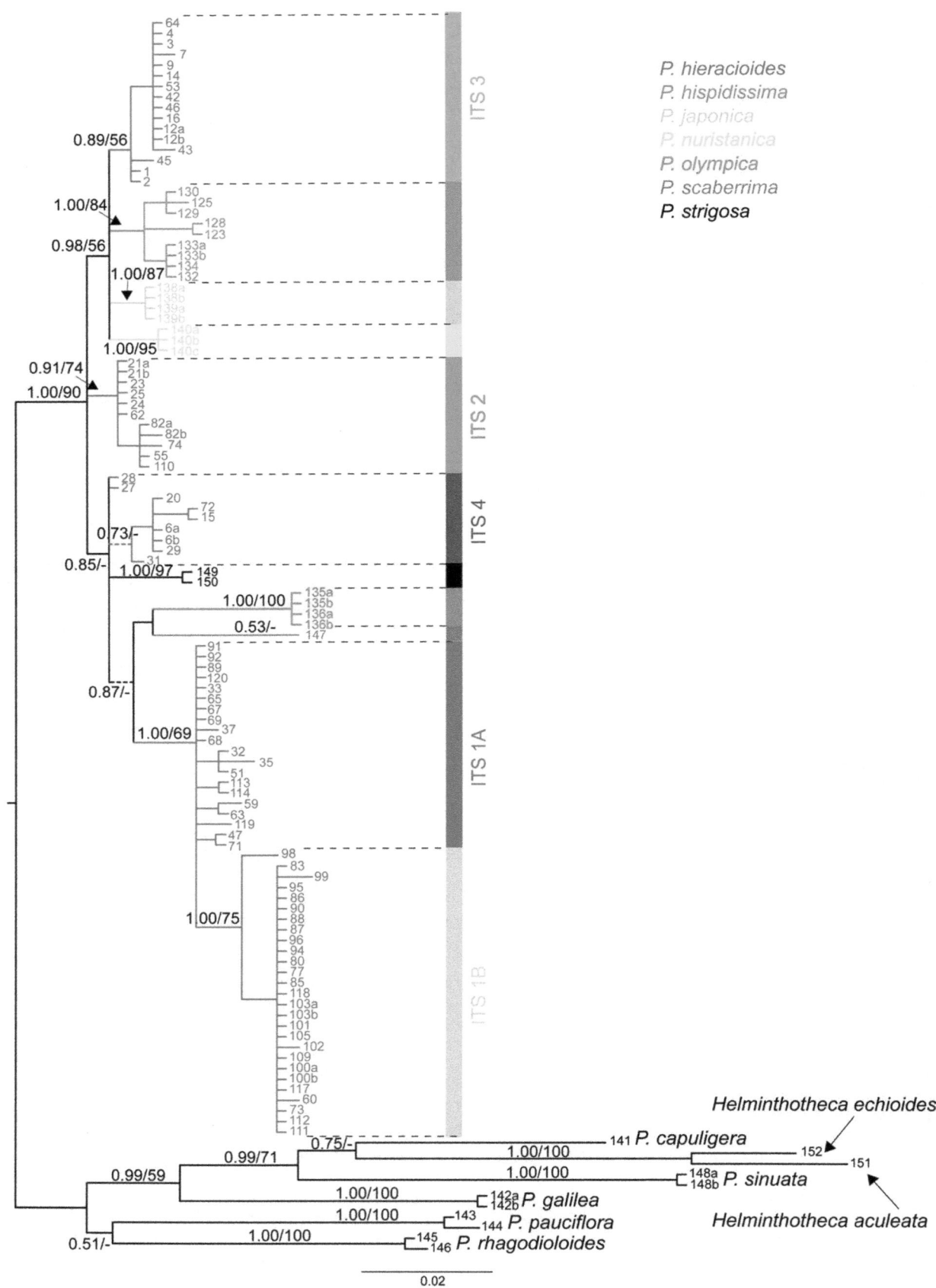

P. hieracioides
P. hispidissima
P. japonica
P. nuristanica
P. olympica
P. scaberrima
P. strigosa

Figure 1. Majority-rule consensus tree of Bayesian inference based on the reduced dataset_ITS_1. The numbers above branches refer to posterior probability values of Bayesian inference/the bootstrap support as inferred for the maximum parsimony analyses (values ≥50% are shown). Support values for terminal branches are not shown. Dashed lines represent branches collapsed in strict consensus tree of the maximum parsimony analyses. Colour line below the tree indicates affiliation of grouping to the ITS genetic lineages as mentioned in the text. Each accession label includes the population code following Table S1.

congruent and in other discordant with respect to the ITS tree (see Figure 1). Although not forming distinct subclades, *Picris japonica* and *P. nuristanica* possessed their own exclusive haplotypes, similarly as they did in the ITS data (see Figure 5). Contrastingly, *P. olympica* and *P. hispidissima* were heterogeneous and shared haplotypes with *P. hieracioides* (Figures 4, 5 and 6). Repeatedly, *P. hieracioides* was heterogeneous and was found in seven genetic cpDNA geographically correlated groups (Figures 4, 5 and 6): unresolved group cpDNA_A – comprising populations of *P. hieracioides* morphologically corresponding to both subspecies and originating from the entire distribution area, *P. hispidissima* from Croatia and Montenegro, and *P. olympica* (cyan in Figures 4, 5 and 6); cpDNA_B – mountain populations of the *P. h.* subsp. *umbellata* morphotype from the Apennines, Alps, and Carpathians, while populations from lowland of northwestern Europe corresponding morphologically rather to *P. h.* subsp. *hieracioides* and finally *P. hispidissima* from Croatia (violet in Figures 4, 5 and 6); cpDNA_C – *P. hispidissima* and lowland populations of the *P. h.* subsp. *hieracioides* morphotype from Croatia and Montenegro (olive in Figures 4, 5 and 6); unresolved group cpDNA_D – including mountain and lowland populations of *P. hieracioides* from Sierra Nevada, the Carpathians and north-western Europe morphologically corresponding to its both subspecies, *P. japonica* and *P. nuristanica* (pink in Figures 4, 5 and 6); cpDNA_E – *P. hispidissima* from Montenegro (grey in Figures 4, 5 and 6); cpDNA_F – populations of both morphotypes of *P. hieracioides* from the Iberian mountain ranges, the western Alps, the Jura Mts, and northwestern Europe (green in Figures 4, 5 and 6); cpDNA_G – lowland populations of the *P. h.* subsp. *hieracioides* morphotype from the Apennine and the Balkan Peninsula, mountain populations of the *P. h.* subsp. *umbellata* morphotype from the Alps and *P. hispidissima* from Croatia (light brown in Figures 4, 5 and 6). The highest density of rare haplotypes was observed on the Balkan Peninsula (nine haplotypes), the Iberian Peninsula mountains (seven haplotypes), and the Apennine Peninsula (five haplotypes) (Figure 6).

The parsimony-based haplotype network revealed 34 very closely related ingroup haplotypes, mostly connected by one or two steps (Figure 5) and the split decomposition diagram displayed a stellar-like structure with numerous haplotypes in an unresolved basal position (data not shown). Similarly as in the ITS dataset, the majority of supporting positions was harboured by outgroup species (data not shown). Both *hieracioides* clades as well as all their sub-clades were supported extremely weakly, in particular, by two positions at the maximum (data not shown).

Concatenated phylogeny and coalescence based species tree

Although the ILD test revealed significant ($p<0.001$) incongruence between both gene trees, we performed concatenated analyses because our split spectrum analyses show that the observed discordances are not strongly supported. Thus, incongruence between both genetic regions could rather reflect paucity of phylogenetically informative signal (Figures 1, 2AB and 4). Concatenated phylogeny also did not unambiguously resolve evolutionary relationships among ingroup taxa, but still depicted several variably supported (sub)clades. Position of two presumed

outgroup taxa, *P. scaberrima* and *P. strigosa*, varied among phylogenetic analyses and remained uncertain (see Figure 7AB and MP/BI phylogenies not shown). Non-*hieracioides* ingroup species, namely *P. japonica*, *P. olympica* and *P. nuristanica*, each clustered in strongly supported sub-clades, repeatedly with *P. olympica* being the most divergent (Figures 7AB). Importantly, although weakly supported, the majority of accessions assigned to morphotypes of both subspecies of *P. hieracioides* as well as to *P. hispidissima* tended to form taxon specific groupings (Figure 7A).

The coalescence-based species tree (Figure 8) showed different topology than the concatenated gene tree and brought more light into the phylogenetic relationships of the analysed taxa: the presumed outgroup taxa, *P. scaberrima* and *P. strigosa*, appeared together in a sister position to the subsection *Hieracioides*; *P. olympica* appeared in a basal position to all taxa from the subsection *Hieracioides*; the Asian *P. japonica* and *P. nuristanica* were clustered together in a sister position to the *P. hispidissima* and *P. hieracioides* clade. Relationships among *P. hispidissima* and *P. hieracioides* were, however, unsupported.

Statistical tree topology tests

Although all tested evolutionary scenarios found in the coalescent species tree (see above and Material and Methods) were not observed in the present gene trees, they cannot be rejected at the significance level of 0.05 by any of the three statistical tree topology tests performed (Table 1). The only exception is the monophyly of *P. hispidissima* and *P. hieracioides*, which can be excluded for the ITS_1 and cpDNA alignment at the significance level of 0.05 only by the AU test but not by the WSH and WKH tests. Thus, the phylogenetic relationships within the subsection *Hieracioides* as unravelled by the Bayesian coalescent-based analysis cannot be very likely excluded also for the gene trees.

Test of hybridization and/or introgression

Only a single case of potential hybridization within the ITS dataset was detected, in between a pair of outgroup taxa, *Helminthotheca aculeata* and *H. echioides* ($p<0.001$). By contrast, 31 cases with $p<0.05$ were found in the cpDNA dataset involving accessions of the corresponding pairs of the following taxa: *P. olympica* vs. *P. hieracioides*, *P. olympica* vs. *P. rhagodioloides*; *P. sinuata* vs. *P. capuligera*; and *H. aculeata* vs. *H. echioides* (Table 3).

Discussion

Intricate phylogenetic relationships within the subsect. *Hieracioides*

Tree-building and network analyses of both the nuclear and plastid markers of the studied *Picris* taxa revealed phylogenies with rather low resolutions and topological discrepancies (see Results), although utilization of these markers has been proposed as one of the most informative approaches even at lower taxonomic levels [35,58]. More light on the evolutionary relationships within the subsection *Hieracioides* was cast by the Bayesian coalescence analysis which generates species instead of gene trees. Although relationships in the present species tree were

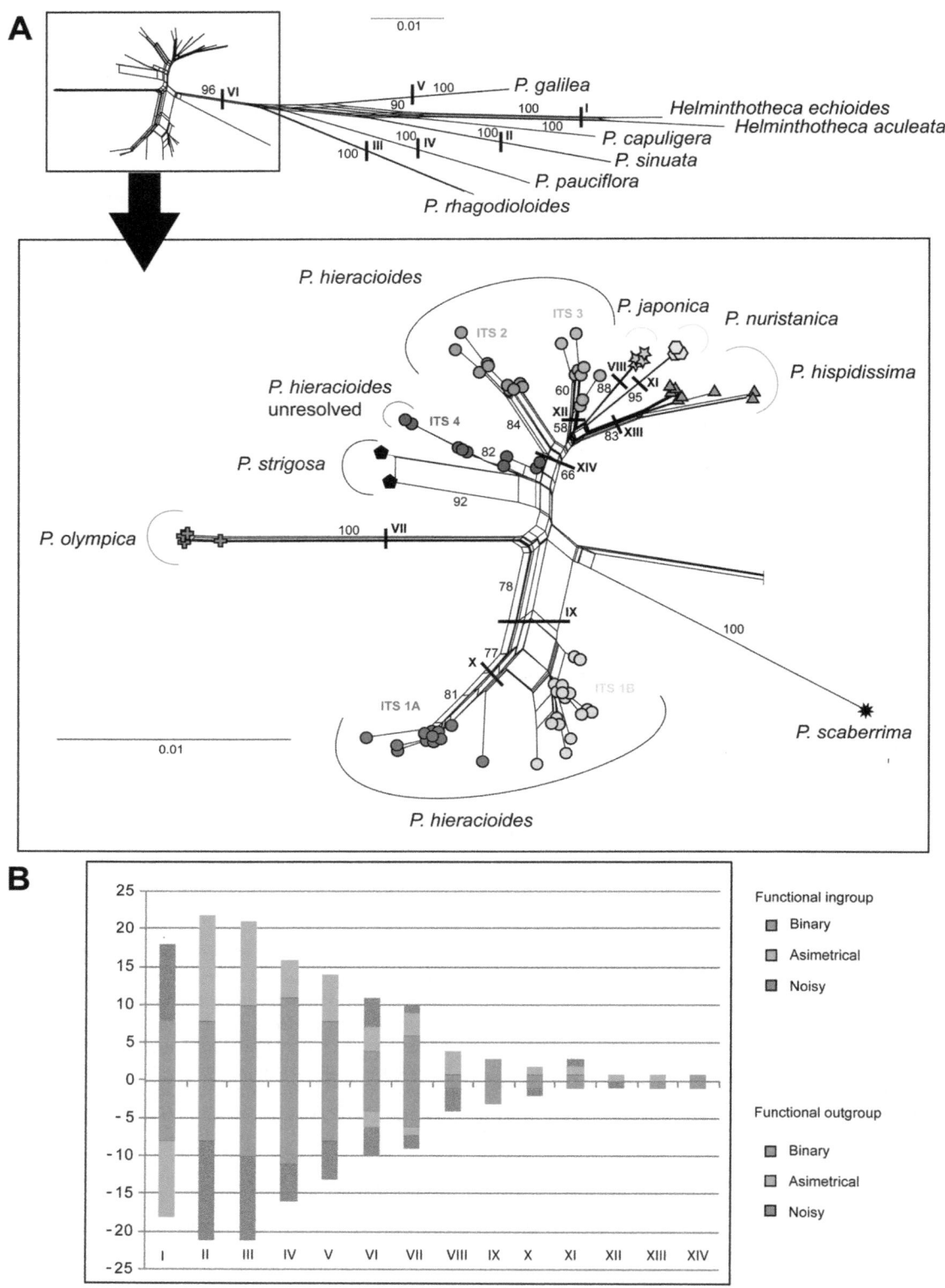

Figure 2. Neighbour-Net diagram (A) and split support spectrum (B) for the ITS_1 dataset. The Neighbour-Net diagram is based on uncorrected *P*-distances. Colour symbols indicate affiliation of accessions to the ITS genetic lineages as mentioned in the text while the shape of symbols indicate affiliation of accessions to taxa. Bootstrap supports of selected important splits are indicated above edges. Column height in the split spectrum represents the number of clade-supporting positions, i.e., putative primary homologies. Column parts above the y-axis represent the ingroup partition while those below the axis correspond to the outgroup partition.

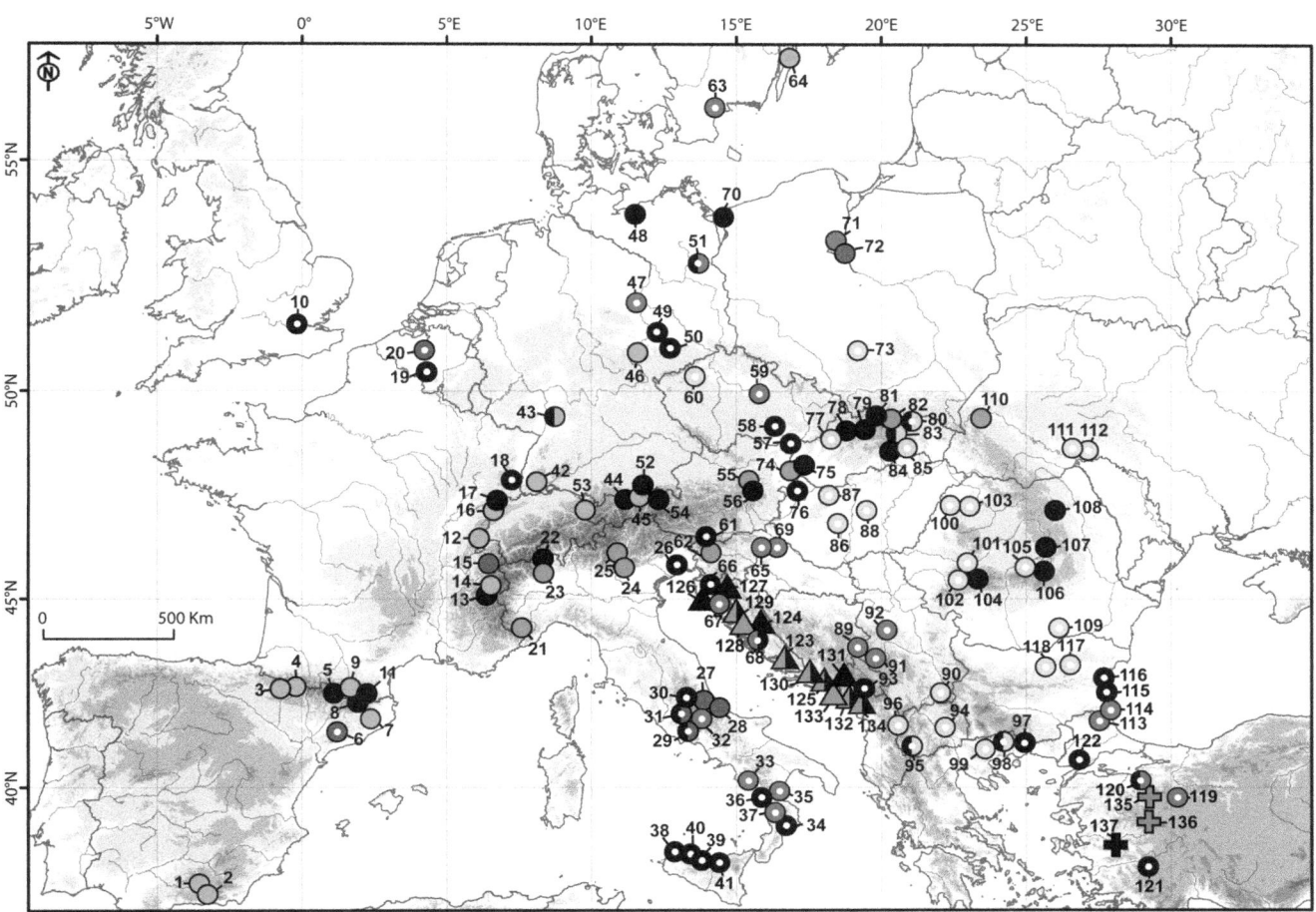

Figure 3. Map displaying geographic distribution of genetic variation in the analysed *Picris* taxa, as inferred from the nuclear ITS sequence data. Coloured symbols indicate affiliation of accessions to the ITS genetic lineages as mentioned in the text. Black symbols indicate populations including individuals with unhomogenised sequences containing additive polymorphic sites. Symbol shapes indicate taxa as follows: *P. hieracioides* – circle, *P. hispidissima* – triangle, and *P. olympica* – square. The circle symbols with empty centres refer to populations morphologically corresponding to *P. hieracioides* subsp. *hieracioides*, while those with solid symbols refer to populations morphologically corresponding to *P. hieracioides* subsp. *umbellata*. The populations of *P. nuristanica*, *P. japonica* and outgroup taxa are not shown. Numbers refer to population codes as denoted in Table S1.

not strongly statistically supported and were not detectable in all gene trees, they could not be rejected by statistical tree topology tests performed on the concatenated dataset (see Results). Thus, in spite of the complex patterns, we might assume the following: (1) the analysed members of the subsection *Hieracioides* have a monophyletic origin, with *P. scaberrima* and *P. strigosa* being their sister taxa; (2) *P. olympica* is basal within the subsection *Hieracioides*, i.e. it is sister to all other members of this subsection; (3) all Asian species are sister to the European ones; and (4) *P. japonica* + *P. nuristanica* are sister to *P. hieracioides* + *P. hispidissima*. Interrelationships among populations of *P. hieracioides* and *P. hispidissima* are left unresolved, however.

Which factor(s) might be responsible for low resolution, varying supports and discordances between gene tree topologies? In the present study, we have shown by the split spectrum analysis that one of the primary and crucial factors responsible for this pattern is a paucity of parsimony informative and node supporting nucleotide positions. Deficiency of phylogenetically informative signal in the datasets analysed was unambiguously evidenced by the low number of primary nucleotide homologies (Figures 2B and 7B). This was indicated also by the presence of short internal

branches in all phylogenetic trees (Figures 1 and 4) and the high number of rare plastid haplotypes in derived positions connected mostly with ancestral haplotypes by only one or two mutational steps (Figure 5). Low level of phylogenetically informative positions in the used markers, which are among the most informative [35,58], strongly indicates a recent diversification of the studied taxa and genetic groups [1,59,60,61,62]. It would be desirable to confirm this by molecular dating. However, in case of insufficient amount of phylogenetic information, such an analysis is assumed to be vague and statistically unsupported.

Furthermore, incomplete lineage sorting represents another evolutionary force that might be responsible for the discordances and complex genetic patterns in the six *Picris* taxa [1,3,4,5]. The Bayesian coalescent-based approach considers the incomplete lineage sorting as a main source of uncertainties in inferred species trees [18]. Thus, the unresolved relationships among genetic lineages of *P. hieracioides* and *P. hispidissima* as well as the low nodal support could signalise either influence of this phenomenon or violation of prior assumption by hybridisation or other processes. Likewise, the disjunctive occurrence of the ancestral plastid haplotypes in populations of *P. olympica* and the spatially

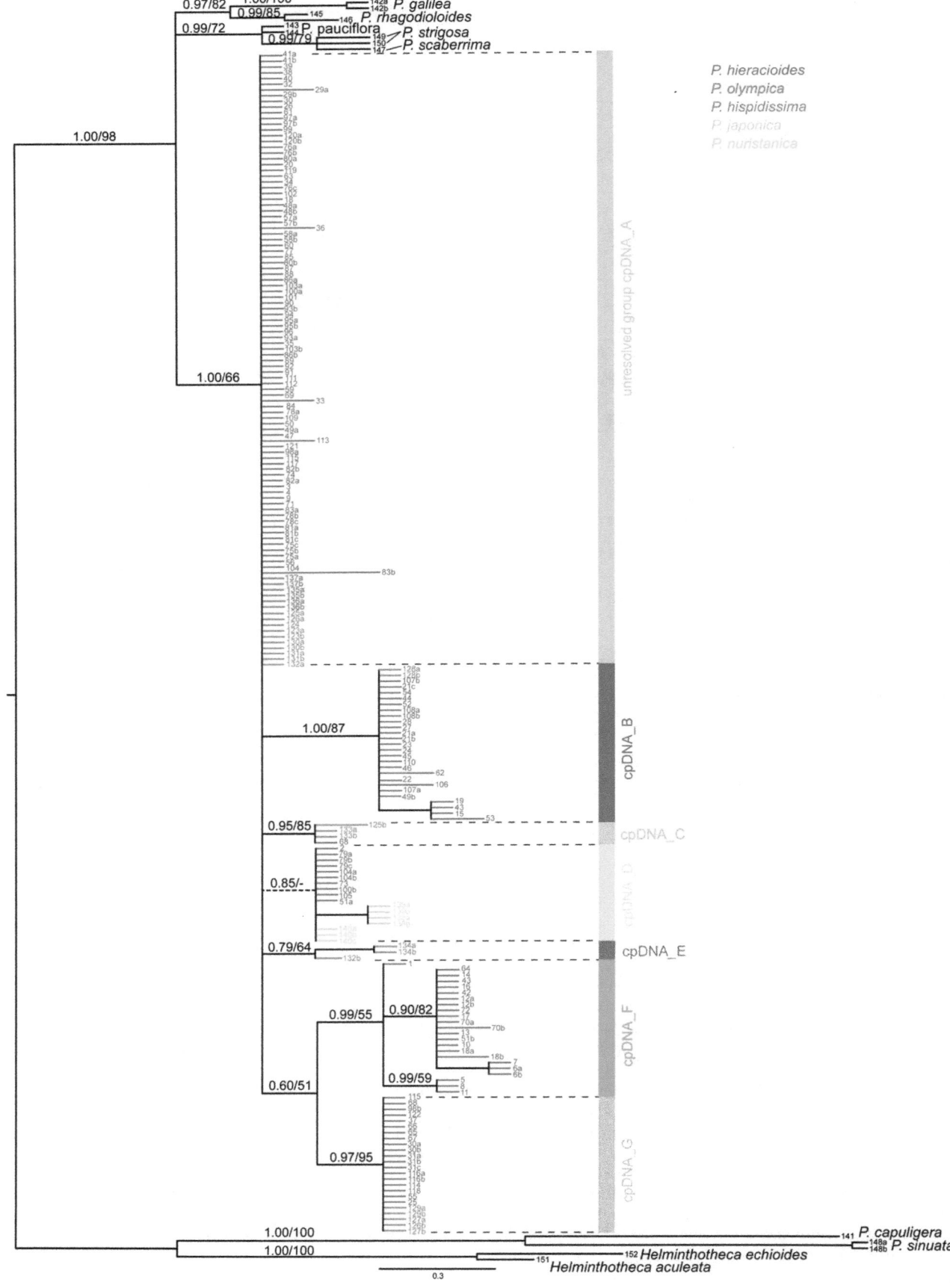

Figure 4. Bayesian majority-rule consensus tree based on the cpDNA intergenic spacer dataset. The numbers above branches refer to posterior probability values of Bayesian inference/the bootstrap support as inferred for the maximum parsimony analyses (values ≥50% are shown). Support values for terminal branches are not shown. Dashed lines represent branches collapsed in strict consensus tree of the maximum parsimony analyses. Colour line below the tree indicates affiliation of groupings to the cpDNA genetic lineages as mentioned in the text. Each accession label includes the population code following Table S1.

distant *P. hieracioides* might be the consequence of incomplete lineage sorting. By contrast, the mosaic pattern detected in the ITS dataset may repeatedly suggest influence of hybridization.

Hybridisation might substantially influence phylogenetic reconstructions at even lower hierarchical levels. Decreased phylogenetic tree resolution, incongruent patterns in different gene trees, and occurrence of alleles transcending taxa or genetic groups are commonly observed in plant phylogenies in which extensive gene exchange between taxa has been documented [2,3,17,41,60, 61,62,63,64,65,66]. Presence of extensive and recurrent gene exchange among investigated *Picris* taxa might be indicated by the following facts. (1) Considerable intra-individual and intra-population ITS sequence heterogeneity and allele transcending clades and taxon borders in *P. hieracioides*, *P. hispidissima*, and *P. olympica*. On the other hand, maintenance of two or more ITS variants within a single genome might be attributed to other molecular-genetic processes [11,67]. (2) Occurrences of heterogeneous ITS sequences predominantly in the contact zones of the studied species or genetic groups, while locations of homogeneous sequences with little or no genetic admixture mostly along the

sampled area's margins inhabited mainly by single lineages or species (see Results, Figure 3). (3) Strict self-incompatibility in *P. hieracioides* and *P. hispidissima*, which has been recently documented ([27]; M. Slovák unpublished data). (4) Last but not least, successful crossing between populations of morphotypes and genetic lineages of *P. hieracioides* as well as with *P. hispidissima* obtained during our preliminary field experiments (M. Slovák, unpublished data).

Posterior predictive checking revealed hybridization/introgression at least in five pairs of the studied species. Four pairs were recorded exclusively by simulations on the cpDNA dataset and one pair was detected by simulations on both the plastid and the nuclear dataset. Since the method of posterior predictive checking is powerful unless the hybridization event has occurred very rapidly after the speciation event [9], it might be hypothesised that hybridizations among the *Picris* taxa are rather recent or have occurred later after speciation events. This hypothesis corresponds well mostly in the case of comparatively distantly related species, *P. olympica* with *P. hieracioides*, *P. olympica* with *P. rhagodioloides*, and *P. capuligera* with *P. sinuata*. Such hybridisation

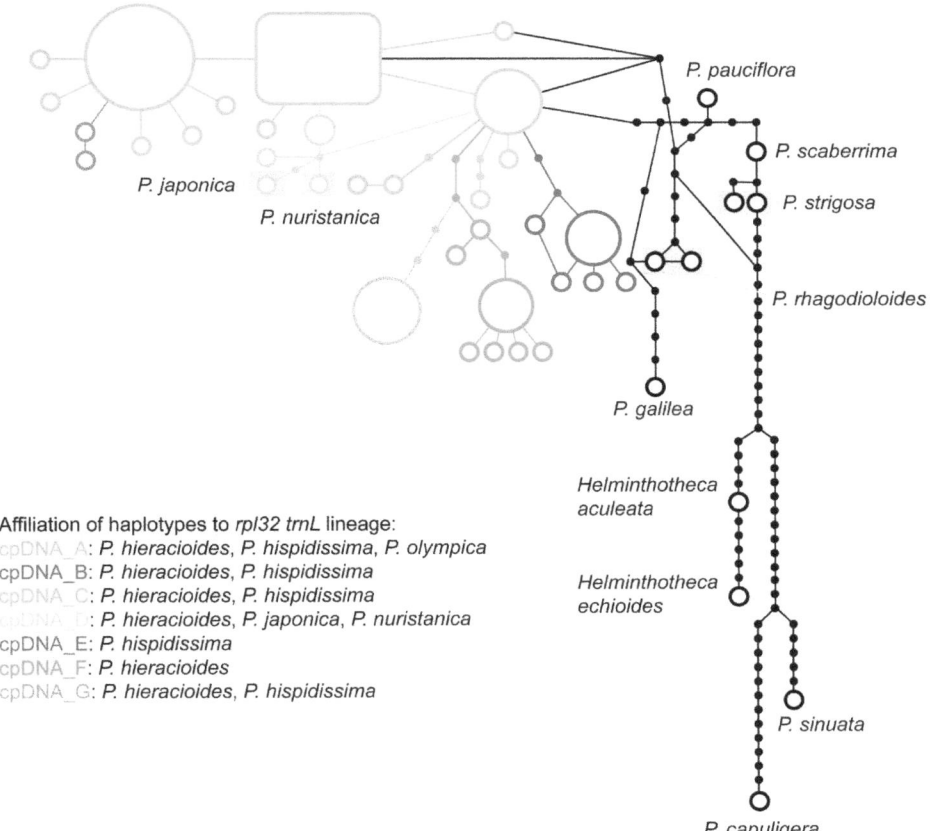

Affiliation of haplotypes to *rpl32 trnL* lineage:
cpDNA_A: *P. hieracioides, P. hispidissima, P. olympica*
cpDNA_B: *P. hieracioides, P. hispidissima*
cpDNA_C: *P. hieracioides, P. hispidissima*
cpDNA_D: *P. hieracioides, P. japonica, P. nuristanica*
cpDNA_E: *P. hispidissima*
cpDNA_F: *P. hieracioides*
cpDNA_G: *P. hieracioides, P. hispidissima*

Figure 5. Maximum parsimony network of the cpDNA haplotypes of *Picris* populations. The symbol sizes are proportional to the haplotype frequencies, the lines represent mutational steps, and black dots are unsampled haplotypes. Colour symbols indicate affiliation of accessions to the cpDNA genetic lineages as mentioned in the text.

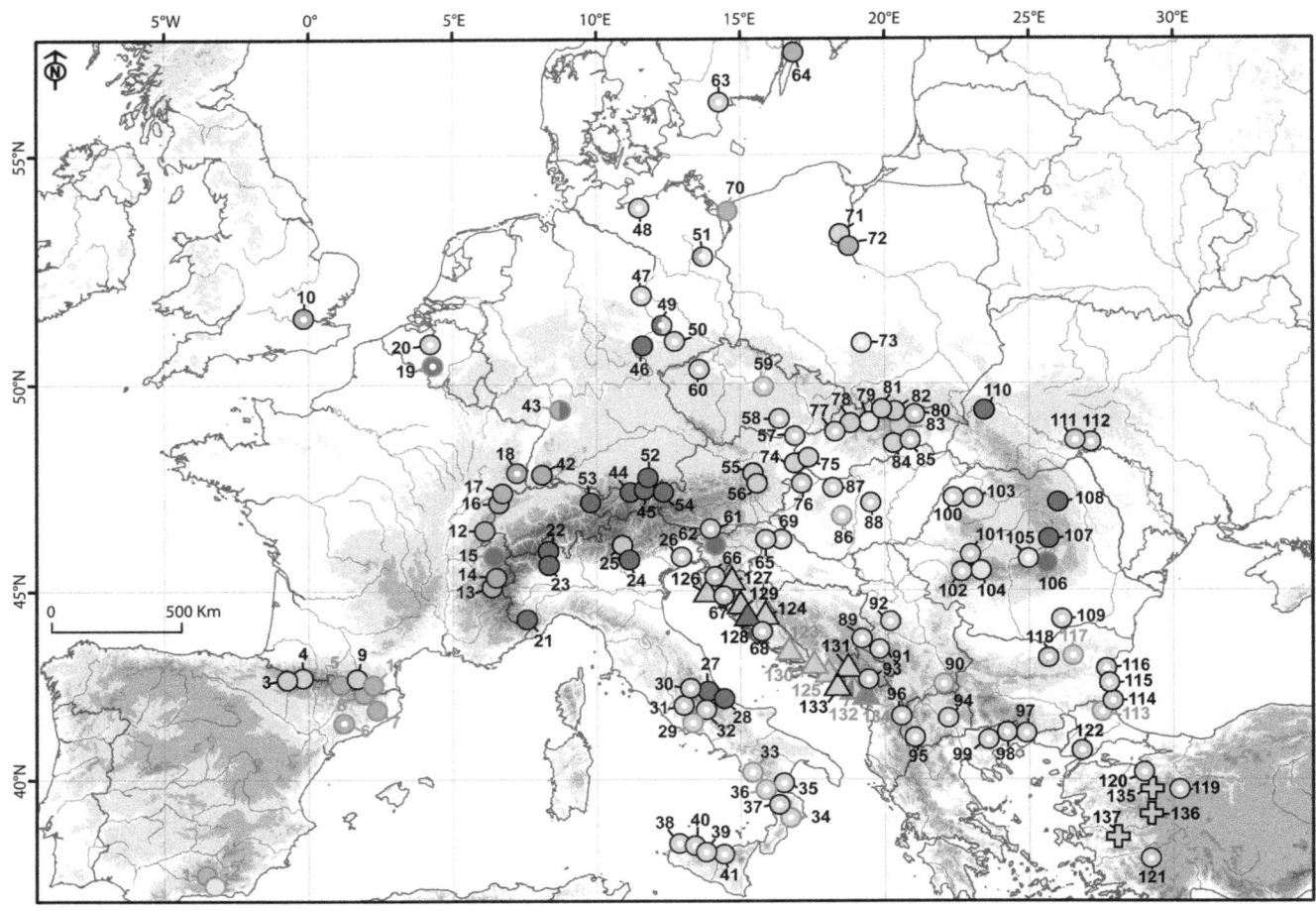

Figure 6. Map displaying geographic distribution of genetic variation in the analysed *Picris* taxa, as inferred from the cpDNA intergenic spacer dataset. Coloured symbols indicate affiliation of accessions to the cpDNA genetic lineages as mentioned in the text. Bi-coloured symbols denote populations possessing two different haplotypes. Symbol shapes indicate taxa as follows: *P. hieracioides* – circle, *P. hispidissima* – triangle, and *P. olympica* – square. The circle symbols with empty centres refer to populations morphologically corresponding to *P. hieracioides* subsp. *hieracioides*, while those with solid symbols refer to populations morphologically corresponding to *P. hieracioides* subsp. *umbellata*. Population of *P. nuristanica* and *P. japonica* and outgroup taxa are not shown. Numbers refer to population codes as denoted in Table S1. The symbols and their numbers highlighted in red indicate populations harbouring rare haplotypes.

events might take part during the Pleistocene glacial/interglacial range shifts, especially, at sites where lowland taxa could have had recurrent contact with mountainous ones [2,68]. Posterior predictive checking, however, did not support hybridisation scenarios among the pairs of closest relative taxa where it would be expected: among subspecies and genetic lineages of *P. hieracioides* as well as between *P. hieracioides* and *P. hispidissima*. One of the most plausible explanations might be that the hybridization events among those taxa happened just after their diversification. Especially in case of the two subspecies of *P. hieracioides*, the divergence of genetic lineages might be shallow causing hybridization to be undetectable [9].

The overall failure to detect hybridization events by simulations on the ITS dataset might be attributed to the presence of recombination and/or concerted evolution within the ITS locus. As stated in [9], some patterns resulting from concerted evolution can potentially bias the estimates of population sizes toward lower values. This causes the loci to coalesce faster and therefore hybridization might be harder to detect. Only one hybridization event was found by both genetic markers, viz., in *H. aculeata* and *H. echioides*. The sample size might be the limitation factor in the

posterior predictive checking approach. Since only a single individual per each aforementioned species (outgroup) was sampled, their population sizes cannot be properly estimated for species tree reconstruction. Therefore we assume this result might be a methodological artefact.

Finally, it remains arguable whether some of the studied taxa, but especially *P. hispidissima*, fit into the reticulate evolution scenario or, alternatively, its large genetic heterogeneity arose from genetic erosion. The latter hypothesis might be explained by recurrent introgression with adjacent populations of *P. hieracioides* after their diversification. We are more inclined to the latter scenario because all taxa are morphologically rather well delimited, inhabit ecologically specific biotopes, and trans-taxon alleles are concentrated predominantly in areas of sympatry or parapatry.

Although methods to potentially distinguish between hybridisation and incomplete lineage sorting have been developed, there is still a high risk of confusing these two phenomena or they cannot be discerned in some cases at all [9,12,13]. In the investigated closely related *Picris* taxa from the subsection *Heiracioides*, it is highly tricky or even impossible to discern between these two

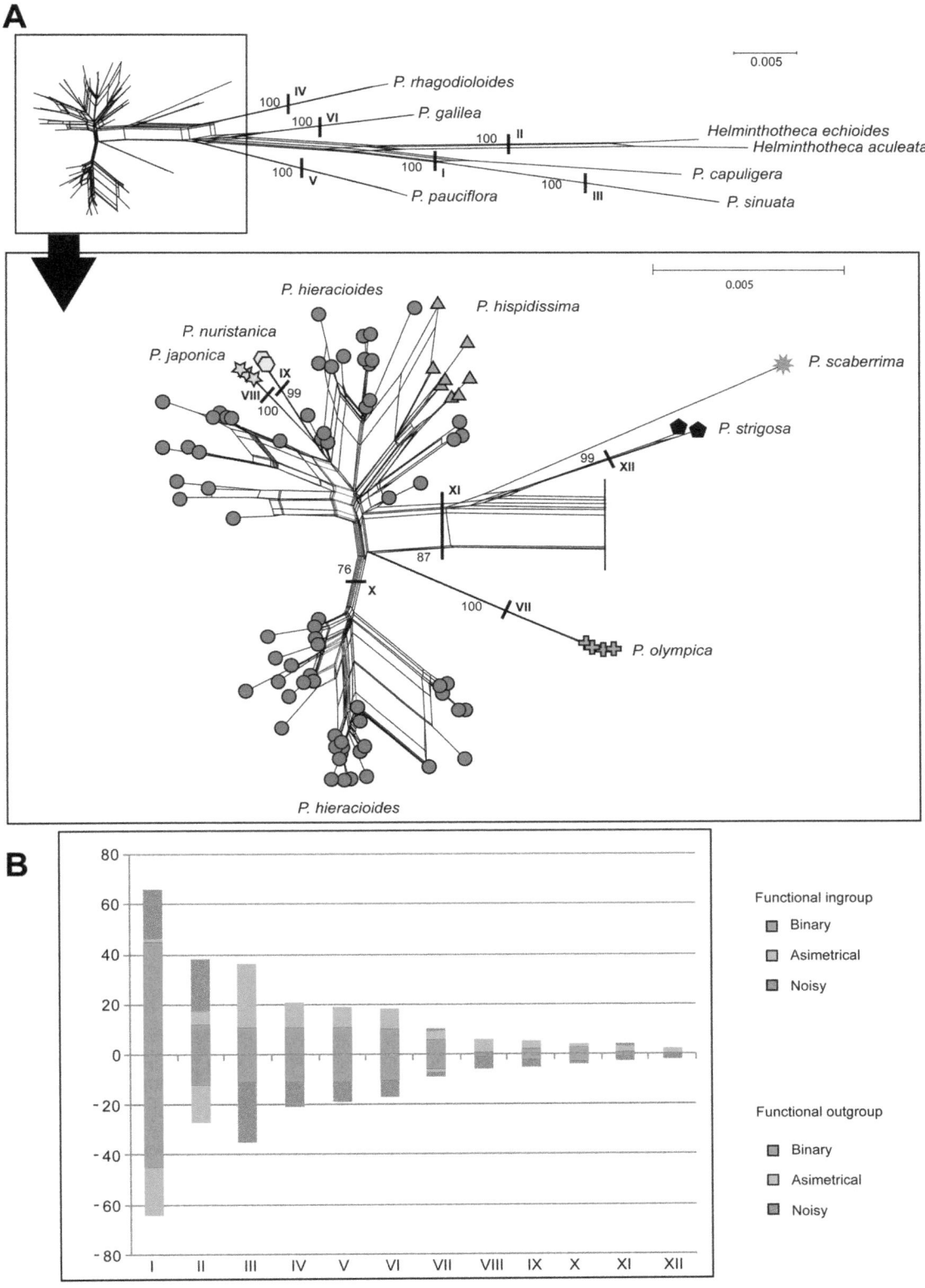

Figure 7. Neighbour-Net diagram (A) and split support spectrum (B) for the concatenated ITS_cpDNA dataset. Coloured symbols indicate affiliation of accessions to taxa. Bootstrap supports of selected important splits are indicated above edges. Column height in the split spectrum represents the number of clade-supporting positions, i.e., putative primary homologies. Column parts above the y-axis represent the ingroup partition while those below the axis correspond to the outgroup partition.

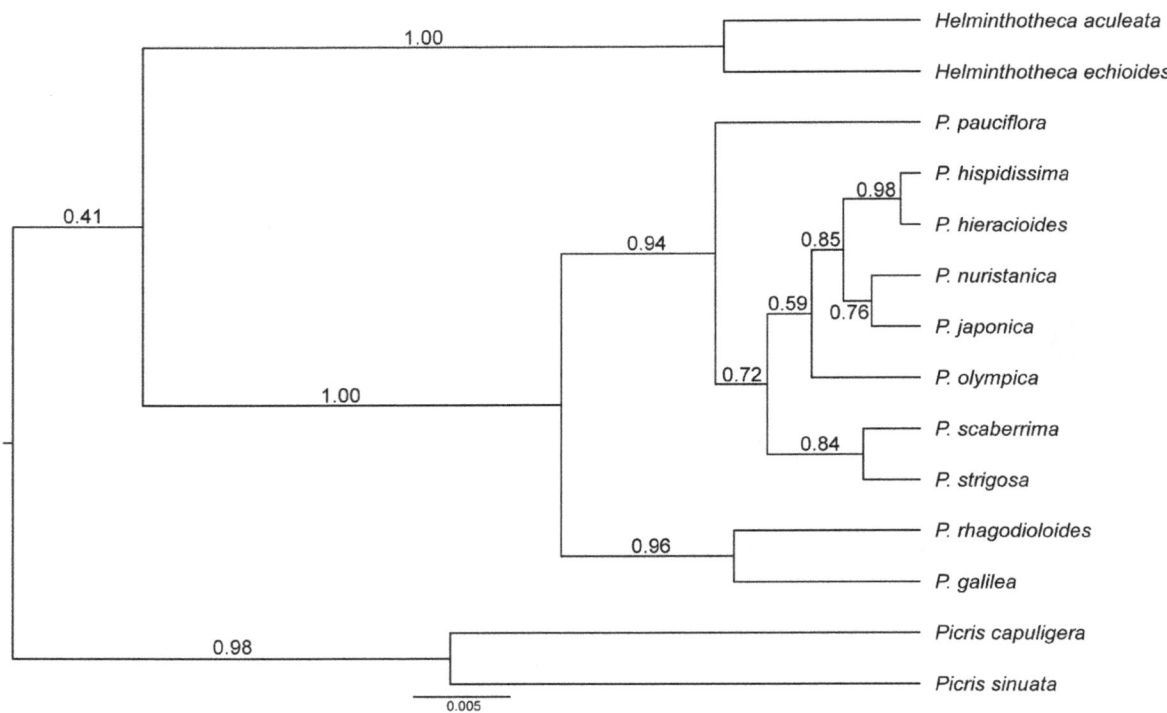

Figure 8. The maximum clade credibility species tree obtained from Bayesian inference in *Beast based on the ITS and cpDNA loci.
The posterior probability values are indicated above the branches.

processes. Reasons for this ambiguity might come from weak genetic divergence of analysed taxa, and importantly from presence of gene transfer via introgression which, moreover, most probably occurred repeatedly during their evolutionary history.

Comparison of genetic patterns with results inferred from AFLP and morphology

Spatial distribution of genetic variation in populations of *P. hieracioides* and *P. hispidissima* varied considerably across phylogenetic trees, with both concordant and discordant patterns being detected (Figures 1 and 4). The ITS data were highly consistent with taxonomic/morphological delimitation of the studied taxa. The genetic ITS groups of all studied species as well as within *P. hieracioides* exhibited high levels of concordance with the results of the AFLP analyses [24]. The congruence of genetic patterns resolved by ITS and AFLP, especially, with regard to the phylogenetic signal, has been already demonstrated for various plant groups [69]. The spatial distribution of genetic variation in populations and genetic lineages of *P. hieracioides* and in *P. hispidissima* was, however, clearly blurred by populations harbouring individuals with non-homogenized ITS sequences. Such a diffuse pattern of genetic variation disrupting geographic structure is not unexpected in the nuclear biparentally inherited genetic markers. In fact, this reflects both seed and pollen-mediated gene dispersal [70], especially, in widespread taxa with unrestricted gene flow [71–73]. Numerous heterogeneous ITS sequences did not, however, display APS for some positions variable in the potentially parental haplotypes (Table S2). This pattern can be attributed to in vivo recombination between ITS ribotypes, or to the eventual combination of minor, undetected alleles with more widespread ones [11,67,74]. Detection of such minor ITS ribotypes within taxa/populations might be prevented by preferential amplification of the major ITS copies or by

quantitative masking of the rare ribotypes [74]. Another highly plausible explanation on the origin of incomplete homogenization of APS involves concerted evolution [2,40,58,63,64,67]. Its presence in our ITS dataset might be one of the crucial reasons responsible for the failure to detect mutual hybridisation among populations and genetic lineages of *P. hieracioides* and *P. hispidissima* (see discussion above).

In contrast, genetic variation inferred from the plastid cpDNA spacer was, however, only partly congruent with that captured by ITS, AFLPs, and morphology. Further, numerous discrepancies in the number and composition of the genetic cpDNA groups were noted (see 'Results'). Thus, the plastid pattern exhibited an apparently mosaic-like spatial structure and two or more species/ genetic groups shared same plastid haplotypes. This was, especially, true for the sympatric/parapatric populations (Figure 6). Complex and incongruent genetic plastid patterns might, beside of the aforementioned hybridisation and incomplete lineage sorting, arise from different attributes of the employed genetic markers such as their evolutionary rate, inheritance mode (bi-parental vs. uni-parental), or number of copies (multiple vs. single copies), which reflects different time horizons of evolution [11]. The nuclear ITS marker evolves faster than plastid ones and therefore enable detection of more recent diversification events [11,75].

Evolution of the studied *Picris* taxa and taxonomic implications

Our results suggested that evolution of the studied *Picris* species was shaped by the interplay of several factors and evolutionary processes. Their diversification was most likely rapid, and may have occurred in an allopatric manner. Since all of these taxa, but especially, *P. hispidissima* and *P. olympica*, have unique habitat and ecology preferences (see Material and Methods), we presume

Table 3. List of the distances among couples of taxa with p-values <0.05, as inferred by posterior predictive distributions.

Gene	Individual 1	Individual 2	Obs. Distance	p-value
cpDNA	*P. olympica1*	*P. h.* subsp. *hieracioides22*	0	0.0288288
	P. olympica 1	*P. h.* subsp. *hieracioides34*	0	0.0288288
	P. olympica2	*P. h.* subsp. *hieracioides22*	0	0.0288288
	P. olympica2	*P. h.* subsp. *hieracioides34*	0	0.0288288
	P. olympica3	*P. h.* subsp. *hieracioides22*	0	0.0288288
	P. olympica3	*P. h.* subsp. *hieracioides34*	0	0.0288288
	P. olympica4	*P. h.* subsp. *hieracioides22*	0	0.0288288
	P. olympica4	*P. h.* subsp. *hieracioides34*	0	0.0288288
	P. olympica1	*P. h.* subsp.*umbellata2*	0	0.0288288
	P. olympica1	*P. h.* subsp.*umbellata4*	0	0.0288288
	P. olympica1	*P. h.* subsp.*umbellata5*	0	0.0288288
	P. olympica1	*P. h.* subsp.*umbellata7*	0	0.0288288
	P. olympica2	*P. h.* subsp.*umbellata2*	0	0.0288288
	P. olympica2	*P. h.* subsp.*umbellata4*	0	0.0288288
	P. olympica2	*P. h.* subsp.*umbellata5*	0	0.0288288
	P. olympica2	*P. h.* subsp.*umbellata7*	0	0.0288288
	P. olympica3	*P. h.* subsp.*umbellata2*	0	0.0288288
	P. olympica3	*P. h.* subsp.*umbellata4*	0	0.0288288
	P. olympica3	*P. h.* subsp.*umbellata5*	0	0.0288288
	P. olympica3	*P. h.* subsp.*umbellata7*	0	0.0288288
	P. olympica4	*P. h.* subsp.*umbellata2*	0	0.0288288
	P. olympica4	*P. h.* subsp.*umbellata4*	0	0.0288288
	P. olympica4	*P. h.* subsp.*umbellata5*	0	0.0288288
	P. olympica4	*P. h.* subsp.*umbellata7*	0	0.0288288
	P. rhagodioloides1	*P. olympica1*	0.00625	0.0405405
	P. rhagodioloides1	*P. olympica2*	0.00625	0.0405405
	P. rhagodioloides1	*P. olympica3*	0.00625	0.0405405
	P. rhagodioloides1	*P. olympica4*	0.00625	0.0405405
	P. sinuata1	*P. capuligera1*	0.0197917	0.0373874
	P. sinuata2	*P. capuligera1*	0.0197917	0.0373874
	H. aculeata1	*H. echioides1*	0.00625	0.000900901
ITS	*H. aculeata1*	*H. echioides1*	0.0306279	0.00135135

that ecological factors also played an important role in their speciation.

The centre of the diversity of the studied species complex, as indicated by the distributions of *P. olympica* and *P. hispidissima* as well as by the accumulation of rare haplotypes, lies in the southern European peninsulas and partly also the Alps and the Carpathians (Figure 6). These regions have been shown to be the most important glacial refugia for the majority of European flora and fauna [76,77]. This might suggest that the evolutionary history of the investigated *Picris* taxa was connected with glacial/interglacial cycles in the Quaternary. This is in concordance with conclusions concerning the origin of the AFLP lineages, as detected within *P. hieracioides* [24].

Genetic data presented herein together with recent distribution [21] indicate that *P. olympica* very likely evolved in alpine levels of high mountain ranges of Anatolia in Asia Minor. Here it very likely survived isolated for a long period, although remote secondary contacts and interaction with other closely related taxa cannot be excluded. Likewise, autecological characteristics and

distribution of *P. hispidissima* suggest that evolutionary history of this taxon has undoubtedly been confined with coastal, calcareous mountain ranges of the western part of the Balkan Peninsula. This region has also been repeatedly proven as important evolutionary hotspot and glacial refugium for other plant groups [32,68].

Furthermore, genetic lineages detected within *P. hieracioides* most likely diversified in allopatry or parapatry in southern European peninsulas and adjacent mountainous regions of the Alps and the Carpathians. The mountain-dwelling populations morphologically corresponding to the *P. h.* subsp. *umbellata* most likely evolved in European mountains or their proximity (the Alps, central Apennines, the Carpathians, Iberian mountain ranges and the Jura Mts.) [24]. If it had existed there already during Pleistocene climatic oscillations, it could potentially survive in the peripheral glacial areas together with other mesophylous herbs [78,79]. As proved by the highest cpDNA haplotype richness, diversification centre of populations morphologically assignable to the lowland *P. h.* subsp. *hieracioides* was located more southwards, especially, in lowland and hilly regions of the Apennine Peninsula and plausibly also in the Balkan Peninsula. Subsequently, they colonised other habitats of their present day occurrence in central and north-western Europe as well as in Asia Minor from these southern refugia.

An important taxonomic question is whether the analysed taxa should be recognised as distinct species/subspecies, or whether they rather represent informal phylogeographic lineages within a single heterogeneous taxon. The latter possibility has been favoured in case of several other complex plant species groups [30,32,80,81]. *Picris hieracioides* has been here evidenced to be highly heterogeneous across both markers. Actually, both plastid and ITS nuclear alleles have transcended borders of both subspecies of *P. hieracioides*, as recently delimited in [24]. Moreover, several populations of *P. hieracioides* displayed transitional morphological variation. In addition, from ecological point of view, *P. hieracioides* is characterised by an exceptionally large range of habitat preferences [22,24]. All these facts might indicate that both currently delimited subspecies of *P. hieracioides* are artificial biological units, representing rather ecotypes evolved recurrently in different evolutionary lineages under similar environmental influences than real taxa.

Ecotype hypothesis, however, strongly contradicts results of our previous study focused not only on investigation of wild populations of *P. hieracioides*, but also on cultivation experiments realised under environmentally homogeneous conditions [22,23,24]. These experiments included more than 300 individuals from 32 populations that belonged to both subspecies of *P. hieracioides*, originating from almost the entire studied area [24]. Subsequent comparative morphological investigations revealed that both subspecies retained morphologically stable under homogeneous environmental conditions. Importantly, these analyses showed that environmental conditions have had only a minute impact on the morphological characters identified as taxonomically diagnostic. Likewise, *P. hispidissima* retained its morphological distinctiveness under environmentally homogenous conditions (M. Slovák, unpublished data). Thus, taxonomically important morphological features of these entities are not a consequence of environmental plasticity, but are genetically inherited instead. All three taxa are well defined morphologically, ecologically and partially also genetically (AFLP's) at least within 'pure' populations untouched by their recent mutual secondary contacts.

Both, subspecies of *P. hieracioides* as well as *P. hispidissima* are recently diversified and thus genetically weakly separated taxa. Additionally, their genetic variation has been significantly affected by their reciprocal multiple secondary contacts, taking place after anthropogenically mediated range expansions. Due to the lack of habitat and breeding isolation barriers, hybridisation and introgression among these taxa occur frequently in contact zones. This process leads to the genetic and morphological homogenisation of affected populations, erasing their distinctness in contact zones. Similar cases of taxon fusion caused by hybridisation preceded by migration of populations enhanced by human activities have been reported for instance in members of the genus *Knautia* [80] and *Cerasus* [82]. For the latter case the term "anthropohybridisation" was coined. Establishment of sound taxonomic concept in such difficult cases is very problematic and usually it is a matter of subjective choice. Therefore, at the present state of knowledge, we prefer to leave both subspecies of *P. hieracioides* at currently

accepted taxonomic ranks [24]. In contrast, *P. hispidissima* should be included under highly variable *P. hieracioides* as separated subspecies: *P. h.* subsp. *hispidissima* (Bartl.) Slovák and Kučera, comb. nova, hoc loco (basionym: *Crepis hispidissima* Bartl. in Bartling & Wendland, Beitr. Bot. 2: 125, 1825).

Although this study does not provide comprehensive and firm taxonomic conclusions, it represents an important background for future investigations on this intricate and dynamically evolving species complex. Last but not least, the present investigation documents rapid process of taxon erosion, happening due to the anthropogenic disturbance of natural biotopes.

Supporting Information

Table S1   Locality details of studied taxa, number of individuals used for the particular molecular analyses, and GenBank accession numbers.

Table S2   Summary of nucleotide site variation for the nuclear ITS region in the studied taxa.

Alignment S1   ITS_1: comprised 117 ITS sequences without APS. The accession labels in alignments follow the population and taxon codes used in Table S1.

Alignment S2   ITS_2: comprised all 206 ITS sequences. The accession labels in alignments follow the population and taxon codes used in Table S1.

Alignment S3   cpDNA: comprised all 219 rpl32-*trn*L[UAG] sequences. The accession labels in alignments follow the population and taxon codes used in Table S1.

Alignment S4   ITS_cpDNA: comprised 117 ITS sequences without APS and corresponding plastid ones. The accession labels in alignments follow the population and taxon codes used in Table S1.

Acknowledgments

We would like to thank the following colleagues for their help in sample collection: Ulla-Grilt Andersson (Sweden), Fabio Conti (Italy), Gianniantonio Domina (Italy), Rolland Douzet (France), Sandrine Godefroid (Belgium), Iva Hodálová (Slovakia), Nicodemo Passalacqua (Italy), Peter Repa (Slovakia), Åke Svensson (Sweden), Luis Villar (Spain) and Torbjörn Tyler (Sweden). The authors also wish to thank The Royal Botanic Gardens, Kew (United Kingdom) and The National Botanic Garden of Belgium (Belgium) who provided *P. hieracioides* and *P. sinnuata* seeds (see details in Table S1). Our special thanks go to Judita Zozomová-Lihová (Slovakia) for her valuable advice and laboratory assistance during this study, and finally to Karol Marhold and all anonymous reviewers for their valuable comments on previous versions of this manuscript.

Author Contributions

Conceived and designed the experiments: JK MS. Performed the experiments: JK MS. Analyzed the data: MS PV EZ. Contributed reagents/materials/analysis tools: JK MS. Wrote the paper: MS PV EZ.

References

1. Comes HP, Abbott RJ (2001) Molecular phylogeny, reticulation, and lineage sorting in Mediterranean *Senecio* sect. *Senecio* (Asteraceae). Evolution 55: 1943–1962.

2. Nieto Feliner G, Gutiérrez Larena B, Fuertes Aguilar J (2004) Fine-scale geographical structure, intra-individual polymorphism and recombination in

nuclear ribosomal internal transcribed spacers in *Armeria* (Plumbaginaceae). Ann Bot (Oxford) 93: 189–200.

3. Willyard A, Cronn R, Liston A (2009) Reticulate evolution and incomplete lineage sorting among the ponderosa pines. Molec Phylogen Evol 52: 498–511.

4. Schmidt-Lebuhn AN, de Vos JM, Keller B, Conti E (2012) Phylogenetic analysis of *Primula* section *Primula* reveals rampant non-monophyly among morphologically distinct species. Molec Phylogen Evol 65: 23–34.

5. Yu WB, Huang PH, Li DZ, Wang H (2013) Incongruence between nuclear and chloroplast DNA phylogenies in *Pedicularis* section *Cyathophora* (Orobanchaceae). PLoS ONE, doi: 10.1371/journal.pone.0074828

6. Rieseberg LH (1997) Hybrid origins of plant species. Ann Rev Ecol Syst 28: 359–389.

7. Wendel JF, Doyle J (1998) Phylogenetic incongruence: Window into genome history and molecular evolution. In: Soltis D, Soltis P, Doyle J, editors.Molecular systematics of plants II: DNA sequencing. Kluwer, Boston, USA. pp. 265–296.

8. Linder CR, Rieseberg LH (2004) Reconstructing patterns of reticulate evolution in plants. Amer J Bot 91: 1700–1708.

9. Joly S, McLenachan PA, Lockhart PJ (2009) A statistical approach for distinguishing hybridisation and incomplete lineage sorting. Amer Nat 174: 54–70.

10. Funk DJ, Omland KE (2003) Species-level paraphyly and polyphyly: Frequency, causes, and consequences, with insights from animal mitochondrial DNA. Annual Rev Ecol Evol Syst 34: 397–423.

11. Small RL, Cronn RC, Wendel JF (2004) Use of nuclear genes for phylogeny reconstruction in plants. Austral Syst Bot 17: 145–170.

12. Holland BR, Benthin S, Lockhart PJ, Moulton V, Huber KT (2008) Using supernetworks to distinguish hybridisation from lineage-sorting. BMC Evol Biol 8: 202.

13. Yu Y, Than C, Degnan JH, Nakhleh L (2011) Coalescent histories on phylogenetic networks and detection of hybridisation despite incomplete lineage sorting. Syst Biol 60: 138–149.

14. Wiens JJ, Hollingsworth BD (2000) War of the iguanas: Conflicting molecular and morphological phylogenies and long-branch attraction in iguanid lizards. Syst Biol 49: 143–159.

15. Duvall MR, Ervin AB. (2004) 18S gene trees are positively misleading for monocot/dicot phylogenetics. Molec Phylogen Evol 30: 97–106.

16. van der Niet T, Linder HP (2008) Dealing with incongruence in the quest for the species tree: a case study from the orchid genus *Satyrium*. Molec Phyl Evol 47: 154–174.

17. Frajman B, Eggens F, Oxelman B (2009) Hybrid origins and homoploid reticulate evolution within *Heliosperma* (Sileneae, Caryophyllaceae) - A multigene phylogenetic approach with relative dating. Syst Biol 58: 328–345.

18. Heled J, Drummond AJ (2010) Bayesian inference of species trees from multilocus data. Molec Biol Evol 27: 570–580.

19. Kubatko L, Carstens B, Knowles L (2009) STEM: species tree estimation using maximum likelihood for gene trees under coalescence. Bioinformatics 25: 971–973.

20. Joly S. (2012) JML: testing hybridization from species trees. Molec Ecol Res 12: 179–184.

21. Lack HW (1974)Die Gattung *Picris* L. sensu lato im ostmediterran-westasiatischen Raum. PhD thesis, University of Vienna, Austria. 184 p.

22. Slovák M, Marhold K (2007) Morphological evaluation of *Picris hieracioides* L. (Compositae) in Slovakia. Phyton (Horn) 47: 73–102.

23. Slovák M, Urfus T, Vít P, Marhold K (2009) Balkan endemic *Picris hispidissima* (Compositae): Morphology, DNA content and relationship to polymorphic *P. hieracioides*. Pl Syst Evol 278: 178–201.

24. Slovák M, Kučera J, Marhold K, Zozomová-Lihová J (2012) The morphological and genetic variation in the polymorphic species *Picris hieracioides* (Compositae, Lactuceae) in Europe strongly contrasts with traditional taxonomical concepts. Syst Bot 21: 258–278.

25. Lack HW (1979) The genus *Picris* (Asteraceae, Lactuceae) in Tropical Africa. Pl Syst Evol 131: 35–52.

26. Samuel R, Gutermann W, Stuessy TF, Ruas CF, Lack HW, et al. (2006) Molecular phylogenetics reveals *Leontodon* (Asteraceae, Lactuceae) to be diphyletic. Amer J Bot 93: 1193–1205.

27. Slovák M, Šingliarová B, Mráz P (2008) Chromosome numbers and mode of reproduction in *Picris hieracioides* s.l. (Asteraceae) with notes on some other *Picris* taxa. Nordic J Bot 28: 238–244.

28. Slovák M, Vít P, Urfus T, Suda J (2009) Complex pattern of genome size variation in a polymorphic member of the Asteraceae. J Biogeogr 36: 372–384.

29. Lack HW (1975) *Picris*. In: Davis PH, editor.Flora of Turkey and the East Aegean Islands 5.Edinburgh, UK: University Press. pp. 678–684.

30. Frajman B, Oxelman B (2007) Reticulate phylogenetics and phytogeographical structure of *Heliosperma* (Sileneae, Caryophyllaceae) inferred from chloroplast and nuclear DNA sequences. Molec Phylogen Evol 43: 140–155.

31. Prentice HC, Malm JU, Hathaway L (2008) Chloroplast DNA variation in the European herb *Silene dioica* (red campion): postglacial migration and interspecific introgression. Pl Syst Evol 272: 23–37.

32. Bardy KE, Albach DC, Schneeweiss GM, Fischer MA, Schönswetter P (2010) Disentangling phylogeography, polyploid evolution and taxonomy of a woodland herb (*Veronica chamaedrys* group, Plantaginaceae s.l.) in southeastern Europe. Molec Phylogen Evol 57: 771–786.

33. Shaw J, Lickey EB, Schilling EE, Small RL (2007) Comparison of whole chloroplast genome sequences to choose noncoding regions for phylogenetic studies in angiosperms: The tortoise and the hare III. Amer J Bot 94: 275–288.

34. Chapman MA, Chang JC, Weisman D, Kesseli RV, Burke JM (2007) Universal markers for comparative mapping and phylogenetic analysis in the Asteraceae (Compositae). Theor Appl Genet 115: 747–755.

35. Shaw J, Lickey EB, Beck JT, Farmer SB, Liu W, et al. (2005) The tortoise and the hare II: Relative utility of 21 noncoding chloroplast DNA sequences for phylogenetic analysis. Amer J Bot 92: 142–166.

36. Francisco-Ortega J, Fuertes Aguilar J, Gómez Campo C, Santos-Guerra A, Jansen RK (1999) Internal transcribed spacer sequence phylogeny of *Crambe* L. (Brassicaceae): Molecular data reveal two old world disjunctions. Molec Phylogen Evol 11: 361–380.

37. White TJ, Bruns T, Lee S, Taylor JW (1990) Amplification and direct sequencing of fungal ribosomal RNA genes for phylogenetics. In: Innis MA, Gelfand DH, Sninsky JJ, White TJ, editors.PCR Protocols: A Guide to Methods and Applications.New York, USA: Academic Press. pp. 315–322.

38. Timme R, Kuehl EJ, Boore JK, Jansen RK (2007) A comparative analysis of the *Lactuca* and *Helianthus* (Asteraceae) plastid genomes: identification of divergent regions and categorization of shared repeats. Amer J Bot 94: 302–313.

39. Hall TA (1999) BioEdit: a user-friendly biological sequence alignment editor and analysis program for Windows 95/98/NT. Nucleic Acids Symp Ser (Oxford) 41: 95–98.

40. Fuertes Aguilar J, Nieto Feliner G (2003) Additive polymorphisms and reticulation in an ITS phylogeny of thrifts (*Armeria*, Plumbaginaceae). Molec Phylogen Evol 28: 430–447.

41. Záveská E, Fér T, Šída O, Krak K, Marhold K, et al. (2012) Phylogeny of *Curcuma* (Zingiberaceae) based on plastid and nuclear sequences: Proposal of the new subgenus *Ecomata*. Taxon 61: 747–763.

42. Farris JS, Källersjö M, Kluge AG, Bult C (1994) Testing significance of congruence. Cladistics 10: 315–319.

43. Swofford DL (2001) PAUP*. Phylogenetic Analysis Using Parsimony (*and Other Methods), v. 4.0 beta 10. Sunderland: Sinauer Associates.

44. Maddison DR, Maddison WP (2000)MacClade4: analysis of phylogeny and character evolution, version 4.0. Sinauer, Sunderland, Massachusetts, USA. 398 p.

45. Ronquist F, Huelsenbeck JP (2003) MrBayes 3: Bayesian phylogenetic inference under mixed models. Bioinformatics 19: 1572–1574.

46. Miller MA, Pfeiffer W, Schwartz T (2010) "Creating the CIPRES Science Gateway for inference of large phylogenetic trees" in Proceedings of the Gateway Computing Environments Workshop (GCE), 14 Nov. 2010, New Orleans, LA : 1–8 pp.

47. Guindon S, Gascuel O (2003) A simple, fast, and accurate algorithm to estimate large phylogenies by maximum likelihood. Syst Biol 52: 696–704.

48. Posada D (2008) ModelTest: Phylogenetic Model Averaging. Molec Biol Evol 25: 1253–1256.

49. Huson DH, Bryant D (2006) Application of phylogenetic networks in evolutionary studies. Molec Biol Evol 23: 254–267.

50. Templeton AR, Crandall KA, Sing CF (1992) A cladistic analysis of the phenotypic associations inferred from restriction endonuclease mapping and DNA sequence data. III. Cladogram estimation. Genetics 132: 619–633.

51. Clement M, Posada D, Crandall K (2000) TCS: a computer program to estimate gene genealogies. Mol Ecol 9: 1657–1660.

52. Wägelle JW, Rödding F (1998) A priori estimation of phylogenetic information conserved in aligned sequences. Molec Phylogen Evol 9: 358–365

53. Rambaut A, Drummond AJ (2007) Tracer v1.4. Available: http://beast.bio.ed.ac.uk/Tracer. Accessed 2014 Jan 21.

54. Silvestro D, Michalak I (2011) raxmlGUI: a graphical front-end for RAxML. Org Divers Evol 12: 335–337.

55. Shimodaira H (2002) An approximately unbiased test of phylogenetic tree selection. Syst Biol 51: 492-508.

56. Shimodaira H (2008) Testing regions with non-smooth boundaries via multiscale bootstrap. J Stat Plan Inference 138: 1227–1241.

57. Shimodaira H, Hasegawa M (2001) Consel: for assessing the confidence of phylogenetic tree selection. Bioinformatics 17: 1246–1247.

58. Baldwin BG, Sanderson MJ, Porter JM, Wojciechowski MF, Cambell CS, et al. (1995) The ITS region of nuclear ribosomal DNA: A valuable source of evidence on angiosperm phylogeny. Ann Missouri Bot Gard 82: 247–277.

59. Hudson RR (1990) Gene genealogies and the coalescent process. Oxford Surv Evol Biol 7: 1–44.

60. Jang CG, Müllner AN, Greimler J (2005) Conflicting patterns of genetic and morphological variation in European *Gentianella* section *Gentianella*. Bot J - Linn Soc 148: 175–187.

61. Flagel LE, Rapp RA, Grover CE, Widrlechner MP, Hawkins J, et al. (2008) Phylogenetic, morphological, and chemotaxonomic incongruence in the North American endemic genus *Echinacea*. Amer J Bot 95: 756–765.

62. Ramsey J., Robertson A., Husband B. (2008) Rapid adaptive divergence in new world *Achillea*, an autopolyploid complex of ecological races. Evolution 62: 639–653.

63. Fuertes Aguilar J, Rosselló JA, Nieto Feliner G (1999) nrDNA concerted evolution in natural and artificial hybrids of *Armeria* (Plumbaginaceae). Mol Ecol 8: 1341–1346.

64. Fuertes Aguilar J, Rosselló JA, Nieto Feliner G (1999b) Molecular evidence for the compilospecies model of reticulate evolution in *Armeria* (Plumbaginaceae). Syst Biol 48: 735–754.

65. Baack EJ, Rieseberg LH (2007) A genomic view of introgression and hybrid speciation. Curr Opin Genet Dev 17: 513–518.

66. Moody ML, Rieseberg LH (2012) Sorting through the chaff, nDNA gene trees for phylogenetic inference and hybrid identification of annual sunflowers (*Helianthus* sect. *Helianthus*). Molec Phylogen Evol 64: 145–155.

67. Álvarez I, Wendel JF (2003) Ribosomal ITS sequences and plant phylogenetic inference. Molec Phylogen Evol 29: 417–434.

68. Kučera J, Marhold K, Lihová J (2009) *Cardamine maritima* group (Brassicaceae) in the amphi-Adriatic area: A hotspot of species diversity revealed by DNA sequences and morphological variation. Taxon 59: 148–164.

69. Koopman WJM (2005) Phylogenetic signal in AFLP data sets. Syst Biol 54: 197–217.

70. Ennos RA (1994) Estimating the relative rates of pollen and seed migration among plant populations. Heredity 72: 250–259.

71. Malm JU, Prentice HC (2002) Immigration history and gene dispersal: allozyme variation in Nordic populations of the red campion, *Silene dioica* (Caryophyllaceae). Biol J Linn Soc 77: 23–34.

72. Tyler T (2002) Geographical distribution of allozyme variation in relation to post-glacial history in *Carex digitata*, a widespread European woodland sedge. J Biogeogr 29: 919–930.

73. Tyler T, Prentice HC, Widén B (2002) Geographic variation and dispersal history in Fennoscandian populations of two forest herbs. Pl Syst Evol 233: 47–64.

74. Rauscher JT, Doyle JJ, Brown AHD (2002) Internal transcribed spacer repeat-specific primers and the analysis of hybridisation in the *Glycine tomentella* (Leguminosae) polyploid complex. Mol Ecol 11: 2691–2702.

75. Rebernig CA, Schneeweiss GM, Bardy KE, Schönswetter P, Villaseñor JL, et al. (2010) Multiple Pleistocene refugia and Holocene range expansion of an abundant southwestern American desert plant species (*Melampodium leucanthum*, Asteraceae). Mol Ecol 19: 3421–3443.

76. Hewitt GM (2004) Genetic consequences of climatic oscillations in the Quaternary. Phil Trans R Soc B 359: 183–195.

77. Médail F, Diadema K (2009) Glacial refugia influence plant diversity patterns in the Mediterranean Basin. J Biogeogr 36: 1333–1345.

78. Kramp K, Huck S, Niketić M, Tomović G, Schmitt T (2009) Multiple glacial refugia and complex postglacial range shifts of the obligatory woodland plant species *Polygonatum verticillatum* (Convallariaceae). Pl Biol 11: 392–404.

79. Huck S, Büdel B, Kadereit JW, Printzen C (2009) Rangewide phylogeography of the European temperate-montane herbaceous plant *Meum athamanticum* Jacq.: Evidence for periglacial persistence. J Biogeogr 36: 1588–1599.

80. Rešetnik I, Frajman B, Bogdanović S, Ehrendorfer F, Schönswetter P (2014) Disentangling relationships among the diploid members of the intricate genus *Knautia* (Caprifoliaceae, Dipsacoideae). Molec Phylogen Evol 74: 97ec P

81. Schönswetter P, Solstad H, Garcia PE, Elven R (2009) A combined molecular and morphological approach to the taxonomically intricate European mountain plant *Papaver alpinum* s.l. (Papaveraceae) – Taxa or informal phylogeographical groups? Taxon 17: 1326–1343.

82. Wójcicky JJ (1991) Variability of *Prunus fruticosa* Pall. and the problem of an anthropohybridisaton. Veröff. Geobot. Inst. ETH, Stiftung Rübel, Zürich 106: 266–272.

17

Local and Regional Scale Genetic Variation in the Cape Dune Mole-Rat, *Bathyergus suillus*

Jacobus H. Visser[1], Nigel C. Bennett[2], Bettine Jansen van Vuuren[1]*

1 Department of Zoology, University of Johannesburg, Auckland Park, South Africa, **2** Mammal Research Institute, Department of Zoology and Entomology, University of Pretoria, Pretoria, South Africa

Abstract

The distribution of genetic variation is determined through the interaction of life history, morphology and habitat specificity of a species in conjunction with landscape structure. While numerous studies have investigated this interplay of factors in species inhabiting aquatic, riverine, terrestrial, arboreal and saxicolous systems, the fossorial system has remained largely unexplored. In this study we attempt to elucidate the impacts of a subterranean lifestyle coupled with a heterogeneous landscape on genetic partitioning by using a subterranean mammal species, the Cape dune mole-rat (*Bathyergus suillus*), as our model. *Bathyergus suillus* is one of a few mammal species endemic to the Cape Floristic Region (CFR) of the Western Cape of South Africa. Its distribution is fragmented by rivers and mountains; both geographic phenomena that may act as geographical barriers to gene-flow. Using two mitochondrial fragments (cytochrome *b* and control region) as well as nine microsatellite loci, we determined the phylogeographic structure and gene-flow patterns at two different spatial scales (local and regional). Furthermore, we investigated genetic differentiation between populations and applied Bayesian clustering and assignment approaches to our data. Nearly every population formed a genetically unique entity with significant genetic structure evident across geographic barriers such as rivers (Berg, Verlorenvlei, Breede and Gourits Rivers), mountains (Piketberg and Hottentots Holland Mountains) and with geographic distance at both spatial scales. Surprisingly, *B. suillus* was found to be paraphyletic with respect to its sister species, *B. janetta*–a result largely overlooked by previous studies on these taxa. A systematic revision of the genus *Bathyergus* is therefore necessary. This study provides a valuable insight into how the biology, life-history and habitat specificity of animals inhabiting a fossorial system may act in concert with the structure of the surrounding landscape to influence genetic distinctiveness and ultimately speciation.

Editor: Alfred L. Roca, University of Illinois at Urbana-Champaign, United States of America

Funding: The work was supported by a Centre for Invasion Biology stipend to BJVV. The funder had no role in study design, data collection and analysis, decision to publish, or preparation of the manuscript.

Competing Interests: The authors have declared that no competing interests exist.

* Email: bettinevv@uj.ac.za

Introduction

The life-history of a species coupled with its ecological requirements and vagility determine its dispersal capability [1]. Together with the structure and composition of the surrounding landscape (e.g., barriers to gene-flow) the dispersal capability impacts on the spatial distribution of genetic variation [2]. The interaction of these factors affects genetic structure and species integrities in various taxa inhabiting niches in riverine [3,4,5], marine [6,7,8], terrestrial [9,10], arboreal [11,12,13] and saxicolous [14,15,16,17] systems. The fossorial/subterranean system, however, remains largely under-investigated. Due to the inconspicuous, subterranean nature of such animals [18,19,20], observational methods of population dynamics, breeding system (reproductive interactions) and dispersal behaviour are difficult or not feasible [21,22,23,24,25,26,27,28].

The structurally simple, constant and predictable subterranean niche [29,30] has led to specialization (narrower niches) in permanently fossorial taxa [25,30,31]. This specialization has resulted in adaptively convergent evolution of both structural and functional traits (low genetic variation, food generalism, cylindrical bodies and anatomical reductions) in fossorial mammals [25,29,30,31]. Consequently, such a specialized morphology decreases vagility [30,32]. In addition, fossorial mammals, especially rodents, exhibit behavioural attributes (a solitary life-style marked by high territoriality and aggressive behaviour [30,33]) which acts to limit dispersal. In solitary species, home ranges are synonymous with territories; these defended areas remain fixed for life, with only minor or incremental boundary changes [30,34,35,36,37] (but see [26] and [28]). As a result, juveniles would have to secure their own territories - a factor influenced by the abundance of the species in the particular habitat. In "saturated" habitats, juveniles would likely traverse larger distances to establish territories. Additionally to these behavioural and biological attributes, suitable habitat patches have a disjunct distribution with restricted food resources [30,32,38,39]. Populations of fossorial mammals are therefore often spatially isolated with patchy distributions [30,32,38]. The genetic patterns evident in such a system, according to Nevo [30], should therefore be dominated by isolation-by-distance with low gene-flow between demes (but see [28]), founder effects, genetic drift and inbreeding.

Intuitively, given these life-history characteristics, the landscape in the form of geographic barriers (e.g., mountains and rivers) should also have a profound influence on the distribution of genetic variation in fossorial systems. Although the effect of such geographic barriers on the genetic structure and evolutionary patterns of subterranean species has been widely suggested [30,32,34,40,41,42,43,44,45,46,47,48,49], the effects of these barriers on processes such as gene-flow remain largely speculative.

The family Bathyergidae offers an enticing opportunity to disentangle the factors which influence the spread of genetic variation in fossorial systems. The bathyergids are a monophyletic group of obligatory subterranean hystricognath rodents endemic to sub-Saharan Africa. Six genera are recognized: *Heterocephalus*, *Heliophobius*, *Bathyergus*, *Georychus*, *Cryptomys* and *Fukomys* [27,31,40,41,44,50,51,52]. Speciation in at least two of the genera is prolific [40] and has been suggested to be linked to population structure and geographic isolation coupled with a labile karyotype [18,32,45,53,54,55,56,57], factors which result in rapid fixation of e.g., chromosomal mutation [30,58].

In spite of these suggested models and the possible influence of behaviour, morphology and the landscape on the distribution of genetic variation in fossorial taxa, previous studies focussed mostly on inter-generic relationships [41,44,45,50,51,59,60,61] or had limited geographic sampling within genera [60]. Few studies have been conducted on geographic variation within either genera or species [41], but more recent investigations have started to disentangle intra-generic relationships (e.g., [31,41,43]).

The solitary mole-rat species have been largely neglected in phylogeographic studies compared to their social counterparts [43,52]. The uncertainty around the intra-generic placement of certain taxa is exemplified by the genus *Bathyergus*, containing two species, *B. janetta* and *B. suillus* [31,50,62]. While *B. suillus* and *B. janetta* are proposed to differ in natural history, chromosome number, allozymes and mitochondrial DNA profiles, no genetic differences in either allozymes or karyotype were found by Janecek *et al.* [51] and Deuve *et al.* [58] respectively. Indeed, these species have been suggested to have a hybrid zone at the border of their distributions (Rondawel, South Africa; [31,58]). As such, any putative genetic differences between these species do not prevent hybridization. In addition, although inter-generic studies have invariably found *B. suillus* and *B. janetta* as sister species [31,50,51,58,62,63], these studies used representatives of *B. suillus* from the west coast area, thereby not allowing for the inclusion of possible geographic variation. Ingram *et al.* [41] used representatives of *B. suillus* from both the south- and west coast and reported *B. suillus* to be paraphyletic with respect to *B. janetta*; this led to the suggestion of higher genetic variation in this genus than previously anticipated.

In this study, we use the Cape dune mole-rat, *Bathyergus suillus*, as a model species to investigate the effect of a fossorial life-style on the distribution of genetic variation in a discontinuous landscape divided by barriers. *Bathyergus suillus* is the largest of the mole-rat species, is solitary and highly aggressive and its size (energetic input of digging; [64,65]) restricts it to the mesic sandy soil areas of the Cape Floristic Region of South Africa characterised by predictable rainfall [27,31,37,51,65,66,67, 68,69,70]. The tunnel system of *B. suillus* is relatively short (< 400 m; [36]) and the morphology, life-history and behaviour of *B. suillus* lends itself to poor dispersal, thus one would expect genetic structure between isolated populations. Despite this intuitive view, very few studies (with the exception of [28]) have to date focussed on the phylogeography and gene-flow of this species.

We therefore tested hypotheses on how the ecology, distribution, life-history and the connectivity of the surrounding landscape have shaped genetic variation across the distribution of *B. suillus*. Our aims were several fold: 1) to test the model of genetic isolation, inbreeding and diminished heterozygosity proposed by Nevo [30], 2) to investigate the effect of geographic barriers on gene-flow patterns in a fossorial system using *B. suillus* as a model, 3) to determine the phylogeographic patterns and intra-generic relationships within *B. suillus* and 4) to compare our results to previous studies on taxa exhibiting similar life-histories and habitat requirements and interpret the impact of a fossorial life-style on the distribution of genetic variation. By using mitochondrial (cytochrome *b* and control region) and nine nuclear (microsatellite) markers, we determined the distribution of genetic variation at local (Sandveld Bioregion - an area divided by the Verlorenvlei and Berg Rivers as well as the Piketberg Mountains) and regional (Cape Floristic Region- a region cleaved by major river systems and mountain ranges) spatial scales through adopting a landscape genetics approach. The influence of connectivity on spatial genetic patterns is the crux of the field of "landscape genetics" [71]. This emerging field offers a framework by which one can isolate the influence of landscape variables and their impact on genetic variation [71,72,73] as well as the identification of barriers to gene flow [74] and integrates ecology, spatial statistics and population genetics to explain evolutionary patterns and processes [71,73,75,76]. Importantly, the movement of an organism is assessed from that organism's perspective [75]; an organism's perception of the landscape differs from the simplistic simulations incorporated in isolation-by-distance (IBD) analyses [72]. This study therefore gives insight into understanding spatial and temporal patterns and processes of biotic diversity across different hierarchical levels in a fossorial system.

Materials and Methods

Sample collection

Sampling was conducted in a hierarchical fashion across the range of *B. suillus*. Tissue samples were collected from five localities (local scale) in the Sandveld region - an area of Quaternary sand deposits and high species diversity situated between the Atlantic Ocean and the western branch of the Cape Fold Mountains [76,77]. Specifically, the Sandveld is divided into three regions by the Berg and Verlorenvlei Rivers namely a northern (Redelinghuys), inner (the area between these rivers) and southern (Vredenburg) region. In total, 10 localities (regional scale) were sampled over the species' range (Figure 1). A total of 20 specimens were sampled per locality with the exception of Stanford (n = 12) and Sedgefield (n = 10). Tail-clippings were taken and stored at room temperature in a saturated salt solution supplemented with 20% dimethyl sulfoxide (DMSO). The protocol was approved by the Ethics Committee of Stellenbosch University (Permit Number: 10NP_VAN01). Handling time was minimized and clipped tails were treated with antibiotics so as to minimize suffering. We are grateful to farmers and landowners in the Western and Southern Cape for access and sample collection (WCNC permit number: AAA-004-00476-0035).

DNA extraction and sequencing

Mitochondrial DNA. Total genomic DNA was extracted from tail-clippings using a commercial DNA extraction kit (DNeasy Tissue and Blood kit; Qiagen) following the manufacturer's protocol. Two mitochondrial DNA segments were amplified using universal primers for the amplifications and sequencing: 928 bp of the protein coding cytochrome b gene (*L14724* and *H15915*; [78,79]) and 897 bp on the 5' side of the hypervariable control region (*LO* and *E3*; [80]).

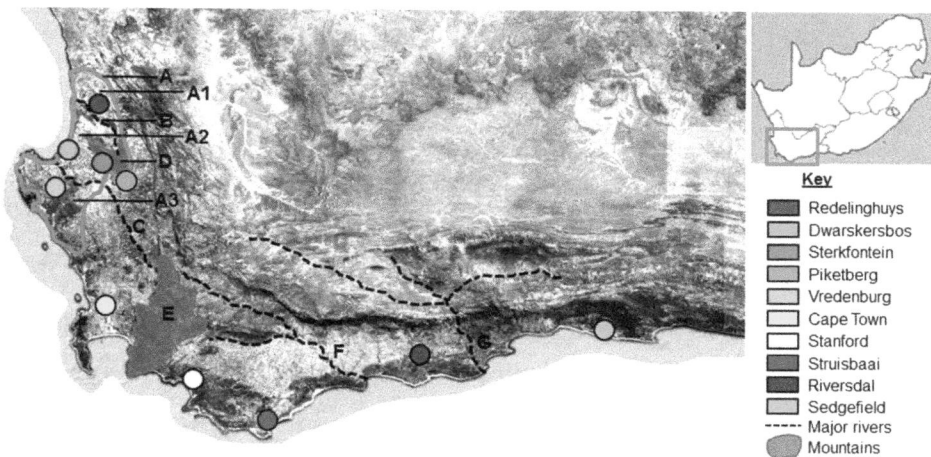

Figure 1. Sampling map of the 10 *B. suillus* populations in this study. Map showing the 10 *B. suillus* collection sites sampled across the Cape Floristic Region of South Africa. The geographic phenomena named in this study are A.) the Sandveld region with A1) the Redelinghuys, A2.) inner-Sandveld and A3.) Vredenburg areas, B.) the Verlorenvlei- and C.) Berg Rivers, D.) the Piketberg mountain, E.) the Hottentots Holland Mountains, F.) the Breede River and G.) the Gourits River. (The sampling map was adapted from ESRI (2007) ArcGIS version 9.3 Media Kit. Redlands, CA: Environmental Systems Research Institute).

PCR amplifications followed standard protocols. In short, amplifications were carried out in a GeneAmp PCR 2700 system (Applied Biosystems) at region-specific annealing temperatures (50°C for cytochrome b and 48°C for the control region) following the protocols outlined in Karsten *et al.* [81]. Successful amplifications were verified on a 1% agarose gel. Sequencing reactions were performed using the protocols outlined in Jansen van Vuuren and Chown [82]. Electropherograms of the raw data were checked manually (Geneious Pro 5.0 software; Biomatters Ltd, New Zealand) and aligned in MacClade version 4.06 for OS X [83].

Microsatellites. Nine microsatellite loci (*DMR1*, *DMR5*, *DMR7*, *CH1*, *Bsuil01*, *Bsuil02*, *Bsuil04*, *Bsuil05*, and *Bsuil06*) were selected from Burland *et al.* [84] and Ingram [43]. These loci were chosen for ease of amplification and polymorphism detected in various populations. The forward primer of each primer pair was 5′-labelled with one of four fluorophores (6-FAM, HEX, VIC and NED). Following primer optimization, all loci were amplified at 60°C; subsequent amplifications were performed in a multiplex at this annealing temperature. For genotyping 1 μl of diluted (1/80) PCR products was combined with 15 μl of deionized formamide and 0.2 μl of the GS500LIZ size standard (Applied Biosystems). Samples were genotyped on an ABI 3170 Prism (Applied Biosystems) and scored using ABI Prism Genemapper version 3.7 software (Applied Biosystems).

Data analyses

Summary statistics and inbreeding. All microsatellite loci across all populations were tested for Hardy-Weinberg equilibrium in Genalex as well as for null alleles in Genepop version 4.0.10 [102,103]. A test value larger than 0.2 suggests that the marker contains null alleles in a particular population [104]. Linkage disequilibrium was also investigated using Genepop version 4.0.10 [102,103] by running Markov chains for 10 000 iterations.

Genetic diversity detected within each sampling locality and summary statistics for the combined mitochondrial DNA analyses (including number of haplotypes and nucleotide (π) diversity) were calculated in Arlequin version 3.5 [97]. Similar measures of genetic diversity resulting from the microsatellite analysis were also calculated; these included allelic diversity indices (total number of alleles and mean number of alleles per locus; FSTAT version

2.9.3.2; [105]), observed as well as expected heterozygosities (Genalex version 6.4; [98]). Inbreeding in each colony was assessed by Wright's F_{IS} (FSTAT version 2.9.3.2).

Fluctuations in population size (based on the combined mitochondrial DNA data) in sampling localities were investigated using Fu's Fs (a statistic that evaluates population equilibrium; [106]) in DnaSP 5.10.01 [96]. Bottleneck version 1.2.02 [107] was used to investigate whether demographic changes were evident in the history of each sampling locality based on the microsatellite data. This programme measures recent effective population size changes. A two-phased model of mutation was employed (recommended by Luikart *et al.* [108] for microsatellite data) and the Wilcoxon sign-rank test value was applied to assess the probability that an excess of heterozygosity existed at a significant number of loci in a population.

Population analyses. To determine whether genetic variation was geographically significantly structured, Φ_{ST} (mitochondrial) and F_{ST} (microsatellites) were calculated between the Redelinghuys-, inner-Sandveld and Vredenburg localities, and between the West- and two South Coast clades. Additionally, pairwise Φ_{ST} and F_{ST} values were calculated between localities. Significance was determined through 9 999 permutations of the data (Arlequin version 3.5; [97]). Isolation-by-distance was evaluated for the combined mitochondrial DNA dataset using a Mantel test as employed in Arlequin and using Genalex version 6.4 [98] for the microsatellites. Geographical distances for each of the datasets were respectively determined "as the crow flies" (i.e., the shortest and most direct route between localities rather than along mountain ranges) and based on the coordinates of each sampling point.

The spatial location of genetic clusters in the microsatellite data within the studied areas was determined using Bayesian assignment approaches implemented in Geneland version 2.0.10 [99]. This programme determines the spatial location of populations (without prior input) from multi-locus genotypes through the simultaneous analysis of both genetic and geographical data [99]. A Reversible Jump (RJ) Markov Chain Monte Carlo (MCMC) algorithm was applied to estimate the number and location of genetic clusters (K) across the landscape [99]. Geneland also outperforms other spatial genetic clustering programmes when F_{ST}

values are high (i.e., when the number of migrants between populations is low) and is efficient at detecting potential contact zones between populations [100]. As allele frequencies were uncorrelated between sampling localities (calculated in Genalex) and gene-flow was expected to be low, the "no admixture" model with "independent/uncorrelated allele frequencies" was selected. Although biologically less relevant to the study species, the "admixture" model with "correlated allele frequencies" was also applied to the data as it is more powerful at detecting subtle population differentiation. We ran 100 000 permutations with a thinning of every 100 trees to search the optimal spatial distribution of markers. Ten chains were run, and the one with the highest likelihood retained.

Gene-flow among sampling localities was estimated in Lamarc version 2.1.6 [101]. Lamarc involves a Markov Chain Monte Carlo coalescent genealogy sampling approach to calculate parameters such as effective population size, growth rate and immigration rate. The programme was run using the Bayesian search strategy under the GTR model for the combined mitochondrial DNA dataset; 10 initial chains were run for 10 000 generations (burnin = 1 000) and 3 final chains of 5×10^6 generations (burnin = 10 000) completed the analysis. In the case of the microsatellite data, the "Brownian" model was selected and 10 initial chains were run for 10 000 generations (burnin = 1 000); two final chains of 1×10^6 generations (burnin = 10 000) completed the analysis.

Genealogical and molecular dating analyses. Genealogical analyses were conducted to search for phylogeographic patterns and date divergence events. As the control region matrix contained indels (insertions or deletions), these were treated as missing data in all analyses. The trees generated from the cytochrome b and control region sequences were congruent and the data were combined into a single sequence for subsequent analyses. Sequences of *B. suillus* were rooted with four *Heterocephalus glaber* (cytochrome b accession numbers: AY 425916.1 and AY 425919.1; control region accession numbers: U 87531.1, U 87534.1 to U 87536.1) and two *Bathyergus janetta* sequences (cytochrome b accession numbers: AY 425916.1 and AY 425919.1; control region accession number: KM 222199) (combined cytochrome b and control region) downloaded from GenBank in the parsimony and Bayesian analyses. For molecular dating, only the cytochrome b data were used and outgroups included *B. janetta*, *Georychus capensis* (Accession numbers: AF 012243.1 and AY 425920.1), *Cryptomys hottentotus* (Accession numbers: AY 425870.1 and AY 425884.1), *Fukomys* spp. (Accession numbers: EF 043513.1 and EF 043514.1), *Heliophobius argenteocinereus* (Accession number: AY 425937.1), *H. glaber* (Accession numbers: U 87522.1 to U 87525.1), *Hystrix cristata* (Accession numbers: FJ 472577.1 to FJ 472579.1), *Mus caroli* (Accession numbers: AB 033699.1 and AB 109795.1), *Rattus rattus* (Accession numbers: AB 033702.1 and AF 295545.1), *Aplodontia rufa* (Accession numbers: JX 420113.1 and JX 420115.1) and *Xerus inauris* (Accession numbers: AY 452689.1 to AY 452690).

Phylogenetic trees were constructed using parsimony and Bayesian Inference approaches. For this, haplotypes were considered as the OTU. Parsimony analyses were executed in PAUP* version 4.0 [85]. Trees were generated with heuristic searches and TBR branch swapping using 100 random taxon additions. Statistical confidence in nodes was determined through 1 000 bootstrap replicates [86]. Bayesian Inference trees were constructed in MrBayes version 3.2 [87]. The best-fit substitution model (GTR+I+G) was selected through Modeltest version 3.7 [88] by using the Akaike Information Criterion (AIC) [89]. The programme was run for 5×10^6 generations with sampling every 100 generations. After discarding the first 25% of the trees as burnin, a majority rule consensus tree with posterior probabilities was constructed. Posterior probabilities >0.90 and bootstraps > 70% were considered acceptable support [90].

To obtain estimates of times of divergence for various clades, a relaxed molecular clock approach was adopted in BEAST version 1.7 [91]. Two fossil calibration points (as published by [92]) were specified, including the divergence between *Mus/Rattus* (12 ± 1 Mya) and *Scuiridae/Aplodontidae* (37 Mya). Runs were continued for 20×10^6 generations sampling every 1 000 generations (burnin = 2 000). Results were visualized in Tracer version 1.5 [93].

Phylogenetic trees are not always sensitive enough to detect variation and relationships below the species level [88]. In addition, several assumptions underpinning phylogenetic tree construction (such as evolution is strictly bifurcating) are violated. As an alternative, a haplotype network was constructed (under a 95% confidence limit) using TCS version 1.21 [94]; see also [95] for network choice). To determine the amount of genetic divergence among groups identified in the phylogenetic analyses, sequence divergences (uncorrected) were calculated in DnaSP version 5.10.01 [96].

Results

Molecular diversity indices

At the local scale, 98 specimens were sequenced (for both the cytochrome b and control region fragments characterized by 54 haplotypes; cytochrome b accession numbers: KJ 866510 to KJ 866607; control region accession numbers: KJ866688 to KJ866785) and genotyped (38 alleles; Table S3). Regionally, we obtained sequence (84 haplotypes; cytochrome b accession numbers: KJ 866510 to KJ 866687; control region accession numbers: KJ 866688 to KJ 866865) and microsatellite (68 alleles; Table S3) data for 178 specimens (Table 1). Less than 5% of the microsatellite dataset comprised missing data (Table S1) with a genotyping error of <0.01 incorrect alleles per genotype. No linkage was detected between loci across the sampled distribution and generally loci did not bear signature of null alleles (Table S1). The exclusion of the loci with higher estimates (>0.1) did not significantly influence analyses and therefore all loci were retained. Several loci were not in Hardy Weinberg equilibrium (HWE) over the landscape (Table S2) with all but one population (Piketberg) showing signs of inbreeding (Table 1).

Fu's Fs values (based on the mitochondrial DNA) indicated that most of the populations were demographically stable (no expansion or contraction); however, the Struisbaai population had experienced a population expansion while the populations of Piketberg and Redelinghuys show signs of population contraction. When the data were pooled, the West Coast (Fu's F = -12.962; p<0.001) and overall (Fu's F = -1.845; p<0.05) distributions showed evidence of population expansions while the South Coast region (Fu's F = 6.245; p<0.01) appears to be demographically contracting. No genetic bottlenecks were evident in the microsatellite data when populations were considered separately (Table 1) or for the combined dataset.

Population analyses

Significant structure was detected in both datasets at local ($\Phi_{ST} = 0.35$, $F_{ST} = 0.13$, p<0.001) and regional ($\Phi_{ST} = 0.82$, $F_{ST} = 0.099$, p<0.001) scales. In addition, significant pairwise differentiation was evident in both datasets between all localities, both at a local and regional scale (Table 2). This genetic structure

Table 1. Genetic diversity of the sampled *B. suillus* populations.

| Locality | Mitochondrial DNA | | | | | Microsatellites | | | | |
	n	Nucleotide diversity	Number of Haplotypes	Haplotype diversity	Fu's F	n	Na	He	F_{IS}	Bottleneck (p-value)
Redelinghuys	19	0.013	8	0.860±0.054	7.191*	19	7.889±0.676	0.793±0.020	0.225	0.473
Dwarskersbos	20	0.013	15	0.963±0.028	−0.549	20	8.667±0.866	0.781±0.033	0.139	0.199
Sterkfontein	20	0.01	13	0.912±0.046	0.573	20	8.778±0.878	0.802±0.014	0.211	0.251
Piketberg	19	0.003	4	0.380±0.134	4.060*	20	4.000±0.408	0.524±0.082	−0.024	0.381
Vredenburg	20	0.016	14	0.916±0.055	1.022	20	8.556±1.107	0.793±0.024	0.094	0.240
Cape Town	18	0.01	11	0.928±0.040	1.219	20	7.111±1.124	0.659±0.081	0.038	0.466
Stanford	12	0.008	7	0.773±0.128	2.414	10	4.444±0.669	0.535±0.090	0.192	0.130
Struisbaai	20	0.003	14	0.963±0.021	−5.635*	20	2.667±0.471	0.393±0.082	0.050	0.299
Riversdal	20	0.009	12	0.921±0.042	1.239	19	6.000±0.764	0.651±0.085	0.022	0.375
Sedgefield	10	0.009	7	0.867±0.107	1.749	10	3.556±0.709	0.403±0.100	0.106	0.073
Total	178	0.043	84	0.979±0.004	1.845	178	6.167±0.336	0.633±0.026	-	-

Genetic diversity for the combined mitochondrial DNA dataset and the microsatellite markers for the 10 *B. suillus* populations sampled across the Cape Floristic Region. In the case of mitochondrial DNA, the number of specimens (n), nucleotide diversity (π), number of haplotypes, haplotype diversity and Fu's F values in each population is given. Fu's F values marked with a "*" are significant at p<0.05. For the microsatellites, the number of specimens (n), average number of alleles per population (Na), expected heterozygosity and F_{IS} values within each population are shown together with the test (p) value indicating whether the population experienced a bottleneck during its evolutionary history.

is likely the result of low (or the absence of) gene-flow between populations (Table 3). Indeed, significant isolation-by-distance was detected over the distribution of *B. suillus* in both datasets on a regional scale (mitochondrial DNA: $r^2 = 0.533$; n = 111; p = 0.001, microsatellites: $r^2 = 0.393$; n = 178; p = 0.001) and in the microsatellites on a local scale ($r^2 = 0.336$; n = 99; p = 0.001). Isolation-by-distance was not found for the mitochondrial data ($r^2 = 0.188$; n = 55; p = 0.165) across the Sandveld (local) region.

There was also a significant partitioning of variance in both datasets ($\Phi_{ST} = 0.49$, $F_{ST} = 0.06$, p<0.001) when the populations were grouped (*a-priori*) into two respective groupings, pertaining to the West Coast (Redelinghuys, Dwarskersbos, Sterkfontein, Piketberg, Vredenburg, Cape Town) and South Coast (Stanford, Struisbaai, Riversdal, Sedgefield). A similar pattern was found locally in the Sandveld (mitochondrial DNA: $\Phi_{ST} = 0.363$, $F_{ST} = 0.503$, p<0.001) when the populations were grouped into the three respective areas constituting this region (Vredenburg, inner-Sandveld and Redelinghuys).

Genealogical analyses - affiliation of *B. janetta*

Bathyergus suillus was found to be paraphyletic with respect to its sister species *Bathyergus janetta*; the latter grouping as a sister taxon (albeit not with strong statistical support in the MP/Bayesian analyses) to the West Coast clade (Figure 2A; Figure S1). Sequence divergence estimates (based on the cytochrome b fragment) between the monophyletic clades are found in Table 4.

Phylogeographic patterns

The trees generated by the various methods were largely congruent with three major clades evident on a regional scale pertaining to the West Coast (including Stanford on the South Coast), Struisbaai and Riversdal/Sedgefield sampling areas respectively (Figure 2A; Figure S1). These three clades are respectively separated by 3.8% (West Coast and Struisbaai) and 3.2% (Struisbaai and Riverdal/Sedgefield) uncorrected sequence divergence for the 928 bp of cytochrome b sequence data. At least four major genetic provinces were evident regionally pertaining to the Sandveld, Cape Town/Stanford, Struisbaai and Riversdal/Sedgefield areas. Of these, the Cape Town/Stanford subclade and Riversdal/Sedgefield clade could further be genetically divided into their constituent populations (Figure 2A; see Table 4 for sequence divergence estimates). On a local scale, there was a signal of three clades (inner-Sandveld, Vredenburg and Redelinghuys) pertaining to the sampled areas on the different sides of rivers. In the inner-Sandveld clade, taxa from different localities were largely paraphyletic. There was, however, an indication of Piketberg- and Sterkfontein subclades within the larger inner-Sandveld clade.

The haplotype network retrieved nineteen haploclades which could not be connected at the 95% confidence level (Figure 2B). There were no shared haplotypes between populations with nearly every population forming a separate haploclade (except those of the inner-Sandveld). Eight haploclades were evident at a local scale that largely consisted of individuals from the Vredenburg, Redelinghuys and the inner-Sandveld populations clustering separately.

Clustering analyses

Invoking the "Admixture" model revealed ten clusters across the landscape thereby confirming the genetic distinctiveness of each sampled population (results not shown). Under the "no admixture" model eight clusters (based on the microsatellite data) were retrieved overall with virtually every sampled population clustering separately (Figure 3A) except for the Redelinghuys/Dwarskersbos/Vredenburg cluster. On a local scale, the afore-

Table 2. Pairwise Φ_{ST}/F_{ST} values between the sampled *B. suillus* populations.

Locality	Redelinghuys	Dwarskersbos	Sterkfontein	Piketberg	Vredenburg	Cape Town	Stanford	Struisbaai	Riversdal	Sedgefield
Redelinghuys	/	0.049	0.051	0.220	0.048	0.168	0.252	0.316	0.162	0.285
Dwarskersbos	0.372	/	0.056	0.230	0.055	0.165	0.254	0.318	0.159	0.272
Sterkfontein	0.419	0.098	/	0.231	0.051	0.163	0.241	0.309	0.139	0.253
Piketberg	0.616	0.249	0.307	/	0.247	0.321	0.414	0.450	0.346	0.443
Vredenburg	0.369	0.246	0.299	0.374	/	0.163	0.249	0.316	0.156	0.282
Cape Town	0.675	0.646	0.691	0.786	0.586	/	0.182	0.383	0.192	0.322
Stanford	0.707	0.676	0.719	0.863	0.601	0.497	/	0.517	0.335	0.454
Struisbaai	0.901	0.895	0.908	0.963	0.862	0.916	0.949	/	0.299	0.433
Riversdal	0.871	0.864	0.879	0.937	0.827	0.889	0.922	0.935	/	0.213
Sedgefield	0.855	0.847	0.867	0.944	0.780	0.876	0.919	0.938	0.741	/

Pairwise Φ_{ST}/F_{ST} values between the 10 *B. suillus* populations sampled across the Cape Floristic Region. Values above the diagonal are based on the microsatellite (F_{ST}) data and those below the diagonal represent the combined mitochondrial sequence data (Φ_{ST}). All values were significant at p<0.001.

Table 3. Gene-flow values (number of individuals per generation) between the sampled *B. suillus* populations.

Sampling localities		Mitochondrial DNA		Microsatellites	
		------>	<------	------>	<------
Dwarskersbos	Redelinghuys	0.883+1.553–0.813*	1.588+1.938–1.335*	0.593+0.304–0.289	0.172+0.220–0.010
Sterkfontein	Dwarskersbos	1.111+2.217–0.997*	1.221+1.544–1.186*	1.739+0.003–0.653*	1.859+0.051–0.045*
Vredenburg	Dwarskersbos	0.002+0.585–0.002	0.388+1.090–0.387*	1.363+0.055–0.059*	1.108+0.048–0.044*
Piketberg	Sterkfontein	0.295+0.996–0.295*	0.108+0.407–0.107	0.434+0.039–0.041	0.424+0.073–0.041
Cape Town	Vredenburg	0.002+0.314–0.002	0.003+0.318–0.002	0.190+0.393–0.063	0.472+0.002–0.001
Stanford	Cape Town	0.196+1.127–0.195*	0.002+0.781–0.002	0.407+0.006–0.268	1.276+0.067–0.037*
Struisbaai	Stanford	0.001+0.510–0.001	0.000+0.187–0.000	0.063+0.120–0.039	0.243+0.127–0.101
Riversdal	Struisbaai	0.009+0.172–0.009	0.013+0.272–0.013	0.398+0.007–0.024	0.503+0.044–0.035
Sedgefield	Riversdal	0.072+0.672–0.071	0.095+0.506–0.095	0.271+0.146–0.248	0.441+0.058–0.072

Gene-flow between the 10 *B. suillus* populations sampled in the Cape Floristic Region. Calculations are based on both mitochondrial DNA and microsatellite data. Values marked with a "*" are gene-flow levels >1 individual per generation (standard error included).

mentioned cluster could be divided into its constituent populations - five populations were revealed across the landscape by both the "admixture" and "no admixture" models (Figure 3B).

Discussion

Genetic patterns

Phylogeography of subterranean taxa is characterised by high inter-population differentiation [20,31,32,45,47,50,51] and *B. suillus* is no exception. Genetic diversity is structured between *B. suillus* populations across both local and regional spatial scales (Table 2; also see [28]) with every sampled locality forming a unique genetic entity (Figures 2 and 3).

Such strong genetic sub-structuring points to the isolated nature of populations of fossorial species. Specifically, the isolated nature of *B. suillus* populations (significant isolation-by-distance based on both datasets) may be linked to habitat fragmentation coupled to low vagility; pertinent factors to the genetic sub-structuring of subterranean taxa [32,34,39,109]. The low vagility of *B. suillus* was previously documented by Bray *et al.* [27] (but see [28]) who found the maximum distance of male gene-flow between two populations to be 2 149 m, resulting in low but significant levels of differentiation (Fst = 0.018). Similarly, Bray *et al.* [28] found low but significant differentiation (Fst = 0.02 and Fst = 0.17) across a one kilometre distance in the Cape Town and Darling areas respectively. Not surprisingly, gene-flow between most populations in the present study was effectively negligibly small (Table 3), although low historical genetic exchange was evident on a local scale. Given the separate genetic profiles of each sampled population (Figures 2 and 3; Table 2), it is likely that a.) these estimates are slightly above the realistic value of genetic exchange in the natural system or b.) if these estimates are indeed correct, the genetic exchange is insufficient to homogenize the genetic profiles of populations. It seems likely that gene-flow over longer distances requires stochastic above-ground movement [27,28,110], therefore the associated cost of such movement results in genetic structure even at fine spatial scales in fossorial taxa (see also [32] and [110]). Genetic structure and differentiation between fossorial populations may therefore occur even in the absence of geographic barriers.

Although we did not directly estimate effective population sizes, inbreeding (F_{IS}) was evident within all but one of the populations (Table 1; also see [28]). Taken together with the gene-flow data (Table 3), the occurrence of non-random mating within populations is likely due to possible small effective population sizes. Genetic sub-structuring may therefore be enforced by a fast accumulation of mutations making genetic drift a pertinent factor in fossorial systems. Accelerated evolution of the mitochondrial genome [50] and specifically the cytochrome b region [42,111] coupled with a short generation time (one year in *B. suillus*; [30]) likely promotes relatively fast genetic differentiation between isolated *B. suillus* populations.

Environmental heterogeneity is low within the subterranean niche, and therefore homoselection was proposed to drive low heterozygosity in fossorial taxa [29,30,50,112,113]. Selection-migration theory also impinges on the genetic population structure of fossorial taxa, hence a negative correlation exists between polymorphism and adult mobility [112]. Low heterozygosity has been demonstrated in a variety of subterranean taxa [30,55], however, these values were based on allozyme data. Levels of expected heterozygosity in *B. suillus* (Table 1) were consequently higher than proposed in the former studies, but were in line with more recent studies which similarly included microsatellite data ([27,28], Konvičková, 2013, unpublished data). A revision and validation of the earlier model by Nevo [30] by using the more variable markers available is therefore necessary. In addition, the model of small founder populations was only supported in two populations (based on the mitochondrial DNA) whilst most populations were stable (Table 1) or even expanding. The demographic decline observed across the South Coast distribution is likely due to agricultural activities which reduces and fragments available habitat.

Barriers to gene-flow

The isolated nature of fossorial populations results in genetic structure. It is therefore intuitive to expect that geographic phenomena may act to further fragment such populations and enforce genetic distinctiveness. As such, both mountains and rivers limit genetic exchange between fossorial taxa, including *B. suillus*.

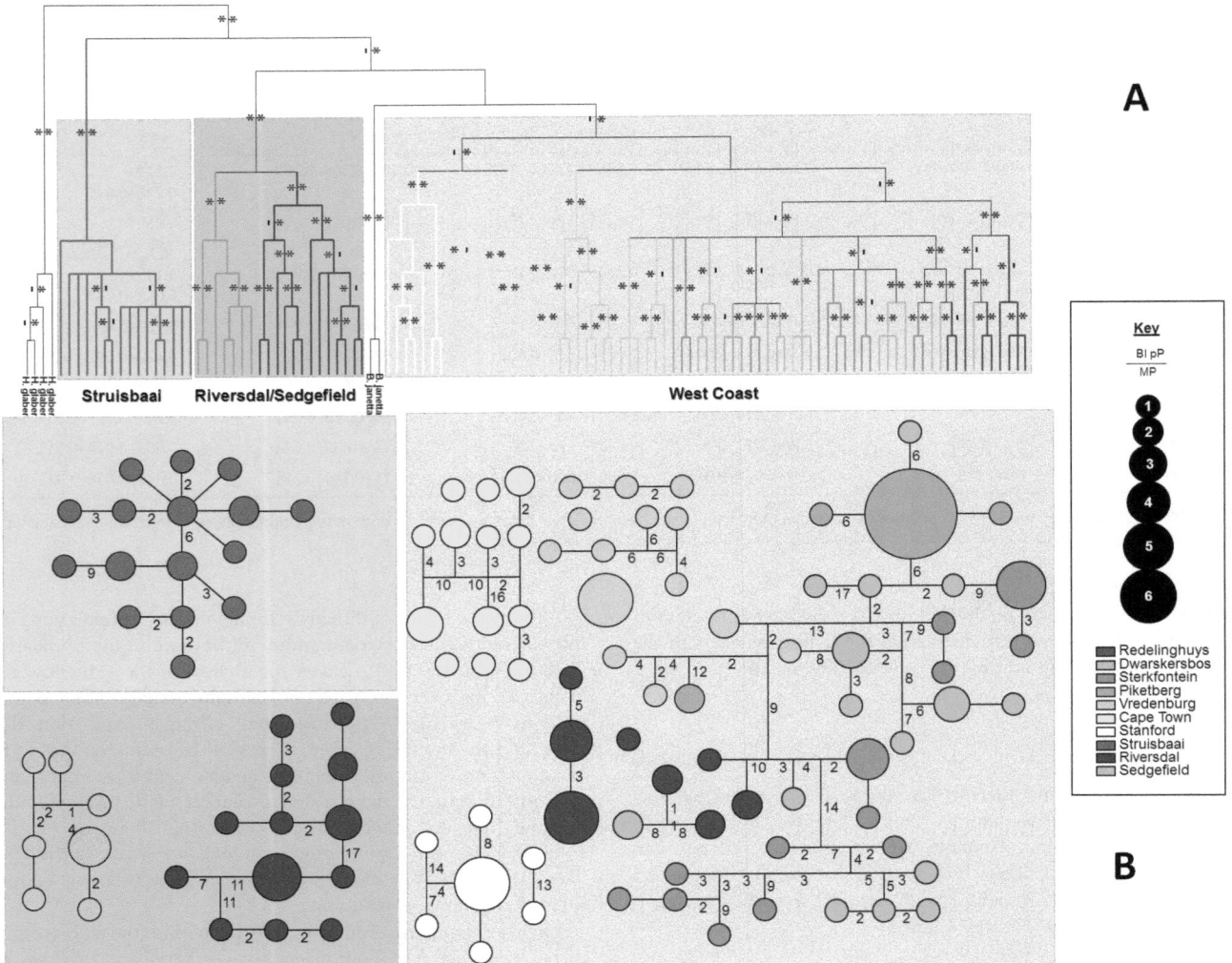

Figure 2. Bayesian phylogram and haplotype network demonstrating the different mitochondrial DNA clades across the sampled distribution. Figure 2A) Bayesian phylogram and B) haplotype networks obtained from the analyses based on combined cytochrome b and control region sequences demonstrating the different mitochondrial DNA clades detected in *B. Suillus* from localities across the Cape Floristic Region. For the Bayesian tree, a "*" above each node represent an acceptable posterior probability (pP>0.90) value derived from the Bayesian inference (MrBayes and BEAST) analyses and those below nodes are the Maximum Parsimony values (MP>70). A "-" indicate that the grouping was not found by the particular analysis. For the haplotype network, the size of each circle reflects the number of specimens with a particular haplotype. Numbers on branches represent the mutational steps separating haplotypes.

Table 4. Uncorrected sequence divergence between the clades retrieved in the genealogical analyses.

Clade A	Clade B	Uncorrected Sequence Divergence (%)
West Coast	*B. janetta*	3.9
West Coast	South Coast	3.9
West Coast	Struisbaai	3.8
Sandveld	Cape Town/Stanford	1.3
Struisbaai	Riversdal/Sedgefield	3.2
Struisbaai	Riversdal	3.3
Riversdal	Sedgefield	0.8

Uncorrected sequence divergence estimates (based on the cytochrome b marker) between the different genealogical clades retrieved within *B. suillus* across its distribution.

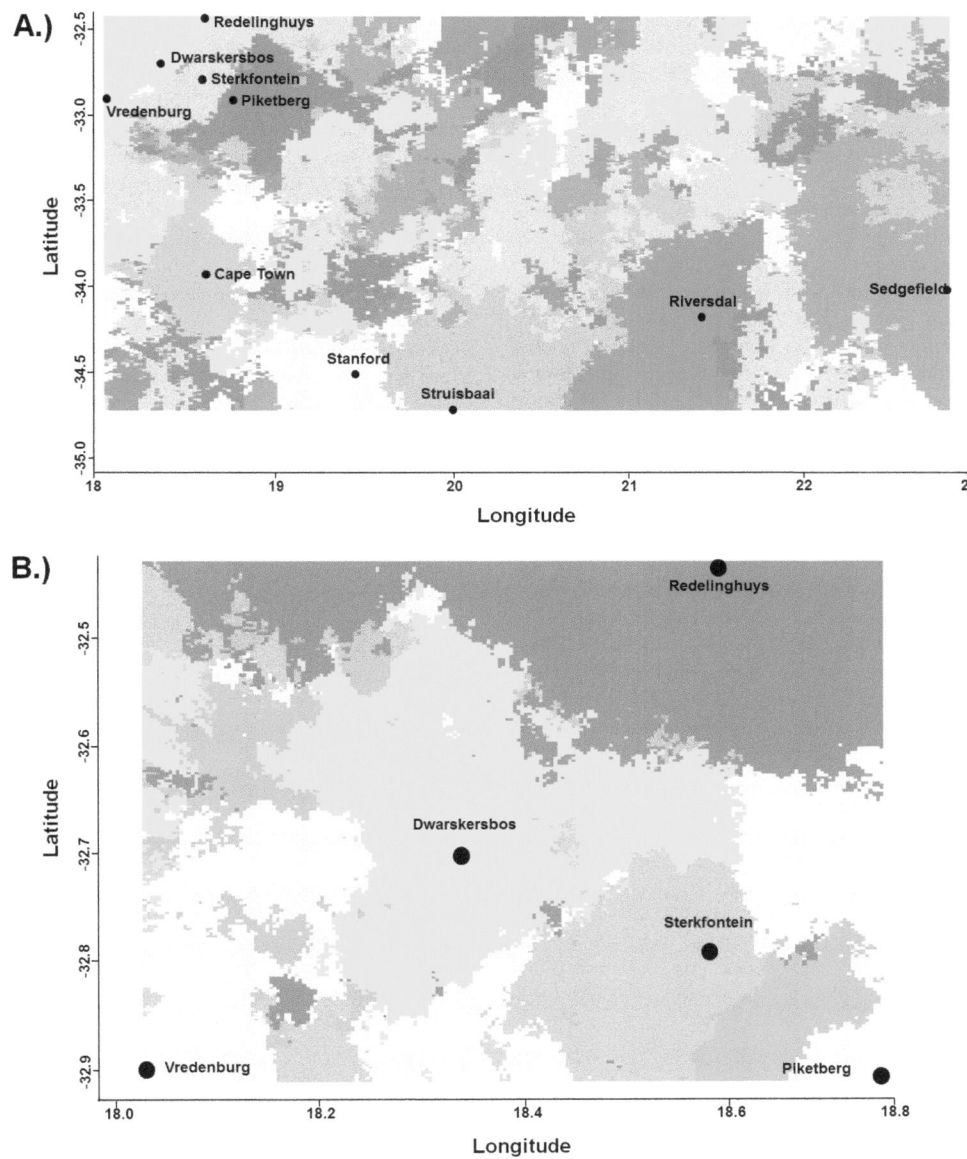

Figure 3. Genetic groupings revealed by the Geneland analysis at regional and local scales. Genetic groupings revealed by the Geneland analysis of the microsatellite data on a A) regional (Cape Floristic Region) and B) local (Sandveld) spatial scale. Dots reflect the location of each population and the colours correspond to each separate genetic grouping.

Mountains act as phylogeographic disruptors between *B. suillus* populations at both local and regional scales. On a local scale, the Piketberg Mountain creates a barrier to gene-flow between the Piketberg and inner-Sandveld populations (Figures 2 and 3). Similarly, the Hottentots Holland Mountains act as a barrier to gene-flow separating two of the major clades (West Coast and Struisbaai; Figures 2 and 3; also see [28]) on a regional scale. Mountains, rifting and volcanism are the drivers of differentiation between populations and speciation within the African Bathyergidae, impacting on genetic structure either directly (as geographic barriers; [31,40,46]) or indirectly (influencing rainfall patterns and therefore vegetation; [31,48]).

When specifically looking at phylogeographic patterns, the African Rift Valley (volcanic uplands and deep valleys) divides the distribution range of the most diverse taxa within the Bathyergidae [40,46] and has influenced the adaptive radiation (speciation) of taxa such as *Fukomys* [46], *Heliophobius* [43,46,69] and *Cryptomys* through geomorphological processes [31,41,46]. While the early divergences within the Bathyergidae were independent of rifting, later divergences, patterns and distributions were mainly influenced by this geological process [31,44]. When more local patterns are taken into account, chromosomal variation in *Thomomys bottae* is maximal in insular montane populations, but minimal in plains populations [30]. In line with this, Patzenhauerová *et al.* [110] also reported that barriers (such as a rocky hill) impeded gene-flow between populations of another mole-rat species, the silvery mole-rat, *Heliophobius argenteocinereus*.

Interestingly, the Cape Town and Stanford areas (that span the Hottentots Holland Mountains) group together in the genealogical analyses (Figure 2A). This grouping may be attributed to a colonization event subsequent to the divergence of the West and South Coast clades (30.65+13.93–10.69 Mya; Figure S2).

According to Siesser and Dingle [114], a marine regression occurred during this period (~25 Mya) which would have opened up a large region of the coastal belt (500–600 m) thereby allowing dispersal and establishment of individuals from the Cape Town area in Stanford. Similar colonization events during major marine regressions have been recorded in *Spalax* and *Nannospalax* [48]. When the sequences of Bray *et al.* [28] (Accession numbers: KC 153980 to KC 153987) were added to our cytochrome b dataset, haplotypes 6 and 8 from the Bray *et al.* [28] study (presumable from the Darling area) grouped together with our Sandveld clade while the rest (haplotypes 1 to 5 and 7) grouped with the Cape Town individuals in the genealogical analyses (results not shown). This genetic discontinuity, also documented by [28], shows historic isolation of these areas across the Cape Flats around 12.32+7.36–5.45 Mya - a period characterized by a major transgression phase (15-4 Mya with the sealevel rising to over 300 m above the current level; [114]).

Drainage systems also act as phylogeographic disruptors in *B. Suillus* at both local and regional scales. On a local scale, the Berg- and Verlorenvlei Rivers form phylogeographic barriers to gene-flow thereby structuring genetic variation (Figure 2; also see [47,49]). Historically, a northward colonization (from the Vredenburg area) of the inner-Sandveld and Redelinghuys areas is evident across these rivers (Figure 2A), therefore stochastic gene-flow events across rivers seems possible. *Bathyergus suillus* prefers lowland sandy areas, especially coastal dunes. These animals therefore occur relatively near the coast and populations would be separated by river mouths (rather than the upper reaches) for most of the time. It is possible that gene-flow and colonization of areas may occur around the upper reaches of these rivers during dry periods (also see [115]). It should be noted that differentiation happened in relative isolation since such colonization events. For instance, the Redelinghuys and Vredenburg haplotypes found to cluster with those of the inner-Sandveld may be attributed to relatively recent colonization of these areas from the same ancestral stock. Therefore, divergence times have been too short to establish monophyly of haplotypes in these areas although distinct genetic profiles are evident.

In contrast, patterns on a regional scale show historic isolation across the larger perennial rivers. The Struisbaai- and Riversdal/Sedgefield clades of the South Coast (Figure 2A; Figure S1) are separated by the Breede River. Similarly, the Riversdal/Sedgefield clade can further be divided into its constituent populations (Figure 2A; Figure S1) showing the action of the Gourits River as a barrier to gene-flow. Indeed, water bodies such as lakes [31,46] and rivers [32,42,43,45,47,49] have impacted on the isolation of fossorial taxa and drainage areas seem to be a hotbed of differentiation due to fragmentation [42,116]. Specifically, genetic structure in mole-rat populations and speciation due to river barriers has been reported within the Bathyergidae e.g., speciation and karyotypic divergence in *Fukomys* [42,45], *Cryptomys* [40,41,42,43] and the separation of *Coetomys* from *Cryptomys* [41,43].

Taxonomic implications

While *B. suillus* and *B. janetta* are definitive species [31,50,51,58,62,63], other studies found contrasting results [41,51,58]. Intergeneric studies that found *B. suillus* and *B. janetta* as sister species invariably included sampling bias (e.g., [31,50,51,58,62,63]). This was, in a sense, rectified by Ingram *et al.* [41] who found *B. suillus* as paraphyletic with respect to *B. janetta* - the latter also forming a sister taxon to the West Coast group. Similar results were obtained in this study (Figure 2A). A dispersal event of animals from the southern West Coast

northward into Namaqualand and Namibia across the Knersvlakte region (via the coastal belt) is therefore the most plausible scenario. The Knersvlakte is a known barrier to gene-flow to even more mobile taxa [15,16,117,118] and divergence between *B. suillus* and *B. janetta* could therefore have occurred in allopatry.

With respect to the genetic structure found within *B. suillus*, the MP and BEAST analyses, indicate that the West Coast clade may be more derived (not found in the Bayesian analysis), suggesting a possible colonization of the West Coast from the South Coast (also see [41,43]). As further evidence, the sister genus to *Bathyergus*, *Georychus*, has a current distribution along the South Coast of South Africa [67]. The most probable scenario thus entails that *B. suillus* colonized the West Coast after the *Bathyergus/Georychus* clade split on the South Coast. This colonization event probably took place during a marine regression in the Miocene [115,119,120] via the coastal belt. The various clades within *B. suillus* show very different evolutionary histories. Additionally, the amount of sequence divergence between *B. Suillus* clades (Table 3) is above the inter-specific threshold (based on cytochrome b) reported for mammalian [111] and fossorial taxa [30]. These findings suggest that the genus *Bathyergus* is in need of a taxonomic revision.

Conclusion

This study exemplifies a holistic approach to investigating the genetic structure within species. The life-history, biology and habitat requirements of fossorial taxa impact on the distribution and isolation of populations - here demonstrated by interpreting phylogeographic and gene-flow patterns in *B. suillus* in the context of previously published biological and ecological information on this species. The model proposed by Nevo [30] to explain the factors (low gene-flow, isolation-by-distance, genetic drift) impacting on the genetic structure of fossorial populations was largely supported by our data; however we could provide no convincing evidence of small founder populations; the comparability of heterozygosity values was compromised by the markers used in the above-mentioned study. Due to the relatively quick generation time of *B. suillus* (a characteristic shared by all rodent taxa), differentiation of isolated populations may happen in a comparatively short period of time. Additionally, geographic barriers to gene-flow (mountains and drainage systems) and geographic distance play a significant role in structuring genetic variation within *B. suillus* - a factor which may partly explain the staggering radiation in fossorial species worldwide. Furthermore, a systematic revision of the genus *Bathyergus* is necessary, given the findings of this study.

Supporting Information

Figure S1 Bayesian phylogram demonstrating the different mitochondrial DNA clades across the sampled distribution. Bayesian phylogram obtained from the analyses based on combined cytochrome b and control region sequences demonstrating the different mitochondrial DNA clades detected in *B. suillus* from localities across the Cape Floristic Region. Values above each node represent posterior probabilities (Pp) derived from the Bayesian inference (MrBayes and BEAST) analyses and those below nodes are the Maximum Parsimony values.

Figure S2 Bayesian phylogram from the BEAST analysis indicating the divergence dates between clades. Bayesian phylogram obtained from the BEAST analyses of the cytochrome b haplotypes among the 10 *B. suillus* sample sites across the Cape Florisitc Region. The values above each node represent the posterior probability (pP) values derived from the Bayesian inference analyses. The populations comprising the two clades evident across the distribution are shown. The divergence dates for four nodes (A–D) are indicated as bars which include the span of the divergence estimate for that particular node.

Table S1 Summary information for the microsatellite loci used in this study. Locus summary information showing the estimated proportion of null alleles (Null), percentage missing data of the total 356 alleles/locus, proportion of missing alleles/locus, the genotyping error per genotype and F_{IS} values for each of the microsatellite loci used in this study.

Table S2 Summary of loci which did not conform to Hardy-Weinberg Equilibrium in each population. Summary of loci which did not conform to Hardy-Weinberg

Equilibrium (marked by an "x") in each sampled population of *B. suillus*.

Table S3 Microsatellite data used in this study.

Acknowledgments

We acknowledge the contribution of the anonymous reviewers to our manuscript.

Author Contributions

Conceived and designed the experiments: JHV BJVV. Performed the experiments: JHV. Analyzed the data: JHV BJVV. Contributed reagents/materials/analysis tools: JHV NCB BJVV. Contributed to the writing of the manuscript: JHV NCB BJVV.

References

1. Smit HA, Jansen van Vuuren B, O'Brien PCM, Ferguson-Smith M, Yang F, et al. (2011) Phylogenetic relationships of elephant-shrews (Afrotheria, Macroscelididae). J Zool 284: 1–11.
2. Avise JC (1994) Molecular markers, natural history and evolution. New York: Chapman and Hall.
3. Katongo C, Koblmüller S, Duftner N, Makasa L, Sturmbauer C (2005) Phylogeography and speciation in the *Pseudocrenilabrus philander* species complex in Zambian Rivers. Hydrobiologia 542: 221–233.
4. Cook BD, Baker AM, Page TJ, Hawcett JH, Hurwood DA, et al. (2006) Biogeographic history of an Australian freshwater shrimp, *Paratya australiensis* (Atyidae): the role life history transition in phylogeographic diversification. Mol Ecol 15: 1083–1093.
5. Hughes JM, Schmidt DJ, Finn DS (2009) Genes in streams: using DNA to understand the movement of freshwater fauna and their riverine habitat. BioScience 59: 573–583.
6. Waters JM, King TM, O'Loughlin PM, Spence HG (2005) Phylogeographical disjunction in abundant high-dispersal littoral gastropods. Mol Ecol 14: 2789–2802.
7. Pelc RA, Warner RR, Gaines SD (2009) Geographical patterns of genetic structure in marine species with contrasting life histories. J Biogeogr 36: 1881–1890.
8. Teske PR, von der Heyden S, McQuaid CD, Barker NP (2011) A review of marine phylogeography in southern Africa. S Afr J Sci 107: 43–53.
9. Montgelard C, Matthee CA (2012) Tempo of genetic diversification in southern African rodents: The role of Plio-Pleistocene climatic oscillations as drivers for speciation. Acta Oecol 42: 50–57.
10. Colangelo P, Verheyen E, Leirs H, Tatard C, Denys C, et al. (2013) A mitochondrial phylogeographic scenario for the most widespread African rodent, *Mastomys natalensis*. Biol J Linn Soc 108: 901–916.
11. Da Silva MNF, Patton JL (1993) Amazonian phylogeography: mtDNA sequence variation in arboreal echimyid rodents (Caviomorpha). Mol Phylogenet Evol 2: 243–255.
12. Da Silva MNF, Patton JL (1998) Molecular phylogeography and the evolution and conservation of Amazonian mammals. Mol Ecol 7: 475–486.
13. Brown M, Cooksley H, Carthew SM, Cooper SJB (2006) Conservation units and phylogeographic structure of an arboreal marsupial, the yellow-bellied glider (*Petaurus australis*). Aust J Zool 54: 305–317.
14. Prinsloo P, Robinson TJ (1992) Geographic mitochondrial DNA variation in the rock hyrax, *Procavia capensis*. Mol Biol Evol 9: 447–456.
15. Matthee CA, Flemming AF (2002) Population fragmentation in the southern rock agama, *Agama atra*: more evidence for vicariance in southern Africa. Mol Ecol 11: 465–471.
16. Smit HA, Robinson TJ, Jansen van Vuuren B (2007) Coalescence methods reveal the impact of vicariance on the spatial genetic structure of *Elephantulus edwardii* (Afrotheria, Macroscelidea). Mol Ecol 16: 2680–2692.
17. Swart BL, Tolley KA, Matthee CA (2009) Climate change drives speciation in the southern rock agama (*Agama atra*) in the Cape Floristic Region. S Afr J Biogeogr 36: 78–87.
18. Nevo E (1999) Mosaic evolution of subterranean mammals: regression, progression and global convergence. Oxford: Oxford University Press.
19. Hart L, O'Riain MJ, Jarvis JUM, Bennett NC (2006) The pituitary potential for opportunistic breeding in the Cape dune mole-rat, *Bathyergus suillus*. Physiol Behav 88: 615–619.
20. Daniels SR, Heideman NJ, Hendricks MGJ (2009) Examination of evolutionary relationships in the Cape fossorial skink species complex (Acontinae: *Acontias meleagris meleagris*) reveals the presence of five cryptic lineages. Zool Scr 38: 449–463.
21. Reichman OJ, Whitham TG, Ruffner GA (1982) Adaptive geometry of burrow spacing in two pocket gopher populations. Ecology 63: 687–695.
22. Rosi MI, Cona MI, Videla F, Puig S, Roig VG (2000) Architecture of *Ctenomys mendocinus* (Rodentia) burrows from two habitats differing in abundance and complexity of vegetation. Acta Theriol 45: 491–505.
23. Spinks AC, Bennett NC, Jarvis JUM (2000) A comparison of the ecology of two populations of the common mole-rat, *Cryptomys hottentotus hottentotus*: the effect of aridity on food, foraging and body mass. Oecologia 125: 341–349.
24. Bishop JM, Jarvis JUM, Spinks AC, Bennett NC, O'Ryan C (2004) Molecular insight into patterns of colony composition and paternity in the common mole-rat *Cryptomys hottentotus hottentotus*. Mol Ecol 13: 1217–1229.
25. Šumbera R, Šklíba J, Elichová M, Chitaukali WN, Burda H (2008) Natural history and burrow system architecture of the silvery mole-rat from Brachystegia woodland. J Zool 274: 77–84.
26. Šklíba JS, Šumbera R, Chitaukali WN, Burda H (2009) Home-range dynamics in a solitary subterranean rodent. Ethology 115: 217–226.
27. Bray TC, Bloomer P, O'Riain MJ, Bennett TC (2012) How attractive is the girl next door? An assessment of spatial mate acquisition and paternity in the solitary Cape dune mole-Rat, *Bathyergus suillus*. PLoS ONE 7(6): e39866. Doi:10.1371/journal.pone.0039866
28. Bray TC, Jansen van Rensburg A, Bennett NC (2013) Overground versus underground: a genetic insight into dispersal and abundance of the Cape dune mole-rat. Biol J of the Linn Soc 110: 890–897.
29. Nevo E (1976) Genetic variation in constant environments. Experientia 32: 858–859.
30. Nevo E (1979) Adaptive convergence and divergence of subterranean mammals. Annu Rev Ecol Syst 10: 269–308.
31. Faulkes CG, Verheyen E, Verheyen W, Jarvis JUM, Bennett NC (2004) Phylogeographical patterns of genetic divergence and speciation in African mole-rats (Family: *Bathyergidae*). Mol Ecol 13: 613–629.
32. Mirol P, Giménez MD, Searle JB, Bidau CJ, Faulkes CG (2010) Population and species boundaries in the South American subterranean rodent *Ctenomys* in a dynamic environment. Biol J Linn Soc 100: 368–383.
33. Jarvis JUM, Bennett NC, Spinks AC (1998) Food availability and foraging by wild colonies of Damaraland mole-rats (*Cryptomys damarensis*): implications for sociality. Oecologia 113: 290–298.
34. Miller RS (1964) Ecology and distribution of pocket gophers (Geomyidae) in Colorado. Ecology 45: 256–272.
35. Andersen DC, MacMahon JA (1981) Population dynamics and bioenergetics of a fossorial herbivore, *Thomomys talpoides* (Rodentia: Geomyidae), in a spruce-fir sere. Ecol Monogr 51: 179–202.
36. Davies KC, Jarvis JUM (1986) The burrow systems and burrowing dynamics of the mole-rats *Bathyergus suillus* and *Cryptomys hottentotus* in the fynbos of the south-western Cape, South Africa. J Zool 209: 125–147.
37. Herbst M, Bennett NC (2006) Burrow architecture and burrowing dynamics of the endangered Namaqua dune mole rat (*Bathyergus janetta*) (Rodentia: Bathyergidae). J Zool 270: 420–428.
38. Cook JA, Lessa EP (1998) Are rates of diversification in subterranean South American tuco-tucos (genus *Ctenomys*, Rodentia: Octodontidae) unusually high? Evolution 52: 1521–1527.
39. Mapelli FJ, Kittlein MJ (2009) Influence of patch and landscape characteristics on the distribution of the subterranean rodent *Ctenomysporteousi*. Landscape Ecol 24: 723–733.
40. Burda H (2000) Determinants of the distribution and radiation of African mole-rats (Bathyergidae, Rodentia). Ecology or geography? In:Denys C, Granjon L, Poulet A, editors. Small African mammals. Paris:Colloques et Seminaires - Editions de l' IRD.

41. Ingram CM, Burda H, Honeycutt RL (2004) Molecular phylogenetics and taxonomy of the African mole-rats, genus *Cryptomys* and the new genus *Coetomys* (Gray, 1864). Mol Phylogenet Evol 31: 997–1014.

42. Van Daele PAAG, Dammann P, Meier JL, Kawalika M, Van De Woestijne C, et al. (2004) Chromosomal diversity in mole-rats of the genus *Cryptomys* (Rodentia: Bathyergidae) from the Zambezian region: with descriptions of new karyotypes. J Zool 264: 317–326.

43. Ingram CM (2005) The evolution of nuclear microsatellite DNA markers and their flanking regions using reciprocal comparisons within the African mole-rats (Rodentia: Bathyergidae). Texas: Unpublished PhD thesis, Texas A&M University.

44. Van Daele PAAG, Faulkes CG, Verheyen E, Adriaens D (2007) African mole-rats (Bathyergidae): a complex radiation in Afrotropical soils. In: Begall S, Burda H, Schleich CE, editors. Subterranean rodents: news from underground. Heidelberg: Springer-Verlag. 357–373.

45. Van Daele PAAG, Verheyen E, Brunain M, Adriaens D (2007) Cytochrome b sequence analysis reveals differential molecular evolution in African mole-rats of the chromosomally hyperdiverse genus *Fukomys* (Bathyergidae, Rodentia) from the Zambezian region. Mol Phylogenet Evol 45: 142–157.

46. Faulkes CG, Mgode GF, Le Comber SC, Bennett NC (2010) Cladogenesis and endemism in Tanzanian mole-rats, genus *Fukomys*: (Rodentia: Bathyergidae): a role for tectonics? Biol J Linn Soc 100: 337–352.

47. Heideman NJL, Mulcahy DG, Sites JW, Hendricks MGJ, Daniels SR (2011) Cryptic diversity and morphological convergence in threatened species of fossorial skinks in the genus *Scelotes* (Squamata: Scincidae) from the Western Cape Coast of South Africa: Implications for species boundaries, digit reduction and conservation. Mol Phylogenet Evol 61: 823–833.

48. Hadid Y, Németh A, Snir S, Pavlíček T, Csorba G, et al. (2012) Is evolution of blind mole rats determined by climate oscillations? PloS ONE 7(1):e30043. Doi:10.1371/journal.pone.0030043

49. Engelbrecht HM, van Niekerk A, Heideman NJL, Daniels SR (2013) Tracking the impact of Pliocene/Pleistocene sea level and climatic oscillations on the cladogenesis of the Cape legless skink, *Acontias meleagris* species complex, in South Africa. J Biogeogr 40: 492–506.

50. Honeycutt RL, Edwards SV, Nelson K, Nevo E (1987) Mitochondrial DNA variation and the phylogeny of African mole rats (Rodentia: Bathyergidae). Syst Zool 36: 280–292.

51. Janecek LL, Honeycutt RL, Rautenbach L, Erasmus BH, Reig S, et al. (1992) AIIozyme variation and systematics of African mole-rats (Rodentia: Bathyergidae). Biochem Syst Ecol 20: 401–416.

52. Bray TC, Bennett NC, Bloomer P (2011) Low levels of polymorphism at novel microsatellite loci developed for bathyergid mole-rats from South Africa. Conserv Genet Resour 3: 221–224.

53. Thaeler CS (1968) Karyotypes of sixteen populations of the *Thomomys talpoides* complex of pocket gophers (Rodentia: Geomyidae). Chromosoma 25: 172–183.

54. Nevo E, Ben-Shlomo R, Beiles A, Hart CP, Ruddle FH (1992) Homeobox DNA polymorphisms (RFLPs) in subterranean mammals of the *Spalax ehrenbergi* superspecies in Israel: Patterns, correlates, and evolutionary significance. J Exp Zool 263: 430–441.

55. Nevo E, Beiles A, Korol AB, Ronin YI, Pavlicek T, et al. (2000) Extraordinary multilocus genetic organization in mole crickets, Gryllotalpidae. Evolution 54: 586–605.

56. Reyes A, Nevo E, Saccone C (2003) DNA sequence variation in the mitochondrial control region of subterranean mole rats, *Spalax ehrenbergi* superspecies, in Israel. Mol Biol Evol 20: 622–632.

57. Kryštufek B, Ivanitskaya E, Arslan A, Arslan E, Bužan EV (2012) Evolutionary history of mole rats (genus *Nannospalax*) inferred from mitochondrial cytochrome *b* sequence. Biol J Linn Soc 105: 446–455.

58. Deuve JL, Bennett NC, Britton-Davidian J, Robinson TJ (2008) Chromosomal phylogeny and evolution of the African mole-rats (Bathyergidae). Chromosome Res 16: 57–74.

59. Allard MW, Honeycutt RL (1992) Nucleotide sequence variation in the mitochondrial 12S rRNA gene and the phylogeny of African mole-rats (Rodentia: Bathyergidae). Mol Biol Evol 9: 27–40.

60. Faulkes CG, Bennett NC, Bruford MW, O'Brien HP, Aguilar GH, et al. (1997) Ecological constraints drive social evolution in the African mole-rats. P R Soc London 264: 1619–1627.

61. Walton AH, Nedbal MA, Honeycutt RL (2000) Evidence from intron 1 of the nuclear transthyretin (prealbumin) gene for the phylogeny of African mole-rats (Bathyergidae). Mol Phylogenet Evol 16: 467–474.

62. Da Silva CC, Tomasco IH, Hoffmann FG, Lessa EP (2009) Genes and ecology: Accelerated rates of replacement substitutions in the cytochrome b gene of subterranean rodents. Open Evol J 3: 17–30.

63. Nevo E, Ben-Shlomo R, Belles A, Jarvis JUM, Hickman GC (1987) Allozyme differentiation and systematics of the endemic subterranean mole rats of South Africa. Biochem Syst Ecol 15: 489–502.

64. Vleck D (1979) The energy costs of burrowing by the pocket gopher *Thommomys bottae*. Phys Zool 52: 122–136.

65. Thomas HG, Bateman PW, LeComber SC, Bennett NC, Elwood RW, et al. (2009) Burrow architecture and digging activity in the Cape dune mole rat. J Zool (Lond) 279: 277–284.

66. Bennett NC, Faulkes CG (2000) African mole-rats: ecology and eusociality. Cambridge: Cambridge University Press.

67. Skinner JD, Chimimba CT (2005) The mammals of the Southern African subregion. Third Edition. Cape Town: Cambridge University Press.

68. Hart L, O'Riain MJ, Jarvis JUM, Bennett NC (2006) Is the Cape Dune mole-rat, *Bathyergus suillus* (Rodentia: Bathyergidae), a seasonal or aseasonal breeder? J Mammal 87: 1078–1085.

69. Faulkes CG, Bennett NC, Cotterill FPD, Stanley W, Mgode GF, et al. (2011) Phylogeography and cryptic diversity of the solitary-dwelling silvery mole-rat, genus *Heliophobius* (family: Bathyergidae). J Zool 285: 324–338.

70. Thomas HG, Bateman PW, Scantlebury M, Bennett NC (2012) Season but not sex influences burrow length and complexity in the non-sexually dimorphic solitary Cape mole-rat (Rodentia: Bathyergidae). J Zool 288: 214–221.

71. Holderegger R, Wagner HH (2008) Landscape genetics. BioScience 58: 199–207.

72. Coulon A, Guillot G, Cosson J (2004) Genetic structure is influenced by landscape features. Empirical evidence from a roe deer population. Mol Ecol 15: 1669–1679.

73. Storfer A, Murphy MA, Evans JS, Goldberg CS, Robinson S, et al. (2007) Putting the 'landscape' in landscape genetics. Heredity 98: 128–142.

74. Dupanloup I, Schneider S, Excoffier L (2002) A simulation annealing approach to define the genetic structure of populations. Mol Ecol 11: 2571–2581.

75. Holderegger R, Wagner HH (2006) A brief guide to landscape genetics. Landscape Ecol 21: 793–796.

76. Conrad J, Nel J, Wentzel J (2004) The challenges and implications of assessing groundwater recharge: A case study - northern Sandveld, Western Cape, South Africa. Water SA 30: 75–81.

77. Low AB, Mustart P, Van der Merwe H (2004) Greater Cederberg biodiversity corridor: provision of biodiversity profiles for management. Project management unit, Greater Cederberg Biodiversity Corridor. 1–78.

78. Kocher TD, Thomas WK, MeyerA, Edwards SV, Paabo S, et al. (1989) Dynamics of mitochondrial DNA evolution in animals: Amplification and sequencing with conserved primers. P Natl Acad Sci USA 86: 6196–6200.

79. Irwin DM, Kocher TD, Wilson AC (1991) Evolution of the cytochrome b gene of mammals. J Mol Evol 32: 128–144.

80. Patterson BD, Velazco PM (2008) Phylogeny of the rodent genus *Isothrix* (Hystricognathi, Echimyidae) and its diversification in Amazonia and the eastern Andes. J Mammal Evol 15: 181–201.

81. Karsten M, van Vuuren BJ, Barnaud A, Terblanche JS (2013) Population Genetics of *Ceratitis capitata* in South Africa: Implications for Dispersal and Pest Management. PloS ONE 8(1): e54281. Doi:10.1371/journal.pone.0054281

82. Jansen van Vuuren B, Chown SL (2007) Genetic evidence confirms the origin of the house mouse on sub-Antarctic Marion Island. Polar Biol 30: 327–332.

83. Maddison DR, Maddison WP (2003) *MacClade version 4.06 for OS X*. Sinauer Associates Inc., Sunderland, Massachusetts.

84. Burland TM, Bishop JM, O'Ryan C, Faulkes CG (2001) Microsatellite primers for the African mole rat genus *Cryptomys* and cross-species amplification within the family Bathyergidae. Mol Ecol Notes 1: 311–314.

85. Swofford DL (2003) PAUP*: Phylogenetic Analysis using Parsimony (* and other methods), Version 4.0b10. Sunderland: Sinauer Associates.

86. Felsenstein J (1985) Confidence limits on phylogenies: an approach using the bootstrap. Evolution 39: 783–791.

87. Ronquist F, Teslenko M, van der Mark P, Ayres D, Darling A, et al. (2011) MrBayes 3.2: Efficient Bayesian phylogenetic inference and model choice across a large model space. Syst Biol 61: 539–542.

88. Posada D, Crandall KA (2001) Intraspecific gene genealogies: trees grafting into networks. Trends Ecol Evol 16: 37–45.

89. Akaike H (1973) Information theory and an extension of the maximum likelihood principle. In: Petrov BN, Csaki F, editors. Second international symposium on information theory. Budapest: Akademia Kiado. 267–281.

90. Mortimer E, Jansen van Vuuren B, Lee JE, Marshall DJ, Convey P, et al. (2011) Mite dispersal among the Southern Ocean Islands and Antarctica before the last glacial maximum. P R Soc B 278: 1247–1255.

91. Drummond AJ, Suchard MA, Xie D, Rambaut A (2007) Bayesian phylogenetics with BEAUti and the BEAST 1.7. Mol Biol Evol 29: 1969–1973.

92. Huchon D, Catzeflis FM, Douzery EJP (2000) Variance of molecular datings, evolution of rodents and the phylogenetic affinities between Ctenodactylidae and Hystricognathi. P R Soc London 267: 339–402.

93. Rambaut A, Drummond AJ (2003) Tracer [computer program] Available: http://evolve.zoo.ox.ac.uk/software/. Accessed 10 October 2013.

94. Clement MD, Posada D, Crandall KA (2000) TCS: A computer program to estimate gene genealogies. Mol Ecol 9: 1657–1660.

95. Joly S, Stevens MI, Jansen van Vuuren B (2007) Haplotype networks can be misleading in the presence of missing data. Syst Biol 56: 857–862.

96. Librado P, Rozas J (2009) DnaSP v5: A software for comprehensive analysis of DNA polymorphism data. Bioinformatics 25: 1451–1452.

97. Excoffier L, Lischer HE (2010) Arlequin suite ver 3.5: a new series of programs to perform population genetics analyses under Linux and Windows. Mol Ecol Resour 10: 564–567.

98. Peakall R, Smouse PE (2006) GENALEX 6: genetic analysis in Excel. Population genetic software for teaching and research. Mol Ecol Notes 6: 288–295.

99. Guillot G, Mortier F, Estoup A (2005) Geneland: a computer package for landscape genetics. Mol Ecol Notes 5: 712–715.

100. Chen C, Durand E, Forbes F, François O (2007) Bayesian clustering algorithms ascertaining spatial population structure: a new computer program and a comparison study. Mol Ecol Notes7: 747–756.

101. Kuhner MK (2006) LAMARC 2.0: Maximum likelihood and Bayesian estimation of population parameters. Bioinformatics 22: 768–770.

102. Raymond M, Rousset F (1995) GENEPOP (version 1.2): population genetics software for exact tests and ecumenicism. J Hered 86: 248–249.

103. Rousset F (2008) Genepop'007: a complete reimplementation of the Genepop software for Windows and Linux. Mol Ecol Resour 8: 103–106.

104. Dakin EE, Avise JC (2004) Microsatellite null alleles in parentage analysis. Heredity 93: 504–509.

105. Goudet J (2001) FSTAT, A Program to Estimate and Test Gene Diversities and Fixation Indices (version 2.9.3.2). Available: <http://www.unil.ch/popgen>. Accessed 15 October 2013.

106. Fu YX (1997) Statistical tests of neutrality of mutations against population growth, hitchhiking and background selection. Genetics 147: 915–925.

107. Cornuet JM, Luikart G (1996) Description and power analysis of two tests for detecting recent population bottlenecks from allele frequency data. Genetics 144: 2001–2014.

108. Luikart G, Allendorf FW, Cornuet JM, William BS (1998) Distortion of allele frequency distributions provides a test for recent population bottlenecks. J Hered 89: 238–247.

109. Jackson CR, Lubbe NR, Robertson MP, Setsaas TH, van der Waals J, et al. (2008) Soil properties and the distribution of the endangered Juliana's golden mole. J Zool 274: 13–17.

110. Patzenhauerová H, Bryja J, Šumbera R (2010) Kinship structure and mating system in a solitary subterranean rodent, the silvery mole-rat. Behav Ecol Sociobiol 64: 757–767.

111. Johns GC, Avise JC (1998) A comparative summary of genetic distances in the vertebrates from the mitochondrial cytochrome b gene. Mol Biol Evol 15: 1481–1490.

112. Nevo E, Beiles A (1991) Genetic diversity and ecological heterogeneity in amphibian evolution. Copeia 1991: 565–592.

113. Nevo E, Kirzhner V, BeilesA, Korol A (1997) Selection versus random drift: long-term polymorphism persistence in small populations (evidence and modelling). Philos T R Soc Lond 352: 381–389.

114. Siesser WG, Dingle RV (1981) Tertiary sea-level movements around Southern Africa. J Geol 89: 523–536.

115. Patton JL, Nazareth M, Da Silva F, Malcolm JR (1994) Gene genealogy and differentiation among arboreal spiny rats (Rodentia: Echimyidae) of the Amazon basin: A test of the riverine barrier hypothesis. Evolution 48: 1314–1323.

116. Cotterill FPD (2003) Species concepts and the real diversity of antelopes. In: Plowman A, editor. Ecology and conservation of mini-antelope: Proceedings of an international symposium on duiker and dwarf antelope in Africa. Füürth: Filander Verlag. 59–118.

117. Matthee CA, Robinson TJ (1996) Mitochondrial DNA differentiation among geographical populations of Pronolagus rupestris, Smith's red rock rabbit (Mammalia: Lagomorpha). Heredity 76: 514–523.

118. Daniels SR, Hofmeyr MD, Henen BT, Baard EHW (2010) Systematics and phylogeography of a threatened tortoise, the speckled padloper. Anim Conserv 13: 237–246.

119. Hendey QB (1983) Cenozoic geology and palaeogeography of the fynbos region. In: Deacon HJ, Hendey QB, Lambrechts NJJ, editors. Fynbos Palaeoecology: a Preliminary Synthesis. Pretoria: CSIR. 35–60.

120. Lambeck K (2004) Sea-level change through the last glacial cycle: geophysical, glaciological geophysical, glaciological and palaeogeographic consequences. C R Geosci 336: 677–689.

18

Determining the Phylogenetic and Phylogeographic Origin of Highly Pathogenic Avian Influenza (H7N3) in Mexico

Lu Lu[1], Samantha J. Lycett[2], Andrew J. Leigh Brown[1]*

1 Institute of Evolutionary Biology, University of Edinburgh, Ashworth Laboratories, Edinburgh, United Kingdom, **2** University of Glasgow, Institute of Biodiversity, Animal Health and Comparative Medicine, Glasgow, United Kingdom

Abstract

Highly pathogenic (HP) avian influenza virus (AIV) H7N3 outbreaks occurred 3 times in the Americas in the past 10 years and caused severe economic loss in the affected regions. In June/July 2012, new HP H7N3 outbreaks occurred at commercial farms in Jalisco, Mexico. Outbreaks continued to be identified in neighbouring states in Mexico till August 2013. To explore the origin of this outbreak, time resolved phylogenetic trees were generated from the eight segments of full-length AIV sequences in North America using BEAST. Location, subtype, avian host species and pathogenicity were modelled as discrete traits upon the trees using continuous time Markov chains. A further joint analysis among segments was performed using a hierarchical phylogenetic model (HPM) which allowed trait rates (location, subtype, host species) to be jointly inferred across different segments. The complete spatial diffusion process was visualised through virtual globe software. Our result indicated the Mexico HP H7N3 originated from the large North America low pathogenicity AIV pool through complicated reassortment events. Different segments were contributed by wild waterfowl from different N. American flyways. Five of the eight segments (HA, NA, NP, M, NS) were introduced from wild birds migrating along the central North American flyway, and PB2, PB1 and PA were introduced via the western North American flyway. These results highlight a potential role for Mexico as a hotspot of virus reassortment as it is where wild birds from different migration routes mix during the winter.

Editor: Florian Krammer, Icahn School of Medicine at Mount Sinai, United States of America

Funding: This work was funded by a China Scholarships Council (www.csc.edu.cn) and the University of Edinburgh (www.ed.ac.uk) Scholarship to LL. SJL was supported by the Wellcome Trust (www.wellcome.ac.uk; grant number 092807). The funders had no role in study design, data collection and analysis, decision to publish, or preparation of the manuscript.

Competing Interests: The authors have declared that no competing interests exist.

* Email: A.Leigh-Brown@ed.ac.uk

Background

Migratory birds are major candidates for long-distance dispersal of zoonotic pathogens and low pathogenicity (LP), avian-origin influenza A viruses (AIVs) are widely distributed in free-ranging water birds [1]. Wild birds spread their viruses to other wild as well as domestic birds as they migrate through an area, allowing extensive reassortment [2]. Once introduced into poultry (especially chickens and turkeys), LPAI may switch to high pathogenic viruses (HPAI) with the introduction of basic amino acid residues into the haemagglutinin cleavage site, which is associated with a high mortality rate in poultry [3,4]. We have recently shown that a higher inter-subtype reassortment rate can be found in wild Anseriformes than domestic Galliformes in the internal segments of Eurasian AIV, indicating the wild bird population was the source of the new reassortants, rather than domestic poultry [5]. Migrating wild birds have been implicated in the spread and emergence of HPAI such as HP H5N1 and H7N3. Viral transmission between wild birds and domestic poultry, and consequent genetic exchange, has contributed to genomic reassortment which confounded disease control efforts [6,7].

Although predictors of such outbreaks have long been sought, surveillance in wild birds in North America has failed to provide a clear early warning signal. Three H7N3 HPAI events in poultry have occurred in North Americas since 2000, and, in one case, it was reported that the outbreak H7N3 AIV were transmitted from poultry to humans [8]. Phylogenetic analyses indicated that each of these H7N3 HPAI strains had a close relationship with LPAI isolated from wild birds sampled in neighbouring provinces [9,10,11].

In June 2012, H7N3 HPAI outbreaks were found in poultry farms in Jalisco state in Mexico, a region of high poultry density [12] and concurrent infections of humans with this HPAI A (H7N3) virus (2 cases) have been confirmed [13]. The outbreak has been affecting broilers, breeders, layers and backyard poultry in the Mexican States of Jalisco, Aguascalientes, Guanajuato and Puebla: the latest outbreak reported by the World Organization for Animal Health (OIE) was on 19th May 2014. Ongoing epidemiological investigations have implicated contact with wild birds as a factor in the outbreaks [12]. However, the specific origin of the novel outbreak strain and its relationship to the previous outbreak strains is not known.

The aim of our study was to investigate the origin of the precursor strain of the Mexico H7N3, using a Bayesian phylogeographic inference framework by reconstructing the spatiotemporal spread of AIV from wild birds in North America.

Results

Phylogenetics of the HPAI H7N3 Mexico with north America AIV

To investigate the origin of the AIV causing the HPAI H7N3 outbreak in Mexico in 2012, an initial phylogenetic analysis using Maximum likelihood was performed for each segment of both the outbreak sequences and a background dataset which comprised all available AIV of North American AIV lineages (Figure 1). The phylogenetic trees of all available H7 segments in north American showed that AIV isolated in recent years have diverged from those before 1990 (Figure 1A). In addition, in the HA segment a sub-lineage mainly composed of H7N2 AIV from domestic birds in New York state is clearly separate from the recent lineage composed of AIV from wild birds, which indicates extensive diversity of LP AIV in wild and domestic birds. Since 2000, the N3 NA segment of North American AIV has split into two separate lineages (Figure 1B). The mechanism for maintenance of this divergence remains unknown as viruses from both lineages co-circulate in geographically overlapping host populations, mainly wild waterfowl.

Diverse reassortment events involving the six internal segments can be inferred from the maximum likelihood phylogenies of 2343 North American AIV. Clades identified in the phylogeny for one segment (e.g., PB2) are not maintained in the phylogenies of other internal segments (Figure S1). In addition, internal segments of AIV viruses isolated in distant locations can be closely related to each other within the same time period, which suggests not just frequent reassortment but also rapid movement of influenza viruses across North America.

Sequences for time-scaled phylogenetic analysis were selected from the closest clades to the novel H7N3 HPAI viruses on the maximum likelihood tree of each segment. This dataset comprised

427 AIV strains collected over a 12 year period (2001 to 2012). The time to the most recent common ancestor (TMRCA) for each segment of the novel HPAI H7N3 in Mexico was estimated from the time-scaled phylogenetic trees. The HPAI H7N3 strains sampled in Mexico shared similar common ancestors among different genes between October 2011 and March 2012, i.e. during the winter of 2011–2012 (Table 1). The common ancestor of the HPAI H7N3 Mexico outbreak and the closest related avian influenza strains existed between 1.1 to 3.9 years ago, which varied among their different genomic segments (Table 1). The difference in unsampled diversity among gene segments suggested that the reassortment of North American AIV lineages which led to the H7N3 Mexico outbreak may have involved several events spread over this time period. This can be seen by comparing the closest related strains in the phylogenetic tree for any segment: they can be quite distant from the H7N3 Mexico strain in the other segments. This result supports our hypothesis of the occurrence of multiple reassortment events.

Co-circulation of multiple H7 clades was observed in HA across North America. Interestingly, the HPAI H7N3 Mexico strains are not related in HA to the HP H7N3 outbreak in British Columbia in 2004 and 2007, but instead are closely related to a subgroup of H7 AIV (H7N3, H7N8 and H7N9) from wild waterfowl isolated from Nebraska, Illinois, Missouri and Mississippi in 2010 and 2011. The mean estimate of the date of the common ancestor is February 2010 (Figure 2). On the other hand, the picture in NA is different: the closest related strain to that of the Mexico outbreak is a subtype H2N3 AIV isolated from a green winged teal in Illinois in 2010 (Figure S2).

In contrast, the three polymerase encoding gene segments PB2, PB1 and PA of the Mexico outbreak strain belong to lineages composed mainly of AIV found in wild waterfowl in California from the beginning of 2012 (Figure S3, S4, S5), with segments PB1 and PA having the same most closely related strain: A/American green-winged teal/California/123/2012 (H1N1). The other internal segments of the Mexico H7N3 strains have a different origin. From the Bayesian phylogenetic tree of the NP segment, the closest AIV strain to H7N3 Mexico is an H11N9 strain isolated

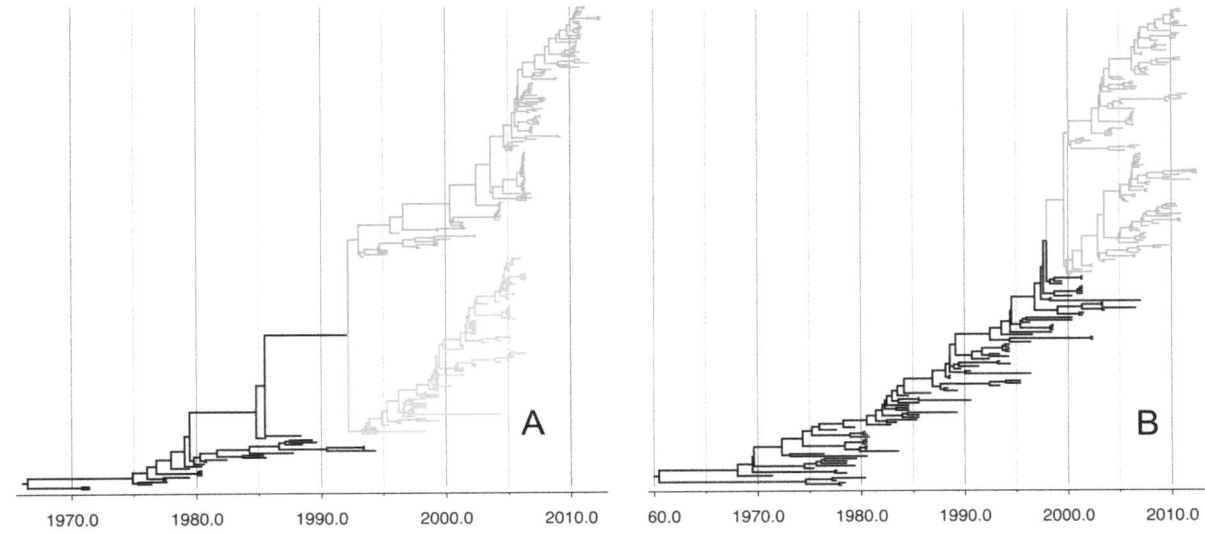

Figure 1. Phylogenies of the H7 and N3 segments of all available North American AIV. A: HA. Sequences in grey are from before 1990; the clade colored blue is composed of H7N2 AIV isolated from a single surveillance in poultry in New York; the clade colored pink was selected for time-scaled phylogenetic analysis. B: NA tree. The uncoloured sequences are from before 2000; the AIV clade in pink was selected for time-scaled phylogenetic analysis.

Table 1. Time of the most recent common ancestors for the Mexico H7N3 virus.

Gene	TMRCA[a]	Most closely related strain[b]
HA	20/3/2012 (29/11/2011, 23/5/2012)	A/north shoveler/Missouri/2010 (H7N3)
NA	4/1/2012 (10/8/2011, 23/4/2012)	A/American green-winged teal/Illinois/2010 (H2N3)
PB2	30/10/2011 (8/6/2011, 25/2/2012)	A/mallard/California/198/2012 (H11N9)
PB1	10/10/2011 (10/10/2010, 28/1/2012)	A/American green-winged teal/California/123/2012 (H1N1)
PA	18/11/2011 (3/7/2011, 26/3/2012)	A/American green-winged teal/California/123/2012 (H1N1)
NP	21/1/2012 (16/8/2011, 5/5/2012)	A/American green-winged teal/Mississippi/2012 (H11N9)
M	2/11/2011 (7/0/2011, 2/5/2012)	A/American green-winged teal/Illinois/2008 (H10N7)
NS	25/10/2011 (25/3/2011, 17/1/2012)	A/American green-winged teal/Illinois/2010 (H10N7)

[a]Time of the most recent common ancestors (TMRCA) of each segment of the novel Mexico H7N3 virus are represented in the order of date/month/year. The values in parentheses represent the 95% HPD intervals.
[b]The strains are identified are those most closely related to the outbreak strains in each tree phylogeny in this study.

from Mississippi in 2012 (Table 1). The NP segment of the H7N3 Mexico outbreak strain is also in the same lineage as a small number of H7N7 AIV strains carried by northern teal in Illinois/Missouri in the fall of 2010; these strains belong to the same lineage in the HA segment as well (see above and Figures 2 and S6). The NS segment was derived from an H10N7 AIV which was also circulating in the same region at that time (Figure S8). In the M segment, however, Mexico H7N3 strains are distinct from all currently available AIV in North America, suggesting a surveillance gap (Figure S7).

These results indicate that the HPAI H7N3 virus that caused the outbreaks in Mexico is not related to any of the previous H7N3 HPAI outbreaks in North America, nor related to other AI outbreaks (HP H5N2 outbreaks in Mexico in 1994–1995, LP H7 outbreaks in Canada in 2009) in domestic birds in recent years [14,15]. In addition, no clear pattern of association among the segments of the Mexico H7N3 strains was observed, indicating multiple segment exchange events occurred among North American influenza strains to give rise to it.

Gene flow of the precursor of the HPAI H7N3 outbreak in Mexico

To further explore the origin of the Mexico outbreak strain, a joint analysis of discrete trait models was performed to estimate the overall genetic transmission process. In this the phylogenetic tree space was sampled independently for each segment, while the transition pattern was jointly estimated in a single analysis as the diffusion parameters being applied in the discrete trait models were assumed to be the same (see methods). Four major factors including seven specific traits were tested by implementing Bayesian stochastic search variable selection (BSSVS): i) host population of AIV (order/species); ii) geographic location of sampled AIV (bird migration flyways/provinces and states of North America); iii) subtype of AIV and iv) virulence (pathogenicity/cleavage sites). For each trait, the evolving process of the HPAI H7N3 in Mexico and closely related AIV can be seen from the reconstructed time-scaled phylogeny of each segment independently (with exception of pathogenicity and cleavage sites which only applied to the HA segment), with the branches colored by the specific trait according to the ancestor trait in the internal nodes (Figure 2 and Figure S2, S3, S4, S5, S6, S7, S8).

Host populations of AIV in this study were first analysed by Order: wild Charadriiformes (cha) such as gulls, wild Gruiformes (gru) such as cranes, wild Passeriformes (pas) such as sparrows,

domestic Galliformes (gal) such chickens and wild Anseriformes (ans) such as mallards, with Anseriformes comprising the majority of our AIV data set (n = 366/427, see table S1. Among all four strongly supported transitions with Bayes Factor (BF) >100 with mean diffusion rate (R) between 0.01 and 0.07, the highest diffusion rate was found between Charadriiformes and Passeriformes. The other three were found between Anseriformes and other bird orders, and the HPAI H7N3 outbreak in poultry (labelled as Galliformes Mexico) is linked to Anseriformes with strong support (R = 0.02, BF>100) (Figure 3A and Table S5). The results confirm that there has been extensive mixing of influenza A virus between different orders of birds, both wild and domestic.

To explore which host species might have been the direct donor of the Mexico outbreak strains, AIV belonging to Anseriformes were further divided into the five predominant species: mallard (Anas platyrhynchos), northern pintail (Anasacuta), northern shoveller (Anas clypeata), blue-winged teal (Anas discors) and green-winged teal (Anas carolinensis) (Table S2). Species that were sampled at relatively low levels were combined as "other Anseriformes" (Table S2). Multiple statistically supported transitions (with R from 0.15 to 2.13, BF from 6 to over 100) were identified among different host species within this Order, and both mallard and green winged teal are linked to 3 other host species (Figure 3B and Table S6). This analysis indicates the Mexican outbreak strains were most likely to have been transmitted from green winged teals (R = 0.15 and BF = 6).

The phylogeographic analysis for each segment of the Mexico HPAI H7N3 strain was summarised by a MCC tree in a geographic context. However, to visualize the evolution process in a spatiotemporal mode we converted the spatial annotated time-scaled phylogeny to an annotated map (Figure 4 and Figure S9). Five segments (HA, NA, NP, M, and NS) of the Mexico HPAI H7N3 strain were introduced directly from different states in central US, while PB2, PB1 and PA were introduced from states in the western region. The introductions of segments from several different geographic locations indicate multiple reassortment events were likely to have been involved in the generation of the novel H7N3 Mexico AIV.

Joint discrete trait analysis of all eight segments indicated frequent gene transfer among locations (states and provinces) where the background AIV sequences were isolated. However, in this initial analysis, no significant support was found between Jalisco (the outbreak state) and any other location, probably due to the large number of possible transitions (26 states, 325 irreversible transition pairs) and the limited number of outbreak strains (3;

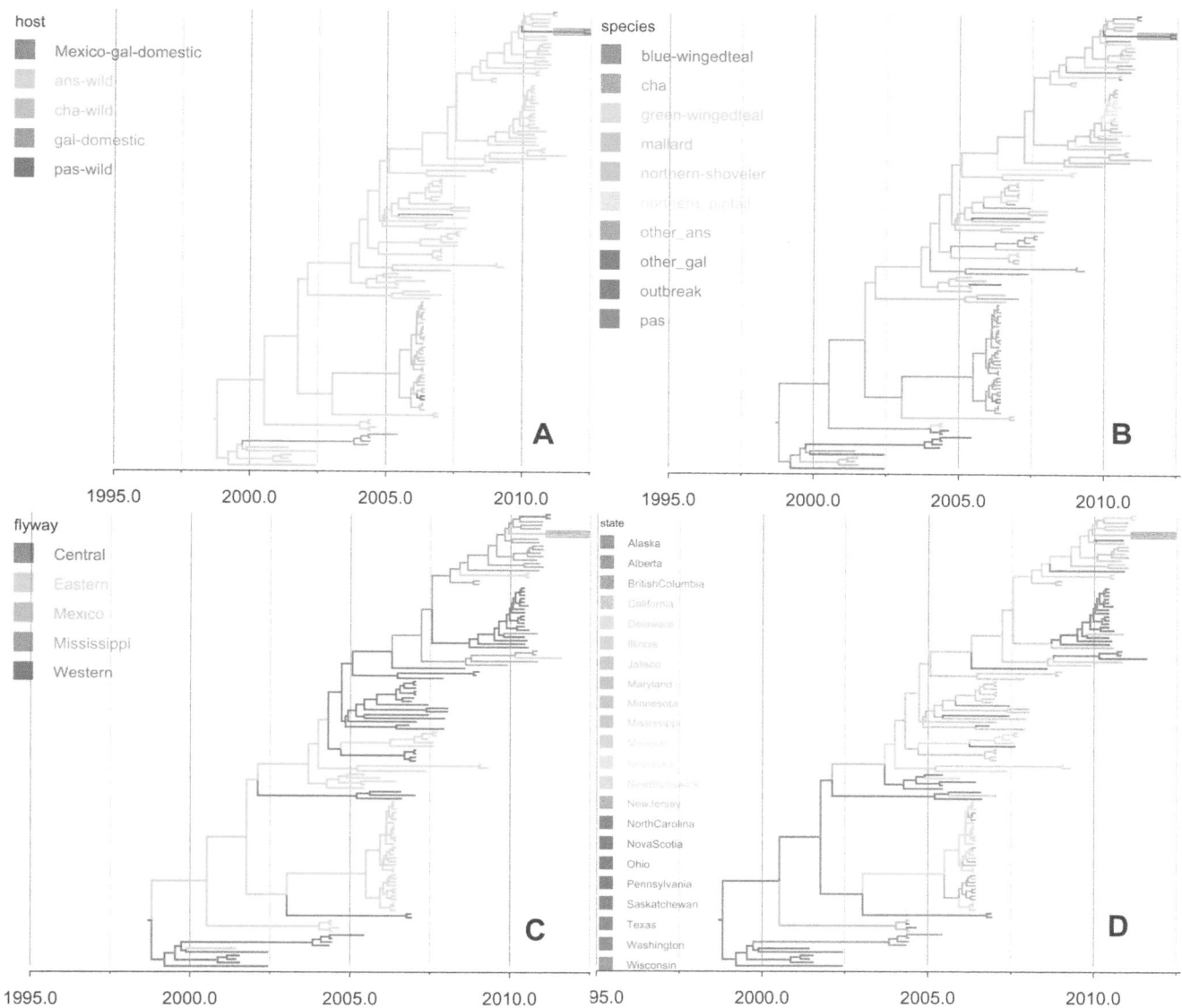

Figure 2. Maximum clade credibility (MCC) phylogenies for the HA segment. Branches are coloured according to the 4 discrete traits (host order, host species, flyway and location) on internal nodes. Mexican outbreak strains are highlighted with pink. A: Host order. Five host orders are labelled on HA tree: wild birds of the order Anseriformes (ans-wild); wild birds of the order Charadriiformes (cha-wild); wild birds of the order Passeriformes (pas-wild); domestic birds of the order Galliformes and Mexico H7N3 outbreak in the order Galliformes (gal-domestic-Mexico). B: Host species. Wild Anseriformes are classified into the five main species and a group comprising the other rarer species of Anseriformes in this study: mallard (Anas platyrhynchos), northern pintail (Anasacuta), northern shoveller, blue-winged teal, green-winged teal and other Anseriformes (other ans); The order Galliformes are shown as "outbreak" (the H7N3 Mexico outbreak) and "other_gal"; The other orders are shown as: Charadriiformes (cha) and Galliformes (gal), Gruiformes (gru) and Passeriformes (pas). C: Flyway. Four specific North American flyways are labelled on the HA tree: the Atlantic, Mississippi, Central, and Pacific. D: State. 22 states and provinces of the viral sample locations are labelled on the HA tree. The original MCC tree files with all taxa names are deposited in Dryad (doi:10.5061/dryad.j5bf8), and trees for the other 7 segments without taxa names can be found in Figure S2, S3, S4, S5, S6, S7, S8.

Table S3). Previous studies have shown that incorporating location greatly improves phylogeographic descriptions of the pattern of virus gene flow [16,17]. Therefore, we enhanced the statistical power of the analysis in two ways: first by decreasing the number of locations by combining states and provinces into major regions (flyways) and secondly by reducing the number of pairwise transitions possible.

To aggregate locations we used the known migration routes, or "flyways": Atlantic, Mississippi, Central, or Pacific (Figure S10). The distributions of avian influenza virus for each flyway are

summarized in Table S4, showing a wide range in rate and statistical support (R = 0.02 to 0.25; BF = 3 to >100). Highly significant links [BF>100, Indicator (I) = 1] were found between major North American flyways, particularly between Atlantic and Mississippi (R = 0.17 exchange/year); Mississippi and Pacific (R = 0.22 exchange/year), Central and Atlantic (R = 0.25 exchange/year) and Central and Pacific (R = 0.18 exchange/year) (Figure 5A and Table S7). Linkages between Atlantic and Pacific flyways, Mississippi and Central flyways were also identified although with weaker support (3<BF<6). The transitions between

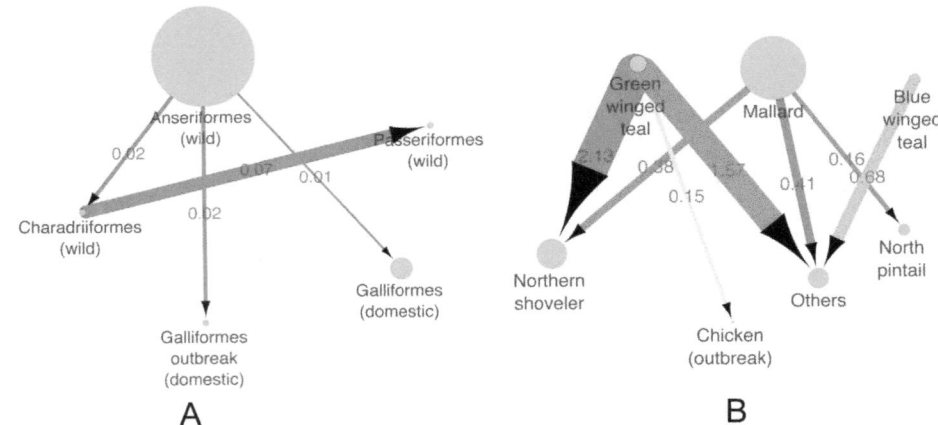

Figure 3. Inferred host transmission networks of Mexican outbreak AIV. A: Host order. Node labels in the nodes are host orders identified following the abbreviations used in the colored phylogenetic trees (Figure 2 and Figures S2, S3, S4, S5, S6, S7, S8): wild birds of the order Anseriformes (ans-wild); wild birds of the order Charadriiformes (cha-wild); wild birds of the order Passeriformes (pas-wild); domestic birds of the order Galliformes and Mexico H7N3 outbreak in the order Galliformes (gal-domestic-Mexico). Arrows show the direction of transmission between two host orders; the arrow weight and the number above each arrow indicates the per capita transmission rate. Node size reflects the number of AIV for each host order (Table S1). Line colours indicate the overall Bayes Factor test support for epidemiological linkage between host orders, Red lines indicate statistical support with BF>100 (very strong support), dark pink lines indicate support with 30<BF<100(strong support), pink lines indicate support with 3<BF<30. B: Host species (Anseriformes only). Wild Anseriformes are further classified into the five main species and a group comprising the other rarer species of Anseriformes in this study: mallard (Anas platyrhynchos), northern pintail (Anasacuta), northern shoveller, blue-winged teal, green-winged teal and other Anseriformes (other ans); As above, arrows show the direction of transmission between two host species; the arrow weight and the number above each arrow indicates the per capita transmission rate. Node size reflects the number of AIV for each host species (Table S3). Line colours indicate the overall Bayes Factor test support for epidemiological linkage between host species, Red lines indicate statistical support with BF>100 (very strong support), dark pink lines indicate support with 30<BF<100(strong support), pink lines indicate support with 3<BF<30.

flyways and the H7N3 HPAI in Mexico are not that strongly supported compared to the between flyway transitions, probably due to the limited number of sequences available from the outbreak AIV, but three direct donors among these flyways to the predecessor of the Mexico outbreak were identified: the Pacific, Central and Mississippi flyways. Among these flows, the transition rate from Mississippi (R = 0.05 exchange/year) is the most strongly supported with BF = 86. In comparison, the link with the Pacific (R = 0.02 exchange/year) and Central (R = 0.04 exchange/year) were weaker (BF = 4; Figure 5B and Table S7). There was no supported link between the Atlantic flyway and Mexico. The results indicated that the HPAI H7N3 in Mexico probably originated from AIV transmitted by wild birds from three different flyways.

Given these findings it appeared likely that the precursor strains were generated somewhere near Mexico as it is the place where birds from the different migration routes meet during winter. To test this hypothesis, we further reduced the number of transitions in the locations transition matrix: transitions between two flyway regions were switched off (by forcing the initial indicator of a given transition pair from 1 to 0, so that this transition pair will not be counted), and those for within flyway transitions and transitions linked to Mexico were maintained. There are 98 non-reversible transition pairs in the new reduced matrix (Table S10). AICM tests (see Methods) revealed that the non-reversible BSSVS model with reduced number of transitions was significantly favoured over the other models with the original matrix (Table 2), indicating the number of transitions has an effect on the performance of discrete trait models. This reduced model has better support than a randomly reduced model with the same number of transition pairs and the same non-reversible BSSVS setting (Table 3); we conclude that gene transitions within flyways and Mexico alone better

explain the gene flow of North America AIV than a model incorporating the between flyway transitions.

Considering individual locations, we found 11 locations were linked to the HPAI H7N3 in Mexico, among which 7 showed significantly strong links (BF>100). These were: Alaska (R = 0.59), Alberta (R = 1.07), California (R = 0.59), Illinois (R = 1.61), Missouri (R = 2.26), Ohio (R = 0.66) and Wisconsin (R = 1.39), belonging to the Pacific, Central and Mississippi flyways. AIV from other four states/provinces (Minnesota, New Brunswick, Quebec and Washington) are also linked to Mexico H7N3 AIV but with weaker support (BF = 21 to 30) and lower rate (0.7 to 0.84) (Figure 5A and Table S8). This result indicates the donor locations of the Mexico outbreak are spread widely across North America.

In addition, an extremely complex pattern of linkage between the 52 ancestral subtypes (Table S9) was identified, which confirmed the extent of the reassortment events which had occurred between, especially in the internal segments (full tree with annotation can be found on the Dryad Digital Repository: doi:10.5061/dryad.j5bf8). However, as for locations, no significant linkage between H7N3 in Mexico and other subtypes was identified due to the large number of candidate donor subtypes.

Mexico H7N3 is confirmed in the phylogenetic trees of the HA segment (Figure S11) to have mutated from a LPAI (low pathogenic avian influenza) virus to HPAI after the ancestral virus was introduced into poultry from wild birds (Figure S11A), rather than being associated with previous HP outbreaks. The virulence of HP avian influenza viruses is associated with the appearance of an insertion of multiple basic amino acids at the cleavage site of the HA protein [18]. Categorizing the cleavage sites in this dataset into three types: 1) Insertion, 2) Partial insertion,3) No insertion (Figure S11 C), we find that the H7N3 Mexico strain has a unique insertion - DRKSRHRR - compared

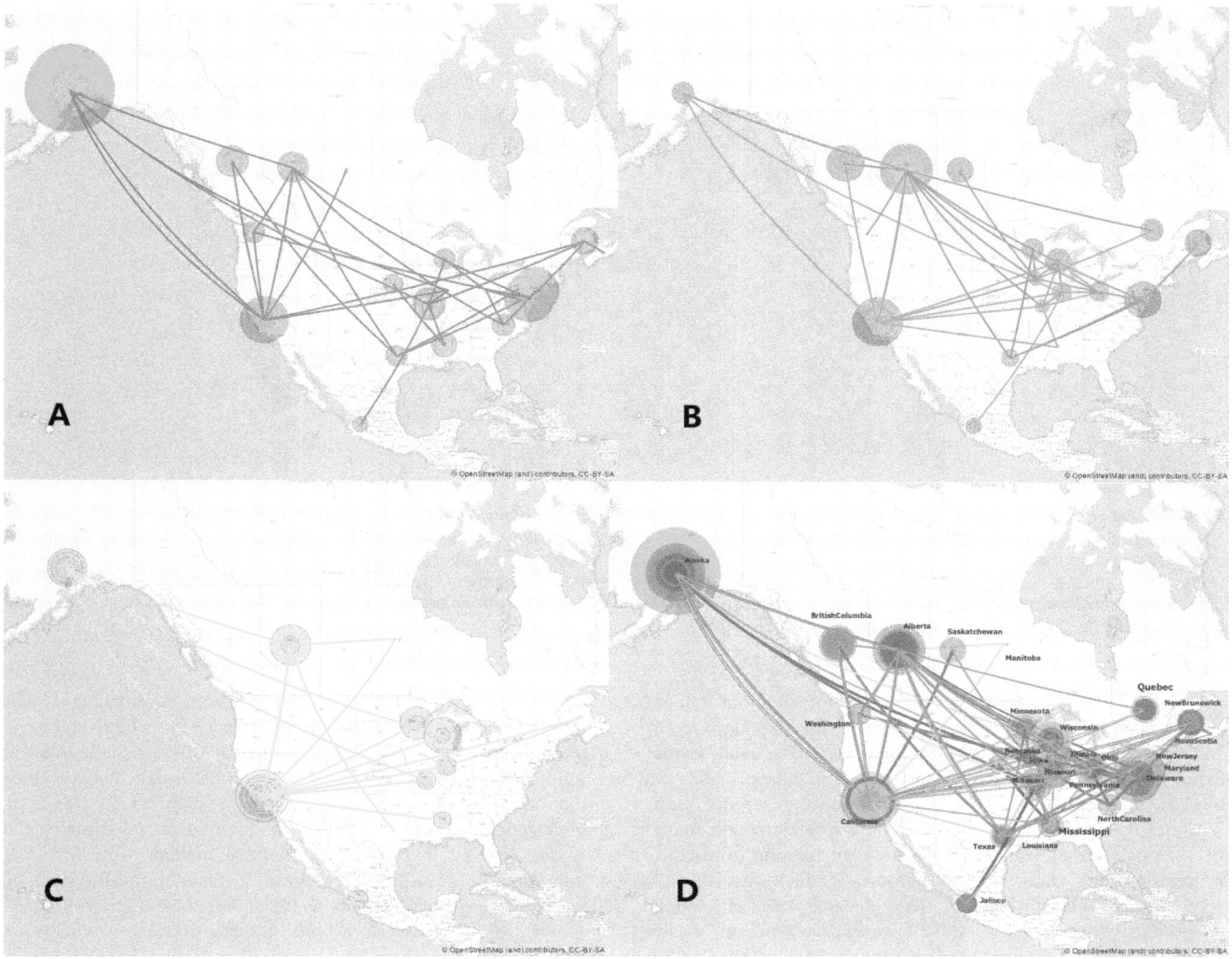

Figure 4. Spatial diffusion of AIV segments of the Mexico outbreak AIV. The first three panels represent three segments separately (A: HA, B: NA, C: PB2) and D represents the spatial transmission of all 8 segments jointly. The plotted lines represent the branches of the MCC trees for different segments, distinguished by color; the size of each circle represents the number of lineages with that location state. The map source for this figure was OpenStreetMap (http://www.openstreetmap.org/). The spatial diffusions of other five segments (PB1, PA NP, M and NS) on the map are shown in Figure S9.

to other HPAI strains (Figure S11 C). We conclude the Mexico H7N3 strains originated from a lineage composed of LP strains with partial insertions in the cleavage sites. Similarly, the H7N3 strains which caused an outbreak of HPAI in Canada in 2004 were also derived from LP strains with the same partial insertions, but from a completely separate HA clade, indicating parallel evolution with respect to the acquisition of the multibasic cleavage site, starting from different lineages (Figure S11 B).

Discussion

We have investigated the origin of the recent highly pathogenic H7N3 outbreaks in Mexico. Our analysis found that the progenitor of the HPAI H7N3 was a reassortant virus with several different origins among the eight segments. We also found that gene segments of AIV in North American wild birds are exchanged at a very high frequency, with no evidence of any restriction which might imply linkage of segments.

Our study confirmed the assumptions of earlier studies based on the HA segment that the outbreak strain derived from wild birds [12,19,20]. We have now shown using powerful Bayesian phylogenetic methods that the origin of the HA segments of the Mexico H7N3 strains can be dated to March 2012, and that they fall into a subgroup of H7 AIV (H7N3, H7N8 and H7N9) from wild waterfowl isolated from Nebraska, Illinois, Missouri and Mississippi with a common ancestor around February 2010. We have extended this analysis to all eight segments, thus obtaining the complete evolutionary history of the outbreak avian influenza viruses.

The predecessors of HPAI H7N3 in Mexico were transmitted from migrating waterfowl in North America. Previous cases of periodic transmission of H7N3 viruses from wild birds to gallinaceous poultry in the Americas suggests that these viruses continuously circulate in wild birds, and their propensity to become highly pathogenic after transmission suggests that they have a gene constellation conducive to generating pathogenic

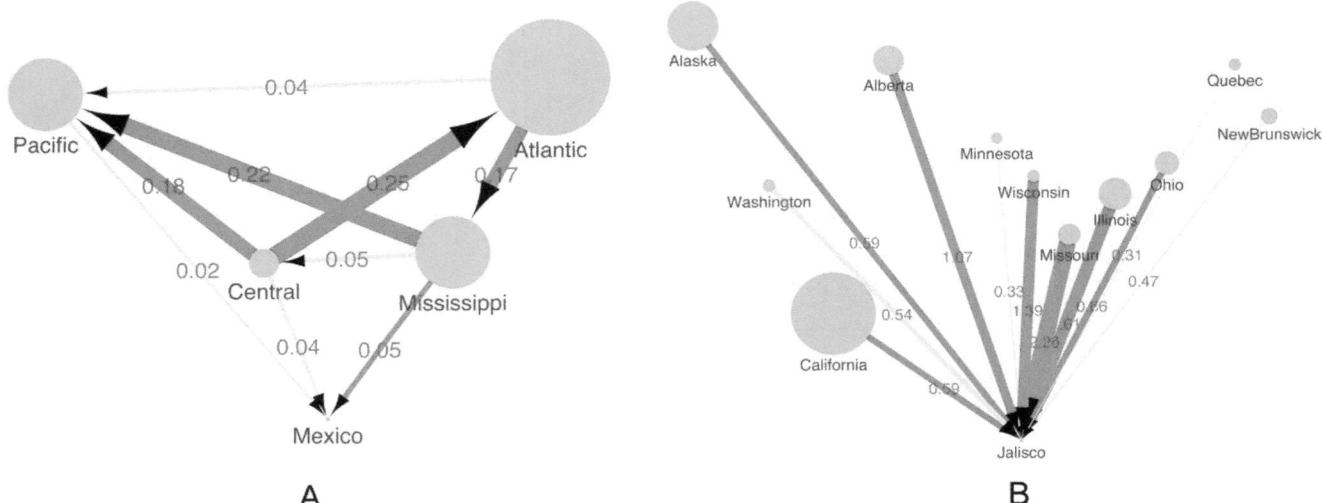

Figure 5. Inferred phylogeographic transmission networks of Mexican outbreak AIV. A: Flyway. AIV transmission among 4 N. American flyways with links to the Mexican outbreak strains. Arrows show the direction of transmission between two flyways; arrow weight and the number above each arrow indicates the per capita transmission rate. Node size reflects the number of AIV for each flyway (Table S6). B: Location. AIV transmission among states/provinces in North America and Jalisco (the Mexican state where the outbreak strains were isolated). Arrows show the direction of transmission between the two states; the arrow weight and the number above each arrow indicates the per capita transmission rate. Node size reflects the number of AIV for each flyway (Table S5).

variants [21]. H7N3 has been responsible for all lethal influenza outbreaks in poultry in the Americas over the past decade [21]. Experimental studies have also indicated that H7 influenza viruses from the North American lineage have acquired sialic acid-binding properties that more closely resemble those of human influenza viruses and have the potential to spread to naive animals [22]. In parallel, H7 influenza viruses from East Asian migratory waterfowl were introduced into domestic ducks in China on several occasions during the past decade and subsequently reassorted with enzootic H9N2 viruses to generate a novel H7N9 influenza A virus, resulting in 44 human deaths in China (WHO reported in Dec.3ʳᵈ 2013) since its first detection in March 2013 [23]. These results indicated that AIV of H7 subtypes carried by wild birds are potential threat to mammalian hosts.

Earlier studies showed that shorebirds and gulls in the Americas are more frequently the source of the potential precursors to HP H5 and H7 avian influenza viruses, while in Eurasia, the precursors of HP influenza viruses are usually from duck species [24,25]. However, we found that wild Anseriformes (ducks and geese) were the origin of the precursor of HPAI H7N3 in Mexico. Anseriformes showed substantial diversity of AIV in North America, and were divided into five avian species groups in our dataset – mallards, northern pintails, northern shovellers, blue-winged teals, green-winged teals, which have specific ranges for breeding, migration and wintering [17]. We found the green winged teal was the species most strongly supported as the direct donor of the predecessor HPAI H7N3 in Mexico. This species nests as far north as Alaska, and migrates along all four flyways. However, the genetic transitions of different segments showed a complicated interaction involving different bird species.

Migrating birds may exchange viruses with other populations at staging, stopover or wintering sites [26]. Many studies have been performed on AIV gene flows in North America during wild bird migration: One such revealed that avian influenza virus exhibits a strongly spatially structured population in North America, and the intra-continental spread of AIV by migratory birds is subject to

major ecological barriers, including spatial distance and avian flyway [17]. Earlier studies suggested that AIV exhibits a strongly spatially structured population in North America, with relatively infrequent gene flow among localities and especially between those that are spatially distant or belong to different flyways using phylogeographic analysis [27]. This hypothesis was supported by studies showed that AIV isolates from mallard were linked by migration between sites in central Canada and Maryland but limited reassortment occurred along the inter-migratory flyway routes [28]. However, more recently, the opposite was seen in a another study, which emphasized that the long-term persistence of the influenza A virus gene pool in North American wild birds may be independent of migratory flyways, and the short-term evolutionary consequences of these ecological barriers may be rapidly erased by East-West virus migration [29].

We also found there are genetic interactions between flyways, using a similar discrete trait model. However, to find the strongest link between the Mexican outbreak and potential precursors we found the model had more power when we switched off the between flyway transitions, keeping only the links within flyways and with Mexico. We found that gene flow from three flyways (Pacific, Central and Mississippi) generated the reassortants which acted as the predecessor of HPAI H7N3 in Mexico, and it is possible that the reassortment events occurred in Mexico or farther afield. Flyway boundaries are not sharply defined and both in the northern breeding grounds and the southern wintering grounds there is overlap to some degree. For example, in Panama parts of all four flyways merge into one (http://www.birdnature.com/flyways.html). Birds that are long-distance migrants typically have ranges that extend from the United States and Canada in the summer to Mexico and further south in the winter and nearly all of the migratory birds of the eastern United States, as well as many western species, use the western Mexican Gulf during migration [30].

While the resolution and detection of migration events has been enhanced through increased surveillance in recent years, critical

Table 2. AICM estimates for the fit of different discrete trait models.

Joint Discrete trait models	AICM[b]	S.E.[c]	Mod1[d]	Mod2	Mod3	Mod4	Mod5	Mod6	Mod7	Mod8
Mod1 Reduce_BSSVS_sym[a]	**111031**	**+/−7.5**	**—**	**107**	**490**	**712**	**863**	**1102**	**4084**	**6582**
Mod2 Reduce_nonBSSVS_sym	111137	+/−9.449	−107	—	384	605	756	995	3977	6475
Mod3 Reduce_BSSVS_asym	111521	+/−11.798	−490	−384	—	221	372	611	3593	6091
Mod4 Reduce_nonBSSVS_asym	111742	+/−7.822	−712	−605	−221	—	151	390	3372	5870
Mod5 Original_BSSVS_sym	111893	+/−6.668	−863	−756	−372	−151	—	239	3221	5719
Mod6 Original_nonBSSVS_sym	112132	+/−13.455	−1102	−995	−611	−390	−239	—	2982	5480
Mod7 Original_BSSVS_asym	115115	+/−14.662	−4084	−3977	−3593	−3372	−3221	−2982	—	2498
Mod8 Original_nonBSSVS_asym	117612	+/−26.584	−6582	−6475	−6091	−5870	−5719	−5480	−2498	—

[a]The names of the 8 models (Mod1-8) in the comparison test.
[b]The estimated AICM score of the posterior: lower values of marginal likelihood indicate a better fit to the data. The model with the best performance is indicated in bold.
[c]The standard error of the AICM estimated using 1000 bootstrap replicates.
[d]The AICM comparisons are shown in the matrix composed of columns 4 to 11. In each row of the matrix, the positive value in a cell represents the support for one model (in column 1) over the other (indicated in the column titles). A difference of AICM = 10 is considered to indicate a strong preference for one model over another.

information for wild bird surveillance remains sparse. Only one AIV in Mexico has been published (A/cinnamon teal/Mexico/2817/2006) and it is not related to the new outbreak virus in any of the eight segments (data not shown). We found the origin of the M segment of the H7N3 outbreak gene was distinct from other North American AIV in the dataset, as seen in the phylogenetic trees. In addition, the relatively greater length of branches preceding the outbreak group in other gene segments (Figures 2 and S2, S3, S4, S5, S6, S7, S8) suggests there may be missing intermediates, possibly through insufficient AIV surveillance in Central and South America. Together with previous phylogenetic studies which also mentioned the importance of filling gaps in viral sampling (with record of sampling time and location) in these regions [31,32], this highlights the need for increased surveillance in those regions.

Overall, by combining the phylogenetic history of AIV with the host distribution and ecology in our analysis, we show the origin of HPAI H7N3 AIV that caused a series of poultry outbreaks in Mexico is an novel reassortant carried by migration of wild waterfowls from different migration flyways in North America throughout the time period studied, and, more importantly, Central America might be a potential hotspot for AIV reassortment events. Our results are useful for identifying the threat of AIV in wild birds and indicate comprehensive surveillance in South and Central America is highly desirable.

Methods

Data preparation

The complete genome of three outbreak strains of H7N3 from Mexico (A/chicken/Jalisco/CPA1/2012; A/chicken/Jalisco/12283/2012; A/Mexico/InDRE7218/2012) and all previously published influenza A virus sequences of North American lineage (complete genome only) were downloaded from GenBank on 1st March 2013. Sequences of each gene segment were aligned using MUSCLE v3.5 [33]. Maximum likelihood (ML) phylogenetic trees for each segment were generated using RAxML v7.04 [34], each employing a GTR GAMMA substitution model with 500 bootstraps. We established a full genome dataset which was composed of the same 2343 North American strains for each segment. The HA and NA segments have extremely high divergence between different subtypes, therefore, we used all available H7 and N3 to generate the raw trees of HA and NA segment respectively; While there are diverse reassortment and interaction among the six internal segments, therefore, we constructed the giant ML trees for the internal segments in order to identify the outbreak strains related strains. Background sequences for further study were selected from the closest clades to the novel H7N3 HPAI viruses on the maximum likelihood tree of each segment. The final dataset of 427 AIV strains collected over a 12 year period (2001 to 2012) is displayed in Table S9. In this table the segments selected for analysis for each strain are indicated. For the majority of strains, only one segment is selected (n = 289), while for others more than one segments is included. There are 131 HA (H7) sequences included in the analysis based on their relationship to the Mexico H7N3 strain (1698 nt); other H7 sequences included in the joint analysis are related to the outbreak strain in other segments. For the other segments the distribution is as follows: NA (N3), n = 100 (1410 nt); PB2, n = 86 sequences (alignment length of 2277 nucleotides); PB1, n = 67 (2271 nt); PA, n = 89 (2148 nt); NP, n = 79 (1494 nt); MP, n = 39 (982 nt); and NS, n = 42 (838 nt). The trait information (host order; host species; location; flyway; state; subtype) of these background AIV sequences are also provided in Table S9.

Table 3. AICM estimates for the fit of different models with reduced number of transitions.

Joint reduced models[a]	AICM[b]	S.E.[c]	Flyway[d]	Ran1	Ran2	Ran3	Ran4	Ran5	Ran6	Ran7	Ran8	Ran9	Ran10
Flyway	**111031**	**+/−5.2**	**—**	**18**	**70**	**108**	**129**	**157**	**158**	**163**	**171**	**521**	**407822**
Ran1	111049	+/−4.4	−18	—	52	90	111	139	139	145	153	502	407804
Ran2	111101	+/−5.9	−70	−52	—	38	59	87	88	93	101	451	407752
Ran3	111138	+/−14.9	−108	−90	−38	—	21	50	50	55	63	413	407714
Ran4	111160	+/−6.5	−129	−111	−59	−21	—	28	28	34	42	391	407693
Ran5	111188	+/−5.6	−157	−139	−87	−50	−28	—	0	6	13	363	407664
Ran6	111188	+/−4.3	−158	−139	−88	−50	−28	0	—	6	13	363	407664
Ran7	111194	+/−12.8	−163	−145	−93	−55	−34	−6	−6	—	8	357	407659
Ran8	111201	+/−8.6	−171	−153	−101	−63	−42	−13	−13	−8	—	350	407651
Ran9	111551	+/−9.9	−521	−502	−451	−413	−391	−363	−363	−357	−350	—	407301
Ran10	518853	+/−3495.6	−407822	−407804	−407752	−407714	−407693	−407664	−407664	−407659	−407651	−407301	—

[a]The names of the 11 models (Flyway model and Ran1-10) in the comparison test: Model1 is the non-reversible BSSVS model without between flyway transitions (Flyway); Ran1-10 are models in which the same number of transitions has been switched off, chosen at random (Ran1-10).

[b]The estimated AICM score of the posterior: lower values of marginal likelihood indicate a better fit to the data. The model with the best performance is indicated in bold.

[c]The standard error of AICM estimated using 1000 bootstrap replicates.

[d]Differences of models are shown in the matrix being composed by column 4 to 13. In each row of the matrix, the positive value in a cell represented the support for one model (in column 1) over the other (the column title). A difference of AICM = 10 is considered a strong preference for one model over another.

Time-scaled phylogeny reconstruction

To estimate the origin in time and space of the HPAI H7N3 outbreak strain in Mexico, models in BEAST v.1.7.3 [35,36] were applied independently to each gene segment (each segment has a different number of AIV sequences). Different combinations of substitution models: general time-reversible (GTR) substitution model+ Γ distributed site-site rate variation [37] and SRD06 [38]; clock models: strict and uncorrelated relaxed lognormal; and population size models: constant size, exponential, skyride models were evaluated by Bayes Factor test. The best fitting model - incorporating a GTR substitution model+ Γ with uncorrelated lognormal relaxed molecular clocks and a constant-population coalescent process prior over the phylogenies was selected. Parameters were estimated using the Bayesian Monte-Carlo Markov Chain (MCMC) approach implemented in BEAST. MCMC chains were run for 100 million states, sampled every 10,000 states with 10% burn-in. MCMC convergence, and effective sample size of parameter estimates were evaluated using Tracer 1.5 (http://beast.bio.ed.ac.uk). Maximum clade credibility (MCC) trees were summarized by using Tree Annotator and visualized by using FigTree v1.4.0 (http://tree.bio.ed.ac.uk/software/figtree/).

A graphical representation of the origin of HPAI H7N3 Mexico was obtained by spatial reconstruction using a Bayesian framework. The SPREAD application [39] was used to convert the estimated divergence times and the spatially-annotated time-scaled phylogeny (by associating each location with a particular latitude and longitude) to a spatiotemporal movement. The mapped objects were exported to keyhole markup language (KML) files and then were visualized by geographic information systems software: ARCGIS (http://www.esri.com/software/arcgis). The map source is OpenStreetMap (http://www.openstreetmap.org/).

Joint analysis of the transition rates between discrete traits

To reduce the effects of sampling bias and identify reassortment events in each individual segment, a joint analysis of all segments was performed using a hierarchical phylogenetic model (HPM) [36,40].

In this hierarchical structure the prior on a parameter may be shared between partitioned datasets in order to increase the efficiency of its estimation [41]. This approach has recently been used in an analysis of the phylogeographic history of Dengue virus where information from multiple phylogeographic datasets was combined in a hierarchical setting using an HPM framework [42]. Here, we use HPMs for discrete trait models where each segment is treated as an independent dataset with an individual relaxed clock model and tree prior, but shares the discrete trait model (subtype, location and host) with all other segments resulting in a joint transition matrix for each discrete trait. Note that is it not necessary to have the same taxa in each of the data partitions within this framework, since the transition rate matrix is estimated using the tip patterns and trees from all of the partitions, when there are missing states and transitions in one partition the estimates of those rates comes from the other partitions where the states and transitions are present.

Bayesian stochastic search variable selection (BSSVS) was employed to reduce the number of parameters to those with significantly non-zero transition rates [43]. A Bayes Factor support (BF)>100 was considered to indicate decisive support, whereas 3≤ BF≤30, and 30≤BF≤100 indicate substantial, and strong statistical support, respectively. The Bayes Factor threshold was estimated from Rate indicators using the indicator BF tool in the BEAST v1.6.2 package. The default Poisson distribution (with

mean equal to the number of states) was used as a prior on the number of rate indicators.

Three factors: host, geographic location and subtype were analysed using discrete trait models in BEAST [44]. Specifically, host population was discretized in terms of bird orders and bird species, while geographic location was discretized in terms of states and migration flyway routes in North America. The hosts of AIV in our study were first categorized into five different bird orders: Anseriformes (ans), Charadriiformes (cha) and Galliformes (gal), Gruiformes (gru) and Passeriformes (pas); The majority of our AIV data set were collected from Anseriformes (366/427), which were further classified as five major species groups: mallard (Anas platyrhynchos), northern pintail (Anasacuta), northern shoveller (Anas clypeata), blue-winged teal (Anas discors), green-winged teal (Anas carolinensis) and other Anseriformes. Each AIV sequence was assigned a discrete geographical state according to its province / state of isolation (in Table S9). In total 26 US states and Canadian provinces were considered in our study. In addition, each state and province was categorised into a specific North American flyway: the Atlantic, Mississippi, Central, or Pacific flyway, following a previous study [17]. Viral sequences were also labelled according to their host species for the analysis of the species contribution to AIV gene flow. The distributions of sequences for each discrete trait (host order, host species, flyway and location) are summarized in Table S1, S2, S3, S4.

These traits were jointly analysed across all eight segments using irreversible substitution models and strict clocks in the discrete trait models. An exponential distribution with mean equal to 1.0 was chosen for the discrete trait clock rate prior in order to favour smaller numbers of transitions across the phylogeny of each segment (especially PB2, M and NS). The parameter values in each trait model were examined using Tracer v1.6. Significant transition rate estimates between discrete traits were calculated using the same methods as in [5] and plotted as a network in Cytoscape v2.8.0 (http://www.cytoscape.org/).

Flyway-restricted transmission

We further modified the discrete transition model specification in the xml configuration files to reflect the hypothesis the gene flow of AIV carried by migration birds are restricted between flyways but can mix in Mexico (and further south). By default, all possible transitions between states are permitted, but we modified the settings to disallow transitions between locations on different flyways (Table S10). The output from this restricted model can then be compared to the original model. The corresponding indicators for BF thresholds were recalculated based on the reduced number of transitions. Eight discrete trait models were further compared based on 3 settings: 1) with BSSVS or without BSSVS implemented; 2) Asymmetric or symmetric transitions model; 3) with reduced or original matrix (The names of the models are: 1. Reduce_BSSVS_asym; 2. Reduce_nonBSSVS_asym; 3. Reduce_BSSVS_sym; 4. Reduce_nonBSSVS_sym; 5. Original_BSSVS_asym; 6. Original_nonBSSVS_asym; 7. Original_BSSVS_sym; 8. Original_nonBSSVS_sym).

The location trait initially had 26 states, resulting in 325 possible transition rates, however to avoid over-parameterising the model, we reduced the number of transitions in the matrix to 98 to increase the power of the analysis. To demonstrate that it was not the case that the reduced flyway matrix performed best only because of the lower number of parameters, we performed a randomization test with the same number of parameters: we randomly switched off (meaning the indicator of a transition pair was forced from 1 to 0, ensuring that it was not included in the model during sampling) the same number of transitions as were disallowed in the between flyway states matrix. We did this for 10 replicates. We then compared the performance to that of the reduced flyway matrix and we found the latter has the best performance, indicating the model with reduced flyway transition is the best model to explain the spatial transmission with location state.

All models were compared using a posterior simulation-based analogue of Akaike's information reiteration (AICM), by measuring AIC from the posterior of each model in a Bayesian Monte Carlo context, with score >10 as a strong evidence in favour of one model over the others [45,46]. The comparisons were conducted in Tracer V 1.6 using a further subsampled log file of each model (>1000 states after burn-in). A recently developed model comparison method, estimating marginal likelihoods using stepping stone (SS) sampling [46,47] was also used to compare the discrete trait models based in order to verify the AICM results. However comparison by SS marginal likelihood estimation was only performed on discrete trait models inferred on the HA segment since the SS method is not yet implemented for the HPM framework. Similar to the joint analysis results by AICM, the SS results also showed the symmetric discrete trait model with reduced matrix in BSSVS is preferred over all the other models with average improvement in marginal likelihood = 10, and the model with between flyway diffusion pairs switched off was favoured over a randomly reduced matrix with an average improvement in marginal likelihood of 52.

The original tree files from this study are available from the Dryad Digital Repository: doi:10.5061/dryad.j5bf8.

Including:

1) Maximum likelihood (ML) trees of six internal segments PB2, PB1, PA, NP, M and NS for the same 2343 North America AIV strains reconstructed using RAxML v7.04. In each ML tree, the Mexico H7N3 strains are colored in red, and the lineage in which H7N3 Mexico fell is colored in blue;

2) Temporally structured maximum clade credibility (mcc) time-scaled phylogenetic trees of all eight segments were generated using Beast V 1.7.3 and annotated with ancestral state (A: host order; B: host species; C: flyway; D: state/province; E: subtype) recovered from the discrete trait analyses. All the tree files can be visualized in FigTree. V 1.4.0.

Supporting Information

Figure S1   Maximum likelihood (ML) phylogenies of the six internal segments of North American AIV. Phylogenetic trees were generated with the same 2343 strains in each of six internal segments. In the PB2 segment, 86 AIV sequences in the same clade as the H7N3 Mexico strains are labelled in red; these strains are widely scattered in the phylogenetic trees of the other internal segments. The distribution of the colored lineages among different segments can be visualized in the original tree files in Dryad mentioned above.

Figure S2   Maximum clade credibility (MCC) phylogenies for the NA segment. Temporally structured maximum clade credibility (mcc) time-scaled phylogenetic tree showing the evolution of NA gene of avian influenza A virus isolated from North American wild birds for each individual gene dataset. Ancestral state (A: host order; B: host species; location; C: flyway; D: state) changes recovered from the discrete trait analyses are indicated by color changes at tree nodes. Mexican outbreak strains are highlighted with pink. A: Host order. Five host orders are

labelled on HA tree: wild birds of the order Anseriformes (ans-wild); wild birds of the order Charadriiformes (cha-wild); wild birds of the order Passeriformes (pas-wild); domestic birds of the order Galliformes and Mexico H7N3 outbreak in the order Galliformes (gal-domestic-Mexico). B: Host species. Wild Anseriformes are classified into the five main species and a group comprising the other rarer species of Anseriformes in this study: mallard (Anas platyrhynchos), northern pintail (Anasacuta), northern shoveller, blue-winged teal, green-winged teal and other Anseriformes (other ans); The order Galliformes are shown as "outbreak" (the H7N3 Mexico outbreak) and "other_gal"; The other orders are shown as: Charadriiformes (cha) and Galliformes (gal), Gruiformes (gru) and Passeriformes (pas). C: Flyway. Four specific North American flyways are labelled on HA tree: the Atlantic (Eastern), Mississippi, Central, and Pacific (Western). D: State. 22 states and provinces of the viral sample locations are labelled on HA tree. The same color code also applies to Figures S3, S4, S5, S6, S7, S8.

Figure S3   MCC phylogenies for the PB2 segment.

Figure S4   MCC phylogenies for the PB1 segment.

Figure S5   MCC phylogenies for the PA segment.

Figure S6   MCC phylogenies for the NP segment.

Figure S7   MCC phylogenies for the M segment.

Figure S8   MCC phylogenies for the NS segment.

Figure S9   Spatial diffusion of AIV segments of the Mexico outbreak. Panels represent eight segments separately (A: HA, B: NA, C: PB2, D: PB1, E: PA, F: NP, G: M, H: NS) and I represents the spatial transmission of all 8 segments jointly. The Map source for this figure was OpenStreetMap (http://www.openstreetmap.org/). Plotted lines are the branches of the MCC trees of different segments, distinguished by color; the size of each circle represents the number of lineages with that location state.

Figure S10   North America states and provinces categorized according to flyways. Provinces and states which belong to Pacific flyways are colored in purple; Central flyways are colored in red; Mississippi flyways are colored in blue; Atlantic flyways are colored in yellow; and Mexico is colored in green. Those colors on the map are the same as the colored branches in the flyway discrete phylogenetic tree in Figure 2 C.

Figure S11   Virulence evolving on the MCC phylogeny of the HA segment. HA cleavage site insertions are classified as follows: 1) Insertion: which are multibasic insertions (insertion) which only being found in HP AIV, including HP H7N3 outbreaks strains in Mexico 2012 and in Canada 2004 and 2007; 2) Partial insertion: LP AIV which has a partial insertion of 2 amino acids; 3) No insertion: other LP AIV with no insertion in the cleavage site. (A): Pathogenicity changes (according to high pathogenicity (HP) and low pathogenicity (LP)) are indicated by color changes in the tree; Cleavage site changes (with insertion, partial insertions and no insertions (C) are indicated by color changes at tree nodes.

Table S1 Host orders distribution of 427 AIV sequences.

Table S2 Host species (Anseriformes only) distribution of 366 AIV sequences.

Table S3 Location (state/province) distribution of 427 AIV sequences.

Table S4 Flyway distribution of 427 AIV sequences.

Table S5 Transmission rates of host orders and Bayes Factor support.

Table S6 Transmission rates of host species and Bayes Factor support.

Table S7 Transmission rates of flyway and Bayes Factor support.

Table S8 Transmission rates of location (state/province) and the Bayes Factor support.

Table S9 AIV sequences being used in this study. The strain name and subtype, host, location, flyway and the segments being selected in the AIV dataset. Details of the 427 North American AIV sequences used in this analysis. Strain name, subtype, host orders, isolated state/provinces and the correspondent flyways are listed.

Table S10 Reduced indicator matrix with between-flyway states turned off. The indicator matrix of non-reversible BSSVS model contains 325 transition pairs between states, which are all initially switched on (indicator = 1). In the reduced matrix, pairs of states belonging to different flyways are switched are switch off (indicator = 0). There are 98 non-reversible transition pairs in the new indicator matrix.

Author Contributions

Conceived and designed the experiments: LL SJL ALB. Analyzed the data: LL. Contributed to the writing of the manuscript: LL SJL ALB.

References

1. Schnebel B, Dierschke V, Rautenschlein S, Ryll M, Neumann U (2007) Investigations on infection status with H5 and H7 avian influenza virus in short-distance and long-distance migrant birds in 2001. Avian Dis 51: 432–433.
2. Webster RG (2002) The importance of animal influenza for human disease. Vaccine 20 Suppl 2: S16–20.
3. Chen J, Lee KH, Steinhauer DA, Stevens DJ, Skehel JJ, et al. (1998) Structure of the hemagglutinin precursor cleavage site, a determinant of influenza pathogenicity and the origin of the labile conformation. Cell 95: 409–417.
4. Stienekegrober A, Vey M, Angliker H, Shaw E, Thomas G, et al. (1992) Influenza-Virus Hemagglutinin with Multibasic Cleavage Site Is Activated by Furin, a Subtilisin-Like Endoprotease. Embo Journal 11: 2407–2414.
5. Lu L, Lycett SJ, Leigh Brown AJ (2014) Reassortment patterns of avian influenza virus internal segments among different subtypes. BMC Evol Biol 14: 16.
6. Duan L, Campitelli L, Fan XH, Leung YH, Vijaykrishna D, et al. (2007) Characterization of low-pathogenic H5 subtype influenza viruses from Eurasia: implications for the origin of highly pathogenic H5N1 viruses. J Virol 81: 7529–7539.
7. Campitelli L, Mogavero E, De Marco MA, Delogu M, Puzelli S, et al. (2004) Interspecies transmission of an H7N3 influenza virus from wild birds to intensively reared domestic poultry in Italy. Virology 323: 24–36.
8. Tweed SA, Skowronski DM, David ST, Larder A, Petric M, et al. (2004) Human illness from avian influenza H7N3, British Columbia. Emerg Infect Dis 10: 2196–2199.
9. Suarez DL, Senne DA, Banks J, Brown IH, Essen SC, et al. (2004) Recombination resulting in virulence shift in avian influenza outbreak, Chile. Emerg Infect Dis 10: 693–699.
10. Hirst M, Astell CR, Griffith M, Coughlin SM, Moksa M, et al. (2004) Novel avian influenza H7N3 strain outbreak, British Columbia. Emerg Infect Dis 10: 2192–2195.
11. Berhane Y, Hisanaga T, Kehler H, Neufeld J, Manning L, et al. (2009) Highly pathogenic avian influenza virus A (H7N3) in domestic poultry, Saskatchewan, Canada, 2007. Emerg Infect Dis 15: 1492–1495.
12. Sherrilyn Wainwright CT, Filip Claes, Moisés VargasTerán, Vincent Martin, Juan Lubroth (2012) Highly Pathogenic Avian Influenza in Mexico (H7N3) - A significant threat to poultry production not to be underestimated. EMPRES WATCH 26: 1–9.
13. Lopez-Martinez I, Balish A, Barrera-Badillo G, Jones J, Nunez-Garcia TE, et al. (2013) Highly pathogenic avian influenza A(H7N3) virus in poultry workers, Mexico, 2012. Emerg Infect Dis 19: 1531–1534.
14. Horimoto T, Rivera E, Pearson J, Senne D, Krauss S, et al. (1995) Origin and molecular changes associated with emergence of a highly pathogenic H5N2 influenza virus in Mexico. Virology 213: 223–230.
15. Pasick J, Berhane Y, Hooper-McGrevy K (2009) Avian influenza: the Canadian experience. Rev Sci Tech 28: 349–358.
16. Wilke CO, Nelson MI, Lemey P, Tan Y, Vincent A, et al. (2011) Spatial Dynamics of Human-Origin H1 Influenza A Virus in North American Swine. PLoS Pathogens 7: e1002077.

17. Lam TT, Ip HS, Ghedin E, Wentworth DE, Halpin RA, et al. (2012) Migratory flyway and geographical distance are barriers to the gene flow of influenza virus among North American birds. Ecol Lett 15: 24–33.
18. Steinhauer DA (1999) Role of hemagglutinin cleavage for the pathogenicity of influenza virus. Virology 258: 1–20.
19. Maurer-Stroh S, Lee RT, Gunalan V, Eisenhaber F (2013) The highly pathogenic H7N3 avian influenza strain from July 2012 in Mexico acquired an extended cleavage site through recombination with host 28S rRNA. Virol J 10: 139.
20. Kapczynski DR, Pantin-Jackwood M, Guzman SG, Ricardez Y, Spackman E, et al. (2013) Characterization of the 2012 Highly Pathogenic Avian Influenza H7N3 Virus Isolated from Poultry in an Outbreak in Mexico: Pathobiology and Vaccine Protection. Journal of Virology 87: 9086–9096.
21. Krauss S, Webster RG (2012) Predicting the next influenza virus. Science 337: 644.
22. Belser JA, Blixt O, Chen LM, Pappas C, Maines TR, et al. (2008) Contemporary North American influenza H7 viruses possess human receptor specificity: Implications for virus transmissibility. Proc Natl Acad Sci U S A 105: 7558–7563.
23. Lam TT, Wang J, Shen Y, Zhou B, Duan L, et al. (2013) The genesis and source of the H7N9 influenza viruses causing human infections in China. Nature 502: 241–244.
24. Krauss S, Obert CA, Franks J, Walker D, Jones K, et al. (2007) Influenza in migratory birds and evidence of limited intercontinental virus exchange. PLoS Pathog 3: e167.
25. Liu D, Shi W, Shi Y, Wang D, Xiao H, et al. (2013) Origin and diversity of novel avian influenza A H7N9 viruses causing human infection: phylogenetic, structural, and coalescent analyses. Lancet 381: 1926–1932.
26. Gunnarsson G, Latorre-Margalef N, Hobson KA, Van Wilgenburg SL, Elmberg J, et al. (2012) Disease dynamics and bird migration–linking mallards Anas platyrhynchos and subtype diversity of the influenza A virus in time and space. PLoS One 7: e35679.
27. Munster VJ, Baas C, Lexmond P, Waldenstrom J, Wallensten A, et al. (2007) Spatial, temporal, and species variation in prevalence of influenza A viruses in wild migratory birds. PLoS Pathog 3: e61.
28. Dugan VG, Dunham EJ, Jin G, Sheng ZM, Kaser E, et al. (2011) Phylogenetic analysis of low pathogenicity H5N1 and H7N3 influenza A virus isolates recovered from sentinel, free flying, wild mallards at one study site during 2006. Virology 417: 98–105.
29. Bahl J, Krauss S, Kuhnert D, Fourment M, Raven G, et al. (2013) Influenza a virus migration and persistence in North American wild birds. PLoS Pathog 9: e1003570.
30. Smith GJ (2005) Migratory Bird Pathways and the Gulf of Mexico USGS Fact Sheet 2005–3069.
31. Gonzalez-Reiche AS, Perez DR (2012) Where Do Avian Influenza Viruses Meet in the Americas? Avian Diseases 56: 1025–1033.
32. Smith GJD, Vijaykrishna D, Bahl J, Lycett SJ, Worobey M, et al. (2009) Origins and evolutionary genomics of the 2009 swine-origin H1N1 influenza A epidemic. Nature 459: 1122–1125.

33. Edgar RC (2004) MUSCLE: multiple sequence alignment with high accuracy and high throughput. Nucleic Acids Res 32: 1792–1797.

34. Stamatakis A, Ludwig T, Meier H (2005) RAxML-III: a fast program for maximum likelihood-based inference of large phylogenetic trees. Bioinformatics 21: 456–463.

35. Drummond AJ, Suchard MA, Xie D, Rambaut A (2012) Bayesian phylogenetics with BEAUti and the BEAST 1.7. Mol Biol Evol 29: 1969–1973.

36. Drummond AJ, Rambaut A (2007) BEAST: Bayesian evolutionary analysis by sampling trees. BMC Evol Biol 7: 214.

37. Lanave C, Preparata G, Saccone C, Serio G (1984) A New Method for Calculating Evolutionary Substitution Rates. Journal of Molecular Evolution 20: 86–93.

38. Shapiro B, Rambaut A, Drummond AJ (2006) Choosing appropriate substitution models for the phylogenetic analysis of protein-coding sequences. Mol Biol Evol 23: 7–9.

39. Bielejec F, Rambaut A, Suchard MA, Lemey P (2011) SPREAD: spatial phylogenetic reconstruction of evolutionary dynamics. Bioinformatics 27: 2910–2912.

40. Edo-Matas D, Lemey P, Tom JA, Serna-Bolea C, van den Blink AE, et al. (2011) Impact of CCR5delta32 host genetic background and disease progression on HIV-1 intrahost evolutionary processes: efficient hypothesis testing through hierarchical phylogenetic models. Mol Biol Evol 28: 1605–1616.

41. Suchard MA, Kitchen CMR, Sinsheimer JS, Weiss RE (2003) Hierarchical phylogenetic models for analyzing multipartite sequence data. Systematic Biology 52: 649–664.

42. Cybis GB, Sinsheimer JS, Lemey P, Suchard MA (2013) Graph hierarchies for phylogeography. Philos Trans R Soc Lond B Biol Sci 368: 20120206.

43. Lemey P, Rambaut A, Drummond AJ, Suchard MA (2009) Bayesian phylogeography finds its roots. PLoS Comput Biol 5: e1000520.

44. Goldberg EE, Igic B (2008) On phylogenetic tests of irreversible evolution. Evolution 62: 2727–2741.

45. Kitakado T, Kitada S, Kishino H, Skaug HJ (2006) An integrated-likelihood method for estimating genetic differentiation between populations. Genetics 173: 2073–2082.

46. Baele G, Lemey P, Bedford T, Rambaut A, Suchard MA, et al. (2012) Improving the Accuracy of Demographic and Molecular Clock Model Comparison While Accommodating Phylogenetic Uncertainty. Molecular Biology and Evolution 29: 2157–2167.

47. Baele G, Li WL, Drummond AJ, Suchard MA, Lemey P (2013) Accurate model selection of relaxed molecular clocks in bayesian phylogenetics. Mol Biol Evol 30: 239–243.

19

Latitudinal Variation of a Defensive Symbiosis in the *Bugula neritina* (Bryozoa) Sibling Species Complex

Jonathan Linneman[1], Darcy Paulus[2], Grace Lim-Fong[2], Nicole B. Lopanik[1]*

1 Department of Biology, Georgia State University, Atlanta, Georgia, United States of America, **2** Department of Biology, Randolph-Macon College, Ashland, Virginia, United States of America

Abstract

Mutualistic relationships are beneficial for both partners and are often studied within a single environment. However, when the range of the partners is large, geographical differences in selective pressure may shift the relationship outcome from positive to negative. The marine bryozoan *Bugula neritina* is a colonial invertebrate common in temperate waters worldwide. It is the source of bioactive polyketide metabolites, the bryostatins. Evidence suggests that an uncultured vertically transmitted symbiont, "*Candidatus* Endobugula sertula", hosted by *B. neritina* produces the bryostatins, which protect the vulnerable larvae from predation. Studies of *B. neritina* along the North American Atlantic coast revealed a complex of two morphologically similar sibling species separated by an apparent biogeographic barrier: the Type S sibling species was found below Cape Hatteras, North Carolina, while Type N was found above. Interestingly, the Type N colonies lack "*Ca.* Endobugula sertula" and, subsequently, defensive bryostatins; their documented northern distribution was consistent with traditional biogeographical paradigms of latitudinal variation in predation pressure. Upon further sampling of *B. neritina* populations, we found that both host types occur in wider distribution, with Type N colonies living south of Cape Hatteras, and Type S to the north. Distribution of the symbiont, however, was not restricted to Type S hosts. Genetic and microscopic evidence demonstrates the presence of the symbiont in some Type N colonies, and larvae from these colonies are endowed with defensive bryostatins and contain "*Ca.* Endobugula sertula". Molecular analysis of the symbiont from Type N colonies suggests an evolutionarily recent acquisition, which is remarkable for a symbiont thought to be transmitted only vertically. Furthermore, most Type S colonies found at higher latitudes lack the symbiont, suggesting that this host-symbiont relationship is more flexible than previously thought. Our data suggest that the symbiont, but not the host, is restricted by biogeographical boundaries.

Editor: Arga Chandrashekar Anil, CSIR- National institute of oceanography, India

Funding: This work was funded in part by the Georgia State University Research Foundation and the Randolph-Macon College Chenery Grant. The funders had no role in study design, data collection and analysis, decision to publish, or preparation of the manuscript.

Competing Interests: The authors have declared that no competing interests exist.

* Email: nlopanik@gsu.edu

Introduction

The biogeographical clines and boundaries that define the geographical distribution of an organism have been studied for many metazoans [1–3]. The importance of these limits to the distribution of symbionts of those metazoans and the partners together (the holobiont), however, has rarely been thoroughly explored. While many symbiotic relationships are considered mutualisms, in which both partners benefit, there can be physiological costs associated with hosting a symbiont [4]. Beneficial defensive symbionts can facilitate the survival of the host against enemies such as pathogens, competitors, or predators [5]. Often, these enemies are ephemeral, or their import varies among populations. In the absence of the selective pressure, the costs of hosting a symbiont may eclipse the potential benefits of a partnership, resulting in an unpartnered host with greater fitness than a partnered host [6–8]. Selection over time would then ultimately result in symbiont loss. Defensive symbionts that are vertically transmitted are thought to represent a significant parental investment. This implies both that the symbionts are

beneficial to the host, and that the host and symbiont have developed mechanisms to prevent loss of the symbiont; therefore, symbiont loss represents an irreversible evolutionary milestone. The increase in aposymbiotic host frequency in the absence of the enemy would then suggest that the symbiont imposes a cost on its host (reviewed in [9]). The interplay of symbiont cost-benefit and biogeography is likely important to host/symbiont partners with widespread distribution, but is not well understood.

The marine bryozoan genus *Bugula* provides an interesting platform for the exploration of the importance of geographic variation to symbiotic interactions. Among commonly studied *Bugula* species, four are known to harbor bacterial symbionts seemingly absent from at least two congeners [10–13]. These closely related symbionts do not appear to have a history of strict cospeciation with their hosts, although vertical transmission is suspected due to their presence in the larvae and larval brood chambers, or ovicells, of colonies [14]. This suspected vertical transmission has also led to a high degree of congruence between the phylogenetic topologies of host and symbiont [11]. Within the genus, the cryptic species complex of the cosmopolitan bryozoan

Bugula neritina allows for even more direct exploration of the ties between symbiosis and host ecology. *Bugula neritina* is known to harbor a γ-Proteobacterial symbiont that is, to date, uncultivated [10,15]. "*Candidatus* Endobugula sertula" is found in all life stages of the host: in larvae, it is found in a circular groove located along the aboral pole called the pallial sinus [10,15]; in adult colonies, it is found in funicular cords, which serve as a vascular system, transporting nutrients and wastes throughout the colony [16] and to a developing embryo in an ovicell [14].

Evidence from antibiotic-curing experiments suggests that "*Ca.* Endobugula sertula" is the source of the bryostatins [17,18], complex bioactive polyketides with medical potential (reviewed in [19]). Feeding assays with extracts of *B. neritina* and purified bryostatins revealed that these symbiotically produced compounds are unpalatable and may serve to protect the vulnerable host larvae from predators, suggesting an association in which bryostatin presence may drive maintenance of symbiosis over generations. Both *B. neritina* larvae and larval extracts are unpalatable to co-occurring invertebrate and vertebrate predators including oysters [20], corals, sea anemones, filefish, and the pinfish, *Lagodon rhomboides* [18,21–23]. Larvae have also shown a marked ability to survive and metamorphose after rejection by predators, with>90% success after rejection by the coral *Oculina arbuscula* [21] and 100% metamorphosing after attacks by *L. rhomboides* and the filefish *Monocanthus ciliatus* [23]. Importantly, the defensive bryostatins appear to be most concentrated in the larval stage of the *B. neritina* life cycle [18,22,23], suggesting that alternative protective means such as structural defenses may be more crucial to adult colonies. The *B. neritina*-"*Ca.* Endobugula sertula" association is one of the most well-characterized examples of a marine system involving defensive natural products that are produced by a microbial symbiont [5]. At least one terrestrial analog has been identified, in which the polyketide pederin, produced by a bacterial endosymbiont, prevents predation on the larval stage of the rove beetle genus *Paederus* [24,25].

In the case of *B. neritina*, while the presence of defensive metabolites suggested that the association with "*Ca.* Endobugula sertula" may provide a distinct advantage for larval survival, some populations of *B. neritina* were shown to lack these endosymbionts and, consequently, the suite of bryostatins associated with them [18,26]. Further genetic characterization over the past several years has identified three cryptic sibling species of *B. neritina* based upon the sequence of the mitochondrial cytochrome *c* oxidase I (COI) gene: (1) Type S ("Shallow") found in temperate environments worldwide, (2) a Type D ("Deep") variety found only on the Eastern Pacific coast, and so named due to its occurring at depths 9 m or greater in southern California, and (3) Type N ("Northern") found only on the Western Atlantic coast, initially in Delaware and Connecticut [12,26,27]. These so-called "sibling species" have recently been proposed as distinct biological species [28].

Types S and D *B. neritina* house closely related bacterial symbionts which differ at just four nucleotide sites (0.4% of the 996-bp shared sequenced region) in the small subunit ribosomal RNA gene (16S rRNA). Interestingly, bryostatin composition has been shown to vary between the Type S and D populations, which could be attributed to either host or symbiont [12]. The *bry* gene cluster, which putatively prescribes bryostatin biosynthesis, shows minimal variation between the sibling species, demonstrating 98% identity and differing most significantly in the genomic location of several accessory genes [29]. Perhaps most notably, Type N populations initially gave no evidence of endosymbionts or of bryostatin production. Although extracts of Type N *B. neritina* were shown to be significantly unpalatable to a predator from

North Carolina, decreasing palatability by nearly 40%, these Type N samples were not as deterrent as extracts with bryostatins, which reduced feeding by 80% [18]. Because of this pattern of palatability and of sibling species location, researchers proposed a biogeographic division on the North American east coast between the ranges of Type N and S *B. neritina*, with the traditional boundary Cape Hatteras [30–32] initially speculated as a point of transition. In this separation of sibling species along the Western Atlantic coast, survival of Type S *B. neritina* in lower-latitude waters would be enabled by maintenance of the symbiosis and associated defensive metabolites in the presence of higher predation pressure [33]. Its spread northward, meanwhile, is limited by environmental tolerance limits of the Type S bryozoan or its symbiont [22,26].

In 2010, we began assessing the genetic and chemical composition of *B. neritina* populations along the United States' east coast with the goal of uncovering differences in local bryostatin composition and symbiont identity. We discovered that *B. neritina* sibling species distribution on both sides of Cape Hatteras was inconsistent with previous reports. Type N *B. neritina* were found in two populations in South Carolina, previously thought to be exclusively Type S, while Type S colonies were located as far north as Maryland. Furthermore, the previously assumed aposymbiotic Type N *B. neritina* was associated with a bacterial symbiont at low latitudes, while Type S colonies north of Cape Hatteras surprisingly lacked this association. These initial findings led us to examine *B. neritina* populations along the Atlantic coast more closely in order to better understand the geographic and phylogenetic relationship of Type N and Type S, as well as to correlate varying symbiotic and bryostatin status with the distribution of both sibling species.

Materials and Methods

Sampling of coastal *B. neritina* populations and extraction of DNA

Adult *B. neritina* colonies were collected from floating docks in coastal locations ranging from Indian River, DE, to St. Augustine, FL (sites listed in Table 1), from 2010–2013. No specific permits were required for the described field studies. The field studies did not involve endangered or protected species. Docks that were used for collections were either private or public marinas. Permissions were sought from the owners or dock masters prior to collecting. All samples were pulled from docks by hand and thus represent animals found close to the water's surface. In many cases, colonies were chosen haphazardly. In some instances, the Type N or S phylotype was targeted for specific investigation, a direction made possible by slight morphological differences observable within multiple mixed *B. neritina* populations (see Fig. 1 for visual representation). A small number of samples were also gathered along the coast of California, at a depth of ~8 m near Catalina Island by SCUBA (GPS coordinates: latitude 33.346°, longitude − 118.331°) and from floating docks in Bodega Bay (GPS coordinates: latitude 38.334°, longitude −123.049°).

For samples subjected to DNA analysis, care was taken after collection to ensure that the gathered zooids represent a single colony rather than multiple associated individuals. Distal zooids were cut from several branches in order to avoid ovicells in more proximal reproductive zooids. In several cases, larvae were also collected as they were released by the colonies in subsequent days (variously for both individual colonies and pooled adult samples). When colonies were not immediately subjected to DNA extraction, zooids were cut and stored in RNALater RNA Stabilization Reagent (Qiagen, Valencia, CA) or 70–100% ethanol at −20°C.

Table 1. *Bugula neritina* symbiont frequency along North American Atlantic coast.

Location	Symbiont occurrence				GPS Coordinates	
	Total Type N		Total Type S			
	Sym+	Sym−	Sym+	Sym−	Latitude (°)	Longitude (°)
Indian River, DE	0	31	-	-	38.612	−75.072
Ocean Pines, MD	0	22	0	2	38.386	−75.130
Chincoteague, VA	0	8	-	-	37.920	−75.405
Wachapreague, VA	0	9	-	-	37.604	−75.687
Oyster, VA	0	60	10	18	37.289	−75.923
Rudee Inlet, VA	0	19	-	-	36.832	−75.976
Beaufort, NC	-	-	49	1	34.717	−76.666
Radio Island, NC	7	4	35	0	34.714	−76.680
Morehead City, NC	3	0	32	0	34.720	−76.707
Salter Path, NC	9	4	0	1	34.690	−76.887
Murrell's Inlet, SC	-	-	11	0	33.557	−79.031
Isle of Palms, SC	-	-	22	0	32.823	−79.730
Beaufort, SC	13	0	22	0	32.466	−80.666
Hunting Island, SC	12	0	2	0	32.346	−80.469
Tybee Island, GA	-	-	12	2	31.991	−80.852
Jacksonville, FL	-	-	10	0	30.299	−81.413
St. Augustine, FL	-	-	10	0	29.946	−81.307

Symbiont occurrence was determined both for colonies collected haphazardly and in targeted sampling (see Materials and Methods); not all haphazardly selected colonies were assayed for symbiont presence. Dashes indicate host phylotypes not found in a location during sampling.

Similarly, larvae not immediately processed through DNA or chemical extraction were stored in RNALater or 100% methanol for later analysis, or were fixed for fluorescence *in situ* hybridization (FISH; see below). For both adult and larval samples, DNA was extracted using either ZR Fungal/Bacterial DNA MiniPrep or MicroPrep (Zymo Research, Irvine, CA), or DNEasy Blood and Tissue Kit (Qiagen), and following the manufacturer's instructions. Quality and quantity were assessed by spectroscopy (NanoDrop 1000, Thermo Fisher Scientific, Wilmington, DE).

Molecular characterization of samples

Polymerase chain reaction (PCR) amplification of the mitochondrial COI gene was carried out using the universal invertebrate primers LCO1490 and HCO2198 [34] or the *B. neritina*-specific primer pair BnCOIf and BnCOIr (Table 2). Restriction fragment length polymorphism (RFLP) analysis was performed on amplicons using the restriction endonucleases *Dde*I, which only digests Type S COI products, or *Hha*I, which cuts amplicons from Type N individuals. In order to identify host phylotypes not documented amongst the individual adult colonies collected, PCR was also performed on DNA from the pooled larvae of a large number of colonies (>200) using primers targeting polymorphisms in the Type N and S COI genes (Table 2). The presence of the symbiont "*Ca*. Endobugula sertula" was assessed in DNA by PCR targeting its 16S ribosomal RNA gene (primers EBn16S_254f and EBn16S_643r, Bn240f and Bn1253r [15]), or the bryostatin biosynthetic gene *bryS* [29] (primers BryS_576f and BryS_774r). PCR amplicons and restriction digested products were visualized after agarose gel electrophoresis.

For selected samples, the COI or 16S PCR amplicon was purified using the GeneJET PCR Purification Kit (Thermo Scientific, Pittsburgh, PA) and sequenced at the Georgia State

Figure 1. Co-occurrence of Type N and Type S *B. neritina*. Colonies collected together in (A) Beaufort, NC, and (B) Oyster, VA. Blue arrows indicate Type N colonies; red indicate Type S.

Table 2. Primers and probes used in this study.

Name	Sequence	Target	Source
LCO1490	GGTCAACAAATCATAAAGATATTGG	Invertebrate COI	[34]
HCO2198	TAAACTTCAGGGTGACCAAAAAATCA	Invertebrate COI	[34]
BnCOIf	ACAGCTCATGCATTTTTA	*B. neritina* COI	This study
BnCOIr	CATTACGATCGGTTAGTAG	*B. neritina* COI	This study
Bn_COI_N_129f	CACCGGTAGAGATAAAAGTAAT	Type N *B. neritina* COI	This study
Bn_COI_N_615r	CGAATTAAGACAACCTGGTAGT	Type N *B. neritina* COI	This study
Bn_COI_S1_129f	CACTGGTAAAGATAAAAGTAAC	Type S *B. neritina* COI	This study
Bn_COI_S1_615r	AGAATTAAGACAACCAGGCAGC	Type S *B. neritina* COI	This study
Bn240f	TGCTATTTGATGAGCCCGCGTT	"*Ca.* Endobugula sertula" 16S	[15]
Bn1253r	CATCGCTGCTTCGCAACCC	"*Ca.* Endobugula sertula" 16S	[15]
EBn16S_254f	TACTCGTTAACTGTGACGTTACTC	"*Ca.* Endobugula sertula" 16S	[57]
EBn16S_643r	ACGCCACTAAATCCTCAAGGAAC	"*Ca.* Endobugula sertula" 16S	[57]
1055F	ATGGCTGTCGTCAGCT	Eubacterial 16S	[58]
1492R	TACGGYTACCTTGTTACGACTT	Eubacterial 16S	[59]
EBn16S_621f	CCTTAGAGTTCCCAGCCAAAC	"*Ca.* Endobugula sertula" 16S (for ITS sequencing)	This study
ITS-23S-r2	TSTGRDGCCAAGGCATCCA	Eubacterial 23S (for ITS sequencing)	After [60]
BryS_576f	CATTGACAGTCAGTTCTTCATTGA	*bryS*	This study
BryS_774r	CTTTTCCAGATTGAGTTTTTAACCA	*bryS*	This study
Eub338	GCTGCCTCCCGTAGGAGT	Eubacterial 16S	[35]

University Core Facility. An additional PCR was performed for a number of samples using the primers EBn16S_621f and ITS-23S-r2 to amplify a contiguous fragment of the ribosomal RNA gene containing the 3′ end of the 16S rRNA gene (for identity confirmation) and the complete internal transcribed spacer (ITS) region leading to the 5′ end of the 23S rRNA gene. These products were purified with the GeneJET kit and sequenced at Yale University's DNA Analysis Facility (New Haven, CT). Additionally, clone libraries of 16S rRNA gene amplicons generated using the universal primers 1055f and 1492r were constructed for four adult *B. neritina* colonies using the Invitrogen TOPO TA Cloning Kit for Sequencing (Life Technologies, Carlsbad, CA), with 164 clones sequenced at Virginia Commonwealth University's Nucleic Acid Research Facility. Colonies included two Type S from Oyster, VA, determined to be aposymbiotic by PCR analysis, one symbiotic Type S from Oyster, and one symbiotic Type N from Salter Path, NC. Clone identities were determined by BLAST searches, and the proportions of "*Ca.* Endobugula sertula" clones obtained from each sample were analyzed using Fisher's Exact Test.

Chemical extraction and analysis of *B. neritina* larvae

Crude extracts were obtained from larvae stored in methanol using a six-step process of exhaustive extraction, with the first and last steps using 100% methanol, and the remaining four using an approximately 1:1 mixture of methanol and dichloromethane as in [18]. Solvents were removed by rotary evaporation and extracts were dissolved in a 6:3:1 mixture of methanol:dichlormethane:water according to the number of larvae collected, based upon counts conducted at the time of collection or immediately after extraction. Extracts were analyzed via high-performance liquid chromatography (HPLC, with photodiode array detection, Shimadzu, Columbia, MD) eluted on a C18 analytical column

(250×4.6 mm Gemini 5 μm, Phenomenex, Torrance, CA) with a gradient time program using water and acetonitrile as solvents. Chromatogram peaks were examined for maximum absorbance at or near 229 nm, a wavelength previously demonstrated to be diagnostic for bryostatins found along the Atlantic coast of the United States [18].

Fluorescence *in situ* hybridization (FISH)

Larvae designated for FISH were fixed for 1 h in 4% paraformaldehyde in MOPS-NaCl buffer, then stored in 70% ethanol at −20°C. Hybridization of larvae followed the protocol previously described [13] using a Cy3-labeled universal eubacterial probe (Eub338, [35]) on a subset of larvae, and a Cy3-labeled "*Ca.* Endobugula sertula"-specific probe (Bn1253r) for the rest of the larvae (Table 2). Briefly, to prepare for hybridization, larvae were rinsed in phosphate buffered saline (PBS) and then transferred to hybridization buffer. Probes were added to obtain a final concentration of 5 ng/μl. After incubation at 46°C for 3–4 h, all samples were rinsed in wash buffer and incubated twice in wash buffer for 20 min at 48°C. Larvae were finally washed in PBS with 0.1% Tween-20 and stored at 4°C until observation. Samples were mounted with Vectashield (Vector Labs, Burlingame, CA) and then viewed using an Olympus FluoView FV1000 confocal microscope, with images composed of stacks of 60–80 optical sections 1 μm apart. All images were captured with identical laser intensity and gain settings.

Results

Distribution of Types N and S *B. neritina*

While *B. neritina* sibling species were largely confined to their initially proposed biogeographic ranges, haphazard colony collection revealed at least four locations in which the atypical species are found among the populations (Fig. 2). To the north of Cape Hatteras, two locations (Ocean Pines, MD, and Oyster, VA) included Type S animals. The Oyster location was extensively sampled (n = 100) and had the largest proportion of Type S animals (17%) among high-latitude locations. These docks were later targeted for Type S colonies, and they were easily found among the Type N animals using observable morphological differences (Fig. 1). Type S COI sequences obtained in this study were identical to the Southern clade of McGovern and Hellberg [26] over the 624 overlapping bases and showed 99.8% identity to the Type S COI sequence of Davidson and Haygood [12], differing by just 1 bp along 483 bases. Type N COI sequences were identical to that of the Northern clade reported in McGovern and Hellberg [26].

As expected, a majority of haphazardly collected *B. neritina* found south of Cape Hatteras were of the Type S sibling species, with only the sites in Beaufort, SC, and Salter Path, NC, including Type N animals. However, collection efforts targeting Type N adult colonies revealed their presence among Type S colonies at a number of other sites: Morehead City, NC; Radio Island, NC; and Hunting Island, SC (Fig. 3). In addition, sibling species-specific PCR using DNA from pooled larvae from a large number of adults (>200) as template revealed the presence of Type N animals as far south as the coast of Florida (Fig. 4).

Distribution of "*Ca.* Endobugula sertula"

A vast majority (94%) of colonies collected north of Cape Hatteras tested negative for the presence of "*Ca.* Endobugula sertula" (Fig. 3). This was especially true for Type N animals, for which no symbiotic colonies were found. In contrast, our collection revealed 95% of colonies to be symbiotic when located south of Cape Hatteras. Just four aposymbiotic Type S colonies were found south of Cape Hatteras (2%). Type N *B. neritina*, similarly, was largely symbiotic when south of the Cape (85%). This study thus reveals two previously undescribed host-symbiotic status pairings of *B. neritina*: symbiotic Type N and aposymbiotic Type S.

The existence of these groups was confirmed by FISH (Fig. 5), as molecular probes specific to "*Ca.* Endobugula sertula" as well as to eubacteria demonstrate the presence of the endosymbiont in the pallial sinus of both Type N and S *B. neritina* larvae which were designated symbiotic by PCR. Similarly, FISH probes failed to reveal the endosymbiont in Type S and Type N animals determined to be aposymbiotic by PCR. Presence of the endosymbiont in larvae collected solely from low-latitude Type N *B. neritina* was additionally demonstrated by PCR targeting the "*Ca.* Endobugula sertula" *bryS* gene, further confirming transmission of symbiont from adult colonies to the next generation (data not shown).

Analysis of 16S rRNA amplicons generated using eubacterial universal primers from adult *B. neritina* colonies (two aposymbiotic Type S, one symbiotic Type S, one symbiotic Type N) revealed

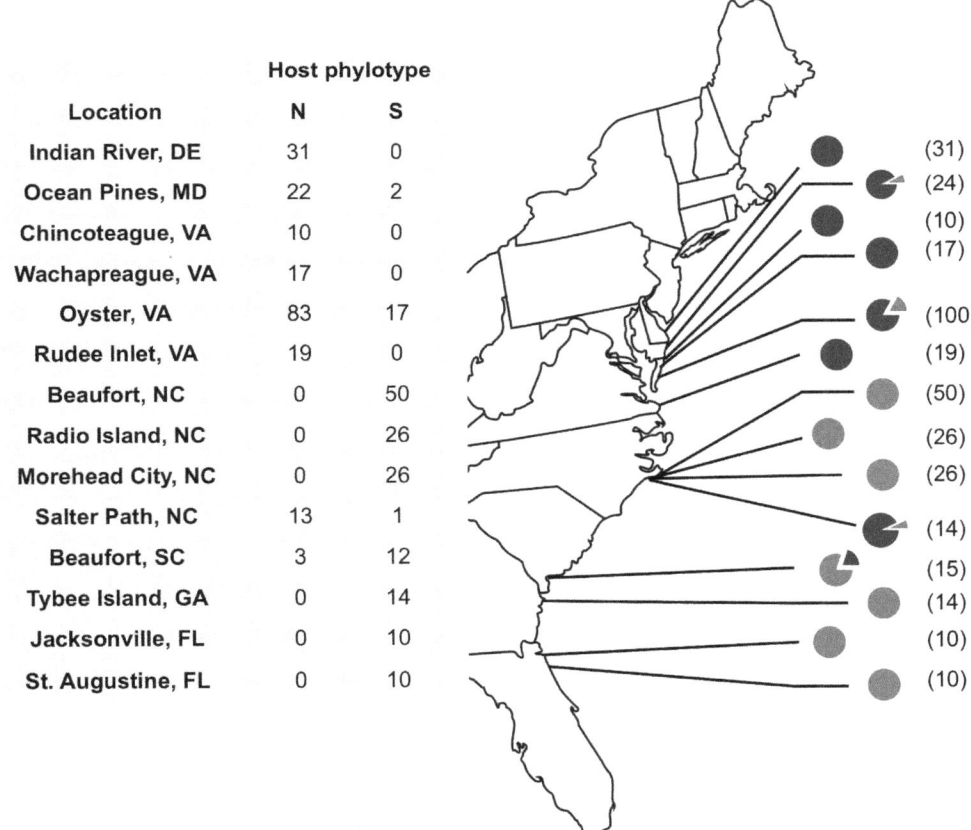

Location	Host phylotype N	Host phylotype S
Indian River, DE	31	0
Ocean Pines, MD	22	2
Chincoteague, VA	10	0
Wachapreague, VA	17	0
Oyster, VA	83	17
Rudee Inlet, VA	19	0
Beaufort, NC	0	50
Radio Island, NC	0	26
Morehead City, NC	0	26
Salter Path, NC	13	1
Beaufort, SC	3	12
Tybee Island, GA	0	14
Jacksonville, FL	0	10
St. Augustine, FL	0	10

Figure 2. *B. neritina* **sibling species collected by haphazard sampling.** Proportions at each site indicated by blue (Type N) and red (Type S) in charts.

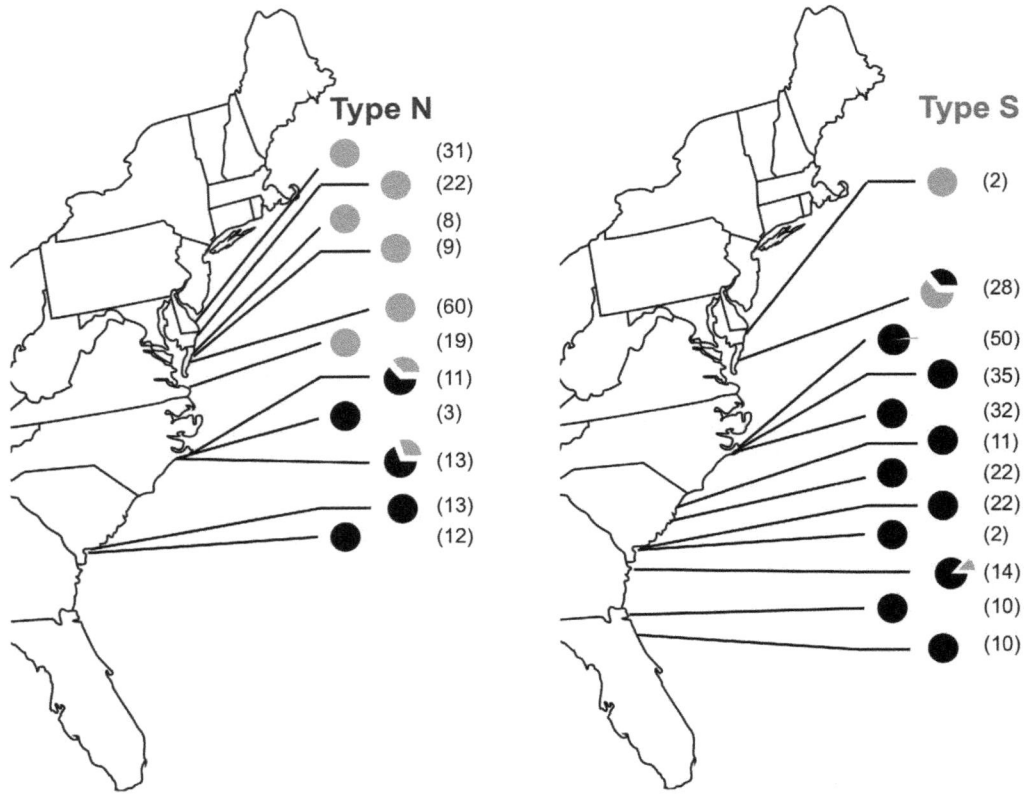

Figure 3. Symbiotic status of sampled colonies. Proportions at each site indicated by black (symbiotic) and gray (aposymbiotic) in charts. Results from both haphazardly collected and targeted samples are included. Number of colonies sampled shown beside each graph.

the presence of "*Ca.* Endobugula sertula" DNA among clones derived from symbiotic animals (8/85 clones), but none among those constructed using aposymbiotic animals (0/79). The proportion of clones matching the endosymbiont were significantly different for animals identified as symbiotic and aposymbiotic by the PCR assay ($p = 0.007$, two-tailed Fisher's Exact Test).

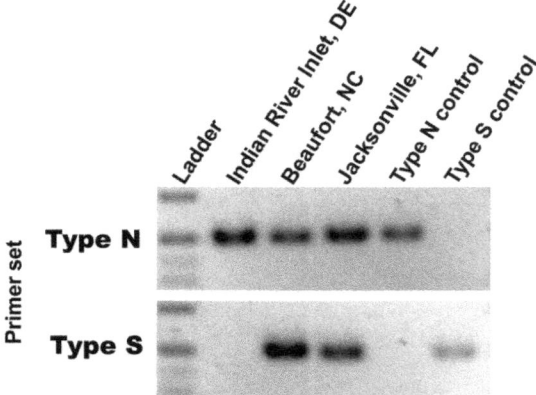

Figure 4. *Bugula neritina* **sibling species-specific PCR.** Reactions were performed with gDNA (10 ng) extracted from pooled larvae collected from >200 adult colonies as template. Control DNA was extracted from adult *B. neritina* colonies of both phylotypes to serve as 100% Type N or S cytochrome *c* oxidase templates.

Bryostatins associated with Type N "*Ca.* Endobugula sertula"

Analysis of bryostatin composition in Type N symbiotic larval extracts by HPLC at 229 nm illustrates a chemical profile that is very similar to that of pooled larvae from largely Type S populations (Fig. 6). Interestingly, larvae from aposymbiotic Type N *B. neritina* found alongside symbiont-protected Type S animals in North Carolina showed few peaks that appear to be bryostatins when analyzed in this manner. While the co-occurrence of these protected and unprotected colonies is somewhat surprising, the stark chromatographic contrast between symbiotic and aposymbiotic Type N animals highlights the relative similarity of symbiotic Type N larvae to Type S larvae from both Beaufort and Morehead City, NC.

Sequence analysis of Type N "*Ca.* Endobugula sertula"

Amplicons of the 16S rRNA gene from Type N symbionts collected from Radio Island, NC, Beaufort, SC, and Hunting Island, SC, had sequences identical to those of Type S symbionts from Hunting Island, SC, and Beaufort, SC. These sequences also matched the 16S rRNA gene sequence reported by both Davidson and Haygood [12] and McGovern and Hellberg [26] for Type S "*Ca.* Endobugula sertula". Sequences obtained using primers starting at position 621 in the 16S rRNA gene and ending in the 5' region of the 23S rRNA gene also matched previous 16S sequences. An ITS region of ~850 bp was amplified and demonstrated to be identical between the Type N and Type S symbionts examined. After verification of contiguity from the "*Ca.* Endobugula sertula" 16S sequence to its newly characterized ITS

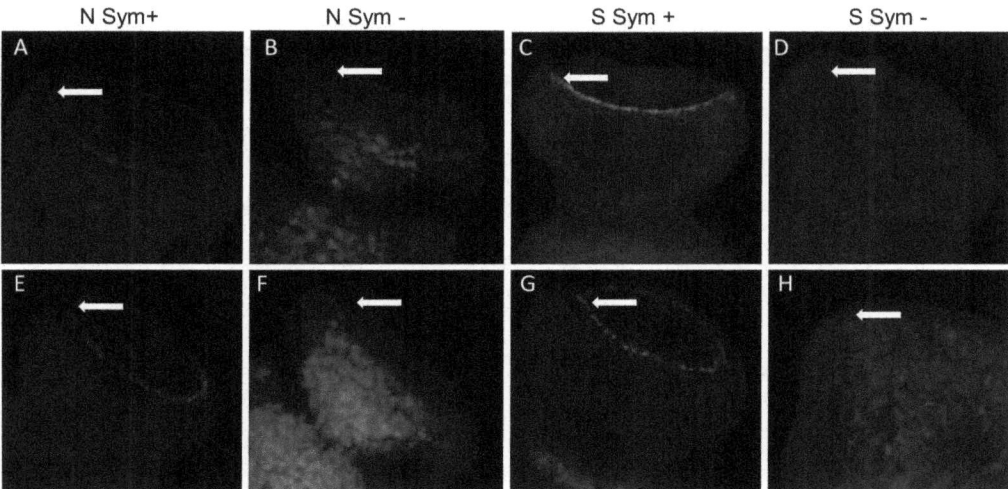

Figure 5. Confirmation of *B. neritina* **symbiotic status via fluorescence** *in situ* **hybridization (FISH).** White arrows indicate the circular pallial sinus on the aboral surface of the larva. (A–D) FISH micrographs of larvae using the Eub338 probe. (E–H) FISH micrographs of larvae using the symbiont-specific Bn1253r probe. Diameter of pallial sinus is approximately 150 µm.

sequence, further sequencing was performed from the 3' end of the amplicon (starting in the 23S region). All six samples (3 Type N, 3 Type S) were identical across 932 bp of shared sequencing.

Significantly, this identity is in stark contrast to Type D ITS sequences obtained from California colonies. Restriction analysis of host COI amplicons with *Msp*I indicated that all California samples from Catalina Island (n = 7) were of the Type D sibling species, with further digestion of 16S rRNA gene amplicons using *Nhe*I [12] revealing the endosymbiont to also be of Type D.

Sequences from two Type D ITS amplicons differed from both the Type N and Type S ITS by 4.0% and 4.3% (Table 3) depending upon the length of quality read. Among *B. neritina* samples taken from Bodega Bay, one proved to be of sufficient quality for DNA characterization. This colony was confirmed to be Type S by COI restriction digestion with *Dde*I and *Msp*I. The ITS sequence amplified from this sample was identical to those of Type S and N "*Ca*. Endobugula sertula" from the Atlantic coast, reducing concerns that the differences between the Type D symbiont ITS

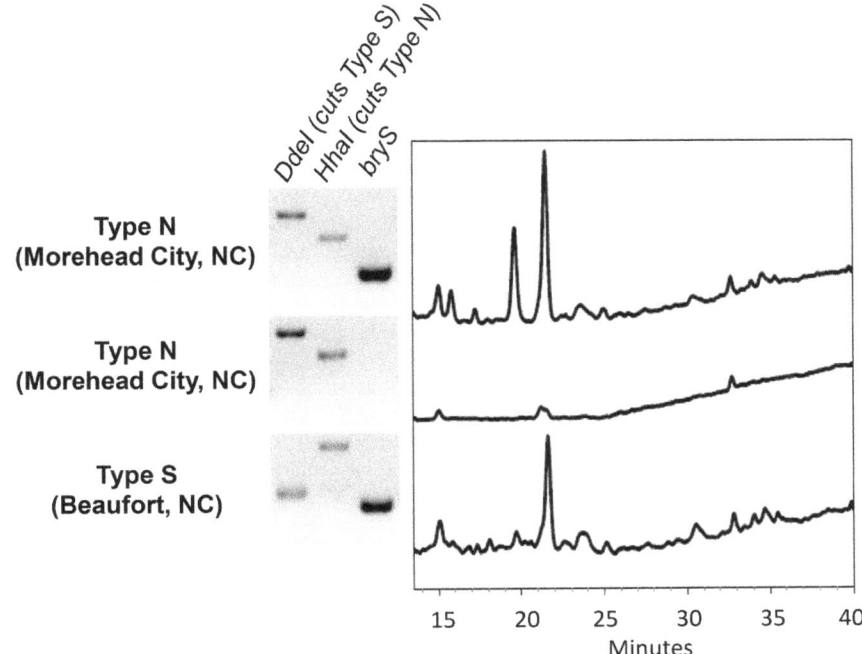

Figure 6. Molecular characterization of *B. neritina* **and HPLC analysis of crude larval extract.** Agarose gel showing *Dde*I and *Hha*I restriction digestion of *B. neritina* adult (top two samples) or pooled larval (bottom sample) cytochrome *c* oxidase I; amplification of *bryS* to demonstrate presence of symbiont; and absorbance at 229 nm (which is diagnostic of bryostatins) of extract collected from larvae represented by this gDNA.

Table 3. Percent identity of marker sequences among *B. neritina* sibling species.

% identity between sibling species	Host COI	Symbiont Small subunit 16S	ITS
D vs. S	91.8	99.6	95.7
D vs. N	89.6	99.6	95.7
N vs. S	88.5	100	100

Types D and S show the greatest COI identity, but Type D differs from both Types S and N in symbiont markers.

and that of Types N and S are a consequence of location. Instead, endosymbiont identity appears to consistently relate to host phylotype.

Discussion

Recent research has revealed that interspecies interactions that are shown to be beneficial in one context may shift to a negative relationship in another where either biotic or abiotic factors vary. For instance, the relationship between cleaning gobies and longfin damselfish is beneficial in high densities of damselfish ectoparasites, but shifts to a more parasitic interaction at low ectoparasite densities [36]. For symbiotic participants that inhabit a wide geographic region, it may be necessary to assess the interaction in different regions to fully understand the dynamic nature of the relationship.

For Western Atlantic *B. neritina* populations, previous sampling had indicated biogeographic separation between two sibling species that varied in their symbiotic status [12,26]. It was proposed that the aposymbiotic Type N *B. neritina*, which was documented only at higher latitudes, did not possess a defensive symbiont because of lower predation pressures, and that the Type S colonies were symbiotic because they inhabited areas of higher predation pressure. During the course of our study, we discovered that the distribution of the host was much broader than previously thought, while that of the symbiont appears to be restricted by biogeography, instead of host phylotype, as previously thought.

Ranges of *B. neritina* sibling species

Both Types N and S *B. neritina* can be found across a much wider range of the United States' Atlantic coast than initially observed [26]. This is the second report of these sibling species co-occurring, as well as of their occurrence on both sides of Cape Hatteras [28]. Both symbiotic Type N and aposymbiotic Type S colonies were unexpected findings in this study. The existence of these previously undescribed types may have been a crucial factor enabling the spread of each sibling species beyond the ranges initially understood. Additionally, while Cape Hatteras is not a strict biogeographic boundary for these sibling species, there is still a marked inversion in relative prevalence occurring around this latitude. Physiological factors likely favor Type N *B. neritina* at higher latitudes and Type S at lower, although population structure may be based both on fitness in and initial settlement of a region. Sequencing completed thus far on conspecific colonies from opposite sides of the Cape has not revealed separate lineages within sibling species that are differentially adapted to varying latitudes. Although *B. neritina* larvae have a very short pelagic phase (0.5–2 hr), suggesting a limited dispersal [37,38], adult colonies are often found on the hulls of ships [39–42] and are thought to be dispersed anthropogenically [27,43]. This transport should result in widespread availability of all haplotypes in many

locations, such that the documented population structure is the result of a marked selection pressure.

Details of *B. neritina* reproduction are not well understood, and the possibility of hybridization between these sibling species has not been adequately addressed. If hybrids exist in the environment, one would expect the sibling species' ranges to be obscured, as some individuals may benefit from adaptations originating within the sibling species different from that of the individual's COI gene. For example, apparently Type N animals may persist at lower latitudes due to genes from the Type S lineage, and vice versa. It should also be noted that if "*Ca.* Endobugula sertula" is exclusively vertically transmitted, it is likely that its inheritance would correlate with mitochondrial transmission. Heritable bacterial symbionts are typically inherited maternally [44,45], though exceptions exist (see [46]). In addition, researchers recently characterized both nuclear and mitochondrial genes from the *B. neritina* species complex worldwide and found no evidence of hybridization, and suggested that these sibling species be regarded as distinct biological species [28]. Thus, it is unlikely that hybridization between Types N and S *B. neritina* is complicating interpretation of the data regarding horizontal transmission of "*Ca.* Endobugula sertula".

Ecology of "*Ca.* Endobugula sertula"

The distribution of the endosymbiont "*Ca.* Endobugula sertula" in both Type S and Type N hosts and the similar chemistry associated with these symbionts suggest that their role in defending *B. neritina* larvae applies to both sibling species. Regardless of host identity, symbiosis is more common at lower latitudes, perhaps reflecting the higher predation pressure in regions nearer to the equator [33,47]. While biogeographic structuring has been noted in marine endosymbionts [48], the latitudinal variation in *B. neritina*-"*Ca.* Endobugula sertula" symbioses may have novel implications for the bryozoan host. For example, symbiosis is more likely to be observed in Type S colonies on both sides of Cape Hatteras, possibly indicating that that they more readily associate with "*Ca.* Endobugula sertula" than do Type N *B. neritina*. It is notable that hybridization may obscure this tendency, as hybrid animals could be falsely identified as strictly one or the other sibling species. However, it is alternatively possible that Type S colonies' stronger association with the endosymbiont is merely a function of greater adaptation to lower latitudes, at which the "*Ca.* Endobugula sertula" is thought to be more beneficial. It is noteworthy that another bryozoan genus, *Watersipora*, also demonstrates both latitudinal variation in genotype [49] and association with putatively defensive symbiotic bacteria [50], but it is unknown if these factors are interrelated as they seem to be for *B. neritina*.

The sympatric association of symbiotic and aposymbiotic colonies at some locations may also be a reflection of the symbiont's impact on host survivorship. The co-occurrence may be briefly observed with the presence of a small number of short-lived colonies, before selection is able to eliminate poorly adapted lineages. However, the occurrence of unexpected associations with "*Ca.* Endobugula sertula," and the lack thereof, may indicate a previously unexplored cost associated with harboring the symbiont. The value of any facultative symbiosis is necessarily context-dependent [51]. Evaluation of cost within the dynamics of the lifestyle of one or more generations of participants may require observation of a host across its lifetime [6], and may be further complicated by genetic variation within the host, as in the association of the protective symbiont *Hamiltonella defensa* with the black bean aphid [4]. These may be additional factors driving the differing association of Type N and Type S *B. neritina* with their bacterial symbiont.

In addition to defining the range of the symbiosis, the cost-benefit variation of the *B. neritina*-"*Ca.* Endobugula sertula" relationship across space may, in fact, be a force behind the bryozoan's abandonment of association with the bacteria. Mutualism breakdown may be a common result of association across a variety of symbiotic systems [52]. A mode of symbiont loss has yet to be demonstrated in this system in nature, leaving open the possibility of mediation by simple selective pressure, perhaps coupled with imperfect inheritance of the bacteria [53]. Alternatively, environmental factors may limit the survival of the endosymbiont in high-latitude waters, with unprotected aposymbiotic animals found only transiently at low latitudes due to anthropogenic dispersal of adult *B. neritina*. The goal of current experiments is to determine the parameters defining successful maintenance of the association with "*Ca.* Endobugula sertula".

Type N "*Ca.* Endobugula sertula"

Analyses using 16S rRNA gene and ITS sequences give no indication that the symbiont associated with Type N *B. neritina* differs from that of Type S colonies. This is surprising if "*Ca.* Endobugula sertula" is subject to strict vertical transmission. Endosymbionts associated with Type S and D *B. neritina* have demonstrated differences of up to 0.5% even in the slowly evolving 16S rRNA gene sequence, with host COI sequences varying by 8.2% [12,26]. In comparison, Type N and Type S COI sequences differ by 11.5%. One would expect this difference to correlate with an equal or greater likelihood of variation in the symbionts' 16S rRNA gene identity, but none is observed. Indeed, multilocus phylogenetic analysis of the *B. neritina* complex indicates that Type N animals form a sister group to the clade containing Types S and D *B. neritina* [28], further suggesting that cospeciation of host and symbiont would lead to greater divergence between Type S and Type N symbionts.

The slow evolution of the 16S rRNA gene, however, makes in-depth comparison difficult. In fact, while the "*Ca.* Endobugula sertula" sequences obtained by McGovern and Hellberg [26] from Type S *B. neritina* differed from those of Haygood and Davidson's [15] Type S sequences by just 2 bp (0.2%) and Type D sequences by 4–5 bp (0.5%), there were an additional 2 bp that were unique to the later Type S sequences and earlier Type D sequences, while not being shared by the first characterized Type S symbionts. The 16S rRNA gene alone may simply be unreliable for elucidating

"*Ca.* Endobugula sertula" phylogeny. In contrast, while 16S rRNA gene sequences may not provide a fast enough clock for contrast, differences might still be seen in ITS sequences, as they have been demonstrated to vary up to 9% among clades even within a single bacterial species [54]. Cytochrome oxidase I variation among *B. neritina* sibling species is well within the expected range for congeneric species of invertebrate phyla, which average 11.3% [55].

Coevolution of host and symbiont would lead to the expectation of associated ITS variation. That these differences have not been observed between Type N and Type S "*Ca.* Endobugula sertula" leaves open the possibility of horizontal transmission of the symbiont between these two sibling species. While potentially not the primary mode of symbiont transfer, such events have been observed in other symbionts demonstrating vertical transmission with a high degree of fidelity [56], and we note that co-analysis of *Bugula* host and symbiont phylogenetic topologies reveals the probability of at least one host-switching event in the lineage's past [11]. The observed similarity between Type N and S symbionts may be the result of a very recent horizontal transmission event, potentially enabled by uptake into the colony's funicular system via the gut. Further investigation of the sibling species' invasion patterns and experimentation with horizontal acquisition of their symbiont may shed more light on the history of this mutualism. In the meantime, our results highlight the flexibility and context-dependence of organismal interactions and the importance of investigating these relationships across scales where abiotic and biotic factors may vary.

Sequence Availability

New sequence data generated in this study are available in GenBank, accession numbers KM246893-KM246900.

Acknowledgments

We thank Niels Lindquist and Nancy Targett for the invaluable use of their lab space and resources at the University of North Carolina at Chapel Hill's Institute of Marine Sciences and the University of Delaware's College of Earth, Ocean, and Environment, respectively. We also thank Vincent Arnone, Kim Beeson, Maggie Benson, Ahron Flowers, Matthew Grober, Amanda Hoppers, Rian Kabir, Colin Maclean, Meril Mathew, and Brittany Wilkerson for their help in the collection and processing of *B. neritina* samples and preparation of figures for this manuscript. This work was funded in part by the Georgia State University Research Foundation and the Randolph-Macon College Chenery Grant. The Olympus microscope was acquired through a National Science Foundation Major Research Instrumentation grant to R-MC. JL is supported by a GSU Molecular Basis of Disease Area of Focus Graduate Research Fellowship, while DP received support from the R-MC Schapiro Undergraduate Research Fellowship.

Author Contributions

Conceived and designed the experiments: JL GL-F NBL. Performed the experiments: JL DP GL-F. Analyzed the data: JL GL-F NBL. Contributed reagents/materials/analysis tools: GL-F NBL. Wrote the paper: JL GL-F NBL.

References

1. Wares JP (2002) Community genetics in the Northwestern Atlantic intertidal. Molecular Ecology 11: 1131–1144.
2. Pennings SC, Silliman BR (2005) Linking biogeography and community ecology: Latitudinal variation in plant-herbivore interaction strength. Ecology 86: 2310–2319.
3. Diaz-Ferguson E, Robinson JD, Silliman B, Wares JP (2010) Comparative Phylogeography of North American Atlantic Salt Marsh Communities. Estuaries and Coasts 33: 828–839.
4. Vorburger C, Gouskov A (2011) Only helpful when required: a longevity cost of harbouring defensive symbionts. Journal of Evolutionary Biology 24: 1611–1617.

5. Lopanik NB (2014) Chemical defensive symbioses in the marine environment. Functional Ecology 28: 328–340.
6. Oliver KM, Campos J, Moran NA, Hunter MS (2008) Population dynamics of defensive symbionts in aphids. Proceedings of the Royal Society B-Biological Sciences 275: 293–299.
7. Polin S, Simon JC, Outreman Y (2014) An ecological cost associated with protective symbionts of aphids. Ecology and Evolution 4: 826–830.
8. Thomas MJ, Creed RP, Brown BL (2013) The effects of environmental context and initial density on symbiont populations in a freshwater cleaning symbiosis. Freshwater Science 32: 1358–1366.
9. Clay K (2014) Defensive symbiosis: A microbial perspective. Functional Ecology 28: 293–298.
10. Woollacott RM (1981) Association of bacteria with bryozoan larvae. Marine Biology 65: 155–158.
11. Lim-Fong GE, Regali LA, Haygood MG (2008) Evolutionary relationships of "Candidatus Endobugula" bacterial symbionts and their Bugula bryozoan hosts. Applied and Environmental Microbiology 74: 3605–3609.
12. Davidson SK, Haygood MG (1999) Identification of sibling species of the bryozoan Bugula neritina that produce different anticancer bryostatins and harbor distinct strains of the bacterial symbiont "Candidatus Endobugula sertula". Biological Bulletin 196: 273–280.
13. Lim GE, Haygood MG (2004) "Candidatus Endobugula glebosa", a specific bacterial symbiont of the marine bryozoan Bugula simplex. Applied and Environmental Microbiology 70: 4921–4929.
14. Sharp KH, Davidson SK, Haygood MG (2007) Localization of "Candidatus Endobugula sertula" and the bryostatins throughout the life cycle of the bryozoan Bugula neritina. ISME Journal 1: 693–702.
15. Haygood MG, Davidson SK (1997) Small-subunit rRNA genes and in situ hybridization with oligonucleotides specific for the bacterial symbionts in the larvae of the bryozoan Bugula neritina and proposal of "Candidatus Endobugula sertula". Applied and Environmental Microbiology 63: 4612–4616.
16. Woollacott RM, Zimmer RL (1975) A simplified placenta-like system for the transport of extraembryonic nutrients during embryogenesis of Bugula neritina (Bryozoa). Journal of Morphology 147: 355–378.
17. Davidson SK, Allen SW, Lim GE, Anderson CM, Haygood MG (2001) Evidence for the biosynthesis of bryostatins by the bacterial symbiont "Candidatus Endobugula sertula" of the bryozoan Bugula neritina. Applied and Environmental Microbiology 67: 4531–4537.
18. Lopanik N, Lindquist N, Targett N (2004) Potent cytotoxins produced by a microbial symbiont protect host larvae from predation. Oecologia 139: 131–139.
19. Trindade-Silva AE, Lim-Fong GE, Sharp KH, Haygood MG (2010) Bryostatins: biological context and biotechnological prospects. Current Opinion in Biotechnology 21: 834–842.
20. Tamburri MN, Zimmer-Faust RK (1996) Suspension feeding: Basic mechanisms controlling recognition and ingestion of larvae. Limnology and Oceanography 41: 1188–1197.
21. Lindquist N (1996) Palatability of invertebrate larvae to corals and sea anemones. Marine Biology 126: 745–755.
22. Lopanik NB, Targett NM, Lindquist N (2006) Ontogeny of a symbiont-produced chemical defense in Bugula neritina (Bryozoa). Marine Ecology-Progress Series 327: 183–191.
23. Lindquist N, Hay ME (1996) Palatability and chemical defense of marine invertebrate larvae. Ecological Monographs 66: 431–450.
24. Kellner RLL, Dettner K (1996) Differential efficacy of toxic pederin in deterring potential arthropod predators of Paederus (Coleoptera: Staphylinidae) offspring. Oecologia 107: 293–300.
25. Kellner RLL (2002) Molecular identification of an endosymbiotic bacterium associated with pederin biosynthesis in Paederus sabaeus (Coleoptera: Staphylinidae). Insect Biochemistry and Molecular Biology 32: 389–395.
26. McGovern TM, Hellberg ME (2003) Cryptic species, cryptic endosymbionts, and geographical variation in chemical defences in the bryozoan Bugula neritina. Molecular Ecology 12: 1207–1215.
27. Mackie JA, Keough MJ, Christidis L (2006) Invasion patterns inferred from cytochrome oxidase I sequences in three bryozoans, Bugula neritina, Watersipora subtorquata, and Watersipora arcuata. Marine Biology 149: 285–295.
28. Fehlauer-Ale KH, Mackie JA, Lim-Fong GE, Ale E, Pie MR, et al. (2014) Cryptic species in the cosmopolitan Bugula neritina complex (Bryozoa, Cheilostomata). Zoologica Scripta 43: 193–205.
29. Sudek S, Lopanik NB, Waggoner LE, Hildebrand M, Anderson C, et al. (2007) Identification of the putative bryostatin polyketide synthase gene cluster from "Candidatus Endobugula sertula", the uncultivated microbial symbiont of the marine bryozoan Bugula neritina. Journal of Natural Products 70: 67–74.
30. McCartney MA, Burton ML, Lima TG (2013) Mitochondrial DNA differentiation between populations of black sea bass (Centropristis striata) across Cape Hatteras, North Carolina (USA). Journal of Biogeography 40: 1386–1398.
31. Briggs JC (1974) Marine zoogeography. New York: McGraw-Hill.
32. Briggs JC, Bowen BW (2012) A realignment of marine biogeographic provinces with particular reference to fish distributions. Journal of Biogeography 39: 12–30.
33. Vermeij GJ (1978) Biogeography and Adaptation: Patterns of Marine Life. Cambridge, MA: Harvard University Press.
34. Folmer O, Black M, Hoeh W, Lutz R, Vrijenhoek R (1994) DNA primers for amplification of mitochondrial cytochrome c oxidase subunit I from diverse metazoan invertebrates. Molecular Marine Biology and Biotechnology 3: 294–299.
35. Amann RI, Krumholz L, Stahl DA (1990) Fluorescent-oligonucleotide probing of whole cells for determinative, phylogenetic, and environmental studies in microbiology Journal of Bacteriology 172: 762–770.
36. Cheney KL, Cote IM (2005) Mutualism or parasitism? The variable outcome of cleaning symbioses. Biology Letters 1: 162–165.
37. Keough MJ (1989) Dispersal of the bryozoan Bugula neritina and effects of adults on newly metamorphosed juveniles. Marine Ecology Progress Series 57: 163–171.
38. Keough MJ, Chernoff H (1987) Dispersal and population variation in the bryozoan Bugula neritina. Ecology 68: 199–210.
39. Gordon DP, Mawatari SF (1992) Atlas of marine-fouling Bryozoa of New-Zealand ports and harbours. Miscellaneous Publication New Zealand Oceanographic Institute 107: 1–52.
40. Floerl O, Pool TK, Inglis GJ (2004) Positive interactions between nonindigenous species facilitate transport by human vectors. Ecological Applications 14: 1724–1736.
41. Hewitt CL (2002) Distribution and biodiversity of Australian tropical marine bioinvasions. Pacific Science 56: 213–222.
42. Carlton JT, Hodder J (1995) Biogeography and dispersal of coastal marine organisms: Experimental studies on a replica of a 16th-century sailing vessel. Marine Biology 121: 721–730.
43. Ryland JS, Bishop JDD, De Blauwe H, El Nagar A, Minchin D, et al. (2011) Alien species of Bugula (Bryozoa) along the Atlantic coasts of Europe. Aquatic Invasions 6: 17–31.
44. Cary SC, Giovannoni SJ (1993) Transovarial inheritance of endosymbiotic bacteria in clams inhabiting deep-sea hydrothermal vents and cold seeps. Proceedings of the National Academy of Sciences 90: 5695–5699.
45. Wilkinson T, Fukatsu T, Ishikawa H (2003) Transmission of symbiotic bacteria Buchnera to parthenogenetic embryos in the aphid Acyrthosiphon pisum (Hemiptera: Aphidoidea). Arthropod structure & development 32: 241–245.
46. Moran NA, Dunbar HE (2006) Sexual acquisition of beneficial symbionts in aphids. Proceedings of the National Academy of Sciences of the United States of America 103: 12803–12806.
47. Manyak-Davis A, Bell TM, Sotka EE (2013) The Relative Importance of Predation Risk and Water Temperature in Maintaining Bergmann's Rule in a Marine Ectotherm. American Naturalist 182: 347–358.
48. Sanders JG, Palumbi SR (2011) Populations of Symbiodinium muscatinei show strong biogeographic structuring in the intertidal anemone Anthopleura elegantissima. Biological Bulletin 220: 199–208.
49. Mackie JA, Darling JA, Geller JB (2012) Ecology of cryptic invasions: latitudinal segregation among Watersipora (Bryozoa) species. Scientific Reports 2: 10.
50. Anderson CM, Haygood MG (2007) α-proteobacterial symbionts of marine bryozoans in the genus Watersipora. Applied and Environmental Microbiology 73: 303–311.
51. Cockburn SN, Haselkorn TS, Hamilton PT, Landzberg E, Jaenike J, et al. (2013) Dynamics of the continent-wide spread of a Drosophila defensive symbiont. Ecology Letters 16: 609–616.
52. Sachs JL, Simms EL (2006) Pathways to mutualism breakdown. Trends in Ecology & Evolution 21: 585–592.
53. Yule KM, Miller TEX, Rudgers JA (2013) Costs, benefits, and loss of vertically transmitted symbionts affect host population dynamics. Oikos 122: 1512–1520.
54. Erwin PM, Thacker RW (2006) Host-specificity of sponge-associated unicellular cyanobacteria, Candidatus Synechococcus spongiarum. Integrative and Comparative Biology 46: E40–E40.
55. Hebert PDN, Ratnasingham S, deWaard JR (2003) Barcoding animal life: cytochrome c oxidase subunit 1 divergences among closely related species. Proceedings of the Royal Society B-Biological Sciences 270: S96–S99.
56. Jaenike J, Unckless R, Cockburn SN, Boelio LM, Perlman SJ (2010) Adaptation via Symbiosis: Recent Spread of a Drosophila Defensive Symbiont. Science 329: 212–215.
57. Mathew M, Lopanik NB (2014) Host Differentially Expressed Genes during Association with its Defensive Endosymbiont. Biological Bulletin 226: 125–163.
58. Ferris MJ, Muyzer G, Ward DM (1996) Denaturing gradient gel electrophoresis profiles of 16S rRNA-defined populations inhabiting a hot spring microbial mat community. Applied and Environmental Microbiology 62: 340–346.
59. Lane DS (1990) 16S and 23S rRNA sequencing. In: Goodfellow M, editor. Nucleic acid techniques in bacterial systematics. New York, NY: John Wiley. pp.115–148.
60. Garcia-Martinez J, Acinas SG, Anton AI, Rodriguez-Valera F (1999) Use of the 16S-23S ribosomal genes spacer region in studies of prokaryotic diversity. Journal of Microbiological Methods 36: 55–64.

20

Strain Classification of *Mycobacterium tuberculosis* Isolates in Brazil Based on Genotypes Obtained by Spoligotyping, Mycobacterial Interspersed Repetitive Unit Typing and the Presence of Large Sequence and Single Nucleotide Polymorphism

Sidra E. G. Vasconcellos[1,8], Chyntia Carolina Acosta[2], Lia Lima Gomes[1], Emilyn Costa Conceição[3], Karla Valéria Lima[3], Marcelo Ivens de Araujo[1], Maria de Lourdes Leite[4], Flávio Tannure[4], Paulo Cesar de Souza Caldas[5], Harrison M. Gomes[1], Adalberto Rezende Santos[1], Michel K. Gomgnimbou[6], Christophe Sola[6], David Couvin[7], Nalin Rastogi[7], Neio Boechat[8,9], Philip Noel Suffys[1]*

1 Laboratory of Molecular Biology Applied to Mycobacteria, Oswaldo Cruz Institute, FIOCRUZ, Rio de Janeiro, Rio de Janeiro, Brazil, 2 Laboratory of Cellular Microbiology, Oswaldo Cruz Institute, FIOCRUZ, Rio de Janeiro, Rio de Janeiro, Brazil, 3 Instituto Evandro Chagas, Section of Bacteriology and Mycology, Belém, Pará, Brazil, 4 Hospital Municipal Rafael de Paula Souza, Municipal Secretary of Health, Rio de Janeiro, Rio de Janeiro, Brazil, 5 Centro de Referência Professor Hélio Fraga, Escola Nacional de Saúde Publica Sergio Arouca, FIOCRUZ, Rio de Janeiro, Rio de Janeiro, Brazil, 6 CNRS–Université Paris–Sud, Institut de Génétique et Microbiologie–Infection Genetics Emerging Pathogens Evolution Team, Orsay, France, 7 Supranational TB Reference Laboratory, Unité de la Tuberculose et des Mycobactéries, Institut Pasteur de Guadeloupe, Abymes, Guadeloupe, France, 8 Multidisciplinary Research Laboratory, University Hospital Clementino Fraga Filho – HUCFF, Federal University of Rio de Janeiro, Rio de Janeiro, Rio de Janeiro, Brazil, 9 Graduate Program in Clinical Medicine, Faculty of Medicine, University Hospital Clementino Fraga Filho, Rio de Janeiro, Rio de Janeiro, Brazil

Abstract

Rio de Janeiro is endemic for tuberculosis (TB) and presents the second largest prevalence of the disease in Brazil. Here, we present the bacterial population structure of 218 isolates of *Mycobacterium tuberculosis*, derived from 186 patients that were diagnosed between January 2008 and December 2009. Genotypes were generated by means of spoligotyping, 24 MIRU-VNTR typing and presence of fbpC[103], RD[Rio] and RD174. The results confirmed earlier data that predominant genotypes in Rio de Janeiro are those of the Euro American Lineages (99%). However, we observed differences between the classification by spoligotyping when comparing to that of 24 MIRU-VNTR typing, being respectively 43.6% vs. 62.4% of LAM, 34.9% vs. 9.6% of T and 18.3% vs. 21.5% of Haarlem. Among isolates classified as LAM by MIRU typing, 28.0% did not present the characteristic spoligotype profile with absence of spacers 21 to 24 and 32 to 36 and we designated these conveniently as "LAM-like", 79.3% of these presenting the LAM-specific SNP *fbpC[103]*. The frequency of RD[Rio] and RD174 in the LAM strains, as defined both by spoligotyping and 24 MIRU-VNTR loci, were respectively 11% and 15.4%, demonstrating that RD174 is not always a marker for LAM/RD[Rio] strains. We conclude that, although spoligotyping alone is a tool for classification of strains of the Euro-American lineage, when combined with MIRU-VNTRs, SNPs and RD typing, it leads to a much better understanding of the bacterial population structure and phylogenetic relationships among strains of *M. tuberculosis* in regions with high incidence of TB.

Editor: Igor Mokrousov, St. Petersburg Pasteur Institute, Russian Federation

Funding: This work was supported by funding by Brazilian funding agencies PAPES/CNPq (407624/2012-0), PROEP/CNPq of the Oswaldo Cruz Institute (IOC), and CNPq PhD grant (142958/2008-5). The funders had no role in study design, data collection and analysis, decision to publish, or preparation of the manuscript.

Competing Interests: The authors have declared that no competing interests exist.

* Email: psuffys@gmail.com

Introduction

Tuberculosis (TB) is an infectious disease with an effective treatment but remains an important cause of morbidity and mortality in many countries. In Brazil, 73.000 new TB cases and 4.600 cases that deceased were registered in 2011. The southeast region has 44.8% of all cases reported in the country and the state of Rio de Janeiro has the highest disease incidence (72.3/100,000) and mortality rate (5.7/100,000) in the country [1].

A couple of years ago, *Mycobacterium tuberculosis* (Mtb) strains designated as RD[Rio] were reported as being a predominant genotype in Rio de Janeiro [2]. These strains have a deletion of 26.3 kb and seem to be restricted to the Latin American-Mediterranean (LAM) family. The RD[Rio] strains have been associated with higher levels of recent transmission and of Multi-Drug Resistance (MDR) but there are contradictory data about their relation with disease severity [3], [4], [5], [6]. Another Region of Difference called RD174 has been described as a co-marker of RD^{Rio} and as a marker for the LAM type [7], [8], [9], [10] and strains with the RD174 deletion were found to have an increased secondary case rate ratio in San Francisco [10].

The most commonly used genotyping methods to characterize *M. tuberculosis* Complex (MTBC) isolates are IS*6110*-RFLP, 24-locus MIRU-VNTR, spoligotyping, detection of Single Nucleotide Polymorphisms (SNPs) and Large Sequence Polymorphisms (LSP). Typing of MTBC by IS*6110*-RFLP was considered the gold-standard during more than a decade [11], but is labor intensive. Currently, the two most commonly used methods are spoligotyp-ing and Mycobacterium Interspersed Repetitive Units – Variable Number of Tandem Repeats (MIRU-VNTR) typing, based on the variability of the direct repeat (DR) locus [12] and of minisatellites [13], [14], respectively. However, for epidemiologic studies, spoligotyping overestimates the proportion of clustered strains and should be used in combination with MIRU-VNTR typing [15]. Spoligotyping alone is also unable to reveal exact phylo-genetic relationships between MTBC strains, particularly for the classification of Euro-American strains, part of the PGG2/PGG3 group [9], [16], [17], [18]. These phylogenetic subtypes include the genotype families LAM, H, T, S and X, frequently observed in Brazil, in South America and elsewhere [2], [4]. [8], [18], [19], [20], [21], [22], [23].

The generalized use of spoligotyping and MIRU-VNTR resulted in the construction of large international databases of genotypes, allowing the study of global distribution and phylo-genetic analysis of the distribution of *M. tuberculosis* worldwide [23], [24], [25], [26], [27], [28], [29]. Combined MIRU-VNTR typing and spoligotyping also helps in revealing epidemiologically meaningful clonal diversity of *M. tuberculosis* strain lineages and is useful to explore internal phylogenetic ramifications [8]. In addition, SNPs and LSPs represent robust markers for inferring phylogenies and for strain classification [30], [31], [32], [33], [34], [35]. Besides the presence of RDs, that of determinate SNPs is also a characteristic of LAM strains, [8], [9], [10] such as *fbpC^{103}* in codon 103 (G to A) of the gene encoding antigen 85 Complex Ag85^c (Rv0129c) resulting in a Glu103Asp amino acid replace-ment in a protein that is involved in biosynthesis of cell wall components of *M. tuberculosis* [7], [36]. Other LAM-specific SNPs are (i) the silent C to G *ligB* mutation at genome position 3426795 [37], validated by the Abadia *et al.* [17] and (ii) the SNP at 240 codon of *mgtC* (C to T) that was discovered by Homolka *et al.* [38].

The objective of this study was to evaluate 24 MIRU-VNTR typing for alternative classification of strain families/lineages by means of phylogenetic tree building and comparison with MIRU databases (MIRU-VNTRPlus), as compared with spoligotypes-based classification (SITVITWEB and SITVIT2), and increase the consistency of the analysis by including a SNP and two LSPs. In addition, the differentiating power of both techniques in a population that is almost exclusively of the Euro-American lineages.

Materials and Methods

Study setting

The study was based on a convenience sampling of TB patients diagnosed between January 2008 and December 2009 at the "Hospital Municipal Raphael de Paula e Sousa", Curicica, Rio de Janeiro, Brazil. Two hundred eighteen clinical isolates were obtained from 188 patients and 28 cases had more than one isolate, including 25 patients with two and three patients with three isolates. Multiple isolates from the same patient were included to evaluate the consistency of the results of genotyping and to verify eventual multiple infections.

Ethics statement

This study was approved by the Research Ethics Committee of the Municipal Health and Civil Defense of the city of Rio de Janeiro Number 160/09 CAAE: 0182.0.314.000-9. Isolates of this study were obtained through bacteriological culture from clinical specimens of patients diagnosed at the Hospital Raphael and Paula Souza as part of routine diagnosis and drug susceptibility testing. The ethical committee stated that, as genotyping of *M. tuberculosis* isolates is complementary to routine diagnosis, there was no need for a written or verbal consent. No information other than that provided for the diagnosis was used.

Culture, DNA extraction and identification of Mycobacterium tuberculosis complex

Sputum samples were cultured on Löwenstein-Jensen (LJ) medium following standard microbiological laboratory procedures as a routine of the Clinical Analysis Laboratory of "Hospital Raphael de Paula e Souza". After cultivation, bacterial mass was re-suspended in 400 µl of sterile distilled water and heat inactivated at 90°C for one hour, followed by DNA extraction and purification using the CTAB method [39].

Spoligotyping

Spoligotyping was performed either as described by Kamerbeek *et al.* (1997) [12], using commercially available kit from Ocimum Biosolutions (Hyderabad, India) according to the manufacturer's instructions or using microbead-based hybridization assay as described by Zhang *et al.* (2009) [40].

MIRU-VNTR typing

Amplification of 24 MIRU-VNTR *loci* was performed by using a commercial typing kit (Genoscreen, Lille, France) and automat-ed MIRU-VNTR analysis performed as previously described [14]. Fragment size of the amplicons was analyzed on a ABI 3730 DNA sequence analyzer (Applied Biosystems, California, USA) and number of copies of each locus was determined by automated assignment using the Genemapper 4.0 software (Applied Biosys-tems, California, USA). In the case of doubtful results, the size of the repeats was double checked by size estimation as compared to a DNA ladder (50 and 100 bp) and the positive control (H37Rv) on agarose gels and by comparing to a reference table as described [13].

PCR-RFLP of *fbpC^{103}*

For characterization of SNP *fbpC^{103}*, we adapted the procedure described by Gibson *et al.* in 2008 [7]. Amplifications were performed in 50 µl reactions containing 40 pmol each of the primers Ag85C103F (5-CTG GCT GTT GCC CTG ATA CTG CGA GGG CCA-3) and Ag85C103R (5-CGA GCA GCT TCT GCG GCC ACA ACG TT-3), 2 mM MgCl2, 0.2 mM dNTPs, 1 U Taq DNA polymerase (Invitrogen, Brazil), buffer (10 mM Tris-HCl, 1, 5 mM MgCl2, 50 mM KCl, pH 8.3), 5% DMSO (v/v) and 10 ng of target DNA. Amplification was performed in a Vieriti Thermal Cycler (Applied Biosystems, Foster City, CA), starting with denaturation at 95°C for 5 min, followed by 45 cycles of 1 min at 95°C, 1 min at 60°C, 4 min at 72°C and final extension for 10 min at 72°C. The amplified products of 519 bp were analyzed on 2% agarose gels after staining with ethidium bromide. Fifteen microliters of the amplified products were subjected to enzymatic digestion with 1 U of *Mnl*I (New England

BioLabs Inc. USA) at 37°C for four hours, generating three fragments of 365 bp, 96 bp and 48 bp for the wild type allele and two bands of 461 bp and 48 bp when the SNP G103A is present. Bands were detected in 3% agarose gel and their size estimated by comparison with a 100 bp DNA ladder (Invitrogen).

RD174 deletion

For detection of RD174, we used the protocol described earlier [7], performing multiplex-PCR using two primers that anneal to the RD174 flanking regions and one to the internal sequence. For amplification, we used 40 pmol of each of the primers RD174F (5′-AGC TGC TCC GGC CGG TCG TCG TCC TTG TC-′3), RD174Fi (5′-TAT GCC GCA GCC CGG GCA TCC GTG ATT A-′3) and RD174R (5′-ATC GTG AAC GCA GCG GTT TCG ACG GCA TCT-′3) in a 50 µl reaction containing 2 mM MgCl2, 0.2 mM dNTP, 1 U Taq DNA polymerase, 1× buffer, 1M of betaine and 10 ng of bacterial DNA. Amplification was performed starting with 5 min incubation at 95°C, followed by 45 cycles of 1 min at 95°C, 1 min at 60°C, 4 min at 72°C and final extension for 10 min at 72°C. To determine the size of the amplicons, 10 µl of PCR product was applied in 2% agarose gel and after electrophoresis, bands of either 300 bp (intact RD174) or 500 bp (deleted) are observed.

RD^{Rio} deletion

Detection of RD^{Rio} was performed using the multiplex PCR-protocol described by Lazzarini *et al.* (2007) [2]. For amplification, we used 20 pmol of each of the primers BridgeF (5′-CAC TCC GGC TGC CAA TCT CGT C′3), BridgeR (5′-CAC GAG GCT CGC GAA TGA GAC C-′3), IS1561F (5′-GAC CTG ACG GCC ACA CTG C′-3) and IS1561R (5′-CAC CTA CAC CGC TTC CTG CC′-3) in 50 µl reactions containing 2 mM MgCl$_2$, 0.2 mM dNTP, 1 U Taq DNA polymerase), 1× buffer, 5% DMSO (v/v) and 10 ng of target DNA. Amplifications were performed starting with incubation for 5 min at 95°C, followed by 45 cycles of 1 min at 95°C, 1 min at 60°C, 4 min at 72°C and final extension of 10 min at 72°C. The amplified products of 1175 bp or 530 bp in the presence or absence of the deletion, respectively were analyzed in 1.5% agarose.

Classification and definition of genotype-based lineages and families

Spoligotype and MIRU patterns were defined according to the definitions in the SITVITWEB database [23], (http://www.pasteurguadeloupe.fr:8081/SITVIT_ONLINE/) in the format of November 20, 2012. In addition, spoligotypes were compared with those in SITVIT2 (proprietary database of the "Institut Pasteur de la Guadeloupe", which is an updated in-house version of the publicly released SpolDB4 and SITVITWEB [23], [24] to be released in 2014. This comparison led to the characterization of families and lineages by spoligotyping and Spoligotype International Types (SIT), MIRU International Types (12-MIT, 15-MIT AND 24-MIT) and 5 *loci* VNTR International Types (VIT). The SIT, MIT and VIT were designated identical patterns when shared by two or more patient isolates, whereas "orphan" patterns were those observed in a single isolate that did not correspond to any of the patterns present in the SITVIT2 database. In order to classify "orphan" patterns, we used SpotClust (http://tbinsight.cs.rpi.edu/run_spotclust.html).

The 24 MIRU-VNTR profiles and spoligotypes were also compared with the MIRU-VNTR*plus* database [28] available at (http://www.miru-vntrplus.org/MIRU/index.faces). The definition of lineages was done on 24 MIRU-VNTR loci by Best-match

analysis and Tree-based identification using the categorical index. The Multi-locus Variable Number Tandem Repeat analysis (MLVA) MtbC 15 and MtbC 12 were determined.

Data analysis

The number and size of genotype clusters were defined initially by introducing numerical values into an Excel table and different criteria for definition of clusters were used, being either individual identical spoligotypes or 15–24 MIRU-based genotypes, or by combining with Spoligotyping both. Excel tables were introduced into BioNumerics software (version 6.1, Applied Maths, Belgium) for construction of similarity matrices and phylogenetic trees, using the Jaccard Index (spoligotypes) and Categorical value (MIRU-VNTR) for construction of the Neighbor Joining (NJ) and Minimal Spanning Tree (MST).

The MIRU-VNTR allelic diversity (*h*) at each of the 24 loci was calculated by the equation described by Graur and Li [41] and the Diversity index was calculated as the Hunter Gaston diversity index (HGDI) [42]. Statistical analysis were performed using chi-square analysis with a confidence interval of 95% using Epi Info.Version 3.51 (Centers for Disease Control and Prevention, Atlanta, GA,USA), by using $\chi2$ test or Fisher exact test for the comparison of proportions. A *p* value<0.05 was considered significant.

To visualize the classification difference observed between Spoligotyping and 24 loci MIRU-VNTRs, we constructed a Confusion Matrix for the frequency of correct and incorrect predictions. From the confusion matrix, the Positive Predictive Value or precision (PPV = TP/TP+FN), Accuracy rate (AAC = TP+TN/TP+FP+TN+FN), True Positive Rate or Sensitivity (TPR = TP/TP+TN), True Negative Rate or Specify (TNR = TN/TP+TN), Negative Predictive Value (NPV = TN/FN+TN) and Error Rate (ER = FP+FN/TP+FP+TN+FN) were calculated. (TP = True Positive, FP = False Positive, TN = True Negative, FN = False Negative).

Results

Classification by spoligotyping

The identification of spoligotype profiles as well as their definition to the family and lineage level was realized by comparison with profiles deposited in the SITVITWEB database and in SITVIT2 and compared with the classification obtained with previous version of the spoligotype database (Table S1).

Based on comparison with SITVIT2, 83 SITs were encountered in 195 isolates (89.4%), while 10 SIT (n = 15) had Unknown profiles and 23 isolates showed to have Orphan patterns. For classification of the latter two pattern types, we used Spotclust (Table S1). Ninety five isolates (43.6%) were classified as belonging to the LAM clade, 76 (34.9%) as belonging to the T clade, 40 (18.3%) as belonging to the Haarlem clade, five to the EAI5 clade (2.29%) and one each to the X clade (0.45%) and to Beijing. Overall, T1 (n = 52/23.8%), LAM9 (n = 32/14.7%) and H3 (n = 24/11%) were the most frequent subclades, while SIT53 (n = 26/50%), SIT42 (n = 18/56.25%) and SIT50 (n = 16/72,7%) were the most frequent SITs.

Table 1 illustrates the difference in frequency of SIT clusters containing three or more isolates (at least 3%) in this study, versus their worldwide distribution (macro-regions and countries), as defined in the SITVIT2 database. We observed a significantly higher frequency of genotypes T1 SIT53 (11.9% vs 5.5%; p = 0,027, X^2 = 3,49) and a lower frequency of LAM9 SIT42 (8.2% vs 12.5%; p = 0,0001; X^2 = 15.82). In addition, SIT 3907 (T1), SIT 2498 (H3) and SIT 3908 (LAM2) have so-far only been observed in Brazil. Presently, we also observed five SITs, 136 (LAM5), 189 (T1), 268 (H3), 2107 (unknown) and 2322 (T2), that had not been reported before in Brazil. When comparing our results with the previous and the present version of the SITVIT database, we noticed some differences in classification (Table S1). Thirty-six isolates could be classified using the new version and among these nine new SITs were observed: 3125 (LAM6), 3504 (LAM6), 3904 (LAM6) 3905 (T1) 3906 (T3), 3907 (T1), 3908 (LAM2; originally designated as SIT2560 Unknown in SITVIT-WEB), 3909 (LAM8) and 3910 (T1). In addition, nineteen orphan profiles were classified and SIT268, SIT1241 and SIT2539, without classification in SITVITWEB, were now classified respectively as H3, LAM9 and LAM1. In addition, some SIT showed changes in clade and/or subclade classification (SITs 73, 102, 159, 451, 742 and 777). Further information is detailed in Table S1.

Classification based on MIRU-VNTR analysis

The 218 isolates were also submitted to 24 loci MIRU-VNTR typing with the objective to confirm or not spoligotype-based classification of existing and new sublineages and to correct eventual miss classification that could have occurred using the latter technique. We observed 208 different patterns and seven clusters with two to four strains (clustering rate = 1.37%) (Figure S1). The allelic diversity of each MIRU-VNTR locus for 218 isolates was evaluated and classified as either highly (HGDI>0.6), moderately ($0.6 \leq 0.3$) and poorly discriminative (HGDI<0.3), according to Sola et al. [15], as summarized in Table 2. Eight loci were highly discriminatory (QUB4156, QUB11, MTUB04, MIRU10, MIRU40, MIRU 23, MIRU26 and MIRU21) ten were moderate (MTUB30, MTUB39, MIRU16, ETR-C, ETR-A, QUB26, MTUB34, MIRU31, MIRU29 AND ETR-B) and six were poorly discriminative (MIRU24, MIRU29, MIRU04, MIRU27, MIRU20 and MIRU2). In the Table 2, we can observe a difference in discriminatory power of some loci according to lineage by Spoligotyping.

The patterns were classified based on the database tool that allows construction of a Neighbor-Joining based phylogenetic tree, visualizing proximity of a particular genotype with that of a set of reference strains to the genotype family level. One hundred thirty-six isolates (62.4%) were classified as LAM (127 patterns), 47 (21.5%) as Haarlem (45 patterns), 21 (9.6%) as T (21 patterns), 11 (5%) as S (11 patterns), and one isolate (0.45%) each as Beijing and EAI. Considerable differences were observed between Spoligotyping and 24 MIRU-VNTR loci-based classifications, even after excluding eventual small typing errors by repeating the assays. The differences observed between the two classifications are presented in the table 3 and in the Confusion Matrix (Figure 1). The precision, accuracy, sensitivity and error rate were respectively 0.74, 0.64, 0.82 and 0.36. The T lineage showed the highest incongruence rate related to classification (sensitivity = 0.26).

Combined analysis of 24 MIRU-VNTR and spoligotyping

The LAM family. For better understanding of the population structure of the Mtb strains, dendrograms of spoligotyping and 24 MIRU-VNTRs, either separately or as combined patterns were constructed using BioNumerics. Among the 136 (62.4%) isolates that were classified as LAM based on MIRU-VNTR analysis, 99 (72.8%) exhibited the characteristic LAM spoligotype profile with absence of spacers 21 to 24 and 33–36. Among these however, four spoligotype isolates were designated to the T clade by SITVIT2 (SITs: 306/n = 3, 1688/n = 1, 3905/n = 1) and one as T2/orphan. Another thirty-seven strains (27.2%) defined as LAM by MIRU-VNTRs did not present the lack of spacers 21–24 and

were therefore designated as "LAM-like". Among isolates initially classified as LAM by spoligotyping (n = 95), 91 were confirmed as LAM (95.8%/TPR = 0.96) by 24 loci MIRU-VNTR (Figure 1). Six spoligotypes referred to as "Unknown" in SITVIT2 (SITs 132/n = 1; 1952/n = 1; 2107/n = 2; 2511/n = 2; 2535/n = 1 and 2548/n = 1) were also classified as LAM. Using SpotClust, SIT132 and SIT2511 were classified as EAI and SIT1952 as Haarlem. All Orphans patterns classified as LAM by SITVIT2 (n = 10) were also classified as LAM according to the MIRU-VNTR based phylogenetic tree.

Signatures based on MIRU copy number for "real" LAM (n = 99) types showed that the majority had two copies of MIRU04 (94; 95%) and two MIRU20 (n = 96; (97%), one copy of MIRU24 (n = 91; 92%), two copies of MIRU39 and ETR-A (n = 93; 94%) and 86 (87%) one copy of MTUB30. This indicates that the 221221 combination of these six alleles is representative for LAM strains, except for LAM3, presenting two copies of MTUB30 (221222).

The "LAM-like" isolates. The thirty-seven isolates that did not present the typical LAM spoligotype signature (absence of spacers 21 to 24 and 33–36) were positioned within the LAM branch in the neighbor-joining tree using MIRU-VNTR*Plus* and were conveniently designated as "LAM-like". These had been classified by spoligotyping earlier as T (n = 24), H (n = 10) and three "Unknown" patterns (SIT2511/n = 2 and SIT1952/n = 1; classified by SpotClust as EAI and H1 respectively). Ninety-one percent of these strains showed one copy of MIRU24 and two copies of MIRU04. Intriguingly, when including both 24 MIRU-VNTRs and spoligotypes for tree building, these isolates are localized among those classified as T and H.

The T family. Among the isolates initially classified as T by spoligotyping, only 26.3% (20/76) was confirmed by MIRU-VNTR 24 loci (TPR: 0.2631) (Table 3 and figure 1). By a Neighbor-Joining based phylogenetic tree, 31 of these isolates (40.9%) were positioned in LAM branch, 16 (21%) in the Haarlem branch and 9 (11.8%) in the S branch. The MIRU-VNTR characteristic signature for the T lineage confirmed by both techniques, was one copy of MIRU24 (95.8%), two copies each of MIRU02 (95.8%), MIRU04 (87.5%), MIRU20 (95.8%), MIRU39 (100%) and three copies each of MIRU27 (87.5%) and MTUB34 (87.5%). Unlike LAM isolates that present a single copy of MTUB30, isolates of the T family presented two or four copies of this allele.

The Haarlem family. Forty-seven isolates were classified as belonging to the H family representing 21.5% of all isolate, 85% of the these isolates initially classified as Haarlem by Spoligotyping were confirmed by MIRU-VNTRs 24 *loci* (TPR = 0.65, TNR = 0,88) (Table 3 and figure 1). Among these isolated 16 (34%) presented spoliotype profile compatible with T family, 3 (6.4%) with LAM family and 2 (2% each) isolates, respectively compatible with X and EAI. The MIRU-VNTRs characteristic for Haarlem are two copies of MIRU02 (100%) and MIRU39 (89.1%) and three copies of MTUB34 (100%), MIRU16 (91.3%), MIRU27 (97.8%), ETR-A (95.6%) and ETR-C (91.3%). Different from LAM isolates having a single copy of MTUB30, H isolates present either three or four copies of this MIRU. Combined analysis of the MIRU-24 and spoligotyping showed that isolates of genotype H1 (absences of spacers 26–31) have two copies of MIRU04, ETR-B and MTUB29 and three copies of MIRU16, MIRU31, ETR-A and ETR-C. Isolates of genotype H3 (absences of spacer 31) share two copies of MIRU02, ETR-C, MIRU4 and MTUB29 and three copies of MIRU27, ETR-A, ETR-B and MTUB34.

Table 1. Description of clusters containing 3 or more isolates in this study and their worldwide distribution in the SITVIT2 database (interrogation made on September 25th 2013).

SIT (Lineage) Octal Number; Spoligotype Description	Number (%) in study	% in study vs. database	Distribution in Regions with ≥3% of a given SITs *	Distribution in countries with ≥3% of a given SITs **
53 (T1) 777777777760771	26 (11.93)	0.41	EURO-W 15.65, AMER-N 13.48, AMER-S 12.45, EURO-S 9.41, EURO-N 7.48, ASIA-W 7.31, AFRI-S 4.97, AFRI-E 4.65, ASIA-E 4.23, AFRI-N 3.52, EURO-E 3.26, AMER-C 3.23	USA 13.2, FXX 7.88, BRA 5.54, ITA 5.33, ZAF 4.86, TUR 3.47, AUT 3.42, CHN 3.09
42 (LAM9) 777777760760771	18 (8.26)	0.54	AMER-S 29.7, AMER-N 11.85, EURO-S 11.31, EURO-W 9.46, AFRI-N 8.57, EURO-N 4.9, AMER-C 3.55, AFRI-E 3.55, AFRI-S 3.16	BRA 12.51, USA 11.85, COL 7.61, MAR 7.05, ITA 6.54, FXX 5.05, ESP 3.34, VEN 3.31, ZAF 3.16
50 (H3) 777777777720771	16 (7.34)	0.47	AMER-S 18.16, EURO-W 17.54, AMER-N 17.54, EURO-S 11.53, EURO-E 5.51, EURO-N 5.45, AFRI-N 4.25, AFRI-S 4.04, CARI 3.48, ASIA-W 3.21	USA 17.51, BRA 7.52, FXX 6.87, AUT 6.07, ITA 5.43, ESP 5.43, PER 4.72, ZAF 4.04, CZE 3.66
SIT64 (LAM6) 777777760750771	6 (2.75)	1.6	AMER-S 49.33, AMER-N 25.07, EURO-W 6.4, EURO-S 4.8, ASIA-W 3.2	BRA 38.4, USA 25.07, GUF 6.13, PRT 3.73
SIT47 (H1) 777777774020771	5 (2.29)	0.32	EURO-W 20.29, AMER-N 15.51, EURO-S 13.47, AMER-S 12.64, EURO-N 10.47, EURO-E 7.21, ASIA-W 4.02, AFRI-N 3.64	USA 15.19, ITA 8.23, AUT 8.04, BRA 7.34, FXX 6.64, FIN 6.0, CZE 3.77, ESP 3.57, SWE 3.38, PER 3.06
SIT60 (LAM4) 777777760760731	5 (2.29)	1.17	AFRI-S 38.08, AMER-S 21.5, EURO-W 7.24, AFRI-N 7.01, AFRI-W 6.31, EURO-S 5.61, AMER-N 4.4	ZAF 38.08, BRA 14.72, FXX 5.14, USA 4.44, MAR 4.44, GMB 3.97, ITA 3.74, VEN 3.51
SIT73 (T) 777737777760731	5 (2.29)	1.94	EURO-S 17.05, EURO-W 13.95, AMER-N 13.95, AMER-S 11.24, AFRI-S 10.85, AFRI-E 7.36, ASIA-E 4.65, AMER-C 3.88	ITA 14.34, USA 13.95, ZAF 10.85, FXX 7.75, BRA 6.59, CHN 4.65, MOZ 3.88, MEX 3.1
4 (Unknown) 000000000760771	4 (1.83)	1.08	EURO-S 16.49, AMER-S 14.32, AMER-N 12.43, EURO-W 10.81, ASIA-W 10.54, AFRI-S 8.11, AFRI-E 8.11, EURO-N 6.22, EURO-E 5.68	USA 9.73, ITA 8.92, BRA 8.92, ZAF 8.11, TUR 7.84, BGR 5.41, ETH 4.32, FXX 4.05, ALB 4.05
SIT17 (LAM2) 677737760760771	4 (1.83)	0.6	AMER-S 59.58, AMER-N 17.95, CARI 8.9, EURO-S 5.58	BRA 28.36, VEN 27.0, USA 17.95, ESP 4.37, HTI 3.32, GLP 3.32
SIT33 (LAM3) 776177760760771	4 (1.83)	0.35	AFRI-S 28.76, AMER-S 25.31, AMER-N 14.07, EURO-S 12.57, EURO-W 7.26, AMER-C 4.78	ZAF 28.76, USA 14.07, BRA 12.04, ESP 7.88, PER 6.02, ARG 5.04, FXX 4.69, ITA 4.07, HND 3.81

Table 1. Cont.

SIT (Lineage) Octal Number; Spoligotype Description	Number (%) in study	% in study vs. database	Distribution in Regions with ≥3% of a given SITs *	Distribution in countries with ≥3% of a given SITs **
SIT3907 (T1) 7770377760631	4 (1.83)	100	AMER-S 100	BRA 100
SIT306 (T1) 777777760156077 1	3 (1.38)	11.54	EURO-W 38.46, AMER-N 26.92, AMER-S 19.23, EURO-N 7.69, AFRI-M 7.69	USA 26.92, BEL 26.92, BRA 15.39, FXX 11.54, GBR 7.69, CAF 7.69, GUF 3.85
SIT828 (LAM4) 377777760731	3 (1.38)	9.38	AMER-S 90.63, EURO-S 3.13, AFRI-W 3.13, AFRI-S 3.13	BRA 84.38, PRY 6.25, ZAF 3.13, ITA 3.13, GNB 3.13,
SIT1711 (LAM2) 677337607760771	3 (1.38)	37.5	AMER-S 75.0, EURO-S 12.5, AMER-C 12.5	BRA 37.5, VEN 25.0, MEX 12.5, ESP 12.5, COL 12.5,
SIT2498 (H3) 777760077620771	3 (1.38)	27.27	AMER-S 100	BRA 100
SIT3908 (LAM2) 677730207760771	3 (1.38)	75	AMER-S 100	BRA 100

*Worldwide distribution is reported for regions with more than 3% of a given SITs as compared to their total number in the SITVIT2 database. The definition of macro-geographical regions and sub-regions (http://unstats.un.org/unsd/methods/m49/m49regin.htm) is according to the United Nations; Regions: AFRI (Africa), AMER (Americas), ASIA (Asia), EURO (Europe), and OCE (Oceania), subdivided in: E (Eastern), M (Middle), C (Central), N (Northern), S (Southern), SE (South-Eastern), and W (Western). Furthermore, CARIB (Caribbean) belongs to Americas, while Oceania is subdivided in 4 sub-regions, AUST (Australasia), MEL (Melanesia), MIC (Micronesia), and POLY (Polynesia). Note that in our classification scheme, Russia has been attributed a new sub-region by itself (Northern Asia) instead of including it among rest of the Eastern Europe. It reflects its geographical localization as well as due to the similarity of specific TB genotypes circulating in Russia (a majority of Beijing genotypes) with those prevalent in Central, Eastern and South-Eastern Asia.
**The 3 letter country codes are according to http://en.wikipedia.org/wiki/ISO_3166-1_alpha-3; countrywide distribution is only shown for SITs with ≥3% of a given SITs as compared to their total number in the SITVIT2 database.

Table 2. Allelic diversity of 24 MIRU-VNTRs loci on 218 *Mycobacterium tuberculosis* strains isolated from pulmonary tuberculosis patients in Rio de Janeiro, Brazil.

| MIRU-VNTR *loci* | Genotype families by Spoligotyping | | | |
	LAM	H	T	All families
MIRU02	0.362	0.148	0.353	0.325
MIRU04	0.061	0.153	0.225	0.139
MIRU10	0.56	0.687	0.692	0.692
MIRU16	0.636	0.243	0.439	0.544
MIRU20	0.081	0.499	0.24	0.27
MIRU23	0.541	0.529	0.584	0.688
MIRU24	0.021	0.101	0.081	0.071
MIRU26	0.723	0.559	0.687	0.639
MIRU27	0.314	0.145	0.105	0.227
MIRU31	0.429	0.357	0.51	0.445
MIIRU39	0.138	0.251	0.027	0.124
MIRU40	0.517	0.623	0.76	0.689
ETR-A	0.14	0.417	0.625	0.537
ETR-B	0.407	0.237	0.396	0.389
ETR-C	0.357	0.447	0.573	0.541
MTUB04	0.612	0.53	0.68	0.716
MTUB21	0.436	0.587	0.673	0.624
MTUB29	0.253	0.432	0.266	0.397
MTUB30	0.269	0.494	0.652	0.591
MTUB34	0.617	0.305	0.617	0.452
MTUB39	0.391	0.573	0.391	0.577
QUB11	0.697	0.702	0.697	0.737
QUB26	0.384	0.63	0.384	0.522
QUB4156	0.842	0.842	0.842	0.84

The S family. Eleven isolates were classified as belonging to the S family and four of these exhibit SIT4, that in SpolDB4 was classified as LAM3-S Convergent, having absence of spacers 1–24 and 33–34 (The characteristic pattern of S family is absence of spacers 9–10, and 33–34). Among the others isolates, we observed SITs 53/n = 1, 102/n = 1, 378/n = 1, 2500/n = 1, 3907/n = 2 and 3909/n = 1; only two of these had the characteristic of S family (SIT3909/LAM8 and SIT2500/Unknown by SITVIT2 and H1 by Spotclust). Isolates of the S family shared their copy number in six loci, being one copy of MIRU24 and MTUB 21, two copies

each of MIRU20 and MIRU39 and three copies MIRU10 and MIRU27.

RD^{Rio}, RD174 and *fbpC*^{103} analysis

The genotypes defined by the presence of RD^{Rio}, RD174 and SNP *fbpC*^{103} were added to the classification based on 24-MIRU-VNTR typing (Table S2). Thirty-six isolates (16.5%) were excluded from the final analysis either because of showing genotypes suggestive for multiple infection of because of failure in at least one of the three genotype procedures. The results of 182

Table 3. Frequencies of strains according to classification by Spoligotyping (SITVIT2) and 24 loci MIRU-VNTR (MIRUVNTRPlus).

Lineage	Spoligotyping (%)		24 loci MIRU-VNTR (%)	
LAM	95	(43.58)	136	(62.39)
T	76	(34.86)	21	(9.63)
H	40	(18.35)	47	(21.56)
S	0	0	11	(5.05)
X	1	(0.46)	1	(0.46)
EAI	5	(2.29)	1	(0.46)
BEIJING	1	(0.46)	1	(0.46)

Spoligotyping[1]								
	LAM	T	Haarlem	S	X	EAI	Beijing	Total
LAM	91	31	11	0	0	3	0	136
T	0	20	1	0	0	0	0	21
Haarlem	3	16	26	0	1	1	0	47
S	1	9	1	0	0	0	0	11
X	0	0	1	0	0	0	0	1
EAI	0	0	0	0	0	1	0	1
Beijing	0	0	0	0	0	0	1	1
Total	95	76	40	0	1	5	1	218

(Left axis label: MIRU-VNTR 24 $loci^2$)

[1] Classification according to the SITIVIT2. For unknown Spoligotypes, we used the SpotClust.
[2] The patterns were classified based on MIRU-VNTRPlus tool that allows construction of a Neighbor-Joining based phylogenetic tree, visualizing proximity of a particular genotype with that of a set of reference strains to the genotype family level.

MIRU-VNTR 24 *loci* vs Spoligotyping	
Positive Predictive Value (PPV) or Precision	0.7433
Accuracy (AAC)	0.6376
True positive rate (TPR) or sencivity	0.8176
Erro Rate	0.3623

	LAM	T	Haarlen
Positive Predictive Value (PPV) or Precision	0.6691	0.9523	0.5531
Accuracy (AAC)	0.7752	0.7385	0.8394
True positive rate (TPR) or sencivity	0.9578	0.2631	0.6500
True Negative Rate (TNR) or Specificity	0.4193	0.9929	0.8820
Negative Predictive Value (NPV)	0.9512	0.7157	0.9181
Positive Predictive Value (PPV)	0.669	0.9523	0.5531
Erro Rate	0.2247	0.2614	0.3500

Figure 1. Confusion Matrix comparing the classifications obtained by Spoligotyping and MIRU-VNTR. [1] Classification according to SITIVIT2. For unknown Spoligotypes, we used SpotClust. [2] Patterns were classified based on a VNTRplus database that allows construction of a Neighbor-Joining based phylogenetic tree, visualizing proximity of a particular genotype with that of a set of reference strains to the genotype family level.

strains are summarized in Table 3 and for simplification of interpretation, we defined three groups, being LAM (n = 77), LAM-like (n = 33) and non-LAM (n = 72) as determined by spoligotyping and MIRU-VNTR typing.

The SNP $fbpC^{103}$ was observed in 107 (58.8%) of the isolates, including 97.3% (74/77) of LAM, 69.6% (23/29) of "LAM-like" and 13.9% (10/72) of non-LAM. The total frequency of RD^{Rio} was 11% (20/182), with 20.7% (16/77) among LAM, 6.0% (2/29) among "LAM-like" and 2.7% (2/72) among non-LAM isolates. The overall frequency of RD174 was 15.3% (28/182), being 33.7% (25/77) in the LAM genotype and 6.8% (2/29) in "LAM-like". All LAM/RD^{Rio} isolates presented the SNP $fbpC^{103}$ (Table 4 and Table 5), but not all were RD174; two isolates classified as LAM2 SIT3908 were not had RD174 (isolated from different patients). The frequency of RD174 (n = 25) was therefore higher than that of RD^{Rio} (32.5% vs 20.7%; p = 0.053 and $X^2 = 2,69$) in LAM. Eleven LAM isolates (14.3%) presented RD174 but not RD^{Rio}, all were positive for the $fbpC^{103}$.

The allelic diversity of the 24 loci MIRU-VNTR loci in LAM/RDRio isolates (Figure S2) showed that the copy number of combining of MIRU04, MIRU20, MIRU24, MIRU31, ETR-A, MTUB21 and MTUB30 loci was a signature of this genotype (2213231) (Table 4). In general, these loci present low variability in LAM, except for MTUB21, is being moderately variable in such isolates (table 2). Upon comparing 24 MIRU-VNTR signatures of LAM, LAM-like and non-LAM strains (Table 4), we observed two isolates that presented the hypothetical ancestral MIRU-VNTR signature (224226153321) for RD^{Rio} that was suggested by Lazzarini et al. (2007) [2]. One isolate (C2009) was classified as LAM1 SIT2539 and the other (C1966) as H3 SIT50 by spoligotyping and "LAM-like" by 24 MIRU-VNTR; both presented SNP $fbpC^{103}$ and were deleted for RD^{Rio} and RD174. Among the LAM/RD^{Rio} strains, frequency of LAM subtypes as defined by spoligotyping using SITVIT2 was 31.5% of LAM2, 25% of LAM9, 12.5% each of LAM6 and LAM5 and 6.3% each of LAM4 and LAM1. Figure 2 is a graphical representation of these three markers in the LAM strains as defined by 24 loci MIRU-VNTRs.

Table 4. Description of RD^{Rio} by 24 loci MIRU-VNTRs, $fbpC^{103}$ and RD174.

Strain	Year	Spoligotype Description	SITVT2	SIT2	Lineage*	RD^{RIO}	RD174	$fbpC^{103}$	24 MIRU-VNTRs — MIRU — 12 MIRU-VNTRs**												Additional 12 MIRU-VNTRs — ETR			MTUB						QUB		
									2	4	10	16	20	23	24	26	27	31	39	40	A	B	C	4	21	29	30	34	39	11	26	4156
C0537A2	2009		LAM9	42	**LAM**	+	+	+	2	2	4	2	2	6	1	5	3	3	2	1	2	1	4	3	3	4	1	5	2	2	7	3
C0996	2008		LAM9	42	**LAM**	+	+	+	2	2	3	2	2	6	1	4	3	3	2	1	2	1	4	3	5	5	1	5	2	2	5	2
C2105	2008		LAM9	42	**LAM**	+	+	+	2	2	2	2	2	6	1	4	3	3	2	1	2	1	4	3	4	4	1	6	2	2	4	2
C1231A2	2008		LAM4	1106	**LAM**	+	+	+	2	2	4	1	2	6	1	4	2	3	2	1	2	1	4	4	4	4	1	3	2	2	1*6	2
C2314	2008		LAM6	64	**LAM**	+	+	+	2	2	3	1	2	6	1	4	2	3	2	1	2	1	4	0	3	4	1	5	2	3	8	2
C0097	2009		Unknow	132	**LAM**	+	+	+	2	2	3	1	2	6	1	4	2	3	2	1	2	1	4	3	3	4	1	4	2	2	8	2
C6689	2008		LAM9	1800	**LAM**	+	+	+	2	2	4	3	2	6	1	5	3	2	1	3	3	2	4	2	4	6	1	3	2	2	7	2
C6582	2008		LAM6	1066	**LAM**	+	+	+	2	2	2	1	2	4	1	4	2	3	2	1	2	1	4	3	4	4	1	5	2	4	7	2
C0817	2008		LAM5	Orphan	**LAM**	+	+	+	2	2	4	2	2	5	1	5	3	3	2	1	2	2	4	3	4	4	1	2	2	3	4	2
C1893	2009		LAM5	1337	**LAM**	+	+	+	2	2	2	2	2	6	1	5	3	3	2	1	2	1	4	3	4	4	1	2	2	2	1	2
C2559	2009		LAM2	17	**LAM**	+	-	+	2	2	4	2	2	5	1	6	3	3	2	1	2	2	5	4	3	4	1	5	2	2	6	2
C3554	2009		LAM2	17	**LAM**	+	-	+	2	2	4	2	2	5	1	6	3	3	2	1	2	2	4	3	4	4	1	3	2	2	6	2
C2020	2009		LAM2	3908	**LAM**	+	+	+	2	2	4	2	2	5	1	5	3	3	2	1	2	2	4	4	3	4	1	2	2	3	4	2
C2046	2009		LAM2	3908	**LAM**	+	+	+	2	2	4	2	2	5	1	5	3	3	2	1	2	2	4	3	4	4	1	2	2	3	4	2
C2017	2008		LAM2	Orphan	**LAM**	+	+	+	2	2	4	2	2	5	1	5	3	3	2	1	2	2	4	3	4	4	1	2	2	3	4	2
C2009	2009		LAM1	2539	**LAM**	+	+	+	2	2	4	2	2	5	1	5	3	3	2	1	3	2	4	1	3	5	1	3	2	2	3	2
C1966	2008		H3	50	**LAM-like**	+	+	+	2	2	4	2	2	6	1	6	3	3	2	1	3	1	4	3	4	4	1	5	2	3	6	2
C0669	2009		T1	Orphan	**LAM-like**	+	-	-	2	2	4	*	2	5	1	5	3	3	2	1	3	1	4	4	4	4	1	*	2	2	8	2

Table 4. Cont.

Strain	Year	Spoligotype Description	SITVIT2	SIT2	Lineage*	RD^Rio	RD174	fbpC^103	12 MIRU-VNTRs**	Additional 12 MIRU-VNTRs	ETR (A, B, C)	MTUB (21, 29, 30, 34, 39)	QUB (11, 26, 4156)
C2142A2	2009	(spoligotype pattern)	H3	50	H	+	−	−	2 2 4 2 2 6 1 5 3 3 2 2	3 2 4 2 2 2 3 2 3 4 2 3	3 2 3	4 1 3 4 2	11 26 4156
C5817	2008	(spoligotype pattern)	Unknow	4	S	+	−	+	3 2 4 3 2 2 5 2 6 3 2 2	3 3 4 2 2 4 3 1 4 2 3 0	3 4 3	4 2 3 3 2	*

*Classification based on construction of phylogenetic tree using the N-J Algorithm and evaluating proximity of particular isolate with a set of 182 reference strains in the MIRU-VNTRPlus. Note: Strains with RD^Rio (LAM, Like-LAM and Non LAM).

**In bold the same number of copy of hypothetic Ancestral RD^Rio as suggested by Iazzarini et al, 2007 [2] and n the probable number of copy of hypothetic Ancestral RD^Rio in the additional 12 loci.

Discussion

Rio de Janeiro is the capital of the state of Rio de Janeiro, located in the southeast of Brazil, has a population of 16 million habitants, six and a half million of these living in the capital, being the second largest city and a major touristic attraction in Brazil (Census 2010 http://cidades.ibge.gov.br/xtras/perfil.php?lang=&codmun=330455 accessed 12/18/2013). According to the Ministry of Health, 11.639 new TB cases were recorded in 2011 in the state with an incidence of 72.3/100.000 habitants and the highest mortality rate (5.1/100.000 habitants) at the national level [1]. Rio de Janeiro city has strong economic and social contrasts, with a large portion of the population living in numerous suburbs, consisting of "communities", urban areas where housing conditions, health, education and security are extremely precarious. These factors are directly related to the number of TB cases detected.

In the present study, we performed spoligotyping and 24 MIRU-VNTR typing and characterized a SNP in $fbpC^{103}$ and the status of RD174 and RD^Rio, to decipher the population structure and the phylogenetic relationships of the MTBC strains of TB cases in this particular population that attended a single reference center in the city of Rio de Janeiro. *Mycobacterium tuberculosis* is classified into six phylogenetic lineages, each of which can be divided into sublineages [32]. Members of the Euro- American lineages represented by the LAM, H, T, S and X Spoligotyping families, which are genetically closely related are the most common in South America [19], [20], [21] and this was confirmed recently also in Brazil [22] and in different states of the country [8], [43], [44], [45], [46], [47] [48], but differences in the frequency of these families are observed, theses difference could also be due to differences in population and immigration history in each region or in period associated genotype frequencies or in differences in sample size. We here confirm the predominance of the Euro-American lineage (also known as lineage 4) with only two isolates classified as Indo-Oceanic lineage (EAI5) and East Asian lineage (Beijing), the first being more prevalent in east Africa, southeast Asia and in south India and the second in east Asia, Russia and South Africa. [32]. Recently, the frequency of such strains in Brazil was described as being less than 1% except for the higher frequency in the state of Pará, north Brazil [22].

The most prevalent families as defined by spoligotyping were LAM (43.6%), T (34.9%) and H (18.3%); however, classification based on 24-MIRU-VNTR and Neighbor-Joining based phylogenetic tree building using the database tool (www.miruvntrplus. org) showed considerable difference with that obtained by spoligotyping, being respectively 62.4%, 9.6% and 21.6% (error rate of 0.36). Among the isolates with discordant results, mostly isolates initially classified as T (n = 31), H (n = 11) and EAI (n = 3) by spoligotyping were classified as LAM by 24 MIRU-VNTRs typing, for isolates that did not show LAM prototype (absence of spacers 21–24 and 33–36), conveniently named LAM-like. Subsequently, 78.3% these isolates confirmed to be LAM by the presence of the LAM-specific SNP $fbpC^{103}$ [7], [38] and when considering the presence of this SNP as an absolute marker for LAM family the frequency of this lineage is 58.9%. We also have preliminary data showing that the supposed LAM-specific marker $ligB^{404}$ was observed in isolates that had been defined by spoligotyping as being T or H (data not shown). Very recently, Mokrousov et al. reported difference in classification of LAM strains as defined by spoligotyping and SNP analysis ($fbpC^{103}$ and $ligB^{404}$) [49] although both SNPs were previously validated in different collection of clinical isolates and reference strains as to be specific for the LAM family [7], [38].

Table 5. The frequency of fbpC^{103}, RD^{Rio} and RD174 in LAM, LAM-like and Non LAM isolates, as designated by spoligotyping and 24 MIRU-VNTR typing.

SNP and RD	LAM (%) (n = 77)	LAM-Like (%) (n = 29)	Non LAM (%) (n = 75)	Total Frequency (%) (n = 182)
fbpC^{103}	74 (97.3)	23 (79.3)	10 (13.3%)	107 (58.8)
RD^{Rio}	16 (20.7)	2 (6.9%)	2 (2.6%)	20 (11)
RD174	25 (32.5)	2 (6.9%)	1(1.3%)	28 (15.4)

It is now common knowledge that spoligotyping has limitations as a tool prediction of the exact phylogenetic relationships between strains of the MTBC, particularly in modern strains (Euro American, East Asian and Indian e East African) [16], [18], [24], [48], [49], [50], [51] [52], mainly due to homoplasy [16], [24]. The accuracy of the phylogenetic grouping by MIRU-VNTR is more exact than that of spoligotyping but depends on the number of loci included in the analysis [49] and classification errors are reduced when analyzing 24 loci [16]. Indeed, several authors have suggested that SNPs are more suitable than spoligotyping and MIRU-VNTR settings for phylogenetic classification [16], [38], [53], [54], [55], [56]. The fbpC^{103} is described as a good marker for the LAM family [7], [38], [49], [57], [58], [59], and our original intention was to evaluate if strains RD^{Rio} was still prevalent in Rio de Janeiro as seen previously by Lazzarini in clinical isolates collected between 2002 and 2003 [2], for this

purpose we believed only a marker that could differentiate LAM and non-LAM associated with the 24 MIRU-VNTR and Spoligotyping could be sufficient, however the scenario was observed more complex.

We here present the first data comparing classification by spoligotyping and MIRU-VNTRs and fbpC^{103} in Euro American lineage prevalent in Rio de Janeiro and Brazil. Among these different types of markers, we observed four major groups: (i) strains classified as LAM by spoligotyping and MIRU-VNTRs 24 loci without the LAM-characteristic SNP fbpC^{103}, (ii) strains not classified as LAM by spoligotyping and MIRU-VNTRs 24 loci but carrying the fbpC^{103} SNP, (iii) strains classified by spoligotyping as non-LAM but as LAM by MIRU-VNTRs 24 loci (LAM-like) and with SNP fbpC^{103} and (iv), strains classified by spoligotyping as non-LAM and LAM by MIRU-VNTRs 24 loci (LAM-like) but not presenting the SNP fbpC^{103}. These different scenarios could be explained by convergent evolution of spoligotypes and of MIRU-VNTRs loci (even including 24 alleles) because a limited number of loci were evaluated that might evolve rapidly and therefore susceptible to pronounced convergence [53] and/or because the existence of ancestral progenitor of Euro- American lineages containing SNP fbpC^{103}. In addition, a possible important limitation of the current classification based on MIRU-VNTRs and their similarity with genotypes present in MIRU-VNTRplus is that, despite including well characterized strains, the database contains a limited number of strains that does not reflect the real genetic diversity of isolates belonging to the MTBC; another limitation of this study is that no additional specific SNPs for H, T, S and X and T were investigated. The Whole Genome Sequencing (WGS) is superior to conventional genotyping for MTBC [60] and has been used in areas not yet studied, from global (phylogeography) for site (transmission chains and diversity of circulating strains), for single patient (clonal diversity) and the bacterium itself (evolutionary studies) [61]. We intend to compare through WGS (developing) these isolates to a better comprehension of the evolution of lineages Euro American, such as the development of new methodologies that allow a more rapid and accurate typing.

Analyzing the data obtained in this study, the spoligotypes of the T family showed the largest number of divergent results when compared to classification by 24-MIRU (TPR = 0,26/ TNR = 0,99). The prototype of the T family is characterized by the absence of spacers 33–36 only [48] and have been observed in almost every country, representing 20% of all isolates deposited in the database SITVITWEB. Despite the high frequency, this genotype family is still considered as "ill-defined" and includes non-monophyletic groups [54] [55]. In South America, the frequency of this family is 26.7% and in Brazil 18.6% (370/ 1991), with the T1 SIT53 subfamily being the most prevalent SIT [22], as observed also in this study. Here, among the 76 isolates classified as T by spoligotyping, only 21 (27.6%) were confirmed

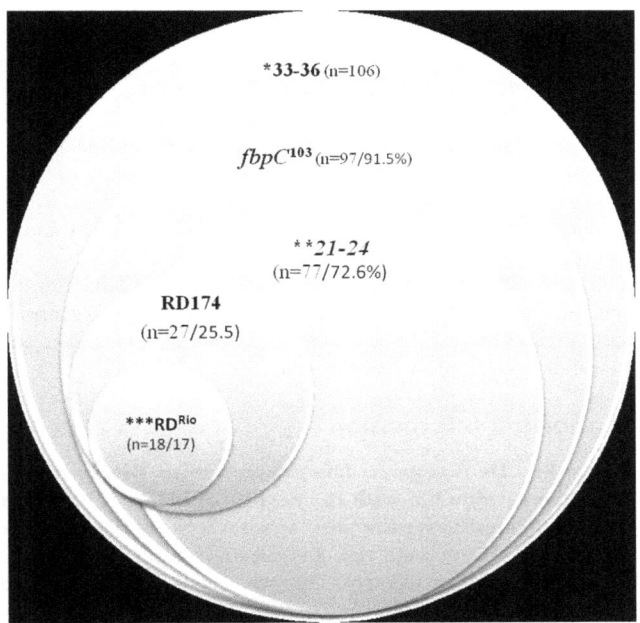

Figure 2. Venn diagram illustrating the different markers in isolated classified as LAM by 24 loci MIRU-VNTR. Notes: The Venn diagram was constructed based on LAM isolates defined by 24 loci MIURU-VNTR. The sizes of the circles is not proportional to the real frequency of these markers. Twenty isolates classified with LAM by MIRU-VNTRs were removed from the final analysis because of showing a mixed genotype or failure in at least one of the three genotype procedures. * absence of spacers 33–36 in the spoligotyping profile, ** absence of spacers 21–24 in the spoligotyping profile, ***Two isolates RD^{Rio} but not RD174 (LAM2 SIT3908).

by 24 MIRU-VNTRs typing, the rest was reclassified as LAM (n = 31), S (n = 9) and H (n = 16). Interestingly, SIT53 was associated with mixed infection in a study conducted in South Africa, a country characterized by a high prevalence of TB [57] and Lazzarini et al. [62], using a computational approach, showed that this is the most frequent false spoligotype derived from mixed infections. However, among isolates with this SIT, we did not observe double signals during 24 MIRU typing indicative for mixed infections. We also observed that spoligotypes, indicative for the T family, were sometimes grouped with spoligotypes of the H family by MIRU-VNTR typing. This could be related with the fact that the prototype spoligotype defining T1 SIT53 and H3 SIT 50 differ only in the absence of the spacer 31. We also observed that in our population, the absence of spacer 31 is not crucial for classification as being either H or T; what differentiates between the two is the number of copies of ETR-A, ETR-B, ETR-C. The H3 subfamily is characterized by the combination 323 of these alleles while T strains present considerable diversity of these loci (one to three copies of ETR-A, two or three copies of ETR-B and three to five copies of ETR-C).

The RD nominated RD^{Rio} was first reported as new *M. tuberculosis* lineage in Rio de Janeiro in 2007, Lazzarini et al. [2] and is a deletion of 26.3 kb restricted to the LAM family and in particular, in subfamilies LAM1, LAM2, LAM4, LAM5, LAM6 and LAM9. This deletion affects 10 genes, including two genes encoding Proline-proline-glutamic Acid Proteins (PPE) [2] and association between RD^{Rio} strains and high prevalence may be related to virulence and/or adaptation specifies the Latin American and European population-based epidemiological and clinical characteristics; however, studies have proven to be inconclusive or contradictory [4], [63], [64]. The lineage RD^{Rio}, was identified in different countries [5], [6], [7], [49] and in Brazil, has been described in different states, including Rio de Janeiro, Rio Grande do Sul, Minas Gerais, Espírito Santo and Rio Grande do Sul [3], [8], [63], [64]. The frequency ranges from 30 to 38% of isolates tested and was associated with MRD-TB [4], [63], [64] and with genotype clustering [5],[63], indicating a higher rate of recent infection and transmissibility. Another deletion, RD174, initially described as a marker for the LAM family and as a co-marker for RD^{Rio} [7], [8] was associated with high transmissibility [10].

The frequency of strains RD^{Rio} in the present study, with an overall frequency of 11% and 20.7% in LAM, is lower than that observed in other studies and even when including the eight isolates that showed mixed RD^{Rio}/WT signals, resulting frequencies of 14.7% and 28.2% still lower than previously observed. This difference could be related with the relative low frequency of LAM1, LAM2, LAM5, LAM6 and LAM9 related subfamilies in our study sample. Earlier studies on classification of RD^{Rio} strains was based on genotypes defined by spoligotyping and/or 12 MIRUs only [2], [5], [6], [7], [8], [63], [64] and again, this is the first study that used 24 MIRU-VNTR typing to conduct a more detailed phylogenetic analysis. We verified the signature of MIRU-VNTR loci for RD^{Rio} and RD174, as compared with that of the hypothetical ancestral of RD^{Rio} as defined by 12 MIRUs [2] and that, besides sharing two copies of MIRU04 and MIRU20, they also share one copy of MIRU24, three copies of MIRU31, two copies of ETR-A, three copies of MTUB21 and one copy of MTUB30, yielding 2213231 as fingerprint for RD^{Rio}. All LAM/RD174 isolates, with or without RD^{Rio}, carried two copies of MIRU20 and ETRA and one copy of MTUB31. We observed that LAM3 has two copies of MTUB30 (2213232) and we propose

that this subfamily that do not carry RD^{Rio}, has evolved independently. We also observed RD^{Rio} in two isolates (2.6%) with a spoligotype not indicative for being LAM; a small number of such cases have also been observed by other research groups [5], [7], [64]. One of these isolates that had been classified by spoligotyping as H3 SIT53 was reclassified by 24 MIRU as being "LAM-Like" and called attention because it had the MIRU signature of the hypothetical ancestor of RD^{Rio} and carried the SNP $fbpC^{103}$ and RD174. Overall, we observed four scenarios: (i) isolates with the RD^{Rio} and RD174, (ii) isolates showing only RD^{Rio} (iii) isolates showing only RD174 and (iv) isolates that did not carry any of the deletions, suggesting that both markers evolved on different time points. This is different from earlier data [7] that claim that RD174 is an absolute co-marker for RD^{Rio} and therefore, studies that use the presence of RD174 to infer the presence of RD^{Rio} [8] may overestimate the frequency of the latter. In 2007, Lazzarini et al. [2] suggested that RD^{Rio} arose by homologous recombination between genes and although all neighboring sequences were identical, such event could have happened more than once. This suggests that the RD174 deletion occurred before RD^{Rio} but this needs to be confirmed as we also observed RD^{Rio} strains that had intact RD174. In Figure S3, we propose a possible evolution of members of the Euro-American lineages but with the limitation that the spoligotype defined lineages S, X and other are not represented and this concerns a sampling only from Rio de Janeiro.

In a recent study, Hill et al. (2012) [53], mentions the difficulty in studying the evolution of the Euro-American lineage (LAM, Haarlem, T, X and S) using spoligotyping due to the large number of IS6110 copies in such strains that may result in IS6110 mediated deletions in the DR locus. This might be the reason why bacterial evolution exclusively based on spoligotyping is not robust and the wide range of profiles reported as unclassified in SITVITWEB. Our approach, combining spoligotyping, MIRU-VNTRs, SNPs and RDs allowed the reclassification of 13 SITs that did not rank in SITVITWEB, allowing definition of 29 new spoligotypes and refine classification of isolates belonging to Euro-American lineage. Our data are also support the idea that absence of spacers 21–24 is not sufficient for classification as LAM and of spacer 31 to differentiate T and H, the latter indicative for subfamilies H3 SIT50 and T1 SIT53. Possible explanations are that ancestral lineages are currently circulating (plesiomorphic state) or that the isolates are suffering homoplasic evolution and reversion into the plesiomorphic state.

Supporting Information

Figure S1 Dendrogram constructed with BioNumerics software version 6.6 with the results of MIRU-VNTRs 24 loci and spoligotyping by similarity coefficient for categorical data and the neighbor-joining algorithm. 1st column (after spoligotypes): number of isolated label; 2nd column: patients who have more than one isolate in the study (n = 27) received numbering 1–27, and the different strains present the same numbering; 3rd column: International Spoligotype Types (SIT); 4th column: classification obtained through SITVITWEB (family and subfamily).

Figure S2 Allelic diversity of the 24 loci MIRU-VNTR loci in LAM/RDRio isolates.

Figure S3 Possible evolution of *M. tuberculosis* lineage Euro-American (LAM, T and H) according to the markers analyzed in this study. The Euro-American Ancestral evolved into two distinct groups: LAM ancestral and T/H ancestral, characterized by absences of spacers 33–36 and one copy of MIRU24 that is common to all Euro-American lineages. The ancestral LAM strains have *fbpC*^{103} and this was the basis of LAM A (LAM9), with additional absence of spacers 21–24, one copy of MTUB30 and two copies of MIRU04 and ETR-A. LAM A on its turn is the basis for two other groups: LAM B (LAM9-LAM4, LAM1-LAM2-LAM5 and LAM6) and LAM C (LAM3). The LAM B evolved from LAM B1, characterized by a deleted RD174 and on its turn to LAM B2, with both deleted RD174 and RD^{Rio}. The H/T Ancestral lineage is the origin of both groups H and T (difference only in MIRU-VNTR copies), showing absence of spacers 33–36; additional loss of spacer 31 led to subtype H A, observed in high frequency in this study.

Table S1 Detailed genotyping results and associated demographic, epidemiologic and nomenclature information on 218 Mycobacterium tuberculosis strains

isolated from pulmonary tuberculosis patients in Rio de Janeiro, Brazil.

Table S2 Detailed genotyping results of *fbpC*^{103}, RD174 and RD^{Rio} on 182 M. tuberculosis strains isolated from pulmonary tuberculosis patients in Rio de Janeiro, Brazil.

Acknowledgments

We thank the Genomic Platform of Network Technology Platforms (PDTIS) Institute Oswaldo Cruz/FIOCRUZ.RPT01D - Analysis of Fragments Oswaldo Cruz Institute in Rio de Janeiro.

Author Contributions

Conceived and designed the experiments: SEGV HMG ARS NB PNS. Performed the experiments: SEGV HMG ARS NB PNS. Analyzed the data: SEGV CCA HMG LLG ECC MIA DC MKG. Contributed reagents/materials/analysis tools: PNS KVL MLL FT PCSC NR CS. Wrote the paper: SEGV CCA PNS NR CS.

References

1. Secretaria de Vigilância em Saúde – Ministério da Saúde, Especial Tuberculose (2012) Boletim Epidemiológico, vol. 43.
2. Lazzarini LC, Huard RC, Boechat NL, Gomes HM, Oelemann MC, et al. (2007) Discovery of a novel Mycobacterium tuberculosis lineage that is a major cause of tuberculosis in Rio de Janeiro, Brazil. J Clin Microbiol 45: 3891–3902.
3. Lazzarini LC, Spindola SM, Bang H, Gibson AL, Weisenberg S, et al. (2008) RDRio Mycobacterium tuberculosis infection is associated with a higher frequency of cavitary pulmonary disease. J Clin Microbiol 46: 2175–2183.
4. Dalla Costa ER, Lazzarini LC, Perizzolo PF, Díaz CA, Spies FS, et al. (2013) Mycobacterium tuberculosis of the RDRio genotype is the predominant cause of tuberculosis and associated with multidrug resistance in Porto Alegre City, South Brazil. J Clin Microbiol 51: 1071–1077.
5. Weisenberg SA, Gibson AL, Huard RC, Kurepina N, Bang H, et al. (2012) Distinct clinical and epidemiological features of tuberculosis in New York City caused by the RD(Rio) Mycobacterium tuberculosis sublineage. Infect Genet Evol 12: 664–670.
6. David S, Duarte EL, Leite CQ, Ribeiro JN, Maio JN, et al. (2012) Implication of the RD(Rio) Mycobacterium tuberculosis sublineage in multidrug resistant tuberculosis in Portugal. Infect Genet Evol 12: 1362–1367.
7. Gibson AL, Huard RC, Gey van Pittius NC, Lazzarini LC, Driscoll J, et al. (2008) Application of sensitive and specific molecular methods to uncover global dissemination of the major RDRio Sublineage of the Latin American-Mediterranean Mycobacterium tuberculosis spoligotype family. J Clin Microbiol 46: 1259–1267.
8. Cardoso Oelemann M, Gomes HM, Willery E, Possuelo L, Batista Lima KV, et al. (2011) The forest behind the tree: phylogenetic exploration of a dominant Mycobacterium tuberculosis strain lineage from a high tuberculosis burden country. PLoS One 6: e18256.
9. Rindi L, Lari N, Garzelli C (2012) Large Sequence Polymorphisms of the Euro-American lineage of Mycobacterium tuberculosis: a phylogenetic reconstruction and evidence for convergent evolution in the DR locus. Infect Genet Evol 12: 1551–1557.
10. de Jong BC, Antonio M, Awine T, Ogungbemi K, de Jong YP, et al. (2009) Use of spoligotyping and large sequence polymorphisms to study the population structure of the Mycobacterium tuberculosis complex in a cohort study of consecutive smear-positive tuberculosis cases in The Gambia. J Clin Microbiol 47: 994–1001.
11. Brudey K, Filliol I, Ferdinand S, Guernier V, Duval P, et al. (2006) Long-term population-based genotyping study of Mycobacterium tuberculosis complex isolates in the French departments of the Americas. J Clin Microbiol 44: 183–191.
12. Kamerbeek J, Schouls L, Kolk A, van Agterveld M, van Soolingen D, et al. (1997) Simultaneous detection and strain differentiation of Mycobacterium tuberculosis for diagnosis and epidemiology. J Clin Microbiol 35: 907–914.
13. Supply P, Mazars E, Lesjean S, Vincent V, Gicquel B, et al. (2000) Variable human minisatellite-like regions in the Mycobacterium tuberculosis genome. Mol Microbiol 36: 762–771.
14. Chacón-Salinas R, Serafín-López J, Ramos-Payán R, Méndez-Aragón P, Hernández-Pando R, et al. (2005) Differential pattern of cytokine expression by macrophages infected in vitro with different Mycobacterium tuberculosis genotypes. Clin Exp Immunol 140: 443–449.

15. Comas I, Homolka S, Niemann S, Gagneux S (2009) Genotyping of genetically monomorphic bacteria: DNA sequencing in Mycobacterium tuberculosis highlights the limitations of current methodologies. PLoS One 4: e7815.
16. Cowan LS, Diem L, Monson T, Wand P, Temporado D, et al. (2005) Evaluation of a two-step approach for large-scale, prospective genotyping of Mycobacterium tuberculosis isolates in the United States. J Clin Microbiol 43: 688–695.
17. Dale JW, Nor RM, Ramayah S, Tang TH, Zainuddin ZF (1999) Molecular epidemiology of tuberculosis in Malaysia. J Clin Microbiol 37: 1265–1268.
18. Kato-Maeda M, Gagneux S, Flores LL, Kim EY, Small PM, et al. (2011) Strain classification of Mycobacterium tuberculosis: congruence between large sequence polymorphisms and spoligotypes. Int J Tuberc Lung Dis 15: 131–133.
19. Abadía E, Sequera M, Ortega D, Méndez MV, Escalona A, et al. (2009) Mycobacterium tuberculosis ecology in Venezuela: epidemiologic correlates of common spoligotypes and a large clonal cluster defined by MIRU-VNTR-24. BMC Infect Dis 9: 122.
20. Cerezo I, Jiménez Y, Hernandez J, Zozio T, Murcia MI, et al. (2012) A first insight on the population structure of Mycobacterium tuberculosis complex as studied by spoligotyping and MIRU-VNTRs in Bogotá, Colombia. Infect Genet Evol 12: 657–663.
21. Gonzalo X, Ambroggi M, Cordova E, Brown T, Poggi S, et al. (2011) Molecular epidemiology of Mycobacterium tuberculosis, Buenos Aires, Argentina. Emerg Infect Dis 17: 528–531.
22. Gomes HM, Elias AR, Oelemann MA, Pereira MA, Montes FF, et al. (2012) Spoligotypes of Mycobacterium tuberculosis complex isolates from patients residents of 11 states of Brazil. Infect Genet Evol 12: 649–656.
23. Demay C, Liens B, Burguière T, Hill V, Couvin D, et al. (2012) SITVITWEB–a publicly available international multimarker database for studying Mycobacterium tuberculosis genetic diversity and molecular epidemiology. Infect Genet Evol 12: 755–766.
24. Brudey K, Driscoll JR, Rigouts L, Prodinger WM, Gori A, et al. (2006) Mycobacterium tuberculosis complex genetic diversity: mining the fourth international spoligotyping database (SpolDB4) for classification, population genetics and epidemiology. BMC Microbiol 6: 23.
25. Filliol I, Driscoll JR, Van Soolingen D, Kreiswirth BN, Kremer K, et al. (2002) Global distribution of Mycobacterium tuberculosis spoligotypes. Emerg Infect Dis 8: 1347–1349.
26. Sola C, Filliol I, Gutierrez MC, Mokrousov I, Vincent V, et al. (2001) Spoligotype database of Mycobacterium tuberculosis: biogeographic distribution of shared types and epidemiologic and phylogenetic perspectives. Emerg Infect Dis 7: 390–396
27. Sola C, Filliol I, Legrand E, Lesjean S, Locht C, et al. (2003) Genotyping of the Mycobacterium tuberculosis complex using MIRUs: association with VNTR and spoligotyping for molecular epidemiology and evolutionary genetics. Infect Genet Evol 3: 125–133.
28. Weniger T, Krawczyk J, Supply P, Niemann S, Harmsen D (2010) MIRU-VNTRplus: a web tool for polyphasic genotyping of Mycobacterium tuberculosis complex bacteria. Nucleic Acids Res 38: W326–331.
29. Allix-Béguec C, Fauville-Dufaux M, Supply P (2008) Three-year population-based evaluation of standardized mycobacterial interspersed repetitive-unit-variable-number tandem-repeat typing of Mycobacterium tuberculosis. J Clin Microbiol 46: 1398–1406.

30. García de Viedma D, Mokrousov I, Rastogi N (2011) Innovations in the molecular epidemiology of tuberculosis. Enferm Infecc Microbiol Clin 29 Suppl 1: 8–13.

31. Goh KS, Rastogi N, Berchel M, Huard RC, Sola C (2005) Molecular evolutionary history of tubercle bacilli assessed by study of the polymorphic nucleotide within the nitrate reductase (narGHJI) operon promoter. J Clin Microbiol 43: 4010–4014.

32. Gagneux S, Small PM (2007) Global phylogeography of Mycobacterium tuberculosis and implications for tuberculosis product development. Lancet Infect Dis 7: 328–337.

33. Huard RC, Fabre M, de Haas P, Lazzarini LC, van Soolingen D, et al. (2006) Novel genetic polymorphisms that further delineate the phylogeny of the Mycobacterium tuberculosis complex. J Bacteriol 188: 4271–4287.

34. Vasconcellos SE, Huard RC, Niemann S, Kremer K, Santos AR, et al. (2010) Distinct genotypic profiles of the two major clades of Mycobacterium africanum. BMC Infect Dis 10: 80.

35. Stucki D, Gagneux S (2013) Single nucleotide polymorphisms in Mycobacterium tuberculosis and the need for a curated database. Tuberculosis (Edinb) 93: 30–39.

36. Musser JM, Amin A, Ramaswamy S (2000) Negligible genetic diversity of mycobacterium tuberculosis host immune system protein targets: evidence of limited selective pressure. Genetics 155: 7–16.

37. Dos Vultos T, Mestre O, Rauzier J, Golec M, Rastogi N, et al. (2008) Evolution and diversity of clonal bacteria: the paradigm of Mycobacterium tuberculosis. PLoS One 3: e1538.

38. Homolka S, Projahn M, Feuerriegel S, Ubben T, Diel R, et al. (2012) High resolution discrimination of clinical Mycobacterium tuberculosis complex strains based on single nucleotide polymorphisms. PLoS One 7: e39855.

39. van Embden JD, Cave MD, Crawford JT, Dale JW, Eisenach KD, et al. (1993) Strain identification of Mycobacterium tuberculosis by DNA fingerprinting: recommendations for a standardized methodology. J Clin Microbiol 31: 406–409.

40. Zhang J, Abadia E, Refregier G, Tafaj S, Boschiroli ML, et al. (2010) Mycobacterium tuberculosis complex CRISPR genotyping: improving efficiency, throughput and discriminative power of 'spoligotyping' with new spacers and a microbead-based hybridization assay. J Med Microbiol 59: 285–294.

41. Graur D, W-H Li (2000) Dynamics of genes in populations, p. 58. In D. Graur and W.-H. Li (ed.), Fundamentals of molecular evolution. Sinauer Associates, Sunderland, Mass.

42. Hunter PR, Gaston MA (1988) Numerical index of the discriminatory ability of typing systems: an application of Simpson's index of diversity. J Clin Microbiol 26: 2465–2466.

43. Noguti EN, Leite CQ, Malaspina AC, Santos AC, Hirata RD, et al. (2010) Genotyping of Mycobacterium tuberculosis isolates from a low-endemic setting in northwestern state of Paraná in Southern Brazil. Mem Inst Oswaldo Cruz 105: 779–785.

44. Mendes NH, Melo FA, Santos AC, Pandolfi JR, Almeida EA, et al. (2011) Characterization of the genetic diversity of Mycobacterium tuberculosis in São Paulo city, Brazil. BMC Res Notes 4: 269.

45. Miranda SS, Carvalho WaS, Suffys PN, Kritski AL, Oliveira M, et al. (2011) Spoligotyping of clinical Mycobacterium tuberculosis isolates from the state of Minas Gerais, Brazil. Mem Inst Oswaldo Cruz 106: 267–273.

46. Luiz Ro S, Suffys P, Barroso EC, Kerr LR, Duarte CR, et al. (2013) Genotyping and drug resistance patterns of Mycobacterium tuberculosis strains observed in a tuberculosis high-burden municipality in Northeast, Brazil. Braz J Infect Dis 17: 338–345.

47. Perizzolo PF, Dalla Costa ER, Ribeiro AW, Spies FS, Ribeiro MO, et al. (2012) Characteristics of multidrug-resistant Mycobacterium tuberculosis in southern Brazil. Tuberculosis (Edinb) 92: 56–59.

48. Sreevatsan S, Pan X, Stockbauer KE, Connell ND, Kreiswirth BN, et al. (1997) Restricted structural gene polymorphism in the Mycobacterium tuberculosis complex indicates evolutionarily recent global dissemination. Proc Natl Acad Sci U S A 94: 9869–9874.

49. Mokrousov I, Vyazovaya A, Narvskaya O (2014) Mycobacterium tuberculosis Latin American-Mediterranean family and its sublineages in the light of robust evolutionary markers. J Bacteriol 196: 1833–1841.

50. Warren RM, Streicher EM, Sampson SL, van der Spuy GD, Richardson M, et al. (2002) Microevolution of the direct repeat region of Mycobacterium tuberculosis: implications for interpretation of spoligotyping data. J Clin Microbiol 40: 4457–4465.

51. Kato-Maeda M, Gagneux S, Flores LL, Kim EY, Small PM, et al. (2011) Strain classification of Mycobacterium tuberculosis: congruence between large sequence polymorphisms and spoligotypes. Int J Tuberc Lung Dis 15: 131–133.

52. Hill V, Zozio T, Sadikalay S, Viegas S, Streit E, et al. (2012) MLVA based classification of Mycobacterium tuberculosis complex lineages for a robust phylogeographic snapshot of its worldwide molecular diversity. PLoS One 7: e41991.

53. Filliol I, Motiwala AS, Cavatore M, Qi W, Hazbón MH, et al. (2006) Global phylogeny of Mycobacterium tuberculosis based on single nucleotide polymorphism (SNP) analysis: insights into tuberculosis evolution, phylogenetic accuracy of other DNA fingerprinting systems, and recommendations for a minimal standard SNP set. J Bacteriol 188: 759–772.

54. Borile C, Labarre M, Franz S, Sola C, Refrégier G (2011) Using affinity propagation for identifying subspecies among clonal organisms: lessons from M. tuberculosis. BMC Bioinformatics 12: 224.

55. Borile C, Labarre M, Franz S, Sola C, Refrégier G (2011) Using affinity propagation for identifying subspecies among clonal organisms: lessons from M. tuberculosis. BMC Bioinformatics 12: 224.

56. Nakanishi N, Wada T, Arikawa K, Millet J, Rastogi N, et al. (2013) Evolutionary robust SNPs reveal the misclassification of Mycobacterium tuberculosis Beijing family strains into sublineages. Infect Genet Evol 16: 174–177.

57. Stavrum R, Mphahlele M, Ovreås K, Muthivhi T, Fourie PB, et al. (2009) High diversity of Mycobacterium tuberculosis genotypes in South Africa and preponderance of mixed infections among ST53 isolates. J Clin Microbiol 47: 1848–1856.

58. Lopes JS, Marques I, Soares P, Nebenzahl-Guimaraes H, Costa J, et al. (2013) SNP typing reveals similarity in Mycobacterium tuberculosis genetic diversity between Portugal and Northeast Brazil. Infect Genet Evol 18: 238–246.

59. Chuang PC, Chen YM, Chen HY, Jou R (2010) Single nucleotide polymorphisms in cell wall biosynthesis-associated genes and phylogeny of Mycobacterium tuberculosis lineages. Infect Genet Evol 10: 459–466.

60. Roetzer A, Diel R, Kohl TA, Rückert C, Nübel U, et al. (2013) Whole genome sequencing versus traditional genotyping for investigation of a Mycobacterium tuberculosis outbreak: a longitudinal molecular epidemiological study. PLoS Med 10: e1001387.

61. Ford C, Yusim K, Ioerger T, Feng S, Chase M, et al. (2012) Mycobacterium tuberculosis–heterogeneity revealed through whole genome sequencing. Tuberculosis (Edinb) 92: 194–201.

62. Lazzarini LC, Rosenfeld J, Huard RC, Hill V, Lapa e Silva JR, et al. (2012) Mycobacterium tuberculosis spoligotypes that may derive from mixed strain infections are revealed by a novel computational approach. Infect Genet Evol 12: 798–806.

63. Von Groll A, Martin A, Felix C, Prata PF, Honscha G, et al. (2010) Fitness study of the RDRio lineage and Latin American-Mediterranean family of Mycobacterium tuberculosis in the city of Rio Grande, Brazil. FEMS Immunol Med Microbiol 58: 119–127.

64. Vinhas SA, Palaci M, Marques HS, Lobo de Aguiar PP, Ribeiro FK, et al. (2013) Mycobacterium tuberculosis DNA fingerprint clusters and its relationship with RD(Rio) genotype in Brazil. Tuberculosis (Edinb) 93: 207–212.

Permissions

The contributors of this book come from diverse backgrounds, making this book a truly international effort. This book will bring forth new frontiers with its revolutionizing research information and detailed analysis of the nascent developments around the world.

We would like to thank all the contributing authors for lending their expertise to make the book truly unique. They have played a crucial role in the development of this book. Without their invaluable contributions this book wouldn't have been possible. They have made vital efforts to compile up to date information on the varied aspects of this subject to make this book a valuable addition to the collection of many professionals and students.

This book was conceptualized with the vision of imparting up-to-date information and advanced data in this field. To ensure the same, a matchless editorial board was set up. Every individual on the board went through rigorous rounds of assessment to prove their worth. After which they invested a large part of their time researching and compiling the most relevant data for our readers.

The editorial board has been involved in producing this book since its inception. They have spent rigorous hours researching and exploring the diverse topics which have resulted in the successful publishing of this book. They have passed on their knowledge of decades through this book. To expedite this challenging task, the publisher supported the team at every step. A small team of assistant editors was also appointed to further simplify the editing procedure and attain best results for the readers.

Apart from the editorial board, the designing team has also invested a significant amount of their time in understanding the subject and creating the most relevant covers. They scrutinized every image to scout for the most suitable representation of the subject and create an appropriate cover for the book.

The publishing team has been an ardent support to the editorial, designing and production team. Their endless efforts to recruit the best for this project, has resulted in the accomplishment of this book. They are a veteran in the field of academics and their pool of knowledge is as vast as their experience in printing. Their expertise and guidance has proved useful at every step. Their uncompromising quality standards have made this book an exceptional effort. Their encouragement from time to time has been an inspiration for everyone.

The publisher and the editorial board hope that this book will prove to be a valuable piece of knowledge for researchers, students, practitioners and scholars across the globe.

List of Contributors

Marcelo Gehara
Division of Evolutionary Biology, Zoological Institute, Technical University of Braunschweig, Braunschweig, Germany
Pós-graduação em Sistemática e Evolução, Centro de Biociências, Universidade Federal do Rio Grande do Norte, Campus Universitário Lagoa Nova, Natal, RN, Brasil

Andrew J. Crawford
Departamento de Ciencias Biológicas, Universidad de los Andes, Bogotá, Colombia
Smithsonian Tropical Research Institute, Panamá, Republic of Panama

Victor G. D. Orrico
Universidade de São Paulo, Instituto de Biociências, Departamento de Zoologia, São Paulo, Brasil

Ariel Rodríguez and Miguel Vences
Division of Evolutionary Biology, Zoological Institute, Technical University of Braunschweig, Braunschweig, Germany

Stefan Lötters
Trier University, Biogeography Department, Trier, Germany

Antoine Fouquet
CNRS-Guyane - USR3456, Immeuble Le Relais - 2, Cayenne, French Guiana

Lucas S. Barrientos
Departamento de Ciencias Biológicas, Universidad de los Andes, Bogotá, Colombia

Francisco Brusquetti
Departamento de Zoologia, Instituto de Biociências, UNESP, Rio Claro, São Paulo, Brasil; Instituto de Investigación Biológica del Paraguay, Asunción, Paraguay

Ignacio De la Riva
Museo Nacional de Ciencias Naturales, Madrid, Spain

Raffael Ernst and Monique Hölting
Museum of Zoology, Senckenberg Natural History Collections Dresden, Dresden, Germany

Giuseppe Gagliardi Urrutia
Peruvian Center for Biodiversity and Conservation (PCRC), Nanay, Iquitos, Peru

Frank Glaw
Zoologische Staatssammlung München, München, Germany

Juan M. Guayasamin
Universidad Tecnológica Indoamérica, Centro de Investigación de la Biodiversidad y el Cambio Climático (BioCamp), Cotocollao, Quito, Ecuador

Martin Jansen
Senckenberg Gesellschaft für Naturforschung, Frankfurt am Main, Germany

Philippe J. R. Kok
Amphibian Evolution Lab, Department of Biology, Vrije Universiteit Brussel, Brussels, Belgium

Axel Kwet
German Herpetological Society (DGHT), Mannheim, Germany

Rodrigo Lingnau
Universidade Tecnológica Federal do Paraná, Francisco Beltrão, PR, Brasil

Mariana Lyra and Celio F. B. Haddad
Departamento de Zoologia, Instituto de Biociências, UNESP, Rio Claro, São Paulo, Brasil

Jiří Moravec
Department of Zoology, National Museum, Prague, Czech Republic

José P. Pombal Jr.
Departamento de Vertebrados, Museu Nacional, Universidade Federal do Rio de Janeiro, Rio de Janeiro, Brazil

Fernando J. M. Rojas-Runjaic
Fundación La Salle de Ciencias Naturales, Museo de Historia Natural La Salle, Caracas, Venezuela

Arne Schulze and Jörn Köhler
Hessisches Landesmuseum Darmstadt, Department of Zoology, Darmstadt, Germany

J. Celsa Señaris
Laboratorio de Ecología y Genética de Poblaciones, Centro de Ecología, Instituto Venezolano de Investigaciones Científicas, Caracas, Venezuela

Mirco Solé
Universidade Estadual de Santa Cruz, Departamento de Ciências Biológicas, Rodovia Ilhéus-Itabuna, Bahia, Brasil

Miguel Trefaut Rodrigues
Universidade de São Paulo, Instituto de Biociências, Departamento de Zoologia, São Paulo, Brasil

Evan Twomey
Department of Biology, East Carolina University, Greenville, North Carolina, United States of America

Yi Liu
The Key Laboratory for Cell Proliferation and Regulation Biology, Ministry of Education, College of Life Sciences, Beijing Normal University, Haidian District, Beijing, China
Department of Bacteriology and Immunology, Beijing Key Laboratory on Drug-resistant Tuberculosis Research, Beijing Tuberculosis and Thoracic Tumor Research Institute/Beijing Chest Hospital, Capital Medical University, Tongzhou District, Beijing, PR China

Xueke Wang, Rongrong Wei, Miao Tian, Tizhuang Ma, Wensheng Li, Yu Xue, Xuxia Zhang, Wei Wang, Tao Wang and Chuanyou Li
Department of Bacteriology and Immunology, Beijing Key Laboratory on Drug-resistant Tuberculosis Research, Beijing Tuberculosis and Thoracic Tumor Research Institute/Beijing Chest Hospital, Capital Medical University, Tongzhou District, Beijing, PR China

Qing Xing, Sumin Wang and Feng Hong
Central Laboratory, Beijing Research Institute for Tuberculosis Control, Xicheng District, Beijing, PR China

Xiaoying Jiang
Clinical Center on TB, China CDC, Beijing Tuberculosis and Thoracic Tumor Research Institute/Beijing Chest Hospital, Capital Medical University, Tongzhou District, Beijing, PR China

Zhiguo Zhang
Beijing Changping Center for Tuberculosis Control and Prevention, Changping District, Beijing, PR China

Junjie Zhang
The Key Laboratory for Cell Proliferation and Regulation Biology, Ministry of Education, College of Life Sciences, Beijing Normal University, Haidian District, Beijing, China

Wen-Hong Chen
Key Laboratory for Plant Diversity and Biogeography of East Asia, Kunming Institute of Botany, Chinese Academy of Sciences, Kunming, Yunnan, China
University of the Chinese Academy of Sciences, Beijing, China

Jun-Bo Yang and Zhi-Rong Zhang
Plant Germplasm and Genomics Center, Germplasm Bank of Wild Species, Kunming Institute of Botany, Chinese Academy of Sciences, Kunming, Yunnan, China

Hong Wang and Yu-Min Shui
Key Laboratory for Plant Diversity and Biogeography of East Asia, Kunming Institute of Botany, Chinese Academy of Sciences, Kunming, Yunnan, China

Kanae Nishii
Science Division, Royal Botanic Garden Edinburgh, Edinburgh, Scotland, United Kingdom

Fang Wen
Guangxi Institute of Botany, Guangxi Zhuang Autonomous Region and Chinese Academy of Sciences, Guilin, Guangxi, China

Michael Möller
Science Division, Royal Botanic Garden Edinburgh, Edinburgh, Scotland, United Kingdom

Elizabeth Cashdan
Department of Anthropology, University of Utah, Salt Lake City, Utah, United States of America

Juanita Olano-Marin, Kamila Plis, Tomasz Borowik, Magdalena Niedziałkowska and Bogumiła JHędrzejewska
Mammal Research Institute, Polish Academy of Sciences, Białowieża, Poland

Leif Sönnichsen
Mammal Research Institute, Polish Academy of Sciences, Białowieża, Poland
Leibniz Institute for Zoo and Wildlife Research, Berlin, Germany

Wei Ji, Gui-Rong Zhang and Kai-Jian Wei
Key Laboratory of Freshwater Animal Breeding, Ministry of Agriculture, College of Fisheries, Huazhong Agricultural University, Wuhan, P. R. China

Wei Ran and Wei-Min Wang
Key Laboratory of Freshwater Animal Breeding, Ministry of Agriculture, College of Fisheries, Huazhong Agricultural University, Wuhan, P. R. China

Gui-Wei Zou
Key Laboratory of Freshwater Biodiversity Conservation, Ministry of Agriculture, Yangtze River Fisheries Research Institute, Chinese Academy of Fishery Sciences, Wuhan, P. R. China

Jonathan P. A. Gardner
Key Laboratory of Freshwater Animal Breeding, Ministry of Agriculture, College of Fisheries, Huazhong Agricultural University, Wuhan, P. R. China
Freshwater Aquaculture Collaborative Innovation Centre of Hubei Province, Wuhan, P. R. China
School of Biological Sciences, Victoria University of Wellington, Wellington, New Zealand

María Quintela
Dept. of Population Genetics, Institute of Marine Research, Bergen, Norway
BIOCOST Research Group, Dept. of Animal Biology, Plant Biology and Ecology, University of
A Coruña, A Coruña, Spain

Hans J. Skaug
Dept. of Population Genetics, Institute of Marine Research, Bergen, Norway
Department of Mathematics, University of Bergen, Bergen, Norway

Nils Øien and Hiroko K. Solvang
Dept. of Marine Mammals, Institute of Marine Research, Bergen, Norway

Tore Haug
Dept. of Marine Mammals, Institute of Marine Research, Tromsø, Norway

Bjørghild B. Seliussen
Dept. of Population Genetics, Institute of Marine Research, Bergen, Norway

Christophe Pampoulie
Marine Research Institute of Iceland, Reykjavik, Iceland

Naohisa Kanda and Luis A. Pastene
Institute of Cetacean Research, Tokyo, Japan

Kevin A. Glover
Dept. of Population Genetics, Institute of Marine Research, Bergen, Norway
Department of Informatics, Faculty of Mathematics and Natural Sciences, University of Bergen, Bergen, Norway

Holly B. Ernest and Michael R. Buchalski
Wildlife Health Center, School of Veterinary Medicine, University of California Davis, Davis, California, United States of America

Wildlife and Ecology Unit, Veterinary Genetics Laboratory, School of Veterinary Medicine, University of California Davis, Davis, California, United States of America

T. Winston Vickers and Walter M. Boyce
Wildlife Health Center, School of Veterinary Medicine, University of California Davis, Davis, California, United States of America

Scott A. Morrison
The Nature Conservancy, San Francisco, California, United States of America

Felix Bast, Aijaz Ahmad John and Satej Bhushan
Centre for Biosciences, Central University of Punjab, Bathinda, Punjab, India

Tetsuya Yanagida, Yasuhito Sako, Minoru Nakao and Akira Ito
Department of Parasitology, Asahikawa Medical University, Asahikawa, Hokkaido, Japan

Jean-François Carod
Institut Pasteur de Madagascar, Antananarivo, Madagascar

Eric P. Hoberg
US Department of Agriculture, Agricultural Research Service, US National Parasite Collection, Animal Parasitic Diseases Laboratory, Beltsville, Maryland, United States of America

Yi Yu, Qiang Fan, Rujiang Shen, Wenbo Liao and Jianhua Jin
Guangdong Key Laboratory of Plant Resources and Key Laboratory of Biodiversity Dynamics and Conservation of Guangdong Higher Education Institutes, School of Life Sciences, Sun Yat-Sen University, Guangzhou, China

Wei Guo
Department of Horticulture and Landscape Architecture, Zhongkai University of Agriculture and Engineering, Guangzhou, China

Dafang Cui
College of Forestry, South China Agriculture University, Guangzhou, China

Natália A. Leite, Alberto S. Corrêa and Celso Omoto
Departamento de Entomologia e Acarologia, Escola Superior de Agricultura ''Luiz de Queiroz'', Universidade de São Paulo, Piracicaba, São Paulo, Brazil

Alessandro Alves-Pereira
Departamento de Genética, Escola Superior de Agricultura "Luiz de Queiroz", Universidade de São Paulo, Piracicaba, São Paulo, Brazil

Maria I. Zucchi
Agência Paulista de Tecnologia dos Agronego´ cios, Piracicaba, São Paulo, Brazil

Mostafa R. Sharaf, Hathal M. Al Dhafer and Abdulrahman S. Aldawood
Plant Protection Department, College of Food and Agriculture Science, King Saud University, Riyadh, Saudi Arabia

Ralf Conrad, Melanie Klose and Peter Claus
Max-Planck Institute for Terrestrial Microbiology, Marburg, Hessen, Germany

Davi Pedroni Barreto
Instituto de Microbiologia Prof. Paulo de Góes, Universidade Federal do Rio de Janeiro, Rio de Janeiro, Brazil

Alex Enrich-Prast
Instituto de Biologia, Universidade Federal do Rio de Janeiro, Rio de Janeiro, Brazil
Department of Water and Environmental Studies, Linköping University, Linköping, Sweden

Diego Cardeñosa and Susana Caballero
Laboratorio de Ecología Molecular de Vertebrados Acuáticos-LEMVA, Departamento de Ciencias Biológicas, Universidad de Los Andes, Bogotá, Colombia

John Hyde
Southwest Fisheries Science Center, National Marine Fisheries Service, La Jolla, California, United States of America

Marek Slovák and Jaromír Kučera
Institute of Botany, Slovak Academy of Sciences, Bratislava, Slovakia

Eliška Záveská
Department of Botany, Charles University, Praha, Czech Republic

Peter Vd'ačný
Department of Zoology, Comenius University, Bratislava, Slovakia

Jacobus H. Visser and Bettine Jansen van Vuuren
Department of Zoology, University of Johannesburg, Auckland Park, South Africa

Nigel C. Bennett
Mammal Research Institute, Department of Zoology and Entomology, University of Pretoria, Pretoria, South Africa

Lu Lu and Andrew J. Leigh Brown
Institute of Evolutionary Biology, University of Edinburgh, Ashworth Laboratories, Edinburgh, United Kingdom

Samantha J. Lycett
University of Glasgow, Institute of Biodiversity, Animal Health and Comparative Medicine, Glasgow, United Kingdom

Jonathan Linneman and Nicole B. Lopanik
Department of Biology, Georgia State University, Atlanta, Georgia, United States of America

Darcy Paulus and Grace Lim-Fong
Department of Biology, Randolph-Macon College, Ashland, Virginia, United States of America

Sidra E. G. Vasconcellos
Laboratory of Molecular Biology Applied to Mycobacteria, Oswaldo Cruz Institute, FIOCRUZ, Rio de Janeiro, Rio de Janeiro, Brazil
Multidisciplinary Research Laboratory, University Hospital Clementino Fraga Filho – HUCFF, Federal University of Rio de Janeiro, Rio de Janeiro, Rio de Janeiro, Brazil

Chyntia Carolina Acosta
Laboratory of Cellular Microbiology, Oswaldo Cruz Institute, FIOCRUZ, Rio de Janeiro, Rio de Janeiro, Brazil

Lia Lima Gomes, Marcelo Ivens de Araujo, Harrison M. Gomes, Adalberto Rezende Santos and Philip Noel Suffys
Laboratory of Molecular Biology Applied to Mycobacteria, Oswaldo Cruz Institute, FIOCRUZ, Rio de Janeiro, Rio de Janeiro, Brazil

Emilyn Costa Conceição and Karla Valéria Lima
Instituto Evandro Chagas, Section of Bacteriology and Mycology, Belém, Pará, Brazil

Maria de Lourdes Leite and Flá vio Tannure
Hospital Municipal Rafael de Paula Souza, Municipal Secretary of Health, Rio de Janeiro, Rio de Janeiro, Brazil

Paulo Cesar de Souza Caldas
Centro de Referência Professor Hélio Fraga, Escola Nacional de Saúde Publica Sergio Arouca, FIOCRUZ, Rio de Janeiro, Rio de Janeiro, Brazil

Michel K. Gomgnimbou and Christophe Sola
CNRS–UniversitéParis–Sud, Institut de Génétique et
Microbiologie–Infection Genetics Emerging Pathogens
Evolution Team, Orsay, France

David Couvin and Nalin Rastogi
Supranational TB Reference Laboratory, Unité de la
Tuberculose et des Mycobactéries, Institut Pasteur de
Guadeloupe, Abymes, Guadeloupe, France

Neio Boechat
Multidisciplinary Research Laboratory, University
Hospital Clementino Fraga Filho – HUCFF, Federal
University of Rio de Janeiro, Rio de Janeiro, Rio de
Janeiro, Brazil

Index